# CHEMISTRY

## An Industry-Based Introduction
## with CD-ROM

**John Kenkel • Paul B. Kelter • David S. Hage**

LEWIS PUBLISHERS

Boca Raton    New York    London    Washington, D.C.

## Library of Congress Cataloging-in-Publication Data

Kenkel, John.
    Chemistry : an industry-based introduction with CD-ROM / John V. Kenkel, Paul B. Kelter, and David S. Hage
        p.   cm.
    Includes bibliographical references and index.
    ISBN 1-56670-303-4 (alk. paper)
    1. Chemistry. I. Kelter, Paul B. II. Hage, David S. III. Title.

QD33.K415 2000
540—dc21
                                                      00-030343

# Preface

How has chemistry shaped your life? What have chemistry professionals done and what are they doing to make your life healthier, more satisfying, and more productive? What specific tasks do chemists and chemistry technicians perform, routine and nonroutine, that help to make your life more comfortable, more fulfilling, and happier? If you answer "I don't know" to these questions, then an introductory chemistry course based on this textbook will provide some answers.

By writing this textbook and creating the accompanying CD-ROM supplement, we seek to provide innovative resources by which an instructor can communicate basic concepts of chemistry while also introducing the workplaces of chemistry professionals to students. This communication reveals special facts concerning this workplace; it reveals the tasks performed and techniques used in this workplace; and it reveals the impact that this workplace has had on their lives as consumers and as citizens. We have accomplished these things by carefully gauging the relevancy of the traditional topics to the lives of the students and to the chemist's real world of work, adjusting the depth of coverage of these topics according to this relevancy, and then tying the subject at hand directly to some aspect of the workplace and/or to the students' lives as consumers. The results are the following unique features.

- Over one hundred "Chemistry Professionals at Work" (CPW) boxes in the textbook (one on about every fifth page), including a box titled "Why Study This Topic" at the beginning of each chapter.
- Homework and class discussion assignments within the CPW boxes that direct the students to obtain and report on or to discuss additional relevant information.
- "For Classroom Discussion and Reports" homework in the end-of-chapter homework sections that provide still further opportunities to discover and communicate the relevancy of topics.
- Innovative "applications" scenarios on the CD-ROM supplement that tie the subject matter to the real world of work through interesting and entertaining industry-based problems to solve.

We feel that these techniques, combined with a clear and readable writing style, will help make introductory chemistry more interesting and less tedious to students and give them a clearer understanding of why principles of chemistry are important by revealing how their lives have been affected by the work of modern chemistry professionals. At the same time, the chemistry professional's workplace will be opened to them, making a chemist's work, and the results of such work, a little less mysterious and a little more appreciated. A separate project has created an industry-based laboratory manual to accompany this textbook and CD-ROM.

We believe the approach we have taken is one answer to the comprehensive and organized efforts undertaken in recent years by two highly visible and respected organizations, the American Chemical Society (ACS) and the National Science Foundation (NSF). With NSF funding, the Division of Chemical Education of the ACS formed the Task Force on the General Chemistry Curriculum which produced a document[1] in 1994 announcing their recommendations. Also, a committee of the NSF itself conducted an "intensive review of the state of undergraduate education in science, mathematics, engineering, and technology (SME&T) in America," which, in 1996, also produced a report.[2] In the ACS Task Force report, a list of goals for a general chemistry course appears. The very first goal reads as follows: "Introduce students to the process of science as it applies to chemistry and *what chemists do*." In the "Recommendations" section of the *Shaping the Future* report, the following is the third recommendation listed for faculty members: "Build into every course inquiry the process of science (or mathematics or engineering), a knowledge of *what SME&T practitioners do*, and the excitement of cutting-edge research." We feel that this textbook and accompanying CD-ROM supplement provide the students of introductory chemistry with the knowledge of what chemistry professionals do, as suggested in these two reports.

This textbook and CD-ROM supplement should be especially useful to chemistry-based technician education programs in the community colleges. The NSF provided the support for our work (see Acknowledgments) for the expressed purpose of producing a relevant curriculum for these programs. We feel the book and CD-ROM may also be useful to high school programs, especially those high school chemistry courses that come under the auspices of "school-to-work" programs, such as Tech Prep.

*John Kenkel*
*Paul Kelter*
*David Hage*

[1]Lloyd, B., Ed., *New Directions for General Chemistry—A Resource for Curricular Change from the Task Force on the General Chemistry Curriculum*, Division of Chemical Education, American Chemical Society, 1994.

[2]*Shaping the Future—New Expectations for Undergraduate Education in Science, Mathematics, Engineering, and Technology*, National Science Foundation, 1996.

# The Authors

**John Kenkel** is a chemistry instructor at Southeast Community College in Lincoln, Nebraska. Throughout his 23-year career at SCC, he has been directly involved in the education of chemistry-based laboratory technicians in a vocational program called Environmental Laboratory Technology. He has also been heavily involved in chemistry-based laboratory technician education on a national scale, having served on a number of American Chemical Society committees including the Committee on Technician Activities and the Coordinating Committee for the Voluntary Industry Standards project. In addition to these, he has served a five-year term on the ACS Committee on Chemistry in the Two-Year College, the committee that organizes the Two-Year College Chemistry Consortium conferences. He was the chair of this committee in 1996.

Mr. Kenkel has authored several popular textbooks for chemistry-based technician education. *Analytical Chemistry for Technicians* was first published in 1988. A second edition of this book was published in 1994. In addition, he has authored two other books, the *Analytical Chemistry Refresher Manual* published in 1992 and *A Primer on Quality in the Analytical Laboratory*, which was published in 2000. All were published through CRC Press/Lewis Publishers.

Mr. Kenkel has been the Principal Investigator for a series of curriculum development project grants funded by the National Science Foundation's Advanced Technological Education Program. The current textbook and laboratory manual, *Chemistry: An Industry-Based Introduction* and *Chemistry: An Industry-Based Laboratory Manual* were produced under these grants. He has also authored or coauthored four articles on the curriculum work in recent issues of the *Journal of Chemical Education* and has presented this work at more than a dozen conferences since 1994.

In 1996, Mr. Kenkel won the prestigious National Responsible Care Catalyst Award for excellence in chemistry teaching sponsored by the Chemical Manufacturer's Association. He has a master's degree in chemistry from the University of Texas in Austin (1972) and a bachelor's degree in chemistry (1970) from Iowa State University. His research at the University of Texas was directed by Professor Allen Bard. He was employed as a chemist from 1973 to 1977 at Rockwell International's Science Center in Thousand Oaks, California.

**Paul Kelter** is Marie Foscue Rourk Professor of Chemistry and Biochemistry at the University of North Carolina-Greensboro (UNCG). This endowed chair in Chemical Education is the first such chair in the United States. Kelter arrived at UNCG in the fall of 1999 after six successful years at the University of Nebraska-Lincoln (UNL). Kelter was honored with a host of teaching and scholarship awards at UNL, including the Student Outstanding Teacher of the Year Award (1996 and 1997, the first two years of the award), induction into the UNL Academy of Distinguished Teachers (1999), and the University of Nebraska Outstanding Teaching and Instructional Creativity Award (1999). These are the latest awards for this nationally known teacher-scholar who has also received the University of Wisconsin Distinguished Teaching Award (1990) and the Wisconsin Society of Science Teachers Regional Award for Excellence in Science Education (1990).

Dr. Kelter has published over 40 journal articles, a popular textbook, *Chemistry: A World of Choices* (McGraw-Hill), several best-selling chemistry study guides (Houghton Mifflin), and a number of chemistry CDs. He is currently writing a textbook, *Chemistry: The Nature of Change* (Houghton Mifflin) on which he is the lead author, and is consulting for a textbook in analytical chemistry (David Hage, lead author, Houghton Mifflin). Kelter has also received over $6 million in competitive grants for his chemical education efforts.

Dr. Kelter currently teaches undergraduate courses in general and analytical chemistry and graduate-level courses in forensic chemistry and chemical education.

**David S. Hage** is Professor of Chemistry at the University of Nebraska-Lincoln. He has received numerous research awards and has published over 65 research articles in the fields of analytical chemistry and biological and environmental testing. He is also the author of several book chapters on analytical techniques for undergraduate chemistry texts. He currently teaches undergraduate courses in quantitative analysis and instrumental analysis, and graduate-level courses in chromatography, advanced chemical equilibria, and data-handling methods.

Dr. Hage is the lead author on a new textbook for analytical chemistry that is in preparation (Houghton Mifflin).

# Acknowledgments

Partial support for this work was provided by the National Science Foundation's Advanced Technological Education program through grant #DUE9553674 and #DUE9751998. Partial support was also provided by the DuPont Company through their Aid to Education Program. Any opinions, findings, and conclusions or recommendations expressed in this material are those of the authors and do not necessarily reflect the views of the National Science Foundation or the DuPont Company.

From the beginning, this project involved many, many people for the purpose of review, advice, field testing, and direct input. While it is not possible to list all who helped, the authors would like to especially thank the following people for their support and contributions to this, the finished product. This list is in alphabetical order.

Mary Adamson, Bayer Corporation
John Amend, Montana State University
Clarita Bhat, Shoreline Community College
George Boggs, Palomar College
Julie Brady, Delaware Technical and Community College
Dale Buck, Cape Fear Community College
Ken Chapman, American Chemical Society (retired)
David Dellar, The Dow Chemical Company
Leticia El Naggar, Bucks County Community College
Dick Gaglione, New York City Technical College (retired)
Denise Gigante, Onondaga Community College
Tom Hager, DuPont
Leslie Hersh, Delta College
Robert Hofstader, Exxon Corporation (retired)
Kirk Hunter, Texas State Technical College
Janet Johannessen, County College of Morris
Richard Jones, Sinclair Community College
Lori Juhl, Eastern Idaho Technical College
Susan Marine, Miami University Middletown
Lynn Melton, University of Texas at Dallas
Ellen Mesaros, DuPont
John Pederson, DuPont
Joseph Pugach, Aristech Chemical Corporation
Joseph Rosen, New York City Technical College
Joan Sabourin, Delta College
Bruce Schuhmacher, Riverland Community College
Kathleen Schulz, Sandia National Laboratory

Robert Smiley, Dixie Chemical Company
Michael Speegle, Amoco
Roy Stein, University of Toledo Community and Technical College
Richard Sunberg, Procter & Gamble
Carol White, Athens Area Technical Institute
Karen Wosczyna-Birch, Tunxis Community-Technical College

Special acknowledgment goes to Susan Rutledge of Southeast Community College. Sue has cheerfully provided important clerical, technical, and academic assistance throughout the project. Special acknowledgment also goes to Sokol Todi, a former student in Paul Kelter's group at the University of Nebraska. Along with a supporting cast of his peers, he drew many of the figures contained in the manuscript. A number of students at the University of Nebraska also worked very hard producing the CD-ROM that accompanies this book. Their names are listed in the acknowledgments section on the CD. In addition, we are grateful to Molecular Arts Corporation of Anaheim, California for granting permission to use their "ChemClip Art 1000" images throughout the text.

Finally, Elizabeth Teles and Frank Settle at the National Science Foundation provided important guidance for this project from the beginning, in order to accomplish the important objectives of the Advanced Technological Education Program as they relate to chemical technology education.

# Table of Contents

# Dedication

To my wife, Lois, and our daughters, Sister Emily, Jeanie, and Laura, for their incredible love and support for everything I do and for recognizing the ultimate Source of all love, Jesus.
—*John Kenkel*

To Barb, the wind beneath my wings.
—*Paul Kelter*

To my wife, Jill, and our children, Ben, Brian and Bethany, for their help and support during this project.
—*David Hage*

1

# Components and Properties of Matter

## 1.1 Introduction

Randy Perry is a chemistry technician employed by the Eastman Chemical Company in Kingsport, Tennessee. He works in a research laboratory developing new laboratory methods, using sophisticated instruments for analyzing important materials utilized and produced by his company.

Edward Cox is a chemistry technician employed by the Procter & Gamble Company in Mason, Ohio. He works in the Oral Care Division of Health Care for this company. He utilizes state-of-the-art analysis equipment to evaluate dental products manufactured by the company, especially investigating the effects these products have on tooth problems associated with tartar, stain, plaque, hot/cold sensitivity, and gum health.

Sharlene Wilson is a photolithography process engineer employed by Kionix, Inc., in Ithaca, New York. She works with a number of photosensitive polymer materials, researching them, checking their resistance to processing, using them as "etch masks," and ultimately using them to transfer photographic images to silicon wafers.

What do these three individuals and hundreds of thousands of other chemistry professionals have in common? They all work to characterize or analyze the composition, structure and properties of material substances or industrial products. They do chemistry!

*Chemistry* can be defined as the study of the composition, structure, and properties of matter and the changes that matter undergoes. Thus, briefly stated, chemistry is the systematic study of matter.

---

*Chemistry Professionals at Work*                                    *CPW Box 1.1*

# WHY STUDY THIS TOPIC?

Chemistry professionals do chemistry. They work with material substances in their laboratories: analyzing them, researching them, and developing them, and all the while making positive contributions to their lives, to their employers' successes, and to the betterment of society. It is critical that they have a viable knowledge of the properties, composition, and structure of these material substances and the changes which they undergo. In short, they must know chemistry.

This chapter represents a beginning. The principles presented here are fundamental. All other aspects of the job of a chemistry professional rest on a sound knowledge of these fundamental principles. Future chapters will build on this foundation.

*For Homework: Interview a chemistry professional in your local area and determine in what manner and for what purpose this person studies the properties, the composition, the structure of matter, or the changes that matter undergoes. Write a report.*

---

What is matter? All basic science textbooks define **matter** as simply the collective aggregation of all material substances that occupy space and have a mass or weight. It follows then that chemistry is the study of the composition, structure, properties, and changes of all material substances that exist.

Chemistry would thus appear to be an extremely complex, far-reaching, and important discipline, and indeed it is all of that and much more. If you simply look around the room in which you are sitting, anything your eyes focus on at any given moment is a material substance whose composition, structure, properties, and changes were studied (or perhaps are routinely being studied) by a chemist or chemistry technician in detail at some point in time. In addition, some complex and high-volume industrial chemical process that was derived from a detailed study of material substances may be responsible for that item even being in the room in the first place. Examples include the plastic materials of which some tables and chairs are composed, the varnish on the wood products, the paint on the walls, the carpets on the floor, or even the clothes on our backs. An example of immense importance to all of us is the study of the composition, structure, properties, and changes that involve the material substances of which our bodies are composed. The fact that life exists on the Earth at all is a result of the composition, structure, properties, and changes of the material substances that are found here. And all of this is just in the room in which you are sitting! When you leave the room, countless material substances are encountered that are the object of the study of chemistry professionals, or were the results of years of study by chemistry professionals. Chemistry is indeed a science of supreme importance to all of us.

How do we begin a study of chemistry? We begin by understanding what is meant by each item listed in the definition. What do we mean by the "composition" of matter? What do we mean by the "structure" of matter? What do we mean by the "properties" of matter? What do we mean by the "changes that matter undergoes?" While these items encompass and pervade the entire discipline we are about to examine in some detail, some simple definitions and examples are in order at this stage.

# 1.2 Matter: Properties and Changes

Let us begin with change. A ***change*** occurs when a material undergoes a transformation in appearance or substance as a result of a process imposed upon it. For example, when water is heated, it boils and is transformed into steam. When iron is exposed to air in the presence of moisture, it rusts. When crystals of sugar are dropped into a container of water, the sugar dissolves. In each of these cases, there has been an observable transformation of some type.

There are two types of change: physical change and chemical change. ***Chemical change*** involves a change of substance while ***physical change*** does not. A change of substance means that one (or more) substance(s) has (have) been converted into something entirely different—a new substance, or something that was not present at the beginning. An illustration from the above examples is the rusting of iron. Iron is a metal with considerable structural integrity (such that it can be used to support large bridges, buildings, etc.). It is a metal that can be described as shiny and is an electrical conductor with a silvery color. When it undergoes the change known as rusting, a reddish (not silvery) solid forms that has virtually no structural integrity, nor is it shiny, nor is it an electrical conductor. An entirely different substance is formed, so a chemical change has occurred.

In contrast, the conversion of water to steam, or the dissolving of sugar in water, does not result in a change of substance. Steam is water in a different physical state, that of a vapor or gas. It can be returned to the liquid state by condensation on a cold surface. The sugar dissolved in water is still sugar; it is just in a dissolved state. The presence of sugar in the water can be confirmed by tasting the water. It will have the sweet taste of sugar. The sugar may be recovered by evaporating the water. All changes of physical state, including melting, evaporation, condensation, and freezing, are physical changes—no transformation of substance occurs. Any change by simply dissolving a substance is likewise a physical change.

Chemical changes, or changes of substance, are very interesting phenomena and are studied by chemists and chemistry technicians to a much greater extent than physical changes. Some additional examples illustrate why.

Most people are familiar with the chemical "formula" for water, $H_2O$. The "H" in this formula represents the chemical substance hydrogen. The "O" represents the chemical substance oxygen. The formula tells us that water is a special combination of hydrogen and oxygen. The formation of water by the combination of hydrogen and oxygen is an example of a chemical change. Hydrogen is a gas (or vapor) at ordinary temperatures. It is colorless, tasteless, and odorless, much like air. However, it is a highly flammable gas. It is so highly flammable that a violent explosion occurs when even a small volume of it is ignited. Oxygen is also a colorless, tasteless, and odorless gas, and, in fact, is a component of air. One of the unique characteristics of oxygen is that it supports combustion. An ordinary fire cannot exist unless there is oxygen present to support it. When these two chemical substances, a *highly flammable gas (hydrogen)* and a *gas that supports burning (oxygen)*, come together, they form water, a *liquid that extinguishes flames*. This is a striking example of a chemical change. An entirely new substance (water) forms, as evidenced by the fact that it has characteristics completely different from the substances that were present before the change.

Another example is ordinary table salt. The chemical formula of table salt is NaCl. The "Na" represents the substance sodium and the "Cl" represents the substance chlorine. Table salt is a special combination of sodium and chlorine. Sodium is a metal, exhibiting many of the characteristics of metals such as those listed earlier for iron (with the exception of structural integrity). It is shiny, silvery, and is a good electrical conductor. It is also highly unstable, undergoing chemical change immediately upon exposure to almost any other chemical substance, and is thus toxic and hazardous. Chlorine is a gas—a greenish yellow gas. It is also quite toxic and has a vile odor. When sodium is exposed to chlorine, a rapid chemical change occurs in which the white crystals of salt form. Salt is not considered to be toxic. Many people season their food with it. Thus, once again we have an example of a chemical change occurring. A new substance forms as a result of the change, the white crystals of salt from two toxic substances, one of which is a silvery metal and the other a foul-smelling gas. We will discuss chemical change, or chemical combination, in quite some detail often throughout this book.

*Chemistry Professionals at Work*                                        *CPW Box 1.2*

# LOOKS ARE DECEIVING

Th</sub>ere is enough iron in our blood to make several nails. Yet, looking at blood with the finest microscope reveals not a trace of iron. There is enough lead in paint chips from an old building to cause serious harm to the health of a child who ingests it. Yet, the finest microscope reveals not a trace of this heavy metal. People eat bananas because they are high in potassium and are therefore good for their health. Yet, the finest microscope does not reveal even a trace of this metallic substance.

The reason for these observations is that the iron, the lead, and the potassium are not in the form we might expect. They have undergone chemical combination with other elements and now exhibit completely different properties.

Chemistry professionals must understand these concepts. Chemists routinely analyze blood for its iron content, environmental samples for their lead content, and natural products for their potassium content. They must understand the chemical principles that are basic to their jobs. They must also be able to communicate them to others.

*For Class Discussion: Think of other examples of materials in which certain elements are in chemical combination and do not exhibit the properties by which they are normally identified.*

---

*Properties* of material substances are simply characteristics by which these substances may be described. For example, a property of water is that it is a liquid; a property of silver is that it is shiny; a property of iron is that it rusts; a property of a pesticide, such as a mosquito repellent, is its disagreeable odor; a property of sugar is that it tastes sweet; a property of salt is that it dissolves freely in water; and a property of the chlorophyll found in green plants is that it is green. There are two kinds of properties: physical properties and chemical properties. *Physical properties* are physical characteristics that might be determined by simple observation or measurement and involve no chemical change. Most of the examples listed above, namely color, odor, taste, physical appearance, physical state (solid, liquid, or gas), and solubility, are physical properties. All of these may be determined by simple observation with one of the five senses, such as sight (color, sheen, physical state, etc.), smell (odor), etc.

There are other physical properties that require a measurement. Examples are melting point, boiling point, and density. Melting point, the temperature at which a solid substance is converted to a liquid (such as ice to water, 32°F), and boiling point, the temperature at which a liquid substance boils (such as water, 212°F) require the measurement of temperature. Because of this, these physical properties have numbers associated with them. They are physical properties because they do not involve a chemical change.

Density is a physical property indicating how heavy a substance is compared to its volume or size. Considering two metal blocks of equal size, one made of aluminum and one made of lead, it is easy to understand the concept. The lead block is much heavier than the aluminum block in spite of the fact that the blocks are the same size. Lead has a greater density, or greater mass or weight per given volume or size. Density is also something that is measured. In a chemistry laboratory, it is easy to measure the weight of a quantity of matter, and it can also be easy to measure volume. Density is calculated by dividing the measured weight by the measured volume. The fact that the density of lead is 11.3 grams per milliliter is a physical property of lead. The fact that the density of aluminum is 2.70 grams per milliliter is a physical property of aluminum.

*Chemical properties* are characteristics that involve chemical change. A chemical property of iron is that it rusts in the presence of air and moisture. A chemical property of hydrogen is that will combine with oxygen to form water. A chemical property of chlorine is that it will combine with sodium to form table salt.

Perhaps the most useful aspect of properties, physical or chemical, is the fact that they can be used for identification purposes. Chemists are sometimes faced with the task of identifying an unknown material. All material substances possess a unique set of properties—no two substances will ever be completely identical in all respects.

For example, suppose you are a chemistry technician faced with the task of identifying an unknown solid substance that someone in your company discovered in his or her work area. Upon close examination, you observe that it has a metallic sheen, and you also observe that it conducts electricity. Such observations would indicate that it is a metal. There are only a limited number of metals in existence, such as iron, lead, aluminum, copper, silver, gold, and sodium, to name a few. You notice that it does not exhibit the color of copper or the color of gold, so those two metals are eliminated from consideration. Next, you notice that it is fairly heavy metal, so you measure its density and discover that it is 11.3 grams per milliliter. You conclude that the metal is lead because it is the only metal that comes close to a density of 11.3 grams per milliliter.

Real-world identification problems are not likely to be as simple as the illustration above. Such problems often demand that chemical properties be determined, or sophisticated instruments that measure more complex properties be used. This example does, however, illustrate the importance of properties in identification processes.

## 1.3 Classification of Matter

### 1.3.1 Introduction

Over 13 million material substances exist. Because of this large number, the study of the composition, structure, properties, and changes of these substances would appear to be an enormously overwhelming task. To help with this problem, chemists attempt to organize their work by placing these substances into various classifications and subclassifications. By doing this, the individual classifications may be studied separately and the problem is simplified.

The number of classifications and subclassifications that have resulted from this approach is fairly large, as one might expect given the large number of substances that must be classified and studied. In many of the chapters of this text, we will encounter a number of these classifications. For the present, however, let us investigate how some initial broad classifications will help us.

One way to classify material substances would be according to physical state. This would mean placing all substances in one of three classifications: *solids, liquids and gases* (or vapors). By this method, substances such as iron, copper, salt, baking soda, plastic, and glass would be solids; substances such as water, alcohol, and gasoline would be liquids; and substances such as oxygen, carbon monoxide, and natural gas would be gases. This method has serious shortcomings, however, because as the study proceeded, it would become clear that temperature and pressure would need to be specified each time. For example, while we may think of water as a liquid, it can be converted to a solid (ice) by simply lowering the temperature, or converted to a gas (steam) by simply raising the temperature. These facts hold true for the vast majority of other substances as well. Because of this problem, further study or classification on this basis would not be practical.

A better method of classification would be according to purity. The vast majority of substances as they are found in nature are not pure. In fact, they are often extremely complex mixtures. Examples are oil, rocks, soil, and ores. Even air and water in different localities can actually be complex mixtures. It is obvious, for example, that ocean water contains many, many substances besides water. Because of this, there are a large number of industrial, household, and laboratory activities

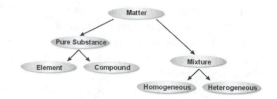

**FIGURE 1.1**    A classification scheme for matter based on purity.

that are dedicated to purifying or separating mixtures, substance by substance. Examples are industrial processes, such as refining oil into component chemicals; household processes, such as the removal of minerals from water with the use of water softeners and distillers; and laboratory processes, such as the use of a sophisticated chromatography instrument to analyze for the quantity of a pesticide residue in river water. In many of these purification schemes, extremely pure substances often result and are available as consumer products or for further industrial or laboratory processing. Thus, while complex mixtures are commonplace in our lives, the same can be said about pure substances, and there is no dependence on temperature or other parameters. Classification according to purity (mixtures vs. pure substances) would thus seem to be a good starting point with our scheme.

All of matter can be classified as either a mixture or a pure substance. All pure substances can be classified as either a compound or an element. All mixtures can be classified as homogeneous or heterogeneous. The flow diagram in Fig. 1.1 shows a pictorial representation of this. We will now define these terms and give additional details and examples.

## 1.3.2    Elements and Compounds

*Elements* and *compounds* consist of just one material substance. There are 109 elements in existence and over 13 million compounds. Many of the elements are familiar to us. Examples are various metals such as iron, copper, aluminum, and lead. Also familiar are elements that are not metals. Examples of nonmetals are helium, hydrogen, oxygen, and iodine. For convenience, all elements have been assigned *symbols*. A symbol consists of one or two letters, the first a capital letter and the second, if used, a lower case letter. These letters are derived from either the English name of the element or the Latin name. For example, "Al" is the symbol for aluminum, and "I" is the symbol for iodine. In these two examples, the symbol is either the first letter or the first two letters of the English name. The symbol "Fe" represents iron, the Latin name for which is "ferrum." Other examples are "Cu" (from "cuprum," the Latin word for copper), "Na" (from the "natrium," the Latin word for sodium), and "Pb" (for "plumbum," the Latin word for lead). All 109 elements are listed in a table that chemists have developed, known as the *periodic table.* The periodic table contains the names and symbols of all the elements as well as some numeric information. This numeric information will become important to us as our study of chemistry proceeds. However, for the present discussion, let us take a look at the periodic table and especially notice the names and symbols. An example of a periodic table is presented in Fig. 1.2.

The periodic table is an arrangement of small squares, each of which contains information about an element. Each square is part of a horizontal row and a vertical column. The large letter or letters in the center of the squares constitute the symbol. Immediately below each symbol is the full name. In the topmost row are the squares for the elements hydrogen (symbol H) and helium (symbol He); hydrogen is on the upper left of the table (first column—Group IA) and helium is on the upper right (last column—Group VIIIA). Notice the number in the upper left corner of each square. This number is known as the *atomic number*. The atomic numbers are in numerical order, left to right, starting with hydrogen (number 1) and continuing to helium (number 2), then to the second row for lithium (number 3), beryllium (number 4), etc., through neon (number 10), and continuing to sodium (number 11) in the third row. The other

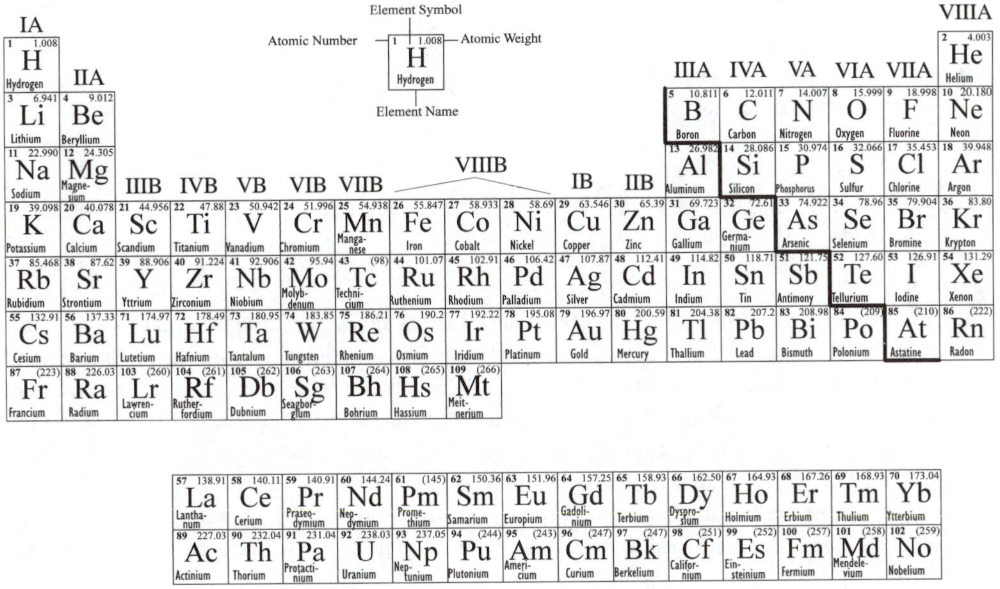

FIGURE 1.2  The periodic table of the elements.

familiar elements we mentioned above can be found as follows: oxygen in the second row (number 8), aluminum in the third row (number 13), iron in the fourth row (Fe, number 26), copper in the fourth row (Cu, number 29), iodine in the fifth row (I, number 53), and lead in the sixth row (Pb, number 82).

Some elements are named after people. Examples are einsteinium, Es, number 99 (after Albert Einstein), and curium, Cm, number 96 (after Marie Curie). Others are named after places. Examples are Californium, Cf, number 98 (after California), and Europium, Eu, number 63 (after Europe). Others have rather odd names, the origins of which are not immediately obvious. Examples are praseodymium, Pr, number 59, and molybdenum, Mo, number 42.

**Compounds** are pure substances that are special combinations of two or more elements. The elements that are present in this combination are shown in a symbolic designation for the compound, which is known as its **chemical formula**, or simply its **formula**. Probably the compound that is most familiar to everyone is water. Water is a special combination of the elements hydrogen and oxygen, which is shown in the formula for water, $H_2O$. Other examples of compounds and their formulas are sodium chloride (formula, NaCl), a special combination of the elements sodium, Na, and chlorine, Cl; calcium carbonate (formula, $CaCO_3$), a special combination of the elements calcium, Ca, carbon, C, and oxygen, O; and potassium chromate (formula, $K_2CrO_4$), a special combination of the elements potassium, K, chromium, Cr, and oxygen, O.

One may ask that if compounds are combinations of two or more elements, why are they not classified as mixtures rather than compounds. The key difference between mixtures and compounds is the nature of this special combination. When two or more elements come together and form a compound, they lose their individual elemental identity and the resulting combination exhibits properties that were not exhibited before the combination. There is a complete transformation of substance, a chemical change, or a **chemical combination**. Water is a prime example. In Section 1.2, we discussed the chemical change that occurs when hydrogen, a highly flammable gas, combines with oxygen, a gas that supports combustion, to produce water, a liquid that extinguishes fires. Because water does not exhibit properties of either hydrogen or oxygen, there has been a complete transformation of substance, and therefore, water is a compound. We can make similar statements about sodium chloride, calcium carbonate, potassium chromate, and the other compounds that exist. The elements in a compound are **chemically combined.**

---

*Chemistry Professionals at Work*                                    CPW Box 1.3

# THE STEEL INDUSTRY

Ordinary steel and stainless steel both have iron as their chief constituent. Ordinary steel contains a small amount of carbon, while stainless steel is an alloy that contains both iron and chromium. Stainless steel can also include the metals nickel and molybdenum. The steel industry is a huge industry because of the wide variety of uses for steel and stainless steel. Over 700 million metric tons of steel are produced annually worldwide. There are dozens of uses for steel and stainless steel that most of us could identify without much help.

Chemistry professionals discovered and developed the methods for extracting iron from iron ores for steel. This had a great deal to do with the rapid development of modern civilization as we know it.

*For Homework: A wide variety of stainless steel alloys exist. In the industry, these are given numbers, such as stainless steel 304, 316, etc. Research various stainless steel types and report on the differences. (Hint: Try this Website:* http://www.steel.org/learning/glossary/s.htm [as of 6/4/00])

---

The elements or substances in a mixture are not. They are simply physically mixed and retain all their properties. They can be separated by physical means, such as filtering.

The fact that there are over 13 million compounds is intriguing. Why so many? The answer is, of course, that there is a very large number of different chemical combinations that can result from the use of the 109 elements that exist. The concept of chemical combination is at the heart of the compositional and structural aspects of our study of chemistry. As our discussions become more and more advanced in this and later chapters, our understanding of this fascinating concept will become more clear.

## 1.3.3  Homogeneous and Heterogeneous Mixtures

As indicated in Section 1.3.1, compounds and elements are most often found in the natural world mixed with other material substances. It is therefore important for us to include these mixtures in our classification scheme. Mixtures may be classified as either homogeneous mixtures or heterogeneous mixtures, as indicated in Fig. 1.1.

*Homogeneous mixtures* are mixtures that are uniform throughout, or mixed so well that we cannot determine visually that there is more than one substance present, even with the most powerful microscope. A good example is a solution of sugar in water. Such a mixture has the same visual appearance as pure water, hence it is not possible to tell, just by looking at it, that it is a mixture. There are, of course, other means by which one can tell that it is a mixture. Such a solution would exhibit a sweet taste, thus indicating that there is some sugar mixed with the water. Also, it is possible to separate the two by a physical change process, that of evaporation. As the water is evaporated from the container, the sugar is left behind. Most *solutions* are homogeneous mixtures. When a solution is prepared, no chemical change takes place. Thus, no transformation of properties nor chemical combination takes place and no compound forms. *Chemical combination* is, of course, the *key difference* between homogeneous mixtures and compounds. The elements making up a compound are chemically combined, while in homogeneous mixtures, they are not.

Homogeneous mixtures can exist as gases and solids, not just liquids. Most mixtures of gases, such as air, are homogeneous. Examples of solid homogeneous mixtures are the metal alloys, such as stainless steel. Stainless steel is actually a homogeneous mixture of iron and chromium. It can also include some nickel and molybdenum.

*Heterogeneous mixtures*, by contrast, are not mixed well, and in fact, one can tell visually that there is more than one substance present. An example would be a mixture of sand and water. If you were to add some sand to a container of water, the sand would not dissolve and it would be visually obvious that what is in the container is a mixture.

# 1.4  Structure and Composition of Matter

At the end of Section 1.3.2, we indicated that chemical combination is at the heart of the structural and compositional aspects of chemistry. Let us now begin to explore the meaning of chemical combination by looking more closely at the structure and composition of matter.

## 1.4.1  Structure and Composition of Elements

Let us imagine a fictitious scenario for the analysis of the structure of matter. Imagine that you hold in your hand a block of silver metal. Now imagine that you also have an extremely sharp cutting tool, one that is capable of cutting through metal. You proceed to cut the block of silver in half with the cutting tool. Now you have two pieces of silver in your hand, both smaller that the original. You lay one piece aside and then proceed to cut the other smaller block in half. Following this, you lay half aside and proceed to cut the other still smaller block in half. Continuing to handle the increasingly smaller blocks and cutting them in half, you reach a point at which the block is too small to see very well. You obtain a magnifying glass and proceed to continue the halving and discarding until again you have a silver piece so small that you cannot see it adequately with the magnifying glass. So, you bring out your microscope and, perhaps with a smaller but sharp cutting tool, you proceed to continue to cut and discard, cut and discard. You then reach a point where you can't see the speck of silver, even with the microscope set on its most powerful setting.

So, with the help of a chemist friend, you gain access to the most powerful microscope in the world and you continue the cutting and discarding process. Finally, you get to the point where you can absolutely go no further. At that point, you have a speck of silver barely visible through this most powerful microscope. The question in your mind is "How far would I be able to go if I could go farther?" And so you start to dream and imagine that you have access to a microscope so powerful that you can see the tiniest objects that can possibly exist. Then you resume the cutting and discarding process and you finally reach the smallest particle of silver that you can have, such that if you proceed further to divide this smallest particle, you discover that it is no longer silver. The question is: "What is this smallest possible particle?" The smallest particle of an element that can exist and still be that element is called an *atom* of that element. The atom is the most fundamental structural unit of all of matter.

All elements are composed of atoms. The atoms of each individual element are unique, which means that an atom of one element is different from an atom of any other element. However, atoms of a given element are very similar, if not identical. Chemists think of atoms as tiny, tiny spheres, or balls. One measure of this uniqueness is the size of this sphere. Hydrogen atoms are the smallest of the spheres, and atoms of francium (Fr, atomic number 87) are among the largest. Francium atoms are over 7 times the size of hydrogen atoms, which is like comparing a ping-pong ball to a basketball.

The utterly small sizes of all atoms, however, cannot be overemphasized. In spite of the fact that francium atoms are relatively large, it would take hundreds of thousands of them, lined up side by side, to make a line visible to the naked eye.

### 1.4.2   Structure and Composition of Compounds

A dividing and discarding process like that described above for the element silver may also be performed on compounds, and the result would be similar: an extremely small fundamental structural unit (visible only through our mythical supermicroscope) that, when divided, would no longer be that compound. The situation is more complex, however, because the nature of this fundamental unit is not the same for all compounds. Some compounds are described as **molecular** or **covalent**; others are described as **ionic**. For compounds described as molecular, the fundamental unit or particle is called a **molecule.** For compounds described as ionic, the fundamental unit is called a **formula unit**, which is composed of a grouping of ions. An **ion** is an atom, or grouping of atoms, that has developed an electrical charge. The subject of electrical charge will be developed further in Sections 1.5, 1.6, 1.7, and 1.8.

The vast majority of the 13 million compounds that exist are molecular. That fact certainly does not diminish the importance of those that are ionic, however, since there are some very common compounds in our daily lives that fall into the ionic category.

Chemical formulas, which we defined in Section 1.3.2 as symbolic designations for compounds, are used for both molecular compounds and ionic compounds. Thus, we can tell what elements are chemically combined to make up a compound, whether it is molecular or ionic by looking at its formula. Examples we presented in Section 1.3.2 included $H_2O$ (water), a combination of hydrogen (H) and oxygen (O); and NaCl (salt, sodium chloride), a combination of sodium (Na) and chlorine (Cl). Water is an example of a molecular compound; while sodium chloride is an example of an ionic compound.

Since compounds are composed of elements in chemical combination, it follows that molecules and formula units are composed of the atoms of the elements in chemical combination. Thus, molecules of water are composed of atoms of hydrogen in chemical combination with atoms of oxygen, and formula units of salt are composed of atoms of sodium and atoms of chlorine in chemical combination. It is the nature of this chemical combination that defines whether a compound is molecular or ionic. This will be detailed further as our study proceeds.

The next question is this: "Can we tell how many atoms of each element are chemically combined to make up a molecule or formula unit?" The answer to this question is "yes." The numbers written as subscripts in the formulas, such as the "2" in the formula for water, $H_2O$, and the "2" and the "4" in the formula for potassium chromate, $K_2CrO_4$, represent the number of atoms of the elements immediately to the left of each subscript in the formula. If there is no subscript immediately to the right of an element symbol, then there is only one atom of that element in the molecule or formula unit. Thus, in a molecule of water, $H_2O$, there are 2 atoms of hydrogen and one atom of oxygen that are chemically combined. In a formula unit of salt, NaCl, there is one atom of each element, sodium and chlorine. In a formula unit of potassium chromate (also an ionic compound), $K_2CrO_4$, there are two atoms of potassium, one atom of chromium, and four atoms of oxygen. Thus, chemical fomulas are reasonable representations of the molecules and formula units of which compounds are composed.

Some formulas utilize parentheses. An example is the formula for calcium nitrate, $Ca(NO_3)_2$. The grouping of elements enclosed by the parentheses is an example of an ion, the nitrate ion, and it is placed in parentheses in order to maintain its identity in the formula. The 2 subscript outside the parentheses means that there are two of these ions in each formula unit of $Ca(NO_3)_2$. This formula would be equivalent to the formula $CaN_2O_6$. The formula with the parentheses is preferred, however, because it gives additional information about this compound. The nitrate ion is a common grouping of atoms in which one nitrogen (N) atom is chemically bound to three atoms of oxygen. Without the parentheses, this small tidbit of information would not be communicated.

We previously indicated that chemists think of atoms as tiny spheres of various sizes. Chemists think of molecules and formula units as entities that have a number of these spheres linked together or "stuck" to each other. The forces that cause them to stick differs depending on whether the compound is ionic or molecular. For example, a chemist's view of a molecule of nitric acid ($HNO_3$) is shown in Fig. 1.3. The sizes of molecules and formula units, then, are most often on the same scale as the incredibly small atoms of which they are composed.

### 1.4.3 Structure and Composition of Mixtures

There are no symbolic designations for mixtures like there are for elements and compounds as discussed above, mostly because the composition of mixtures is infinitely variable. With compounds, the composition is fixed. With a homogeneous mixture (solution), such as sugar and water, for example, one can have many, many differing amounts of sugar and water mixed, but each would still be described as a homogeneous mixture. It would be impossible to assign a symbolic designation for each and every possibility. We will look at this problem further in Chapter 10.

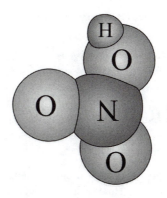

**FIGURE 1.3** A chemist's visualization of a molecule of nitric acid, $HNO_3$.

However, a look at homogeneous and heterogeneous mixtures through our supermicroscope that would make atoms, molecules, and formula units visible would be a useful activity in order to fully understand the nature of these categories of matter and would provide a more useful definition of a homogeneous mixture.

We defined a homogeneous mixture in Section 1.3.3 as a mixture in which the components are mixed so well that one cannot visually determine that it is, in fact, a mixture, even with the most powerful microscope. Through our supermicroscope, however, we can see the atoms, molecules, and ions of which they are composed and recognize that there are indeed two or more substances present. Thus, an *alternate definition* of **homogeneous mixture** would be a mixture in which the atoms, molecules, and/or ions of the component substances are mixed as well as they can be mixed such that the composition is the same throughout.

Heterogeneous mixtures viewed through our supermicroscope are of some interest. Segments of such a mixture that consist of a pure element or compound, such as inside a grain of sand in the sand/water mixture, would, of course, provide views of the same atoms and molecules or formula units that the pure substance provides. However, at the interface of two segments, such as at the surface of a grain of sand in the sand/water mixture, we would be able to see a clear dividing line or boundary showing a change from the atoms or molecules or formula units of one substance to the atoms or molecules or formula units of the other substance. There are some studies of chemistry in which chemists theorize about chemical activity occurring at such a boundary and what it would look like. We will study a few of these as our knowledge of chemistry develops.

## 1.5 A Quick Look Inside the Atom

We have stated that the atom is the smallest particle of an element that can exist and still be that element. We have also stated that if an atom is split in an attempt to make a smaller particle of an element, we would no longer have that element, but would instead have something else. We would have a collection of "subatomic" particles called protons, neutrons, and electrons. These are the major particles of which all atoms are composed.

Scientists discovered the existence of these particles and their properties in the late 1800s and early 1900s when some interesting experiments were performed. A brief historical account of this work will be presented in Chapter 3. The modern theory describing these particles, especially the electrons, is complex and highly mathematical. The tenets of this theory will also be presented in Chapter 3. For our present discussion, however, a very brief physical description of these particles is necessary.

---

*Chemistry Professionals at Work*                                    *CPW Box 1.4*

# MICROSCOPY

Some aspects of modern chemical research require chemists to be able to look at extremely small things. This work is known as "microscopy." Ordinary light microscopes are not powerful enough. The electron microscope, which utilizes a stream of electrons instead of light, has been very helpful, especially when combined with sophisticated elemental analysis techniques. Its use has become routine in real-world scientific studies, such as in observing contamination in substandard materials, foreign substances in cracks in structural metals, and small fragments in or on forensic samples.

A recent development in microscopy is a device called the *scanning tunneling microscope*. This microscope maps a surface by sweeping a tiny probe across it and measuring a small electrical current between the probe tip and the surface. The probe tip is the size of a single atom. Small peaks and valleys that are due to atoms on the surface can be seen. It is the closest chemists have come to seeing atoms.

*For Homework: Interview a forensic scientist in your area and ask him/her about his/her use of the electron microscope. Write a report.*

---

Let us begin with the mass of these particles. The term **mass** refers to the amount of a material. This amount of a material is measured on the earth's surface by measuring the effect of the pull of gravity on it, which is determined by measuring its weight. References to mass in this discussion can thus be thought of as references to weight. The masses of a proton and a neutron are approximately the same. The mass of an electron is significantly less.

The problem with discussing the masses of these particles is that they are so tiny and weigh so little that any numerical values that we state for the mass (in ounces, for example) would likewise be so small that they would have little meaning to us. Scientists recognized this and decided to create a new unit for mass, the **atomic mass unit**, or amu, and arbitrarily defined it in terms of something very small so that we could compare one particle with another. The *amu* is officially defined as one-twelfth the mass of a particular kind of atom of the element carbon, the so-called carbon-12 atom. With the amu defined this way, a proton weighs approximately 1 amu, a neutron weighs approximately 1 amu, and an electron weighs approximately 0.00055 amu.

Next, we describe the so-called electrical **charge** of these particles. The concept of electrical charge is still not well understood by scientists. However, we are familiar with the effects of charge. For example, lightning is the movement of charge through the atmosphere. Electrical current is the movement of charge through a conductor. The "shock" we feel when we touch a grounded metal object after sliding our shoes across a carpeted floor on a dry day is the movement of charge from our body to the object. This static charge can be a hazard in a laboratory because it can ignite flammable substances, causing fires.

In describing the charge of the subatomic particles that are part of the structure of all matter, and in order to explain their behavior, we assign each proton a charge of "+1," and each electron a charge of "−1."

**TABLE 1.1** The Subatomic Particles in an Atom

| Particle | Mass | Charge | Location |
|----------|------|--------|----------|
| proton | 1 amu | +1 | In the nucleus |
| neutron | 1 amu | None | In the nucleus |
| electron | 0.00055 amu | −1 | Outside the nucleus |

They are thus equal, but opposite in their charge, which means that they attract each other, like the opposite poles of a magnet, and can neutralize each other's charge upon contact. Neutrons have no charge.

How are these particles arranged within the atom? Much of the work of scientists in the early 1900s was focused on this question. Again, the historical account will be presented in Chapter 3. Our modern theory of the atom states that all the protons and neutrons are located in an incredibly small core situated at the very center of the atom, and the electrons are located in the space outside this core. Obviously the core contains most of the mass of the atom, since the protons and neutrons possess most of the mass. It is a very dense agglomeration of particles, often thought of as a single particle itself. Scientists have named it the ***nucleus***. To give a reasonable perspective on the nucleus in relation to the electrons, imagine that if the atom is a ball with a diameter of a football field, the nucleus would then be the size of the period at the end of this sentence and the electrons would take up the remaining space.

Table 1.1 summarizes the information presented in this section concerning protons, neutrons, and electrons.

# 1.6    The Number of Subatomic Particles

One very important component of the study of the subatomic particles in atoms is the study of the number of each of these particles within an atom and of how one element's atoms differ from another element's atoms in terms of this number. In Section 1.4.1, we indicated that the atoms of different elements are indeed different from each other, noting, for example, that the size of a francium atom compared to the size of a hydrogen atom is like comparing a basketball to a ping-pong ball. Besides size, they also differ in terms of the number of protons, neutrons, and electrons that each has.

## 1.6.1    Protons and Electrons

Let us refer back to the periodic table presented in Fig. 1.2. We have noted that each of the elements in this table has an identifying "atomic number," which is the number located in the upper left corner of each square. For example, hydrogen is number 1, nitrogen is number 7, calcium is number 20, and gold is number 79, etc. We can define the ***atomic number*** of an element as the number of protons in the nucleus of an atom of that element. All elements differ in terms of this number of protons. In fact, the atomic number, since it is unique to a given element, is a primary characteristic by which an element is identified. If you do not know the identity of an element, but you know the number of protons in the nucleus (its atomic number), you can look in the periodic table to determine its identity.

In an electrically neutral atom, or an atom in which the number of positive charges equals the number of negative charges, the number of protons must equal the number of electrons. Thus, for a neutral atom, the atomic number is not only the number of protons in an atom, but it is also the number of electrons. An atom may gain and lose electrons, which is why the number of electrons is not an identifying characteristic for an element like the number of protons is. An atom cannot lose or gain protons. The particle that results from the gain or loss of electrons is called an ***ion***. The term ion was first defined in Section 1.4.2. Additional discussion can be found in Section 1.8.

## 1.6.2   Neutrons and Isotopes

The number of neutrons within an atom is quite variable, even for different atoms of the same element. An atom of oxygen, for example, has eight protons and eight electrons, but it can have either eight, nine, or ten neutrons. Similarly, an atom of neon has ten protons and ten electrons, but it can have either ten, eleven, or twelve neutrons. Atoms of a given element that differ because they contain a different number of neutrons in the nucleus are called ***isotopes.*** The varying number of neutrons does not change the chemical identity of an element. It is oxygen or neon, for example, regardless of how many neutrons its atoms have. It also does not change the electrical neutrality because the neutron is not charged. It does, however, change the mass of the atom, since neutrons are important contributors to this mass, and it also changes a characteristic number called the mass number.

The ***mass number*** is the sum of the number of protons and the number of neutrons in an atom:

$$\text{mass number} = \text{\# of protons} + \text{\# of neutrons} \tag{1.1}$$

Each isotope of a given element has the same atomic number but a different mass number.

### Example 1.1

What is the mass number of an isotope of magnesium if it has 13 neutrons?

### Solution 1.1

The mass number is the sum of the number of protons and the number of neutrons. The number of neutrons is given as 13. The number of protons is the atomic number of magnesium, which can be found in the periodic table. The atomic number of magnesium is 12. Therefore,

$$\text{mass number} = 12 + 13 = 25$$

### Example 1.2

Name the element described by the following: mass number = 19, number of neutrons = 10.

### Solution 1.2

The atomic number (the number of protons) determines an element's identity. Since the mass number is given and the number of neutrons is given, the number of protons may be calculated:

$$\text{number of protons} = \text{mass number} - \text{number of neutrons}$$

$$= 19 - 10$$

$$= 9$$

From the periodic table, we see that the element with atomic number of 9 is fluorine.

Some symbolism may be used to distinguish one isotope from another. One symbol that is frequently used is to simply write the name of the element, followed by a hyphen and the mass number. Thus, we have "oxygen-17," to symbolize the oxygen isotope with 9 neutrons, and "neon-22," to symbolize the neon isotope with 12 neutrons. In Section 1.5, we referred to the fact that the "carbon-12" isotope is used to define the atomic mass unit. This would be the isotope of carbon with 6 neutrons.

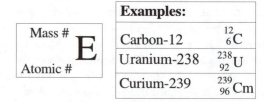

**FIGURE 1.4** Symbolism for isotopes. The mass number is a superscript and the atomic number is a subscript to the left of the element symbol.

Chemists often use another representation for isotopes. This representation utilizes the symbol for the element. It has the mass number as a superscript and the atomic number as a subscript, each immediately to the left of the element symbol. This is as indicated in Fig. 1.4 for an element with symbol "E." Some examples of real representations are also shown in this figure.

## 1.6.3   The Atomic Weight

Isotopes always occur in nature in the same proportions. For example, consider boron, the element with an atomic number equal to 5. All boron atoms contain five protons, five electrons, but either five or six neutrons. Thus, two isotopes of boron exist. The isotope with 5 neutrons constitutes approximately 20% of all boron atoms on Earth, while the isotope with 6 neutrons constitutes the other 80%. The source of the boron, whether it be in the U.S., Europe, or Africa, etc., generally makes no difference. The ratio of one boron isotope to the other is almost always 20/80.

The mass or weight of an atom is the sum of the weights of all the particles of which it is composed: protons, neutrons, and electrons. However, since the electron's weight is negligible compared to the proton and neutron, the mass or weight of an atom in amu is nearly equal to the mass number. The more neutrons there are, the heavier the atom is. Since all isotopes that occur naturally are always present in the same proportions, and since it is not practical or useful to attempt to separate one isotope from another in most cases, the *atomic weight* is almost always expressed as the average weight of all the isotopes that exist. This average always reflects the proportion by which these atoms are found in nature. The atomic weight of boron, for example, is 10.811, not 10 ( the mass number of an atom with 5 neutrons), and not 11 (the mass number of an atom with 6 neutrons). It is also not 10.5, or the average of 10 and 11. Rather, it is 10.811, a number closer to 11 than to 10. This reflects the fact that most boron atoms on Earth (80%) have a mass number of 11, and so the average is "weighted" toward 11. The atomic weight for any element is never a whole number.

It is important to recognize that *the atomic weight and the mass number are not the same*. The *mass number* is the sum of the protons and neutrons in a given atom, while the *atomic weight* is the average weight of all the naturally occurring isotopes. The atomic weight is found in the periodic table. It is the number in the upper right of each square in the periodic table presented in Fig. 1.2. *The mass number is not in the periodic table.* There is no single mass number for any element.

## 1.6.4   Important Isotopes

Probably the most important set of isotopes are those for carbon. There are three main isotopes of carbon, one with six neutrons (mass number 12, carbon-12), one with 7 neutrons (mass number 13, carbon-13), and one with 8 neutrons (mass number 14, carbon-14). Carbon-12 is the one that is used to define the atomic mass unit, as we have mentioned, and hence this isotope is of considerable importance. Approximately 99% of all carbon atoms are carbon-12. The carbon-14 isotope is also of considerable importance. It is used in determining the age of materials found in archeological discoveries.

The element hydrogen has three isotopes, hydrogen-1, hydrogen-2 (also known as deuterium), and hydrogen-3 (also known as tritium). Deuterium may be isolated from naturally occurring hydrogen and is used in research studies in which hydrogen atoms change positions in molecules. The movement of hydrogens in chemical change processes can therefore be traced.

Finally, there are a number of isotopes that are *radioactive*, such as uranium isotopes. This means that their atoms spontaneously break apart into simpler atoms with time. This breaking apart is accompanied by dangerous radiation that is given off from the materials in which these elements are found, and special precautions must be taken by technicians who work with them.

# HANDLING RADIOACTIVE MATERIALS

The Nuclear Regulatory Commission (NRC) requires that the radioactive materials found in nuclear power plants be frequently sampled and handled by the chemistry technicians that work there. To take the samples without endangering the health of the sampling technician or others, protective clothing must be worn. This clothing includes a lab coat, glove liners, rubber gloves, and special safety glasses. In addition, a radiation monitoring meter is required.

Following the sampling, the protective equipment must be disposed of according to well-defined safety protocols. Strict adherence to the proper procedures is crucial to the health and safety of the workers and to the quality of the environment.

*For Homework: Check out the Website for the Nuclear Regulatory Commission (www.nrc.gov) (as of 6/4/00) and find information about the handling and disposal of radioactive waste. Write a report.*

## 1.7 Subclassifications of Elements

We mentioned in Section 1.3 that chemists simplify the task of studying the enormous quantity of different material substances that exist by placing them into classifications and subclassifications. Let us now take another look at the periodic table and see what subclassifications chemists have used for the elements.

The arrangement of the elements in the periodic table may seem odd. In Chapters 3 and 4 we will explain the scientific reasons for this arrangement. It will be apparent from the following discussion, however, that they are grouped according to their properties. This is especially true of the vertical columns, such as Groups IA, IIA, etc. The properties of the elements within a vertical column are so similar that vertical columns of elements have come to be known as *families.* Before we discuss some individual families, let us look at a few larger groupings.

The majority of elements exhibit the properties of *metals.* These properties are: metallic sheen, electrical conductivity, malleability (ability to be flattened into foil), ductility (ability to be formed into wires), and high tensile strength (ability to withstand forces that would pull it apart). Atoms of metals tend to lose electrons to form positively charged ions (see Section 1.8). Metals are located to the left of the bold stair-step line that is shown extending diagonally across the right side of the periodic table in Fig. 1.5 (see also Fig. 1.2).

Two other classifications are the *nonmetals* and the *metalloids* (see Fig. 1.5). Nonmetals exhibit none of the properties listed above for metals. Many are gases. One is a liquid (bromine). Those that are solids (e.g., carbon, phosphorus, and sulfur) are electrical insulators, do not have a metallic sheen, are neither ductile nor malleable, and do not have a high tensile strength. Nonmetals tend to take on electrons and form negatively charged ions (see Section 1.8). Nonmetals are located to the right of the bold diagonal line in the periodic table in Fig. 1.5. Some of the elements positioned adjacent to the bold dividing line exhibit properties of both metals and nonmetals. These are the metalloids. The metalloids are boron, silicon, germanium, arsenic, antimony, tellurium, polonium, and astatine. These are shown within the dark shaded region of Fig. 1.5.

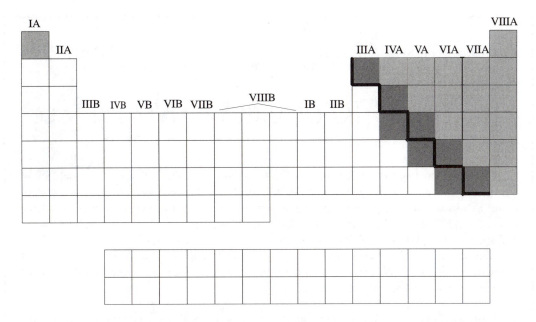

**FIGURE 1.5** An outline of the periodic table showing the locations of the metals (white squares), the nonmetals (light gray squares), and the metalloids (dark gray squares). The stair-step dividing line is also shown.

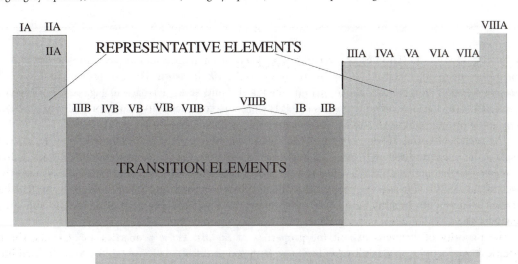

**FIGURE 1.6** An outline of the periodic table showing the locations of the representative elements, the transition elements, and the inner-transition elements.

Another classification scheme divides the periodic table into three parts: the ***representative elements,*** the ***transition elements*** (also known as transition metals), and the ***inner-transition elements*** (also known as inner-transition metals or rare earth elements). These divisions are shown in Fig. 1.6. The inner-transition elements are set apart (below and slightly to the right) from the rest of the elements. The two horizontal rows (***periods***) that constitute this group were actually removed from the last two rows of the transition elements to conserve space. If this were not done, the table would be needlessly long from left

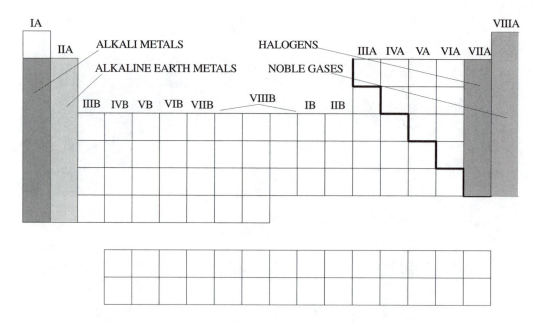

**FIGURE 1.7** An outline of the periodic table showing the locations of the alkali metals, the alkaline earth metals, the halogens, and the noble gases.

to right. Again, the scientific reason for these elements being grouped together will become clear in Chapter 4.

Finally, four of the families of elements have been given names and these names deserve to be mentioned here. The metals in Group IA are known as the **alkali metals.** These include lithium, sodium, potassium, rubidium, cesium, and francium (see Fig. 1.7 and also refer back to Fig. 1.2). Hydrogen, a Group IA element, is excluded from this list. It is a gas and considered to be a nonmetal. The scientific reasons for assigning it to Group IA will become clear as our discussion progresses.

The metals in Group IIA are known as the **alkaline earth** metals. These include beryllium, magnesium, calcium, strontium, barium, and radium. The nonmetals in Group VIIA are known as the **halogens.** These include fluorine, chlorine, bromine, iodine, and astatine. The elements in Group VIIIA are known as the **noble gases.** These include the nonmetals helium, neon, argon, krypton, xenon, and radon. Refer to Figs. 1.7 and 1.2 to clearly see the positions of these groups and these elements in the periodic table.

## 1.8  Formation of Ions

As we've indicated previously, atoms are capable of gaining or losing electrons so that charged particles called ions form. Ions are charged because they have an imbalance of protons and electrons, which are oppositely charged particles. It is possible for single atoms to develop a charge, but it is also possible for groups of atoms to collectively develop a charge. When single atoms gain or lose electrons, a **monatomic ion** forms. When a collective group of atoms has a charge, it is called a **polyatomic ion.** The prefix "mono-" refers to one atom and the prefix "poly-" refers to many atoms. The number of atoms a polyatomic ion has is variable. Some examples are discussed below.

The gain or loss of electrons by an atom is not a random event. An atom gains or loses one or more electrons to achieve a more stable state—a state in which the atom possesses a certain number of electrons. The most stable of states is exemplified by the noble gases. Thus, the number of electrons representing

---

*Chemistry Professionals at Work* *CPW Box 1.6*

# SLEUTHING OUT THE UNKNOWNS AT DOW CORNING

"I am Tina Leaym. I am a chemistry professional at Dow Corning in Midland, Michigan. Sleuthing out unknowns, growing crystals, and finally seeing the pink end point of a tricky titration are a few of my best memories from school. I enjoyed these labs and my co-op experience at Dow Corning Corporation so much that I quickly chose chemistry as my major. That was in 1988, and I've never been sorry (or bored). I began my chemistry career as a Technician, and after several years of addi-

tional classes, I successfully interviewed for a chemist's position. Juggling a full-time technician career with family obligations and night classes was a struggle, but a career in "the central science" has helped block out the pain! I'm still sleuthing out those unknowns, currently in a resin development lab at Dow Corning's headquarters in Midland, Michigan. I recommend a career in chemistry to anyone who loves a good mystery."

*For Homework: Check out the Internet site for Dow Corning and write a report on this company.*

http://www.dowcorning.com/ (as of 6/4/00)

---

stable states is the number of electrons found in the atoms of noble gases, 2 (helium), 10 (neon), 18 (argon), etc. Atoms of elements that have an electron count close to those of the noble gases will gain or lose one or more electrons (and become a monatomic ion) in order to have the same count as the nearest noble gas. A metal, since it is located on the left side of the periodic table, tends to lose one or more electrons to obtain an electron count equal to the noble gas that precedes it. A nonmetal, since it is located on the right side of the periodic table, tends to gain one or more electrons to obtain an electron count equal to the noble gas to its right. Thus, given the opportunity, chlorine, for example, will gain an electron to become like argon; sodium will lose an electron to become like neon, etc. The result is a "chloride ion," an ion with a −1 charge because the chlorine atom has gained an electron (symbolized as $Cl^-$), and a "sodium ion," an ion with a +1 charge because a sodium ion has lost an electron (symbolized as $Na^+$). All alkali metals tend to lose one electron and become a +1 ion ($Li^+$, $Na^+$, $K^+$, etc.) and all halogens tend to gain one electron and become a −1 ion ($F^-$, $Cl^-$, $Br^-$, etc.). Additionally, all alkaline earth metals (Be, Mg, Ca, etc.) tend to lose 2 electrons to achieve the electron count of the nearest noble gas. Thus, we have the monatomic ions $Be^{2+}$, $Mg^{2+}$, and $Ca^{2+}$, etc. Similarly, the monatomic ions $O^{2-}$ and $S^{2-}$ form

**TABLE 1.2** "The Big Six"
Polyatomic Ions

| FORMULAS | NAME |
| --- | --- |
| $NO_3^{1-}$ | Nitrate |
| $SO_4^{2-}$ | Sulphate |
| $PO_4^{3-}$ | Phosphate |
| $CO_3^{2-}$ | Carbonate |
| $OH^-$ | Hydroxide |
| $NH_4^+$ | Ammonium |

due to the gain of two electrons by the elements oxygen and sulfur to become like their nearest noble gas. For representative elements, *the number of electrons that can be gained or lost by an atom is equal to the number of families the element is away from a noble gas.*

The number of electrons that transition metal and inner-transition metal atoms can lose when forming monatomic ions is not as easily predicted. The number of electrons that can be lost varies from one to four. Thus, the charges on such ions vary from +1 to +4. In many cases, a given transition or inner transition metal can form two or more different monatomic ions. The two monatomic ions of iron, for example, are $Fe^{2+}$ and $Fe^{3+}$. Possible reasons for this behavior will be discussed in Chapter 3.

The formation of polyatomic ions is also not random, but the explanation as to why they form is complex and will not be addressed here. It is important to know some of the common polyatomic ions that do exist, however. Table 1.2 presents the formulas and names of six of the most common ones. For the purpose of future reference, let us call these "The Big Six" polyatomic ions. Others will be discussed in Chapter 4.

## 1.9   Homework Exercises

1.  With what four aspects of matter is the science of chemistry concerned?
2.  How do chemical properties of matter differ from physical properties? Give two examples of each.
3.  Which of the following represent physical properties and which represent chemical properties? (a) Iron rusts.   (b) Naphthalene smells like moth balls.   (c) Potassium permanganate is purple.   (d) The density of gold is 19.3 grams per milliliter.   (e) Silver tarnishes.   (f) Hydrogen reacts with oxygen to form water.
4.  What is meant in chemistry by the term "physical change?" Give some examples of physical changes.
5.  What is meant in chemistry by the term "chemical change?" Give some examples of chemical changes.
6.  Consider the change that occurs in each of the following. Tell whether it is a physical change or a chemical change and why.
    (a)  Sodium, a silvery metal, changes to a white solid in the presence of chlorine, a yellow-green gas.
    (b)  A few crystals of sugar dissolve in the water in a laboratory beaker.
    (c)  Alcohol, a liquid, changes to a vapor when heated.
7.  Which of the following represent a chemical change and which represent a physical change?
    (a) Water is converted to steam by boiling. (b) Carbon, a black powder, is converted to a colorless gas when reacted with oxygen. (c) Salt completely dissolves in water to form a solution that looks like ordinary water. (d) When the chemical mercury oxide is heated in a test tube, a silvery substance condenses near the top of the test tube. (e) When acetic acid is cooled to below 16.7°C, it changes to a solid.
8.  What is a "homogeneous mixture"? Give one example.
9.  What is a "heterogeneous mixture"? Give one example.

10. Tell whether each of the following is an element, a compound, a homogeneous mixture, or a heterogeneous mixture.
    (a) Water. (b) A multicolored rock. (c) A sample of stainless steel. (d) $CO_2$. (e) Bromine. (f) A solution of sugar in water. (g) Fe. (h) Sand and water mixed together in a container.

11. Fill in the blank with the appropriate letter selected from the choices that follow. Some selections may be used more than once while others not at all.
    ___ Consists of two or more substances which are plainly visible.
    ___ An example of this is the melting of ice.
    ___ Gold.
    ___ Consists of more than one substance in chemical combination.
    ___ An example of this is density.
    ___ An example of this is the dissolving of sugar in water.
    ___ This can be found in the periodic table.
    ___ Ferric chloride.
    ___ Even though this consists of substances that are only physically mixed, this cannot be determined even with the most powerful microscope.
    ___ This has a "formula" rather than a "symbol."
    (A) Compound
    (B) Chemical property
    (C) Heterogeneous mixture
    (D) Physical change
    (E) Element
    (F) Homogeneous mixture
    (G) Physical property

12. What is the difference between a compound and a homogeneous mixture?

13. If compounds contain more than one element, why do we classify them as pure substances rather than as mixtures?

14. When two elements come together and chemically combine,
    (a) what happens to their properties?
    (b) what happens to their identities?
    (c) what happens to their atoms?
    With reference to each of these, give examples.

15. What is the meaning of the subscript "2" to the right of the H, and the lack of a subscript to the right of the O in the formula for water?

16. For each of the following formulas of compounds, indicate what elements (their names, not symbols) are present in the compound and also tell how many atoms of each element are present in one formula unit or molecule of the compound.
    Example: $NaNO_3$ sodium, 1 atom; nitrogen, 1 atom; oxygen, 3 atoms
    (a) $Na_3PO_4$  (b) $CCl_4$  (c) $Al_2(SO_4)_3$  (d) $(NH_4)_2CO_3$  (e) $Ba(OH)_2$

17. Define each of the following terms.
    (a) atom  (b) molecule  (c) ion

18. State whether each of the following is an atom, molecule, or ion.
    (a) Se  (b) $CCl_4$  (c) $HCO_3^-$
    (d) $C_6H_6$  (e) Pb  (f) $Br_2$

19. What is a formula unit? How is it used in chemistry?

20. What is the difference between an atom and a molecule?

21. Define each of the following terms.
    (a) mass
    (b) atomic mass unit
    (c) atomic weight
    (d) atomic number

22. Complete the following table.

| Particle | Relative Mass | Charge | Location |
|----------|---------------|--------|----------|
| Neutron  |               |        |          |
|          | 0.00055 amu   |        |          |
|          |               |        | In the nucleus |

23. Complete the following table.

| Particle | Relative Mass | Charge | Location |
|----------|---------------|--------|----------|
|          |               | −1     |          |
| Neutron  |               |        |          |
|          | 1 amu         |        |          |

24. What is an isotope? How does one isotope of an element differ from another isotope of the same element?

25. Fill in the blanks:

The symbol for the element with atomic number 54 is _____.

The number of electrons in a neutral atom of zinc is _____.

The particle inside the atom that does not have a charge is the _____.

The number of neutrons in an atom of titanium that has a mass number of 41 is _____.

The number of protons in an atom of molybdenum is _____.

The mass number of an atom of sodium that has 13 neutrons is _____.

The atomic number represents the number of _____ in any atom of that element.

The atomic number of an atom of calcium that has 18 neutrons is _____.

The symbol for the element that has a mass number of 14 and a total of 8 neutrons is _____.

The particle inside the atom that has a negative charge is the _____.

26. Fill in the blanks.

The symbol for the element with atomic number 47 is _____.

The number of protons in an atom of iodine is _____.

The number neutrons in an atom of chromium that has a mass number of 52 is _____.

The number of electrons in a neutral atom of calcium is _____.

The mass number of an atom of cadmium that has 47 neutrons is _____.

The atomic number of an atom of lead that has 85 neutrons is _____.

The symbol of the element that has a mass number of 27 and has 14 neutrons is _____.

Of atomic number, mass number, and atomic weight, the one that is not found in the periodic table is _____.

27. Fill in the blanks:

The number of neutrons in an atom of bismuth that has a mass number of 210 is _____.

The atomic number of an atom of cadmium that has 56 neutrons is _____.

The symbol for the element with atomic number 35 is _____.

The symbol of the element that has a mass number of 52 and a total of 28 neutrons is _____.

The mass number of an atom of aluminum with 14 neutrons is _____.

Of atomic number, mass number and average atomic weight, the one that is not in the periodic table _____.

The number of electrons in a neutral atom of an element with 13 protons and 12 neutrons _____.

The number of protons in an atom with atomic number 28 and mass number 55 is _____.

The number of protons in an atom of rubidium is _____.

28. Give the atomic number, the mass number, and the number of neutrons for each of the following:
    (a) carbon-14   (b) iron-58   (c) bismuth-209
29. Give the atomic number, the mass number, and the number of neutrons in each of the following:
    (a) $^{239}_{93}Np$   (b) $^{235}_{92}U$   (c) $^{206}_{82}Pb$   (d) $^{87}_{37}Rb$
30. What is the name of the isotope indicated by $^2H$?
31. An atom with a nucleus of 8 protons and 9 neutrons is an isotope of oxygen. Is a atom with 9 protons and 8 neutrons another isotope of oxygen? Explain your answer.
32. Why are the mass number and the atomic weight not the same? Why are they approximately the same? Which is found in the periodic table?
33. The two most common isotopes of the element chlorine are chlorine-35 and chlorine-37. But the atomic weight that is given in the periodic table for chlorine is 35.453. Explain why this atomic weight is not the same as the weight of either of these isotopes.
34. It was stated earlier that the element hydrogen has three isotopes: hydrogen-1, hydrogen-2, and hydrogen-3, but the atomic weight that is given for hydrogen in the periodic table is 1.008. Based on this, which hydrogen isotope do you think is the most abundant in nature? Explain the reasons for your answer.
35. Iodine-125 is a radioactive isotope that is used as a label in some types of analytical methods. How many protons, neutrons, and electrons are in one atom of this isotope?
36. Why are radioactive materials dangerous to a person's health?
37. List the various "families" of chemicals that were discussed in this chapter. Where does each of these families occur in the periodic table?
38. Describe some properties for each of the following types of elements.
    (a) metal   (b) nonmetal   (c) metalloid
39. Identify the following elements as metal, nonmetal, or metalloid: B, Si, F, H, Pu.
40. Place each of the following in as many of the listed subgroups of the periodic table.
    List: Representative Element, Transition Element, Inner-Transition Element, Metalloid, Alkali Metal, Alkaline Earth Metal, Halogen, Noble Gas.
    (a) Cr   (b) Na   (c) U   (d) Ca   (e) Cl   (f) S   (g) Ar   (h) Ge
41. Name an element found in each of the following groups: halogen, inner-transition element, representative element, noble gas, alkali metal, alkaline earth metal, transition metal.
42. What is a "monatomic ion"? Give two examples.
43. Explain why the elements in Group VIIA do not lose electrons readily.
44. Why is it easier for a nonmetal to gain electrons than to lose electrons?
45. Name 3 elements that have only 2 electrons in their outer level.
46. The usual ion of Mg has a +2 charge. Why do magnesium ions with charges of +1 and +3 not form?
47. What is a "polyatomic ion?" Give two examples.
48. (a) Write down the formula of the nitrate ion. _____
    (b) Write down the name of this ion: $PO_4^{3-}$. _____
    (c) Write down the formula of the hydroxide ion. _____
    (d) Write down the name of the following: $SO_4^{2-}$. _____

## For Class Discussion and Reports

49. For homework and class discussion: What are some recent news stories you have read or heard that require some basic knowledge of chemistry for a full understanding? Write a report.
50. For homework: Research and write a report on the process used for extracting iron from iron ore.
51. For homework: Using an Internet search engine, search "scanning tunneling microscopy" and report on some interesting aspect of this technique.
52. For class discussion: What are some issues regarding nuclear power generation and safety in your city/state?

# 2

# Compounds and Reactions

## 2.1   Introduction

As we discussed in Chapter 1 (Section 1.4.2), all of the over 13 million chemical compounds that exist consist of elements in chemical combination. Compounds that are "covalent" or "molecular" have molecules as their fundamental particles. Compounds that are "ionic" have formula units as their fundamental particles. Molecules are composed of various numbers of different atoms in covalent chemical combination; formula units are composed of various numbers of different ions.

Ionic and covalent are words that describe the nature of the chemical combination involved. In this chapter, we examine these concepts in more detail in order to characterize the nature of chemical combination and the reasons why elements come together to form compounds in the way that they do. We also look at the schemes for naming common compounds and at the symbolism used for expressing chemical change in the formation and destruction of compounds.

## 2.2   Formulas of Ionic Compounds

Positively charged ions are called *cations* (pronounced "cat-eye-uns") and negatively charged ions are called *anions* (pronounced "ann-eye-uns"). Cations and anions come together to form formula units because oppositely charged ions attract each other electrostatically. Cations attract anions, and vice versa, in a manner similar to the way in which opposite poles of a magnet attract each other. If the cation has

*Chemistry Professionals at Work*                                    *CPW Box 2.1*

# WHY STUDY THIS TOPIC?

Chemical laboratory workers handle commercial chemicals, especially compounds, in all phases of their work. Standard laboratory procedures refer to these chemicals by name and formula. Names and formulas of chemicals are found on labels identifying the contents of vials and bottles in the laboratory and lab workers must be able to recognize these names and formulas in order to handle chemicals safely and to accurately perform laboratory duties. Names and formulas of chemicals are utilized in scientific publications. When ordering chemicals to restock storage room shelves and when taking inventory of stock chemicals, a laboratory worker must be familiar with names and formulas.

Chemistry professionals perform chemical reactions as part of their jobs. The standard and accepted methods of representing chemical reactions on paper, such as in notebooks and scientific communications, must be a part of a chemist's knowledge base.

A familiarity with names and formulas of compounds and the accepted techniques of representing chemical reactions are thus essential to the job of a chemical technician.

*For Homework: Look at the list of ingredients on the label of a consumer product found in your home or garage. Select one chemical found there and research it by name. Report on its formula, properties, hazards, cost, and other descriptive information.*

a charge of +1 (as opposed to +2 or +3), it attracts and exactly neutralizes one anion that has a charge of $-1$. With these two ions electrostatically stuck to each other, the result is a formula unit that has neither a positive charge nor a negative charge, but is neutral, or has no net charge. An ionic compound, with a set formula, has thus formed. For example, one sodium ion, $Na^+$, and one chloride ion, $Cl^-$, by virtue of their equal but opposite charges, attract each other forming the formula unit depicted in Fig. 2.1(A). The result is a compound with the formula NaCl.[1] A similar one-to-one combination occurs between any cation with a charge of +1 and any anion with a charge of -1. Examples include the ammonium ion or any alkali metal ion (+1) combined with the nitrate ion, the hydroxide ion, or any monatomic ion formed from a halogen atom ($-1$). These latter ions are called the "halide" ions: fluoride ($F^-$), chloride ($Cl^-$), bromide ($Br^-$), and iodide ($I^-$). We thus have compounds with such formulas as LiBr, KI, $NaNO_3$, $NH_4F$, $NH_4OH$, etc.

One-to-one combinations also occur when the cation has a +2 charge and the anion has a $-2$ charge, since such a combination would also result in a neutral or uncharged unit. Examples of this include any alkaline earth metal ion (+2) combined with the sulfate ion and carbonate ions, and also the oxide ion ($O^{2-}$), or any other monatomic ion of the oxygen family (Group VIA), such as the sulfide ion ($S^{2-}$). Thus, we have compounds with such formulas as CaO, $MgSO_4$, $BaCO_3$, etc. Figure 2.1(B) depicts a formula unit of CaO as an example.

---

[1]The positive ion is always listed first in the formula of an ionic compound. Thus, it is NaCl, in this example, and not ClNa.

Two-to-one combinations occur when the cation has a +1 charge while the anion has a −2 charge. This is because two of the +1 charges are required to neutralize the −2 charge. Examples of this include the ammonium ion or any alkali metal ion combined with the carbonate ion, the sulfate ion, the oxide ion, or any other monatomic ion of the oxygen family. This gives us compounds with such formulas as $Na_2SO_4$, $K_2CO_3$, $(NH_4)_2S$, etc. Figure 2.1(C) shows $Na_2SO_4$ as an example. The reason for using parentheses in a formula now becomes a bit clearer (see Chapter 1, Section 1.4.2). A formula with parentheses highlights the polyatomic ion that is present in the formula unit.

One-to-two combinations occur when the cation has a +2 charge while the anion ion has a −1 charge. In this case, two of the −1 charges neutralize the +2 charge. Examples of this include any alkaline earth metal ion combined with the nitrate ion, the hydroxide ion, or any halide ion. Thus we have compounds with such formulas as $Ca(NO_3)_2$, $Mg(OH)_2$, and $BaCl_2$, etc. $Ca(NO_3)_2$ is shown in Fig. 2.1(D) as an example.

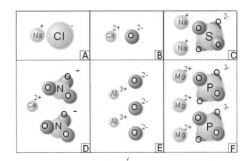

**FIGURE 2.1** Examples of formula units produced by the combination of various positive and negative ions. In each case, the net positive charge must be equal (but opposite) to the net negative charge. (A) a sodium ion with a chloride ion, NaCl; (B) a calcium ion with an oxide ion, CaO; (C) two sodium ions with a sulfate ion, $Na_2SO_4$; (D) a calcium ion with two nitrate ions, $Ca(NO_3)_2$; (E) two aluminum ions with three oxide ions, $Al_2O_3$; (F) three magnesium ions with two phosphate ions, $Mg_3(PO_4)_2$.

The phosphate ion has a −3 charge. Also, an aluminum atom has three electrons beyond that of neon, and therefore would form a +3 ion. The compound formed between aluminum and the phosphate ion would thus consist of a one-to-one combination, $AlPO_4$, since that is what would produce a neutral unit. Similarly, aluminum combined with a −1 ion would produce a one-to-three unit, and a +1 ion combined with the phosphate ion would produce a three-to-one unit. Thus we have compounds with such formulas as $AlCl_3$, $Al(NO_3)_3$, etc., and $Na_3PO_4$, $(NH_4)_3PO_4$, etc.

The existence of +3 ions and −3 ions such as the $Al^{3+}$ ion and the $PO_4^{3-}$ ion means that combinations of these ions with −2 and +2 ions, respectively, present some difficulties. For example, what combination of $Al^{3+}$ with the sulfate ion, $SO_4^{2-}$, or an alkaline earth metal ion with the phosphate ion, $PO_4^{3-}$, would produce a neutral unit? In the case of the aluminum ion and sulfate ion, it would take two aluminum ions and three sulfate ions, since the net charge from the aluminum would then be +6 and the net charge from the sulfate would be −6. Thus, we have the formula $Al_2(SO_4)_3$. In the case of an alkaline earth metal ion and the phosphate ion, it would take three alkaline earth metal ions and two phosphate ions, since the net charge from the alkaline earth metal ion would be +6 and the net charge from the phosphate would be −6. Thus we have the formula $Mg_3(PO_4)_2$ (see Examples (E) and (F) in Fig. 2.1).

It may be noted here that the value of the charge on the positive ion becomes the subscript for the negative ion, and vice versa, as depicted below for $Mg_3(PO_4)_2$.

$$Mg^{2+}PO_4^{3-} \Rightarrow Mg_2^{2+}PO_{4_3}^{3-} \Rightarrow Mg_3(PO_4)_2$$

One technique to determine the correct formula of a compound if we know the ions involved, then, consists of transposing the value of the charge of each ion to the subscript of the other. Such a technique will work for all formulas except where the values of the charges are both equal to 2, 3, 4, etc. An example of the latter is the formula of the compound formed between calcium and oxygen. Transposing the charges, as described above, would result in the formula $Ca_2O_2$, when actually the correct formula is CaO, since a one-to-one ratio would produce the required neutral unit. We should also mention that it is only the *value* of the charge that is transposed, and not its sign. For negative ions, such as −1, −2, −3, etc., the numbers 1, 2, 3, etc., are transposed.

## Example 2.1

What is the formula of the compound formed between barium and bromine?

## Solution 2.1

Barium is an alkaline earth metal and it loses 2 electrons when forming an ion. Bromine is a halogen and gains one electron when forming an ion. Thus, we have $Ba^{2+}$ and $Br^{1-}$. It would take 2 bromine (bromide) ions to form a neutral unit with the +2 barium ion, giving $BaBr_2$. Alternatively, we could use the technique of transposing the values of the charges. In this case, the subscript of the barium becomes 1 while the subscript of the bromine becomes 2, again giving the formula $BaBr_2$.

## Example 2.2

What is the formula of the compound formed between the ammonium ion and the carbonate ion?

## Solution 2.2

The ammonium ion, $NH_4^{1+}$, has a +1 charge, while the carbonate ion, $CO_3^{2-}$, has a −2 charge. Therefore, it takes two ammonium ions to form a neutral unit with the carbonate ion and the formula would be $(NH_4)_2CO_3$. Utilizing the technique of transposing charge values, the subscript for the ammonium becomes 2, and the subscript for the carbonate becomes 1. This means that $(NH_4)_2CO_3$ is the formula.

Some less common monatomic anions not mentioned in the previous discussion are hydride ($H^{1-}$), carbide ($C^{4-}$), nitride ($N^{3-}$), phosphide ($P^{3-}$), and selenide ($Se^{2-}$). Like the other monatomic ions discussed, these ions result from the elements gaining electrons so as to have an electron count equal to that of the noble gas atoms (see also Chapter 1, Section 1.8). Formulas of ionic compounds that include these anions can be determined by using the same schemes as those discussed earlier.

The charges on transition metal cations and cations of other metals, and the formulas of compounds in which they are found, are not as predictable, and can, in fact, be variable. Iron (Fe), for example, forms cations that have either a +2 charge or a +3 charge, neither of which is predictable from iron's position in the periodic table. These cations will be discussed in greater detail in Section 2.5 and in Chapter 4.

## 2.3   Naming Ionic Compounds

The naming scheme for ionic compounds is quite simple. We first state the name of the positive ion, which is the name of the metal if the positive ion is a metal ion, then state the name of the negative ion. The values of the subscripts do not enter into the naming scheme. Thus, the name of the compound with the formula NaCl is sodium chloride; the name of the compound with the formula $Ca(NO_3)_2$ is calcium nitrate; and the name of the compound with the formula $(NH_4)_3PO_4$ is ammonium phosphate. Several other examples are listed in Table 2.1.

One curious point is how the names of the monatomic anions are determined. The names are obviously derived from the names of the corresponding elements. For example, the monatomic ion derived from chlorine is "chloride"; the monatomic ion derived from oxygen is "oxide"; etc. The system for deriving these names is as follows: **stem of element name followed by "ide."**

**TABLE 2.1**   Examples of Ionic Compounds

| Formula | Name |
| --- | --- |
| $MgBr_2$ | Magnesium bromide |
| $K_2O$ | Potassium oxide |
| $Al_2(SO_4)_3$ | Aluminum sulfate |
| $NH_4OH$ | Ammonium hydroxide |
| CaS | Calcium sulfide |

For the halogens, the stem is the name of the element without the "ine" ending. So for fluorine, it is "fluor"; for chlorine, it is "chlor"; for bromine, it is "brom"; and for iodine it is "iod." This gives fluoride, chloride, bromide, and iodide as the names of the monatomic anions of the halogens.

For other nonmetals, there is no set system for determining the stem. For oxygen, it is "ox," and the anion is oxide. For sulfur, carbon, hydrogen, and nitrogen, the stem is the first four letters of the element name. Thus, for sulfur, it is "sulf" and the anion is sulfide. For carbon, it is "carb," and the anion is carbide. For nitrogen it is "nitr" and the anion is nitride. For hydrogen it is "hydr" and the anion is hydride. For selenium, the stem is the first five letters of the name, and for phosphorus, it is the first six letters of the name. Thus we have "selen" as the stem for selenium and "phosph" as the stem for phosphorus, and the anions are selenide and phosphide.

## Example 2.3

What is the name of the compound with the formula $MgI_2$?

## Solution 2.3

This is named by stating the name of the metal followed by the name of the ion. The name of the ion is "iod" followed by the "ide" ending. The name of the compound is magnesium iodide.

## Example 2.4

What is the name of the compound with the formula $(NH_4)_2CO_3$?

## Solution 2.4

The name of the cation is given first, followed by the name of the anion. The name of the compound is ammonium carbonate.

## Example 2.5

What is the name of the compound with the formula $Ca_2C$?

## Solution 2.5

The name of the metal followed by the name of the anion is the scheme used here. The name of the anion is "carb" followed by "ide." The name of the compound is thus calcium carbide.

## Example 2.6

What is the formula of the compound with the name potassium hydroxide?

## Solution 2.6

The symbol for potassium (K) is given first, followed by the formula of the hydroxide ion ($OH^-$). Remember that subscripts may be involved. The charge on a potassium ion is +1 while the charge on the hydroxide ion is $-1$. One potassium ion with one hydroxide ion produces a neutral unit. Thus, the formula of potassium hydroxide is KOH.

## Example 2.7

What is the formula of the compound with the name aluminum hydride?

## Solution 2.7

The symbol for aluminum (Al) is given first followed by the symbol for hydride (H). The appropriate subscripts are also included. Hydride is the monatomic ion of hydrogen and is a $-1$ ion. Aluminum forms a +3 ion. Thus, aluminum hydride has one aluminum with three hydrogens, or $AlH_3$.

You may have noticed in the previous discussions, that the cations in ionic compounds are almost always metal ions. The only exception to this that we have seen is the ammonium ion, $NH_4^+$. The reason for this, as stated in Chapter 1, Section 1.8, is that metals tend to lose electrons to become like the noble gas that precedes them. You may also have noticed that the anions in ionic compounds are always either ions of nonmetals or polyatomic ions containing only nonmetals. The reason for this, as stated in Section 1.8, is that nonmetals tend to gain electrons to become like the noble gas located to their right in the periodic table. Briefly stated, metals form cations and nonmetals form anions. While there are some important compounds in which metal atoms are a part of negatively charged polyatomic ions (as discussed in Chapter 4), the above description covers almost all ionic compounds.

What about covalent compounds? As stated previously, covalent compounds are nonionic, molecular compounds. For the most part, they contain only nonmetal atoms. A huge number of such compounds exist and the naming schemes for these are much more involved than what we have seen for ionic compounds. The schemes we will study in this chapter are limited to compounds that contain only two nonmetals and compounds that are "acids." Compounds, whether ionic or covalent, that contain just two elements are called **binary compounds**. Compounds, whether ionic or covalent, that contain three elements are called **ternary compounds.** The naming scheme for binary covalent compounds is discussed in Section 2.6. The naming scheme for acids, both binary and ternary, is discussed in Section 2.7.

## 2.4   Oxidation Numbers

There are many examples in which two elements combine in different atom ratios, forming two or more different compounds. The naming schemes we have seen thus far do not differentiate between such compounds. For example, iron forms two different compounds with chlorine, $FeCl_2$ and $FeCl_3$. Using the naming schemes we have seen earlier, both of these would be called "iron chloride." They are, however, two distinctly different compounds, each with unique properties. Thus, each must be given a unique name.

This situation occurs in ionic compounds of transition metals, inner-transition metals, and with metals located in Groups III(A), IV(A), and V(A) of the periodic table. This situation also occurs in covalent compounds, such as the compounds of carbon with oxygen, $CO$ and $CO_2$. The naming scheme for ionic compounds involves a concept known as the oxidation number of elements. It is also possible to name covalent compounds by this scheme.

The **oxidation number** of an element in a compound is a number assigned to that element that reflects how its electrons are involved in making up the compound. For example, if an atom of an element has lost one electron and formed a +1 cation, the oxidation number of that element in a compound is +1. If an atom of an element has gained one electron and formed a −1 ion, the oxidation number is −1. This may also be referred to as the **oxidation state** of an element. It is a sort of bookkeeping concept that is not only used in naming schemes, but is also used to track electrons in reactions that involve electron transfer and to predict chemical properties of compounds. Thus, it is a concept of some importance.

While the examples of a +1 or a −1 oxidation number may appear simple enough (being simply the charge on the ion), there are many cases in which the determination of an oxidation number is more involved. For this reason, a list of rules is used to help determine the oxidation number. These rules are listed in Table 2.2.

Let us make a few observations about these rules before we look at some examples of their use. First, in Rule 1 the *free elemental state* refers to the state of the element when it is not combined with any other element. For most elements, this is shown by the use of the symbol of the element, as found in the periodic table. For example, the oxidation numbers of silver metal (Ag), radon gas (Rn), and mercury liquid (Hg) would be zero. However, there are some elements whose free elemental state refers to *diatomic molecules*, or molecules that consist of two atoms of the element that are covalently combined. This list includes hydrogen gas ($H_2$), fluorine gas ($F_2$), nitrogen gas ($N_2$), oxygen gas ($O_2$), chlorine gas ($Cl_2$), bromine liquid ($Br_2$), and iodine solid ($I_2$). Thus, whenever these diatomic symbols are observed, these substances are in their free elemental state and the correct oxidation number to be assigned would be zero.

---

*Chemistry Professionals at Work*                    *CPW Box 2.2*

# THE FERTILIZER INDUSTRY

Chemical fertilizer is one of the most important products of the chemical industry. Ionic compounds containing nitrogen, phosphorus, and potassium (especially ammonium and potassium compounds, nitrate compounds, and phosphate compounds) are among the ingredients of fertilizers.

The two most important nitrogen-bearing fertilizers are ammonia ($NH_3$) and ammonium nitrate ($NH_4NO_3$). Ammonia ranked 6th and ammonium nitrate ranked 15th in total chemical production worldwide in 1995, according to the April 8, 1996 issue of *Chemical and Engineering News*. Over 85% of all ammonia produced in the world is used for fertilizer, either directly or in the manufacture of ammonium compounds, such as ammonium nitrate.

On most bags and containers of fertilizers, the nitrogen, phosphorus, and potassium content are indicated with three numbers separated by dashes, such as "6-12-6." The first number is the percentage of "total nitrogen," the second is the percentage of "available phosphorus," calculated by using the chemical formula $P_2O_5$, and the third is the percentage of "soluble potash" which has the chemical formula $K_2O$. Chemical process operators are responsible for mixing these ingredients in the proper ratios and chemical laboratory technicians are responsible for quality control to assure that the fertilizers do indeed conform to the numbers shown on the containers.

*For Homework: Visit with a local fertilizer dealer and ask him/her to interpret the label on the fertilizer container for you. Ask him/her some key chemistry questions, like: "What does 6-12-6" refer to?" and "What is potash?" Report on his/her knowledge of chemistry.*

---

**TABLE 2.2**   The Rules for Determining the Oxidation Number of an Element

---

1. Any element in its free elemental state has an oxidation number of zero.
2. Any element that exists as a monatomic ion has an oxidation number equal to the charge on the ion of that element.
3. Any alkali metal in the combined state has an oxidation number of +1.
4. Any alkaline earth metal in the combined state has an oxidation number of +2.
5. Any halogen in a binary compound with a metal or hydrogen has an oxidation number of −1.
6. Oxygen in the combined state has an oxidation number of −2, except in compounds called peroxides, in which it has an oxidation number of −1.
7. Hydrogen in the combined state has an oxidation number of +1 unless it is in a binary compound with a metal. In that case it has as an oxidation number of −1.
8. Aluminum in the combined state has an oxidation number of +3.
9. For any element not covered by the above rules, the oxidation number is determined as follows: (a) the oxidation numbers of all atoms in a neutral molecule or formula unit must add together to give a total net charge of zero, and (b) the oxidation numbers of all atoms in a polyatomic ion must add together to give the charge on the ion.

Second, in Rule 5, note that the $-1$ oxidation number for a combined halogen atom is strictly limited to binary compounds with a metal or hydrogen. We will see other oxidation numbers for halogens in Chapter 4. Third, Rule 6 mentions peroxides. A *peroxide* is a polyatomic ion that consists of 2 oxygen atoms that have a net charge of $-2$, or $O_2^{2-}$.

## Example 2.8
What is the oxidation number of nitrogen in $N_2$?

## Solution 2.8
The formula $N_2$ represents nitrogen in the free elemental state. Thus, according to Rule 1, the oxidation number is zero.

## Example 2.9
What is the oxidation number of barium in $BaSO_3$?

## Solution 2.9
Barium is an alkaline earth metal, and according to Rule 4, it has an oxidation number of $+2$ when it is in the combined state.

Rule 9 is needed in examples such as Cu in CuCl. Cu is not mentioned in Rules 1 through 8. Its oxidation number is determined by knowing that it must add to the oxidation number of Cl to give zero. The oxidation number of Cl is $-1$ (Rule 5) and so the oxidation number of Cu is $+1$ in CuCl.

Rule 9 can be a source of confusion in the sense that often, more than one atom of an element is present in the same formula. Examples are the potassium in $K_2S$, the oxygen in $NO_3^{1-}$, etc. In these cases, the rule should be interpreted to mean that the oxidation numbers of *all* atoms must add together to either give zero or the charge on the ion. Thus, in $K_2S$, because there are two potassium atoms, each with an oxidation number of $+1$ (Rule 3), the total contribution from potassium is $+2$, not $+1$; thus, sulfur's oxidation number must be $-2$ in order for the sum to be zero. In $NO_3^{1-}$, the total contribution from oxygen is $-6$ ($-2$ for each of the three oxygen atoms in the formula—see Rule 6); thus, the oxidation number for nitrogen must be $+5$ in order for the sum to be $-1$, the charge on the ion.

## Example 2.10
What is the oxidation number of sulfur in $Li_2SO_4$?

## Solution 2.10
Sulfur is not covered by Rules 1 through 8, so its oxidation number must be determined by Rule 9. The sum of the oxidation numbers of all the atoms in $Li_2SO_4$ must be zero. Lithium is an alkali metal and its oxidation number is $+1$. Since there are two lithium atoms in the formula, its total charge contribution is $+2$. Oxygen in the combined state, according to Rule 6, has an oxidation number of $-2$ and since there are four oxygen atoms in the $Li_2SO_4$ formula, the total charge contribution from oxygen is $-8$. In order for all oxidation numbers to add to zero, sulfur's oxidation number must be $+6$, since $+2$, $+6$, and $-8$ add to zero.

## Example 2.11
What is the oxidation number of phosphorus in the phosphate ion, $PO_4^{3-}$?

## Solution 2.11

Phosphorus is not covered directly by Rules 1 through 8, so its oxidation number is determined by Rule 9. The sum of the oxidation numbers of all the atoms in $PO_4^{3-}$ must be $-3$. All oxygen atoms have an oxidation number of $-2$ (Rule 6), and thus the total charge contribution from oxygen is $-8$. If the sum is to be $-3$, then the oxidation number of phosphorus must be $+5$, since $+5$ and $-8$ add to $-3$.

One final point about Rule 9 involves a situation in which there is more than one element in a formula that cannot be determined by Rules 1 through 8, such as when one is a transition metal and the other is a nonmetal that is part of a polyatomic ion. In this case, the charge on the polyatomic ion can be used to determine the oxidation number of the metal. The oxidation number of the nonmetal can then be determined as in Example 2.10.

## Example 2.12

What are the oxidation numbers of Fe and S in $FeSO_4$?

## Solution 2.12

The oxidation number of iron is $+2$ because the charge on the sulfate ion is $-2$ and they must add to zero. The oxidation number of oxygen is $-2$, giving a total charge contribution of $-8$ for four oxygen atoms. If iron is $+2$, sulfur must be $+6$ in order to add to zero ($+2 + 6 - 8 = 0$).

## Example 2.13

What are the oxidation numbers of Sn and P in $Sn_3(PO_4)_2$?

## Solution 2.13

The phosphate ion has a $-3$ charge and, since there are two phosphate groups in the formula, the total charge from the phosphate is $-6$. This means that in order to add to zero, the total charge of Sn must be $+6$ and, since there are three Sn ions in the formula, each individual Sn must have an oxidation number of $+2$. Similarly, the P and O oxidation numbers must add to $-3$, the charge on the phosphate ion. Oxygen is $-2$ for a total of $-8$, so phosphorus must be $+5$ ($+5 - 8 = -3$).

In some rare instances, an element may seem to have a *fractional oxidation number*. For example, in applying Rule 9, iron in $Fe_3O_4$ appears to have an oxidation number of $+8/3$, or $+2\frac{2}{3}$. Since it does not make sense to say that an atom has given up part of an electron (2/3), chemists have concluded that two of the Fe atoms are in the $+3$ state while one is in the $+2$ state. Because of this we say that the iron in $Fe_3O_4$ has a *mixed oxidation state*.

## 2.5 Naming Ionic Compounds of Metals that Have More Than One Oxidation Number

At the beginning of Section 2.4, we mentioned that more than one combining ratio can occur for certain elements. Compounds between iron and chlorine, $FeCl_2$ and $FeCl_3$, were cited as examples. The reason for this is that the cations of certain transition metals and other metals can have variable charges, or oxidation numbers (see the end of Section 2.2.). Nonmetals may also have various oxidation numbers. Naming schemes must be devised to deal with this situation.

Complicating the situation is the fact that two different (but related) naming schemes for this are in use. One is a classical, or traditional naming scheme, while the other is a newer scheme that was approved by a conference of the International Union of Pure and Applied Chemistry (IUPAC) organization in the 1920s.

---

*Chemistry Professionals at Work*          *CPW Box 2.3*

# HYDROGEN PEROXIDE

Rule #6 in Table 2.2 states that oxygen in peroxides has an oxidation number of "−1." This is an unstable oxidation number for oxygen since an oxidation number of "−2" would provide the stable electron count of the noble gases. This means that peroxides, such as hydrogen peroxide, $H_2O_2$, sodium peroxide, $Na_2O_2$, and organic peroxides are very reactive compounds because they take electrons from, or oxidize, other chemicals. The most common of all peroxides is hydrogen peroxide. It can be purchased as a 3% solution in any pharmacy. This solution has household uses as an antiseptic.

The most popular method of manufacturing hydrogen peroxide is by the reaction of a solution of anthraquinone, a complex organic molecule, with hydrogen in the presence of a catalyst. When the resulting product reacts with oxygen, the original anthraquinone and the hydrogen peroxide are formed.

Hydrogen peroxide has significant application in environmental water treatment because its reaction with microbial and chemical contaminants produces harmless oxygen gas and water. It is also used in the textile and paper industries as a bleaching agent.

*For Homework: Check out the following Internet site and other sources and write a report on the uses of hydrogen peroxide.*

http://www.h2o2.com/intro/overview.html (as of 6/4/00)

---

The first scheme utilized stems from the Latin names of the elements along with different suffixes, while the second utilizes Roman numerals to designate the oxidation numbers.

Let us look at three examples involving compounds of iron, copper, and tin to illustrate the two naming methods. Iron, in most of its compounds, is either in the +2 or +3 oxidation state. Copper, in most of its compounds, is either in the +1 or +2 oxidation state. Tin, in most of its compounds, is either in the +2 or +4 oxidation state. Notice that in each case there are predominantly only two oxidation numbers. We can designate one as the *lower* oxidation number and the other as the *higher* oxidation number. The traditional naming scheme utilizes the stem of the Latin name followed by "ous" for the lower oxidation number and "ic" for the higher oxidation number. This is then followed by the name of the anion, as in the normal scheme for binary ionic compounds. This method is illustrated in Table 2.3 for the chlorides of iron, copper, and tin.

The IUPAC naming scheme utilizes the English names and simply inserts a Roman numeral (in parentheses) between the name of the metal and the name of the anion, with the Roman numeral indicating the oxidation number. This is illustrated in Table 2.4 for the chloride compounds of iron, copper, and tin. Additional examples of the method are $Fe_2O_3$, which would be named either "ferric oxide" or "Iron (III) oxide"; $Cu_2CO_3$, which would be named either "cuprous carbonate" or "copper (I) carbonate"; and $Sn(NO_3)_4$, which would be named either "stannic nitrate" or "tin (IV) nitrate."

**TABLE 2.3**   Examples of the "Traditional" Scheme of Naming Ionic Compounds of Metals that Can Have Several Possible Oxidation Numbers

| Element | Latin Name | Stem | Compound | Oxidation Number of Metal | Traditional Name |
|---|---|---|---|---|---|
| Iron, Fe | Ferrum | Ferr | $FeCl_2$ | +2 | Ferrous chloride |
| | | | $FeCl_3$ | +3 | Ferric chloride |
| Copper, Cu | Cuprum | Cupr | $CuCl$ | +1 | Cuprous chloride |
| | | | $CuCl_2$ | +2 | Cupric chloride |
| Tin, Sn | Stannum | Stann | $SnCl_2$ | +2 | Stannous chloride |
| | | | $SnCl_4$ | +4 | Stannic chlorie |

**TABLE 2.4**   Examples of IUPAC Naming Schedule for Ionic Compounds of Metals that Have Various Oxidation Numbers

| Compound | Oxidation Number of Metal | Roman Numeral | Name |
|---|---|---|---|
| $FeCl_2$ | +2 | II | Iron (II) chloride |
| $FeCl_3$ | +3 | III | Iron (III) chloride |
| $CuCl$ | +1 | I | Copper (I) chloride |
| $CuCl_2$ | +2 | II | Copper (II) chloride |
| $SnCl_2$ | +2 | II | Tin (II) chloride |
| $SnCl_4$ | +4 | IV | Tin (IV) chloride |

## 2.6   Binary Covalent Compounds

Learning names of binary covalent compounds may also be troublesome because of the variety of naming schemes that exist. Covalent compounds are those that are formed between two or more nonmetals. Like the compounds of the metals that were discussed earlier, nonmetals can exist in a variety of oxidation numbers. Thus, naming schemes have also been devised to distinguish between two or more different compounds formed between the same two nonmetal elements. Examples are the compounds formed between carbon and oxygen, $CO$ and $CO_2$.

Probably the most useful naming scheme for this class of compounds is one that uses prefixes to indicate the number of each type of atom in the formula. These prefixes and the corresponding number of atoms they represent are given in Table 2.5. Utilizing this scheme, $CO$ would be named carbon monoxide and $CO_2$ would be named carbon dioxide. Notice in these examples that the element that is shown first in the formula is also named first. The second word utilizes the stem of the second nonmetal and the "ide" ending, as in the scheme for ionic binary compounds. However, a prefix is placed in front of the stem to indicate the number of atoms of the second nonmetal in the formula ("mono," in the case of one atom, and "di" in the case of two). Actually, a prefix is also used in front of the name of the first element if there is more than one atom of this element indicated in the formula. Thus, the compound $N_2O$ would be named dinitrogen monoxide, for example. The general scheme then is as follows. The *first word* in the name consists of the prefix (if more than one atom), followed by the name of the first element. The *second word* in the name consists of the prefix for the second element (in all cases) followed by the stem of the second element and the suffix "ide." Examples are presented in Table 2.6.

In addition to this scheme, the "ous" - "ic" and the Roman numeral schemes are also used. For example, $N_2O$ is also known as nitrous oxide and as nitrogen (I) oxide. While the Roman numeral scheme would provide a name for all possibilities, the "ous" - "ic" scheme is limited to only two compounds, the lower and higher of two oxidation numbers. Some nonmetal-nonmetal combinations can exist in multiple combining ratios. For example, there are seven oxides of nitrogen, $N_2O$, $NO$, $NO_2$, $NO_3$, $N_2O_3$, $N_2O_4$, and $N_2O_5$. Nitrous oxide is the name assigned to $N_2O$ and nitric oxide is the name assigned to $NO$. The rest of the oxides of nitrogen must be named by the other scheme.

*Chemistry Professionals at Work*                                    *CPW Box 2.4*

# OXIDATION NUMBERS AND "SHELF LIFE"

Not all oxidation numbers represent stable states for elements. Comparing $Fe^{2+}$ with $Fe^{3+}$, for example, $Fe^{3+}$ is much more stable. Contact with ordinary oxygen in the air is sufficient to convert $Fe^{2+}$ to $Fe^{3+}$. Ground water may contain iron in the "+2" state, but upon exposure to air it is converted to the "+3" state, which then exhibits undesirable properties. These include the formation of rust on household fixtures in contact with water, such as toilets and showers.

This type of problem is also encountered by chemistry laboratory professionals in their work and by consumers. A chemical in use in a laboratory may degrade due to unstable oxidation numbers and be rendered unusable after a period of time. This time period is called the "shelf life." In addition, pharmaceutical formulations often have an expiration date, past which the product should not be used. Chemists need to be aware and dispose of chemicals whose storage time has exceeded the shelf life. Consumers also need to be aware and dispose of expired pharmaceutical products.

*For Class Discussion: What do you think would happen if you were to consume a pharmaceutical product, such as a cough medicine, after its expiration date has passed? How do you think chemists determine what the expiration date is?*

**TABLE 2.5**    Prefixes for Naming Binary Covalent Compounds

| Number of Atoms | Prefix |
| --- | --- |
| 1 | Mono |
| 2 | Di |
| 3 | Tri |
| 4 | Tetra |
| 5 | Penta |
| 6 | Hexa |

**TABLE 2.6**    Examples of the Naming of Binary Covalent Compounds Using Prefixes

| Compound | Name |
| --- | --- |
| $CO$ | Carbon monoxide |
| $CO_2$ | Carbon dioxide |
| $N_2O$ | Dinitrogen monoxide |
| $NO_3$ | Nitrogen trioxide |
| $CCl_4$ | Carbon tetrachloride |
| $P_2O_5$ | Diphosphorus pentaoxide |
| $SF_6$ | Sulfur hexafluoride |

## 2.7 Identifying and Naming Acids

One subclassification of covalent compounds is that of **acids**. Acids have unique properties that are related to the fact that they are a source of hydrogen ions, $H^{1+}$. The details of these properties are presented in Chapter 12. Since they are a source of hydrogen ions, it follows that acids would have one or more hydrogens in their formulas. Also, since it is the hydrogens that give them their unique properties, the hydrogens are presented first in these formulas. Compounds whose formulas have one or more hydrogens that are listed first are acids. Common examples are the binary acids HF, HCl, HBr, HI, and $H_2S$. Common examples of ternary acids are those in which the hydrogens are bonded to the familiar polyatomic ions, $H_2SO_4$, $HNO_3$, $H_2CO_3$, and $H_3PO_4$. These are often called **oxyacids** because the third element of the "ternary" formula (besides hydrogen and a nonmetal) is oxygen.

Not all compounds that have hydrogen atoms covalently bonded to other atoms are acids. That is to say, not all compounds that have hydrogens in their formulas are sources of hydrogen ions. Some compounds with covalently bonded hydrogens have a potential to release hydrogen ions and some do not. Those that do release hydrogen ions have hydrogen listed first in the formula, while those that do not release hydrogens may have hydrogen given elsewhere in the formula. An example of a compound that has hydrogens, none of which can become hydrogen ions, is ammonia, which has the formula $NH_3$. This formula indicates that ammonia is *not* an acid. Those hydrogen atoms that have a potential for becoming hydrogen ions are called **acidic hydrogens**. There are some compounds that have both acidic and nonacidic hydrogens; these compounds have hydrogens in two different locations in the formula. An example of such an acid that is fairly common is acetic acid, which has the formula $HC_2H_3O_2$. Acetic acid, and other related acids will be further discussed in Chapters 4, 6, 11, and 12.

Let us examine the schemes for naming binary and ternary acids. All *binary acids* are named using the following scheme. The first word in the name consists of the prefix "hydro" followed by the stem of the nonmetal and the ending "ic." The second word in the name is the word "acid." For the *ternary acids* that we will look at here (we will look at others in Chapter 4), the first word consists of the stem of the nonmetal, followed by the ending "ic." The second word in the name is word "acid." The difference in these two schemes is the use of the prefix "hydro." This prefix is only found in the names of binary acids. Examples of names of acids are given in Table 2.7.

You may have noticed that the stem of the elements sulfur, carbon, and phosphorus that are used in the names of acids are different from the stems used for these elements in Section 2.3. In the cases of sulfur and carbon, the stems used in the names of acids are the complete names of the elements rather than "sulf" and "carb." In the case of phosphorus, the stem for an acid is "phosphor" rather than "phosph."

One final comment about acids is that binary acids that are not dissolved in water are gases. These gases can be named as if hydrogen were an alkali metal, which means that they can be named like the ionic compounds in Section 2.3. Using this method, HCl would be named "hydrogen chloride," HBr

**TABLE 2.7**  Formulas and Names of Common Binary and Ternary Acids

| Acid | Name |
|------|------|
| HF | Hydrofluoric acid |
| HCl | Hydrochloric acid |
| HBr | Hydrobromic acid |
| HI | Hydroiodic acid |
| $H_2S$ | Hydrosulfuric acid |
| $H_2SO_4$ | Sulfuric acid |
| $HNO_3$ | Nitric acid |
| $H_2CO_3$ | Carbonic acid |
| $H_3PO_4$ | Phosphoric acid |

would be named "hydrogen bromide," $H_2S$ would be named "hydrogen sulfide," etc. This is *not* true of *ternary* acids, however. For example, nitric acid, $HNO_3$, does not exist as a gas when it is not in a water solution and would *not* have "hydrogen nitrate" as an alternate name.

There are other common acids that have not been mentioned here. These will be named and dealt with in Chapter 4.

## 2.8    Bases and Salts

Another subclassification of compounds that is important is that of **bases**. *Bases* are the opposite of acids in the sense that they react with hydrogen ions rather than supply them. Examples of bases are compounds containing the hydroxide polyatomic ion, $OH^{1-}$. Thus, NaOH, KOH, $NH_4OH$, etc., are known as bases. They react with hydrogens because the hydrogens and the hydroxides together will form water, $H_2O$ (or HOH).

When water forms upon the contact of an acid with a base, another substance that forms is a salt. Salt in this context does not only refer to table salt (NaCl), but rather to any ionic compound (except for the hydroxides, which are bases). A *salt* can be thought of as a compound that forms as a result of the replacement of the hydrogen in an acid with a metal or the ammonium ion. For example, NaCl can form from HCl, $K_2SO_4$ from $H_2SO_4$, $NH_4NO_3$ from $HNO_3$, and so forth, and each of these products is a salt. This phenomenon will be discussed again in Chapters 4 and 12.

## 2.9    Introduction to Chemical Equations

Symbols and names of elements, as well as formulas and names of compounds, are fundamental to the description and expression of chemical principles and chemical processes. To describe the formation of water from the chemical combination of hydrogen and oxygen would be difficult to do without having the name "water" assigned to the clear, colorless, and odorless liquid that is involved in this reaction, or without having the names "hydrogen" and "oxygen" assigned to the colorless, odorless, and invisible gases that are involved. Additionally, knowing that the formulas of hydrogen and oxygen are $H_2$ and $O_2$, respectively, and knowing that the formula for water is $H_2O$ reduces the time and space needed to describe the process. In a laboratory notebook, a chemist might state that $H_2$ and $O_2$ combine to form $H_2O$, rather than writing the actual names of these substances. It is very important to have symbols and formulas available for the substances involved in chemical processes, in addition to the names. These symbols and formulas provide a type of "chemical shorthand" that chemists have found invaluable because of their convenience. Even if you have found the task of learning symbols, formulas, and names to be laborious, this work ultimately reaps the important reward of convenience.

The use of symbols and formulas is especially helpful in the expression of chemical change. Another term for chemical change, or the transformation of one or more substances with known properties into one or more other substances with completely different properties, is ***chemical reaction***. The chemical combination of hydrogen and oxygen to form water is an example of a chemical reaction. We say that hydrogen and oxygen "react" to form water, or that hydrogen "reacts" with oxygen to form water. A statement of a chemical reaction that utilizes the convenient chemical shorthand of symbols and formulas is known as a ***chemical equation***. In a chemical equation, the symbols and/or formulas of the substances present before the reaction takes place are written to the left and separated by a "+" sign. An arrow pointing to the right signifies a chemical change. Symbols and/or formulas of the elements or compounds that form as a result of the reaction appear on the right. These are also separated by "+" signs. Thus, for the water formation reaction we would write:

$$H_2 + O_2 \rightarrow H_2O \qquad\qquad (2.1)$$

*Chemistry Professionals at Work* CPW Box 2.5

# CHEMICAL EQUATIONS IN ANALYTICAL LABORATORIES

The jobs of virtually all chemistry professionals involve the use of chemical reactions. In order to represent and document these reactions, chemical equations are used in written analytical procedures, in laboratory notebooks, in scientific papers, and in encyclopedias and reference works. The ability to write and understand chemical equations and the ability to communicate via chemical equations is therefore very important. Beyond the qualitative, however, it is also necessary to write and understand chemical equations in terms of the measurement and expression of the quantities of chemicals involved. This means the ability to balance equations and to understand what a balanced equation represents must be a part of a chemist's knowledge base. Often the outcome and results of a task performed in the laboratory depend on it.

*For Homework: Find a chemical equation in a scientific paper, an encyclopedia, or reference work. Write a short paragraph explaining the value of this equation to the work being described. Also indicate whether it was important for this equation to be balanced in order to prove the chemist's point.*

The substance or substances to the left of the arrow, $H_2$ and the $O_2$ in the above example, are called the **reactants**. The substance or substances to the right of the arrow, water in this example, are called the **products** of the reaction. Thus, we have reactants and products separated by the arrow.

$$\text{reactants} \rightarrow \text{products} \tag{2.2}$$

The information in a chemical equation is verbalized by using the word "plus" or "and" or "reacts with" for the "+" signs on the left; the words "yields" or "reacts to produce" or "to form" for the arrow; and the word "plus" or "and" for the "+" signs on the right. Thus, the water formation reaction above might be verbalized as follows: "Hydrogen reacts with oxygen to form water," or "Hydrogen and oxygen react to produce water," or "Hydrogen plus oxygen yields water."

Examples in which there is only one reactant but two or more products, or two or more reactants with two or more products are verbalized similarly. For example, the equation

$$CaCO_3 \rightarrow CaO + CO_2 \tag{2.3}$$

may be verbalized as "Calcium carbonate yields calcium oxide plus carbon dioxide." And the equation

$$NaBr + Cl_2 \rightarrow NaCl + Br_2 \tag{2.4}$$

may be verbalized as "Sodium bromide reacts with chlorine to produce sodium chloride and bromine."

## 2.10   Balancing Chemical Equations

### 2.10.1   Quantitative Information from Chemical Equations

Chemical equations also can provide us with quantitative information, which means that it is possible to tell how much of a reactant or a product is involved. In order to do that, we assume that each symbol and formula present in the equation represents exactly one atom, one molecule, or one formula unit of the element or compound. It is possible to then indicate two or more atoms, molecules, or formula units by placing coefficients in front of each of the symbols or formulas, such that the number of total atoms of each element is the same on both sides. Such an equation is said to be "balanced." Utilizing our water formation example, the following represents a balanced equation:

$$2H_2 + O_2 \rightarrow 2H_2O \qquad\qquad (2.5)$$

We could now verbalize this equation as follows: "Two molecules of hydrogen and one molecule of oxygen react to produce two molecules of water."

Since two molecules of hydrogen gas, $H_2$, are represented on the left side, there are a total of four hydrogen atoms represented on the left (two molecules of $H_2$ with two hydrogen atoms each). Since there are two molecules of water represented on the right side, there are also four atoms of hydrogen on the right (two hydrogen atoms in each of two molecules of water). Also, since there is one molecule of oxygen on the left side (no coefficient means "one"), there are two atoms of oxygen on the left (two atoms per molecule); and, since there are two molecules of water on the right side, there are also 2 atoms of oxygen represented on the right (one atom per molecule). Thus, the equation is said to be balanced. The equation now indicates *quantities* of chemicals in terms of the number of molecules that are involved.

All chemical equations can be balanced. In Section 2.10.3, we will discuss the process of how simple chemical equations may be balanced. First, we must understand why we must balance these equations.

### 2.10.2   The Law of Conservation of Mass

One of the most basic of all chemical principles is the *Law of Conservation of Mass* which states that in a chemical reaction, mass can be neither created nor destroyed. For the purpose of this discussion, consider mass to be the same as weight. This means that the total mass of the products formed in a chemical reaction must equal the total mass of the reactants that were used. It is a law that is easily proven in the laboratory. If an enclosed container is used so that no material can escape (or be added) during the reaction, and if the reaction is not initiated until the mass of the container with reactants is measured, you should find that when the reaction is initiated and allowed to proceed, the mass of the container does not change.

A balanced chemical equation is a demonstration of the Law of Conservation of Mass. While in a typical chemical reaction atoms lose their identity because they enter into chemical combination, are released from chemical combination, or enter into a different mode of chemical combination, they do *not* lose their identity in terms of the number of protons in the nucleus.[2] Thus, for example, an atom of iron is an atom of iron in all instances, whether it is chemically combined or not. Recognizing this, we can restate the Law of Conservation of Mass to say that in a chemical reaction, the total number of atoms of each element present in the reaction container must be the same before the reaction begins and after the reaction is complete. The act of balancing a chemical equation, then, puts this important law into practice. The total number of atoms of each element on the left side of the equation must equal the total number of atoms of each element on the right side.

---

[2]The exception to this are "nuclear reactions" or reactions in which the nuclei of atoms are either split or fused. Nuclei and atoms of other elements are thus formed.

## 2.10.3 The Process of Balancing Equations

Many equations can be balanced "by inspection." This is a trial-and-error method by which we try various coefficients on both sides, going back and forth, from the left side to the right side, and then to the left side again, etc., until the equation is balanced. This will be the process described here.

Equations are balanced by placing coefficients in front of the individual symbols and formulas. The equation representing the water formation reaction

$$H_2 + O_2 \rightarrow H_2O \qquad (2.6)$$

is balanced as shown below.

Starting with the first element encountered as you look at the reaction equation from left to right, we see that the hydrogens are already balanced, with two being on the left and two on the right. Proceeding through the equation from left to right, we next encounter oxygen. We see that we need more oxygen on the right and that a 2 in front of the $H_2O$ would balance the oxygens:

$$H_2 + O_2 \rightarrow 2H_2O \qquad (2.7)$$

While the oxygens are now balanced, the hydrogens have become unbalanced. However, we see that placing a 2 in front of the $H_2$ would balance the hydrogens and this does not affect the oxygens.

$$2H_2 + O_2 \rightarrow 2H_2O \qquad (2.8)$$

The equation is now balanced.

As a general rule, the coefficients used should be whole numbers and not fractions. For example, this equation is balanced if the fraction $1/2$ would be used as the coefficient for $O_2$:

$$H_2 + \tfrac{1}{2}O_2 \rightarrow H_2O \qquad (2.9)$$

However, fractions in general should be avoided. Also, the coefficients used should be the smallest possible numbers. For instance, this same equation is balanced as follows

$$4H_2 + 2O_2 \rightarrow 4H_2O \qquad (2.10)$$

but the coefficients can be smaller and still give a balanced equation.

There are two very important rules that we must introduce here. These are (1) equations cannot be balanced by changing the subscripts in the formulas, and (2) equations cannot be balanced by placing coefficients in the middle of formulas.

Changing the subscripts in the formulas is incorrect because this changes the identity of the chemical that is involved in the reaction. For example, the water formation reaction can be balanced by placing a 2 as a subscript to the O on the right

$$H_2 + O_2 \rightarrow H_2O_2 \qquad \text{(incorrect)} \qquad (2.11)$$

but even though this equation is balanced, the result is incorrect because hydrogen peroxide ($H_2O_2$) would be listed as the product instead of water, which makes this an equation representing a different reaction from the original water formation reaction. We have changed the equation; we have not balanced it.

In the case of the second rule, we might be tempted to balance the equation by placing a 2 between the H and the O in the formula for water. Thus, we would have the following:

$$H_2 + O_2 \rightarrow H_2 2O \qquad \text{(incorrect)} \qquad (2.12)$$

This is incorrect because the formula on the right is not a legitimate formula.

Additional examples of the trial-and-error inspection process are given below; these examples provide some hints that will be helpful in certain cases.

## Example 2.14

Balance the following equation.

$$KClO_3 \rightarrow KCl + O_2$$

## Solution 2.14

Proceeding from left to right, we notice that the potassiums and the chlorines are already balanced with one of each on each side of the equation. The oxygens, however, are not balanced. We need more oxygens on the right. Let us place a 2 in front of the $O_2$.

$$KClO_3 \rightarrow KCl + 2O_2$$

This gives us four oxygens on the right but only three on the left. Let us place a 2 in front of the $KClO_3$.

$$2KClO_3 \rightarrow KCl + 2O_2$$

This gives us six oxygens on the left, but only four on the right. We need more oxygens on the right again. If we place a 3 in from of the $O_2$, we will have six oxygens on the right and this matches the six on the left.

$$2KClO_3 \rightarrow KCl + 3O_2$$

The potassium and the chlorine are not balanced now, but, placing a 2 in front of the KCl balances the potassium and the chlorine and does not change the balance of the oxygens.

$$2KClO_3 \rightarrow 2KCl + 3O_2$$

The equation is now balanced.

## Example 2.15

Balance the following equation:

$$Pb(NO_3)_2 + AlCl_3 \rightarrow Al(NO_3)_3 + PbCl_2$$

## Solution 2.15

The nitrate polyatomic ion ($NO_3^{-2}$) is seen in this equation on both sides. One helpful hint when you have the same polyatomic ion on both sides is to balance this ion as a group rather than doing the component elements individually. Keeping this in mind, let us balance the equation, starting with the first element on the left and proceeding to the right, as before. There is one lead on both sides, so they are already balanced, with one on each side. Next is the nitrate ion. There are two on the left but three on the right. If we place a 3 in front of the $Pb(NO_3)_2$ and a 2 in front of the $Al(NO_3)_3$, the nitrate ion will be balanced, with six on each side.

$$3Pb(NO_3)_2 + AlCl_3 \rightarrow 2\,Al(NO_3)_3 + PbCl_2$$

Proceeding to the aluminum, we need a 2 in front of the $AlCl_3$ to balance the two aluminums on the right. This gives us six chlorines on the left; and these are balanced by placing a 3 in front of the $PbCl_2$ on the right.

$$3Pb(NO_3)_2 + 2AlCl_3 \rightarrow 2Al(NO_3)_3 + 3PbCl_2$$

Rechecking the lead, we see that these are now balanced, as are all the other elements in the equation.

## Example 2.16

Balance the following equation:

$$NaOH + H_2SO_4 \rightarrow Na_2SO_4 + H_2O$$

## Solution 2.16

As in Example 2.15, this example has a polyatomic ion on both sides, in this case the sulfate ion, $SO_4^{-2}$. We will proceed to balance the sulfate as a group. This is also an example of the reaction of an acid with a base, with NaOH being the base (recognized as a base because of the hydroxide polyatomic ion) and $H_2SO_4$ being the acid (recognized as an acid because there is an H given first in its formula). In these so-called neutralization reactions, water is always a product. One trick that helps in balancing this equation is to consider the $H_2O$ as HOH so that the hydroxide can be balanced as a group. Proceeding now to balance the equation and starting with sodium, we see that we need a 2 in front of NaOH so that there are two sodiums on each side.

$$2NaOH + H_2SO_4 \rightarrow Na_2SO_4 + HOH$$

Proceeding to the right, we balance the hydroxide groups next. We need a 2 in front of the HOH so that we will have 2 hydroxides on both sides.

$$2NaOH + H_2SO_4 \rightarrow Na_2SO_4 + 2HOH$$

Continuing to the right, we see that the hydrogens that are not part of the hydroxide groups are already balanced, with two on the left (in the $H_2SO_4$) and two on the right (the first H in HOH). The sulfates are also already balanced with one on each side. All elements are thus balanced. It is now desirable to convert the HOH back to $H_2O$. The final balanced equation is shown below.

$$2NaOH + H_2SO_4 \rightarrow Na_2SO_4 + 2H_2O$$

# INDUSTRIAL CATALYSIS

Catalysis, or the speeding up of a chemical reaction using a catalyst, is a very important concept in the chemical industry. A number of industrial chemical processes utilize catalysts. An example is the manufacturing of sulfuric acid, $H_2SO_4$, which far outpaces all chemicals in terms of quantity produced by industry. One of the steps in the manufacturing process is the reaction of sulfur dioxide with oxygen to form sulfur trioxide.

$$SO_2(g) + O_2(g) \rightarrow SO_3(g)$$

This step uses an oxide of vanadium, $V_2O_5$, as a catalyst. It is an example of "heterogeneous catalysis," in which a solid material, the $V_2O_5$, is the catalyst for a process that otherwise involves only gases. (See also CPW Box 12.2)

*For Homework: The Ostwald Process for manufacturing nitric acid utilizes platinum as a catalyst. Look up information on the Ostwald Process and write a full report, including the balanced chemical equations involved, the role of the catalyst, and other details.*

## 2.11   Optional Symbols

Besides identifying what substances are the reactants and products of a chemical reaction and how many atoms, molecules, and/or formula units are involved, chemical equations can provide additional information if the writer so chooses. Symbols may be used to provide some qualitative information about the substances involved or about the reaction in general. However, such symbols are usually optional. Symbols giving specific information about the reactants or products are placed in parentheses immediately to the right of the chemical symbol or formula in the equation. Symbols giving information about the reaction in general are usually placed above the arrow.

Three examples of symbols that are placed over the arrows are the symbol for heat ($\Delta$), the symbol or formula for a catalyst (such as "Pt" for platinum), and the abbreviation "Elect" to indicate that electrolysis is required. A **catalyst** is a material that helps speed up a reaction but that is not chemically altered in the process. **Electrolysis** is a process whereby a chemical reaction is caused by the passage of electrical current through a solution. Some examples of equations in which these optional symbols are used are presented in Table 2.8.

To indicate the physical state of a substance, the symbols "s" (solid), "$\ell$" (liquid), and "g" (gas) are used. To indicate that a substance is dissolved in water, the symbol "aq" is used. This latter symbol is an abbreviation of the word aqueous, which means "dissolved in water." The word aqueous is derived from the Latin word "aqua," which means "water." To indicate that a product of the reaction is a gas that forms in a reaction medium that is not otherwise gaseous, we use an "up arrow" ($\uparrow$). This symbol refers to the fact that such a gas often escapes the reaction medium and is lost to the atmosphere. To indicate that a product of the reaction is a solid material that forms under the surface of a liquid (a precipitate), we use a "down arrow" ($\downarrow$). This symbol refers to the fact that such a solid often settles to the bottom of the container. Figures 2.2 and 2.3 show drawings of a gaseous product and a precipitate as described here and should help clarify these latter two symbols.

**TABLE 2.8**   Examples of Equations That Can Use the Special Symbols
Discussed in the Text. See the Text for the meaning of these Symbols

$$2H_2(g) + O_2(g) \rightarrow 2H_2O(\ell)$$
$$K_2SO_4(aq) + Ba(NO_3)_2(aq) \rightarrow BaSO_4(\downarrow) + 2KNO_3(aq)$$
$$Zn(s) + 2HCl(aq) \rightarrow ZnCl_2(aq) + H_2(\uparrow)$$
$$2KClO_3(s) \xrightarrow{\Delta} 2KCl(s) + 3O_2(\uparrow)$$
$$2H_2O(\ell) \xrightarrow{\text{elect}} 2H_2(\uparrow) + O_2(\uparrow)$$

**FIGURE 2.2**   Hydrogen gas bubbles up and out of the
solution when zinc metal is in contact with a solution of
hydrochloric acid. The "($\uparrow$)" symbol would be used with
the symbol for hydrogen in the equation for this reaction.

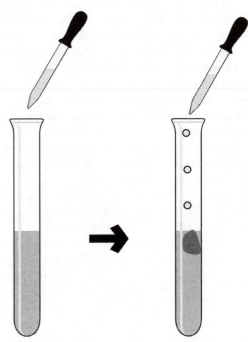

**FIGURE 2.3**   A solid material, or precipitate, can
form when an ingredient of one solution reacts with
an ingredient in another. The "($\downarrow$)" symbol is placed
to the right of the formula for the precipitate in the
equation for this reaction.

# WHO CHEMISTRY TECHNICIANS REALLY ARE

My name is Ed A. Teamer. I am a chemistry technician employed at the E.I.DuPont DeNemours agricultural chemical plant in LaPorte, Texas. I would like to give my perspective on the role that chemistry techinicians play in the modern workplace.

The image of Igor, Dr. Frankenstein's trusted assistant, has left a lasting, albeit inaccurate impression on the minds of many people. Today's chemical analysts, laboratory technicians, and chemist's assistants are much more than hunchbacked, mindless gophers who cater to the chemist's beck and call. As we embark on a new millennium, it is evident that laboratory professionals must be prepared to fill a variety of roles that will be necessary to ensure the continued success of scientific discovery.

As more and more chemists see their duties being expanded to entail administrative, customer-related, and business functions, technicians will be expected to fill the gap. Besides the normal duties of setting up chemical process simulations, synthesizing and analyzing compounds, and documenting results, technicians must also be able to perform repairs on instrumentation, calibrate equipment, and set up and maintain quality control systems. Additionally, they must be astute observers of any unexpected or extraordinary phenomena that might occur while performing routine chemical reactions. Many of our greatest discoveries have been the result of a technician's watchful eye.

Perhaps most importantly, technicians must be prepared for the enormous responsibility the workplace poses. An aptitude for mathematics, chemistry, mechanics, and computer usage is absolutely necessary in today's highly technical, electronic environment. Such proficiencies, coupled with an inquisitive spirit, will ensure that the stereotypical image of Igor from a bygone era is replaced with the reality of the new technician for the new millennium.

*For Homework: Research material from the American Chemical Society (ACS) national offices and also from the ACS Division of Chemical Technicians. Your instructor may be able to suggest specific publications. Write a report on the ACS's perspective on who chemical technicians are.*

## 2.12   Homework Exercises

1. What is a "cation"?
2. What is an "anion"?
3. What are the formulas of the compounds formed between each of the following?
   (a) calcium and oxygen
   (b) calcium and bromine
   (c) potassium and the phosphate ion
   (d) magnesium and the nitrate ion
   (e) aluminum and sulfur
   (f) the ammonium ion and the carbonate ion
4. What are the formulas for the compounds formed between each of the following?
   (a) potassium and oxygen
   (b) the ammonium ion and oxygen
   (c) rubidium and the sulfate ion
   (d) sodium and bromine
   (e) aluminum and chlorine
   (f) calcium and the nitrate ion
   (g) magnesium and the phosphate ion
   (h) barium and sulfur
5. Give the formula for the compound formed between each of the following.
   (a) calcium and fluorine
   (b) aluminum and oxygen
   (c) potassium and sulfur
   (d) magnesium and oxygen
   (e) rubidium and the carbonate ion
   (f) lithium and the phosphate ion
   (g) strontium and the hydroxide ion
   (h) the ammonium ion and chlorine
6. Could the compound $Ba_2SO_4$ exist? Why or why not?
7. Tell which of the following chemicals are ionic and which are covalent: $SrCO_3$, $P_2O_5$, $MgBr_2$, $NH_4NO_3$, $CCl_4$, $Al_2(SO_4)_3$, $C_3H_8O$, $NCl_3$, $FeBr_2$, $K_2O$, $MgCl_2$, $IBr$, $CO_2$.
8. What is a "binary compound"?
9. What is a "ternary compound"?
10. What is a "hydride"?
11. Define the term "oxidation number." Explain why this number is useful to chemistry professionals.
12. What is meant by the "free elemental state" of an element?
13. Write the free elemental state for each of the following elements.
    (a) Gold                      (b) Helium
    (c) Oxygen                    (d) Chlorine
14. Write the normal oxidation numbers for each of the following elements when they are combined with different elements in chemical compounds.
    (a) Oxygen                    (b) Hydrogen
    (c) Sodium                    (d) Calcium
    (e) Aluminum                  (f) Chlorine
15. What is the oxidation number for hydrogen in a hydride?
16. What is a "peroxide"? What is the oxidation number of oxygen in a peroxide?
17. What is the oxidation number of nitrogen in each of the following compounds or ions?
    (a) $N_2O$  (b) $NO_2^-$  (c) $NaNO_3$  (d) $N_2$  (e) $Ba_3N_2$  (f) $NH_4^{1+}$  (g) $NCl_3$

18. What is the oxidation number of sulfur in the following compounds or ions?
    (a) $Na_2S$   (b) $H_2SO_3$   (c) S   (d) $SO_4^{2-}$
19. What is the oxidation number of phosphorus in the following compounds or ions?
    (a) $P_2O_5$   (b) $PO_3^{-3}$   (c) $H_3PO_4$   (d) $PCl_3$
20. What is the oxidation number of each of the following?
    (a)  calcium in any compound
    (b)  sodium when it is not chemically combined with anything
    (c)  chlorine in NaCl
    (d)  manganese in $MnO_2$
    (e)  oxygen in $H_2O$
    (f)  iron in $FeCO_3$
    (g)  copper in $CuNO_3$
    (h)  tin in $Sn(SO_4)_2$
    (i)  copper and phosphorus in $Cu_3PO_4$
    (j)  tin and nitrogen in $Sn(NO_3)_2$
21. Give three examples of metals that can have more than one oxidation number when in the combined state.
22. Define the following terms. Explain how they are related to each other.
    (a) acid   (b) base   (c) salt
23. Identify each of the following as an acid, base, or salt.
    (a) $CaCl_2$   (b) $H_2SO_4$   (c) KI   (d) $NH_4OH$   (e) $NaC_2H_3O_2$   (f) $Ba(OH)_2$
24. Name the following compounds:
    (a) HBr   (b) CaO   (c) $P_2O_5$   (d) KI   (e) $FeCl_2$   (f) $Na_2S$   (g) $SO_3$
    (h) $Mg(OH)_2$  (i) $CCl_4$   (j) HCl   (k) $CaBr_2$   (l) FeO   (m) $CaSO_4$   (n) $FeBr_2$
    (o) $SF_6$   (p) HI   (q) $BaCl_2$   (r) $H_2S$   (s) MgO   (t) $CuCl_2$   (u) $PCl_3$
25. Give the formula of the chemicals with the following names.
    (a) carbon tetrachloride   (b) cuprous bromide   (c) ammonium chloride   (d) hydrosulfuric acid
    (e) sodium fluoride   (f) iron(III) oxide   (g) dinitrogen trioxide
    (h) stannous sulfate   (i) cupric iodide (j) sulfur hexafluoride (k) hydrofluoric acid
    (l) tin (IV) chloride (m) potassium oxide   (n) sodium hydroxide   (o) cupric oxide
    (p) hydrochloric acid   (q) diphosphorus pentoxide   (r) magnesium chloride
    (s) iron (II) sulfide   (t) dinitrogen pentoxide   (u) stannous fluoride   (v) aluminum bromide
26. For each name, select the correct formula from those given.
    (a)  magnesium bromide       $MgBr$, $MgBr_2$, or $Mg_2Br$
    (b)  phosphoric acid         $H_3PO_4$, $H_2PO_4$, $H_3PO_3$, or $H_2PO_3$
    (c)  ferric iodide           $FeI_2$, $FeI_3$, $Fe_2I$, or $Fe_3I$
    (d)  potassium sulfate       $KSO_3$, $K_2SO_3$, $KSO_4$, or $K_2SO_4$
    (e)  copper (I) carbonate    $CuCO_3$, $Cu_2CO_3$, $CuCO_4$, or $Cu_2CO_4$
    (f)  sulfur trioxide         $SO_3$ or $SO_3^{2-}$
    (g)  hydrosulfuric acid      $HSO_4$, $H_2SO_4$, HS, $H_2S$
    (h)  dinitrogen monoxide     NO, $NO_2$, $N_2O$, or $N_2O_3$
27. For each formula, select the correct name from those given.
    (a)  $HNO_3$                 hydronitric acid or nitric acid
    (b)  $CCl_4$                 carbon chloride or carbon tetrachloride
    (c)  $AlBr_3$                aluminum bromide or aluminum tribromide
    (d)  $SnO_2$                 stannic oxide, stannous oxide, tin oxide, or tin dioxide
    (e)  $(NH_4)_2CO_3$          ammonium carbonate, diammonium carbonate, or ammonium tricarbonate
28. Describe what is meant by a "chemical reaction."
29. Describe what is meant by a "chemical equation." How is this used by chemistry professionals?

30. Define the following chemical terms.
    (a) reactant   (b) product
31. What is the "Law of Conservation of Mass"? Why is this important to chemistry professionals?
32. Balance the following equations.
    (a) $Fe + O_2 \rightarrow FeO$
    (b) $P + O_2 \rightarrow P_2O_5$
    (c) $SO_2 + O_2 \rightarrow SO_3$
    (d) $H_3PO_4 + NaOH \rightarrow Na_3PO_4 + H_2O$
    (e) $BiCl_3 + H_2O \rightarrow BiOCl + HCl$
    (f) $S + O_2 \rightarrow SO_3$
    (g) $FeO + O_2 \rightarrow Fe_2O_3$
    (h) $Ag + H_2S + O_2 \rightarrow Ag_2S + H_2O$
    (i) $NO_2 + H_2O \rightarrow HNO_3 + NO$
    (j) $HgO \rightarrow Hg + O_2$
    (k) $Mg + H_2O \rightarrow H_2 + Mg(OH)_2$
    (l) $N_2 + (O_2 \rightarrow N_2O)$
    (m) $PCl_5 + H_2O \rightarrow POCl_3 + HCl$
    (n) $NaOH + H_3PO_4 \rightarrow Na_3PO_4 + H_2O$
33. State which of the reactions represented by the equations in Problem #32 are acid-base reactions and explain why.
34. State the meaning of each of the following chemical symbols.
    (a) (s)              (b) (g)              (c) ($\ell$)
    (d) ($\uparrow$)     (e) ($\downarrow$)   (f) ($\Delta$)
35. Consider the following chemical equation:

$$CaCO_3(s) + 2HCl(aq) \rightarrow CaCl_2(aq) + CO_2(\uparrow) + H_2O(\ell)$$

    (a) Write the formula for a chemical that is a reactant.
    (b) Write the formula for a chemical that is a liquid.
    (c) Write the formula for a product that is a gas.
36. A chemist observes the following reaction in her laboratory: magnesium metal (Mg) and hydrobromic acid (HBr) are the reactants; magnesium bromide ($MgBr_2$) and hydrogen gas ($H_2$), which bubbles up out of the solution, are the products. The HBr and the $MgBr_2$ are in water solution. Given this information, write a balanced chemical equation incorporating all the symbolism appropriate for this reaction.
37. Consider the following chemical equation:

$$Zn(s) + 2HCl(aq) \rightarrow ZnCl_2(aq) + H_2(\uparrow)$$

    Explain what the symbols (s), (aq), and ($\uparrow$) mean in this equation.
38. Given the following information, write a balanced chemical equation that includes all the appropriate formulas and symbols in their proper places. The products are $PbCrO_4$ and $KNO_3$. The reactants are $Pb(NO_3)_2$ and $K_2CrO_4$. The reaction requires heat in order to proceed. The $Pb(NO_3)_2$, $KNO_3$, and $K_2CrO_4$ are in water solution. The $PbCrO_4$ is a precipitate.

## For Class Discussion and Reports

39. Stannous fluoride has been an ingredient of most toothpastes for many years. Do a library and Internet search and write a report on the merits of stannous fluoride as an ingredient of toothpaste. Explain the mechanism by which tooth decay is prevented by this ionic compound.

40. Write a report on the presence of iron in well water. Is this a health hazard? What properties does it give the water? What changes in oxidation number occur before and after water containing iron is pumped from the ground? How does a chemist analyze for iron in the water?

41. Select one compound that is either a salt, a base, or an acid and find as much information about it as possible (for example, how it is manufactured, how much of it is produced annually, its uses in the chemical industry, where it appears in consumer products, safety issues concerning the compound, etc.). Write a report. Use a library, Internet service, or personal contacts for references.

42. Disposal of waste acids and bases is often an issue in chemistry labs. Visit a local industry and discover how disposal of acids and bases is handled in both the manufacturing area and, if applicable, in the laboratory. Identify one specific acid or base and report on specific methods used in the disposal of waste that contains that acid or base. If a chemical reaction is used to neutralize the waste, write and balance the equation.

# 3

# Atomic Structure, Light, and Spectroscopy

## 3.1  Introduction

One of the most fascinating topics in our study of chemistry is the concept of chemical combination. What is it about chemical combination that converts a highly flammable gas (hydrogen) and a gas that supports combustion (oxygen) into a liquid that puts out fires (water)? What is it about the process of two atoms of hydrogen and one atom of oxygen chemically combining to form a molecule of water that results in the formation of a whole new substance that in no way resembles the substances that went into it? And that's just one example! There are innumerable examples of this phenomenon in our world. With this chapter, we begin to unravel the mystery of chemical combination by studying the structure of the smallest unit of matter, the atom, in more detail.

You will recall that we previously defined chemistry as the study of the structure, composition, and properties of matter and the changes that matter undergoes. With an examination of the phenomenon of chemical combination comes a need for the consideration of the structure and composition portions of this definition. Specifically we need to look at the characteristics of that ultrasmall structural unit that is at the heart of the structure of all matter—elements, compounds, and mixtures alike—the atom. We presented some basic information about the atom in Chapter 1. We will now go into greater detail in describing its structure.

# WHY STUDY THIS TOPIC?

Atomic structure is at the heart of our study of the basic structure and composition of all matter. While this obviously important fact alone warrants a detailed study of the makeup of the atom, the basic concepts of atomic structure, in particular the energy of electrons, are fundamental to understanding some popular techniques that are used by chemists and laboratory technicians in their jobs. These techniques come under the general heading of "atomic spectroscopy" and have been very useful for the analysis of laboratory samples for metals. Examples include the analysis of environmental water samples for heavy metals such as mercury, the analysis of food samples for sodium, and the analysis of vitamin preparations for iron.

*For Class Discussion: For what other consumer products or environmental samples would an analysis for metals be useful? Why?*

## 3.2   History of Theories and Experiments

The structure of matter on a submicroscopic scale is something that has intrigued scientists for years. Modern theories of atomic structure had their origin in 1808 when **John Dalton**, an Englishman, published a theory based on certain facts and experimental evidence that were known at the time. Not only did he propose the existence of the atom, but he went on to precisely describe some attributes of the atom that we still hold to be true today. The listing of these attributes is called **Dalton's Atomic Theory** and consists of the following:

1. All matter is made up of tiny indivisible particles called atoms and the atom is the smallest structural unit of an element.
2. In terms of their mass and chemical and physical properties, atoms of the same element are all very similar and atoms of different elements are very different.
3. Atoms chemically combine with each other in simple whole number ratios to form molecules.
4. In chemical reactions, atoms of elements retain their identity, but rearrange, reorganize, and reunite in the formation of other substances.

Thus, important facts relating to the structure of matter have been known and accepted for some time. The mystery and intrigue since Dalton's time has mostly related to what is *inside* the atom. Indeed, if we are to understand more clearly the whys and hows of chemical combination (i.e., what exactly happens when two atoms, or an atom and a molecule, or two molecules, come together and undergo chemical change), then an examination of the interior of this mysterious little particle is in order.

Two of the particles we find inside the atom, the proton and the electron, were discovered in the late 1800s and early 1900s by scientists working with instruments and devices that were fairly sophisticated for the time. The electron was discovered and characterized by **William Crookes, J. J. Thomson,** and **Robert Millikan** as a result of independent experimentation occurring between 1879 and 1909.[1]

---

[1]Thomson (in 1906) and Millikan (in 1923) won Nobel Prizes in Physics.

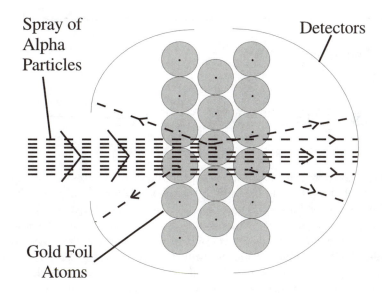

**FIGURE 3.1** Rutherford's gold foil experiment.

Some of the work involved the use of a device called a cathode ray tube. An electric current (which we now know to consist of moving electrons) passing between two metal plates in an evacuated glass tube could be made to "bend" when a magnet was brought near. Thomson measured the ratio of the electron's charge to its mass, and Millikan, through an experiment that has come to be known as the "Millikan Oil Drop Experiment," measured the charge of an electron by measuring the velocity of a tiny electrically charged oil droplet as it moved through a varied electric field. The mass of the electron was then calculated.

After the discovery and characterization of the electron, Thomson proposed a model for the atom in which the electrons were thought of as particles embedded in a "cloud" of positive charge. This model came to be known as the *"plum pudding" model* because the electrons were pictured like plums embedded in plum pudding. A famous experiment performed by a team of scientists led by the Englishman **Ernest Rutherford** in 1911 disproved this model, however.[2]

Rutherford used a spray of positively charged particles called alpha ($\alpha$) particles to probe the interior of the atom to determine the arrangement of the particles. Alpha particles are one of the three kinds of radiation that are discharged by radioactive materials. The others are beta ($\beta$) particles and gamma ($\gamma$) rays. An alpha particle consists of two neutrons and two protons in a single particle (a helium nucleus). Thus they have a +2 charge. Beta particles are electrons, and gamma rays are a very high-energy form of light. Alpha particles are positively charged (and very heavy, comparatively speaking) and were used by Rutherford for his experiment.

The atom Rutherford chose to probe was a gold atom—he used an extremely thin gold foil. He set up an experiment in which the gold foil was suspended directly in the path of the moving alpha particles, as pictured in Fig. 3.1, such that the alpha particles would strike the foil at a right angle for a direct hit. In front of and around the back of the gold foil he placed detectors that were sensitive to alpha particles so that the alpha particles could be detected wherever they ended up. The results produced a model for the atom that we still subscribe to today, almost a century later.

Rutherford's gold foil was mostly transparent to the alpha particles, that is, the vast majority of alpha particles passed through the foil and were observed by the detectors directly behind it, apparently unaffected by the presence of the foil in their path. It was as if the foil was not even there. A small number of the particles, while not completely blocked by the foil, were found by the detectors on the back side, deflected from their path a

---

[2]Rutherford won the Nobel Prize in Chemistry in 1908 for other related work.

---

*Chemistry Professionals at Work*                  *CPW Box 3.2*

# ALL IN THE FAMILY

A very famous scientist made her mark on history by winning two Nobel Prizes, one for physics in 1903 and one for chemistry in 1911. Her name was Marie Curie. The physics prize was awarded jointly with her husband, Pierre, for their study of radioactivity, a term she herself introduced. The chemistry prize was awarded for being the first to isolate the metal radium and measure its atomic weight. Pierre and Marie Curie's daughter, Irene, also won a Nobel Prize, jointly with her husband, Frederic Joliot. It was the chemistry prize in 1935 and it was awarded for their discovery that radioactive elements could be artificially prepared from stable elements.

The sad part of the story is that both Marie Curie and her daughter died of leukemia thought to have been caused by their exposure to dangerous levels of radioactivity. Marie died in 1934 and Irene in 1956.

*For Homework: Select one of the distinguished scientists mentioned here or in the text and write a report on his or her life.*

---

little above and a little below the main stream of particles. In addition, an extremely small number were found on the detectors on the front side, indicating that they had bounced off the foil and did not pass through it.

The model derived from these observations is called the **Rutherford model**, or the **nuclear model**, as shown by the shaded circles in Fig. 3.1. The scientists concluded that most of the mass of the atom had to be concentrated into an extremely small volume of space inside the atom which they called the nucleus (the dots in the figure). That was the only way to explain the small number of alpha particles that were found reflected off the front side of the foil. In addition to this small volume of space where most of the mass is found, they concluded that the nuclei of neighboring atoms were surprisingly distant from one another and that the masses of any particles found there had to be extraordinarily small. This was the only way to explain finding the vast majority of the particles directly behind the foil from the point where they would have struck it.

To explain the particles that were deflected from their path and found a little above or a little below the main grouping of particles, the scientists concluded that the nucleus must have a charge that would repel the alpha particles, causing them to slightly alter their course if they came near it. Thus, the nuclear model was born. In this model, the space outside and between the nuclei is viewed as being mostly empty space which is occupied by the electrons, the extremely small and light particles of negative charge.

In order to explain some cathode ray phenomena observed 25 years earlier by other scientists, Rutherford proposed in 1914 that small positively charged particles were contained within the nucleus. He named these particles protons. Neutrons in the nucleus were proposed by the British scientist **James Chadwick** in 1932 as a result of additional experiments.[3]

Two other models for the atom are important. One was proposed by a Danish scientist by the name of **Neils Bohr** in 1913.[4] This model was called the "***solar system model***" because he proposed that the

---

[3]James Chadwick won the Nobel Prize in Physics in 1935 for his discovery of the neutron.
[4]Neils Bohr won the Nobel Prize in Physics in 1922 for his solar system model.

TABLE 3.1    Names and Years for Noteworthy Achievements Relating to the Development of Atomic Theory

| Year(s) | Scientist(s) | Achievement |
|---------|--------------|-------------|
| 1808 | Dalton | Atom |
| 1879–1909 | Crookes, Thomson, Millikan | Electron characterized |
| 1909 | Thomson | Plum pudding model |
| 1911 | Rutherford | Nuclear model |
| 1913 | Bohr | Solar system model |
| 1926 | Schrodinger | Quantum mechanical model |

electrons orbit the nucleus much like the planets orbit the sun. The other was proposed by Austrian physicist **Erwin Schrodinger** in 1926.[5] The Schrodinger model is called the **"*quantum mechanical model*"** and is the model that we use today to explain and predict atomic behavior. Each of these two models is explained more fully in the sections to follow. A summary of the history of the development of atomic theory is given in Table 3.1.

## 3.3   The Modern Theory of Light

In order to have a clear understanding of the atomic theories espoused by Bohr and Schrodinger, and therefore a clear understanding of the current models of atomic structure, we must first briefly explore the modern theory of the nature of light, which is a form of energy.

### 3.3.1   What is Energy?

*Energy* is defined as the capacity for causing matter to move due to an applied force. Energy can take many forms. First, there is *mechanical energy*, also called *kinetic energy*, or the energy of motion. An example of this is the energy of a moving bowling ball as it speeds down a bowling lane. The energy it possesses is its capacity to move the pins. The kinetic energy a moving object has depends on its mass and velocity. The faster it moves and the heavier it is, the more energy it has. The mathematical relationship to describe this is:

$$\text{Kinetic Energy} = \tfrac{1}{2}(mv^2) \tag{3.1}$$

in which m is the mass and v is the velocity of the object.

There is also *potential energy*, or the energy an object has because of its position rather than its motion. An example of this is the energy of an object sitting on the edge of a table. Given a nudge, it falls from the table and its potential energy is released as kinetic energy. But as long as it sits at the edge, its energy is stored as potential energy.

This latter example illustrates an important point—one form of energy can be converted to another form and anything that can *potentially* cause an object to move is a form of energy. This means that we can add light energy, electrical energy, chemical energy, heat energy, and nuclear energy to the list because all have the *potential* to cause an object to move. The electrical energy possessed by a battery causes the Energizer™ Bunny to move. When the battery is put to use, the chemical energy possessed by the chemicals present inside the battery is converted to electrical energy and ultimately to kinetic energy. Light energy from the sun can be converted to electrical energy via the solar energy cell in a solar calculator. Heat energy heats our homes in the winter. Nuclear energy possessed by certain isotopes held

---

[5]Schrodinger won the Nobel Prize in 1933 for his quantum mechanical model.

as the "fuel" in a nuclear power plant can be converted to electrical energy to drive a variety of devices in our homes. The present discussion will be limited to light energy, however. We need to have a basic sense of what light is so that we can discuss the relationship of light to the modern theory of atomic structure.

### 3.3.2   What is Light?

The modern theory of light is that it has a dual nature, a rather complex and abstract notion. First, some properties of light are best explained if we think of it as a stream of particles emanating from the source of the light (light bulb, sun, etc.). These particles are called ***photons***. This is the ***particle theory***. Other properties of light are best explained if we consider it to be a wave-like disturbance that has both an electrical component and a magnetic component. This is the reason light is sometimes called electromagnetic radiation. This is the ***wave theory***. The wave theory is the most useful to us for our present purpose.

In the wave theory, light is thought of as a continuous series of repeating electromagnetic waves that travel through space and matter. As an analogy, you may picture a single wave as what you observe when a rock is dropped into a pond of water. A "water wave" is created that travels outward from where the rock contacted the water. The differences are (1) the water wave is a "mechanical" wave, as opposed to light, which is an electromagnetic wave and, (2) the water wave is just one wave, whereas electromagnetic radiation is a continuous series of waves, repeating one after the other, flowing from the light source. A mechanical wave requires matter to exist. Electromagnetic waves do not require matter to exist because they are electrical and magnetic in nature. That is why we are able to see light from the sun. The void (absence of matter) between the Earth and the Sun is no problem for light as it travels to the Earth.

With electromagnetic radiation, the waves are repeating and there is a finite distance from the crest of one wave to the crest of the next wave. This distance is called the ***wavelength***. Some wavelengths are extremely short while others are extremely long. A wavelength can be shorter than the radius of the smallest atom, or it can be as long as several miles. See Fig. 3.2 for the concept of wavelength and for drawings of two waves with different wavelengths. Wavelengths are expressed in units of length. Extremely short wavelengths are expressed in nanometers (1 billionth of a meter) or angstroms, which are on the order of the size of an atomic radius. Extremely long wavelengths are expressed in kilometers. (See Chapter 7 for a basic discussion of the units of length.)

The energy of light depends on this wavelength. The shorter the wavelength, the greater the energy. Thus, very short wavelengths of light (angstroms) possess a great deal of energy—enough to do harm to the human body (e.g., gamma rays and x-rays). Very long wavelengths, like radio and TV waves (kilometers), are not harmful at all.

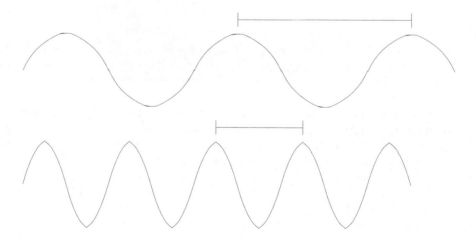

**FIGURE 3.2**   Drawings of two waves with different wavelengths. Wavelength is the distance from the crest of one wave to the crest of the next, as indicated.

---

*Chemistry Professionals at Work* *CPW Box 3.3*

# USING LIGHT FOR LABORATORY ANALYSIS

How much light is absorbed by a solution can be a measure of how much of a particular constituent is present in that solution. Such an analysis may tell a chemist how much iron is present in a sample of well water, for example. Other kinds of samples, such as food, soil, and manufactured products may also be analyzed in this way. This involves the use of an instrument called a spectrophotometer. This instrument, such as the one pictured below, has a light bulb inside that is positioned in such a way that a light beam is directed through a transparent sample holder (called a "cuvette") filled with the sample solution. A particular wavelength of light is first isolated from the beam so that only this one wavelength is directed at the cuvette. There is a device inside for isolating this wavelength. A Spectronic 20 spectrophotometer is shown. A chemist is placing the cuvette with the test solution into the instrument.

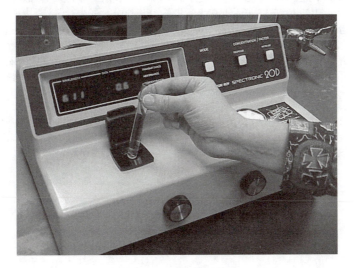

*For Homework and Class Discussion: Visit with a chemist or chemistry technician in a laboratory in your locale and ask about his/her use of a spectrophotometer. Report back to the class on what type of sample was being analyzed, what the constituent was, how the sample was prepared, how the data were evaluated, and why the spectrophotometer was the tool of choice.*

---

The human eye is sensitive only to a narrow range of wavelengths, the so-called "visible" wavelengths. Hence, these are the only wavelengths we can see. Visible light wavelengths are seen as colors. While visible light, such as from the sun or from a light bulb, is "white," this light is actually a mixture of all the colors, or of many different wavelengths. Under certain conditions, these wavelengths can be separated from each other, such as by water droplets in the atmosphere when a rainbow is visible or by a prism which is a solid piece of glass that has a triangular shape. In addition, the colors that various forms of matter appear to have are due to the fact that some wavelengths get absorbed by matter and some do not. For example, if the cover of the notebook you use for taking class notes is red, it means that the red wavelengths are being reflected off (transmitted by) the cover while all other wavelengths are being absorbed. The concept of absorption and transmission of light by matter is of considerable importance as we will see as our study continues.

The wavelength range of visible light is very narrow compared to the range that encompasses all light. There are wider ranges of wavelengths that are much shorter, for example gamma rays, which have wavelengths on the order of a fraction of an atom's diameter, and ultraviolet waves, which are just outside the visible range on the short side. There are also wider ranges of wavelengths that are much longer, for example radio and TV wavelengths, which are on the order of miles in length, and infrared wavelengths, which are just outside the visible range on the long side. Blue light represents the shorter wavelengths in the visible range while red light represents the longer wavelengths. Green and yellow are in between, as in the rainbow. The wavelength range for visible light is about 350–750 nanometers. While these wavelengths are extremely short (the same as about 16,000 hydrogen atoms lined up end to end), their energies are not harmful under normal circumstance. In fact, we depend on visible light in order to view everything we see. Blue light, possessing the shorter wavelengths in this range, has the highest energy. Red light, possessing the longer wavelengths, has the lowest energy. Green and yellow are in between.

As mentioned, ultraviolet wavelengths are shorter than visible wavelengths and infrared wavelengths are longer than visible wavelengths. Thus, of the three, ultraviolet has the greatest energy (e.g., causes sunburn) while infrared has the least energy. Ultraviolet, visible, and infrared wavelengths are used routinely in a chemistry laboratory for chemical analysis purposes as we will see later in this chapter. The complete range of all light, from the short wavelength gamma rays to the long wavelength radio and TV waves, is called the ***electromagnetic spectrum***. See Fig. 3.3 for a graph of the electromagnetic spectrum summarizing the points made in this section.

Wavelength is symbolized by the Greek letter, lower case lambda, $\lambda$. Thus we might refer to a visible wavelength of 534 nanometers as $\lambda = 534$ nm, or an ultraviolet wavelength of 256 nanometers as $\lambda = 256$ nm (where "nm" is the abbreviation for nanometers).

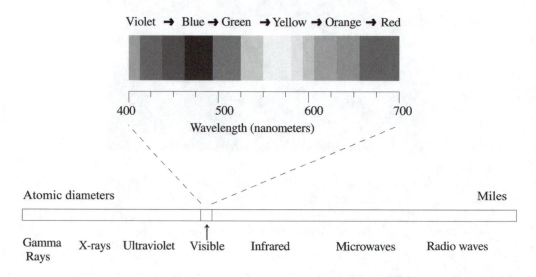

**FIGURE 3.3** The electromagnetic spectrum showing the visible range as a very narrow range, (below) expanded to show the various colors and corresponding wavelengths expressed in nanometers (above).

Wavelength can also be described by its energy or its frequency. Wavelength and frequency (symbolized by the Greek letter lower case nu, $\nu$) are related mathematically as follows:

$$c = \lambda\nu \tag{3.2}$$

where c is the speed of light. Energy (E) and frequency are related by:

$$E = h\nu \tag{3.3}$$

where h is a proportionality constant called "Planck's constant." Thus, by combining Eqs. (3.2) and (3.3), energy and wavelength are related by:

$$E = h\left(\frac{c}{\lambda}\right) \tag{3.4}$$

Equation (3.4) indicates that energy and wavelength are inversely proportional, meaning that the shorter the wavelength, of light, the greater the energy of light will be. The longer the wavelength, the less the energy of the light will be.

Concepts of light and energy will be revisited and mathematically treated in Chapters 6 and 13.

## 3.4 The Bohr Model

### 3.4.1 Introduction

With the protons and neutrons in the nucleus and buried deep inside the atom, it is logical to think that it is the electrons that play the most important role as far as contact with other atoms is concerned. As represented in Fig. 3.4, it is logical to assume that it is the outer edges of the atomic spheres, and thus the electrons, that interact when there is a close approach between the atoms ultimately resulting in chemical combination.

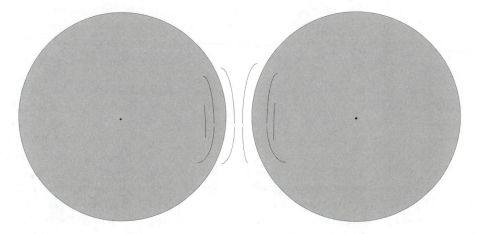

**FIGURE 3.4** The outer edges of the atomic spheres, where the electrons are located, interact when there is chemical combination.

---

*Chemistry Professionals at Work*                                          **CPW Box 3.4**

# FIREWORK DISPLAYS

As has been mentioned in the text, the colorful light we see in firework displays is due to the line spectra emitted by the elements contained in the fireworks. People who design and assemble the shells used in fireworks are called "pyrotechnicians." The actual design of a shell involves a casing, several fuses, including time-delay fuses, black powder propellant, bursting charges, and "stars." These stars contain the actual light-emitting compounds.

The pyrotechnicians use many different chemical compounds in the stars. Compounds of strontium and lithium are used to produce red colors; compounds of barium to produce greens; compounds of copper to produce blues and compounds of sodium and calcium to produce yellows and oranges.

The energy produced by the exploding charges are absorbed by the atoms of the metals and the electrons are elevated to a higher energy level. When they drop back to the original level, the energy that is lost is emitted in the form of light. The different colors correspond to different amounts of energy lost, which in turn depends on how far the electrons drop back. (See Section 3.8 for more information about specific energies involving specific elements.)

*For Homework: Research this topic and write a more detailed report on how fireworks are designed. If available, use the following reference: Graham, "Colors bursting in air," ChemMatters, American Chemical Society, October 1998, pages 7–9 .*

---

In the time immediately following Rutherford's experiment, the electrons and the space they occupy still remained a mystery for scientists. They still had to consider important questions. Exactly how do we picture the electrons? If they are so very small, exactly where are they located in this vast empty space? Since positive charges attract negative charges, why don't these small negatively charged particles fall into the positively charged nucleus? Exactly what is the role of electrons when chemical combination takes place? The next step appeared to be the development of a model that would fit the known data and answer these questions.

An explanation as to why the electron doesn't fall into the nucleus is that each electron travels in an orbit, much like the planets travel around the sun, such that they are influenced by a force directed outward (centrifugal force) that would counterbalance the force of attraction for the nucleus. This would be similar to a string tied to a ball that you would twirl around with your hand. While there is an attractive force between your hand and the ball (represented by the pulling force exerted on the string), there is an equal but opposite force, due to the ball's motion, that keeps the ball from crashing into your hand. This solar system model became popular, and indeed is still popular today. The idea of electrons as particles moving around the nucleus in orbits much like planets moving around the sun is a very common picture of electrons. A rather crude drawing of this may be familiar to you. An example is shown in Fig. 3.5.

According to this model, electrons, as moving particles, have kinetic energy. In order for an electron not to fall into the nucleus, it must have sufficient energy to keep it aloft. The further the electron is from the nucleus, the more energy it must have—the longer the string, the faster the ball must move to keep the string taut. It is clear with this model that an electron's energy is associated with its speed. The further it is from the nucleus, the faster it must move, and thus the more energy it must have [recall Eq. (3.1)]. While these ideas did not answer all the questions, it was a significant step forward. There was also the matter of light emission by chemical substances under certain conditions that Bohr wanted to address.

## 3.4.2 The Bohr Model and Light

All matter tends to want to be in its lowest possible energy state. When an object on the edge of a table is gently pushed off the edge, it falls to the floor—it loses energy. The Bohr theory states that when an electron is in an orbit of high energy, it will drop to an orbit of lower energy if a vacancy for it exists in the lower orbit. This explained a strange phenomenon observed at the time—atomic spectra.

Bohr had observed that under certain conditions, elements can emit light. Examples of this emission that we might observe today include fireworks and colored flames. Bohr's explanation was that when an atom is energized, such as in a fireworks display or a flame, an electron gains energy and moves from an orbit of lower energy to one of higher energy. Because this new orbit is not the lowest possible energy state for the electron to be in, and since there is a vacancy in a lower orbit, the electron subsequently loses this gained energy and drops back to the lower orbit. The energy that is lost is the energy of light, and the light that is created is seen as a particular color (wavelength). What color it is depends on exactly how much energy was lost.

**FIGURE 3.5** A drawing of electrons orbiting the nucleus of an atom.

In most cases, more than one wavelength is emitted due to more that one orbit transition occurring in the atom, creating a pattern of wavelengths called a line spectrum. See Fig. 3.6 for

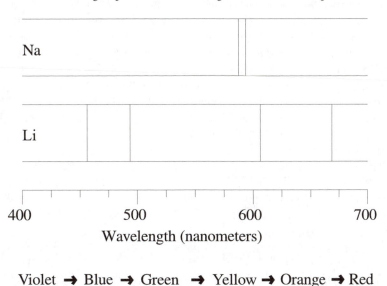

**FIGURE 3.6** Line spectra of sodium and lithium. The sodium emission is bright yellow. The lithium emission is bright crimson.

representations of the line spectra of sodium and lithium. Notice that even though the lithium emission includes light ("lines") in the blue, green, yellow, and red regions, we perceive the combined light as a crimson color. The particular pattern of light emitted is unique to the element causing the emission. A **line spectrum** is a display of lines of light of different wavelengths (colors), with each line representing a distinct wavelength or amount of energy that corresponds to a particular energy transition in an atom.

Thus, the acceptance of the concept of the energy of the electron and the fact that part of this energy can be lost in the form of light, is key to the acceptance of this Bohr model. In the language of modern chemistry, we say that the energy of the electron is "quantized," which means that the Bohr orbits represent levels of energy between which the electron can move and that the electron can only have those energies that are represented by these levels. In other words, the electron cannot be found anywhere but in an orbit, meaning it cannot be positioned between orbits.

The Bohr model was important because, while not the correct model, it led to the modern model of the atom, the model that is widely accepted today as the true model of atomic structure.

## 3.5   The Schrodinger Model

The model of electronic structure that is widely accepted by scientists today changes one basic premise of the Bohr Model and, in doing so, makes the entire picture much more abstract and more difficult to understand. The one basic premise that the modern theory dismisses is that the electron is a particle traveling in an orbit.

The modern theory of the electron is very similar to the modern theory of light. As complex as it might seem, scientists today believe that the electron has a dual nature. Some of its properties are best explained if we assume that an electron is a particle, while other properties are best explained if we assume that the electron is an electromagnetic wave. The modern theory of the electron, while maintaining that the electron has energy, says that the electron is not a particle traveling in well-defined orbits, as specified in the Bohr Model. Rather, the electron is an entity with the same electromagnetic characteristics as light, but which is, with a very high probability, confined to a particular region of space inside an atom.

The modern theory described above was initially introduced by Erwin Schrodinger in 1926, as mentioned in Section 3.2. It is popularly known as the **quantum mechanical model**. This model is based on some very complex mathematics, with which we will not concern ourselves here. The essence of the model is that electrons exist in **principal (or "main") energy level**s (sometimes called **shells**), in energy **sublevels** (sometimes called **subshells**) within these principal levels, and in regions of space called **orbitals** within the sublevels. These electrons also have a particular spin direction. Light emission by elements, or atomic spectra, is explained by nearly the same statements as given by the Bohr model. The energy entity we call the electron can jump from one level to another. When it falls from a higher level to a lower one, energy is lost and this energy may be given off in the form of light, often visible light as in fireworks or colored flames.

Principal energy levels for electrons are characterized by a number assigned to that level. This number is symbolized by the letter $n$. The principal level closest to the nucleus is the first level and is referred to as the $n = 1$ principal level. Proceeding out from the nucleus, the next principal level is the $n = 2$ level, the next is the $n = 3$ level, etc. Figure 3.7 presents a simplified view of the first few principal levels.

The values that $n$ can take on (1, 2, 3, 4, 5, etc.), correspond to the first, second, third, fourth, fifth, etc., levels. The larger the number of the level, the further from the nucleus. Theoretically, the number of such levels is limitless. Practically, however, no atom that we currently know of has more than seven principal levels. Thus, the electron configuration of all elements in the periodic table can be described by using just seven principal levels. Since there are over 100 elements in the periodic table, it follows that the average number of electrons that can be found in each level is relatively large. In addition, the number

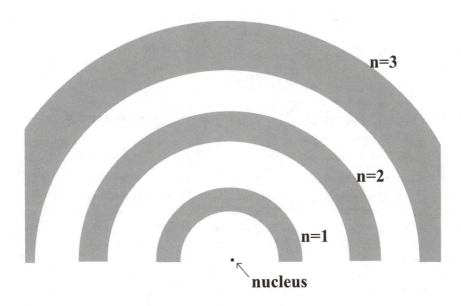

**FIGURE 3.7**   A diagram showing a simplified picture of several principal energy levels for electrons around a nucleus.

of electrons possible in each principal level varies with the level number. A maximum of only two electrons can be found in the $n = 1$ level, 8 in the $n = 2$ level, 18 in the $n = 3$ level, 32 in the $n = 4$ level, etc. A simple equation can be used to compute the maximum number of electrons that a principal level, $n$, can hold:

$$\text{maximum number of electrons possible in level } n = 2n^2 \qquad (3.5)$$

It is not sufficient to simply utilize the $n$ value to describe the configuration of the electrons around the nucleus since there are so many. The electrons within each principal level are not equal in energy. In fact, the modern theory says that each is unique. To describe each one, we must explore the concepts of energy sublevels within the principal levels, orbitals with the sublevels, and electron spin direction.

The different sublevels that exist are designated by the four letters *s, p, d,* and *f.* Thus, there are four possibilities for different sublevels within the principal energy levels for electrons. Not all are found in every principal level. In the $n = 1$ level, there is only an *s* sublevel. In the $n = 2$ level, there are only *s* and *p* sublevels, etc. As the *n* value increases, more sublevels are required. This is because as *n* increases, more electrons are found, as is determined from calculations performed with Eq. (3.5). Again, theoretically, the number of sublevels is unlimited; however, all electrons in all known elements can be accounted for by only utilizing four sublevels—the four symbolized by the letters *s, p, d* and *f.* We often refer to the *s, p, d,* and *f* sublevels (the "kind" of sublevel within the principal level), to the *s, p, d,* and *f* orbitals (the orbitals that are within the sublevels), and call the electrons they contain the *s, p, d,* or *f* electrons.

The maximum number of electrons that a sublevel can have depends on which sublevel it is:

| Sublevel | Maximum Number of Electrons |
|----------|-----------------------------|
| *s*      | 2                           |
| *p*      | 6                           |
| *d*      | 10                          |
| *f*      | 14                          |

---

*Chemistry Professionals at Work*                                    *CPW Box 3.5*

# FLAME EMISSION SPECTROSCOPY

One source of line spectra is a flame in which a solution of an element has been introduced. The study of the light emitted by such a flame is called *flame emission spectroscopy* (FES). FES has been used by chemists for many years to determine quantities of elements dissolved in laboratory samples. It is especially applicable to the analysis of solutions for sodium, potassium, calcium, lithium, and strontium and has had particular application in clinical laboratories in which the determination of calcium and sodium in body fluids has been important. The flame photometer, as the laboratory instrument is often called, measures the intensity of one of the emitted lines for these elements and the chemist relates this intensity to the element's concentration.

*For Homework: Do a library or Internet search, or contact a health professional to determine what body fluids are analyzed for sodium and calcium. What concentration levels are expected in these fluids? How are the samples prepared? Write a report.*

---

Electrons within sublevels exist in finite regions of space that are called orbitals. The term "orbital" is not to be confused with the electron orbit that was discussed for the Bohr model (the so-called "Bohr orbit"). The Bohr orbit is a specific path on which the electron, assumed to be a particle, travels, like a planet around the sun. Remember, the quantum mechanical model does not recognize the electron as a particle, and so there is no such thing as a "Bohr orbit." Instead, the *orbital* is viewed as a region of space around the nucleus in which the entity we call the electron is found.

By specifying an orbital, we come pretty close to uniquely describing each electron in an atom. We can say that a particular electron is in a particular principal level, a particular sublevel, and in a particular orbital. Any given orbital can only hold two electrons. In order to complete this unique description, we only need to differentiate between the two electrons in the orbital. The quantum mechanical model states that these two electrons have opposite spins.

Since there can be a maximum of two electrons in any orbital, the number of orbitals contained within a principal level of an atom is half the number of electrons, or,

$$\text{number of orbitals in principal level } n = n^2 \tag{3.6}$$

Compare Eq. (3.6) above with Eq. (3.5). Notice that the word "maximum" is found in Eq. (3.5) and not in Eq. (3.6). This is because all possible orbitals exist in all energy levels even though some may not have any electrons in them. So there is no maximum number. Therefore, some orbitals are empty, some may have one electron in them, and some may be "filled" with two electrons. The number of electrons, on the other hand, is limited by the element under consideration. Thus, the number of electrons in a given level varies from one element to another and the maximum number is what a level will have if all orbitals are completely filled. We will diagram a few examples in Section 3.7.

# 3.6 Shapes, Orientations, and Relative Sizes and Energies of Orbitals

For consideration of the chemical combination principles to be discussed in Chapter 5, it is important to discuss the physical appearance and shapes of the regions of space that are called orbitals. We will only study the *s* and *p* orbitals in detail here.

As indicated above, there is only one *s* orbital in any *s* sublevel, and hence only one *s* orbital in any principal level. Thus, since any orbital can hold a maximum of two electrons, the maximum number of *s* electrons that can be found in an *s* sublevel in any principal level is two. The physical appearance of an *s* orbital in any principal level is that of a sphere or ball, as shown in Fig. 3.8. The boundary of the sphere encloses the region of space that has the greatest probability (about 90%) of the location of the electrons in that orbital. As we progress from lower principal levels to higher principal levels, the ball simply gets larger. For

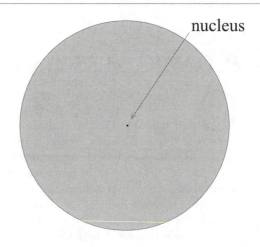

**FIGURE 3.8**  A drawing of an *s* orbital, a sphere where the nucleus is at the exact center.

perspective, consider the *s* orbital in the *n* = 1 level as being like a ping-pong ball, the *s* orbital in the *n* = 2 level as being like a baseball, and the *s* orbital in the *n* = 3 level as being like a softball, etc. All *s* orbitals are concentric, meaning that the center point is the same for each orbital. Thus, the ping-pong ball is "inside" the baseball, the baseball is inside the softball, etc., and at the exact center of each is the atomic nucleus.

The situation with *p* orbitals is very different. First, as indicated in the previous section, there are three *p* orbitals in each *p* sublevel rather that just one (as in the *s* sublevels), and a maximum of six electrons, as opposed to just two. Second, the shape is very different from the *s* orbitals. Each *p* orbital is like a figure 8 that has been rotated around a vertical axis as shown in Fig. 3.9(a)—a 3-dimensional figure 8 with the nucleus again being in the middle. The shape is similar to that of a bowling pin. The top and bottom halves of the 3-dimensional figure 8 are referred to as "lobes".

Since there are three of these 3-dimensional figure 8's in each *p* sublevel and in each principal level, we must consider how they are different from each other, something we did not need to do with the *s* orbital. The quantum mechanical model states that the three *p* orbitals are all the same shape, all the same size, but oriented differently in space. They are positioned in such a way that each is perpendicular to the other two. The usual description is that each is aligned along one of the axes in 3-dimensional space, the *x*-axis, the *y*-axis, and the *z*-axis, as shown in Fig. 3.9(b).

The boundaries of the 3-D figure 8 shapes, as with the boundaries of the spheres, enclose the regions of highest probability for the electrons'

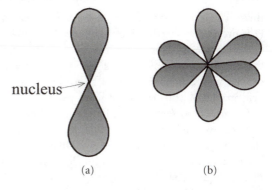

(a)                    (b)

**FIGURE 3.9**  (a) A single *p* orbital, (b) the orientations of the three *p* orbitals within a given *p* sublevel.

location. The "ping-pong ball–baseball–softball" analogy applies to the *p* orbitals in the sense that the set of three *p* orbitals gets progressively larger in size as we progress from lower principal levels to higher ones. The set of three *p* orbitals first occurs in the *n* = 2 principal level. In addition, some of

---

*Chemistry Professionals at Work*                                    *CPW Box 3.6*

# ATOMIC ABSORPTION SPECTROSCOPY

Closely related to the FES method described in CPW Box 3.5 is **atomic absorption spectroscopy** (AAS). In AAS, the sample is contained in a flame, but rather than measuring the *emission* of light as in FES, the chemistry professional measures how much light is *absorbed* by the atoms in the flame in a manner similar to what was described in CPW Box 3.3. The electrons that exist in the orbitals within the energy levels of atoms can absorb the specific wavelengths represented in the line spectrum of an element. The amount of light absorbed is related to the concentration of the element in a solution. The following story relates how such a technique may be used in an industrial laboratory.

Chemistry technicians were using AAS to monitor the concentration of chromium in the wastewater from their company's chromium electroplating facility. One day, the results of the AAS work indicated that the waste contained a very high level of chromium. It was apparent that the plant would not pass an inspection by their state's environmental protection division, and they alerted the process operators in their electroplating facility of the problem. The operators discovered that one of the plating tanks was leaking solution into the waste stream. They repaired the leak and the laboratory technicians, after analyzing the waste stream a day later by AAS, reported a normal and safe chromium level.

*For Homework: Find a reference on atomic absorption spectroscopy in a library and report on the details of this technique.*

---

the space taken up by the *s* orbital spheres is shared with some of the space taken up by the *p* orbital 3-D figure 8s. These orbitals overlap, which is also true of the *d* and *f* orbitals. The physical appearance and orientations of the *d* and *f* orbitals are beyond our scope here. However, it is important to remember that the *d* sublevel has 5 orbitals with a maximum of 10 electrons possible, while the *f* sublevel has 7 orbitals with a maximum of 14 electrons possible. The set of five *d* orbitals first occurs in the $n = 3$ principal level, while the set of seven *f* orbitals first occurs in the $n = 4$ principal level.

It is also important to consider the energy of these sublevels and orbitals compared to each other. This is so that we can know which sublevels within a principal level (or which orbitals within a sublevel) have electrons and which don't, in the event that the principal level (or the sublevel) is not filled with electrons. If a principal level is not filled with electrons, the sublevels with lower energy will have electrons, while those at higher energies will not. Levels with lower energy get electrons before those with higher energy. Electrons tend to seek their lowest possible energy state. The order of increasing energy of sublevels within the same principal level is as follows:

$$s \rightarrow p \rightarrow d \rightarrow f$$

We are now nearly ready to uniquely describe the electron picture within the atoms of each individual element. As we move through the periodic table from hydrogen to sulfur to cobalt and beyond, within any principal level, the *s* sublevel gets electrons before the *p*, which gets electrons before the *d*, which gets electrons before the *f*. The only question remaining is in what order do the *p* orbitals within a *p* sublevel fill with electrons, or in what order do the *d* orbitals within a *d* sublevel fill, etc. The answer is that all orbitals within a given sublevel are equal in energy. Thus, deciding which orbital gets an electron is arbitrary. However, ***all orbitals within a given sublevel must get one electron before any get two.*** This is a statement of what has become known as Hund's Rule. The application of **Hund's Rule** is visualized in the next section in the use of orbital diagrams. When an orbital has two electrons, the electrons are said to be "paired." When an orbital has just one electron, that electron is said to be "unpaired."

## 3.7   Orbital Diagrams

One way to diagram the sublevels, orbitals, and electrons is to use small squares ("boxes") to symbolize orbitals, and to use arrows inside the boxes to symbolize electrons. The use of arrows for electrons is valuable because we can symbolize the fact that the electrons within a given orbital have opposite spin directions by drawing two arrows pointing in opposite directions, one up and one down. Thus, we have the following symbol for an orbital with paired electrons:

The designation of the spin direction is the final step in the characterization of an individual electron. No two electrons in an atom will have the same four such designations (principal level, sublevel, orbital, and spin) and hence each electron is unique.

Let us look at some examples of the "boxes and arrows" orbital diagrams. Let us begin with hydrogen and proceed through the first ten elements in the periodic table. In order to label each sublevel, we will place symbols such as 1*s*, 2*s*, 2*p*, etc., under the boxes. This will identify each set of orbitals as being in the $n = 1, 2, 3$, etc. principal level, and in the *s*, *p*, *d*, or *f* sublevel. Remember that the electrons will go into their lowest possible energy state. The principal levels increase in energy in the order $n = 1, 2, 3$, etc. and the sublevels within the same principal level increase in energy in the order *s*, *p*, *d*, *f*. Only the *s* sublevel occurs in the $n = 1$ principal level (the first period of the periodic table), so for hydrogen and helium we have:

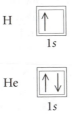

Helium represents a situation in which the $n = 1$ principal level is filled and the *s* sublevel is also filled. When we proceed to lithium and beryllium in the second period of the periodic table, the $n = 2$ principal level and the *s* sublevel within the $n = 2$ principal level must be used for the third and fourth electrons:

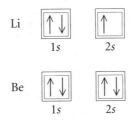

---

*Chemistry Professionals at Work*                                    *CPW Box 3.7*

# HEAVY METALS POLLUTION

In environmental chemistry, a ***heavy metal*** refers to a metal that has an atomic weight in the range of the first-row transition metals or higher. The term is used especially with respect to metals that are toxic to humans and tend to accumulate in the environment. On the list as being particularly troublesome are chromium, nickel, cadmium, antimony, mercury, and lead. These are often found in solid wastes and in the dissolved state in environmental water. Some chemistry professionals are involved directly in reducing the accumulation of heavy metals in the environment through safety and awareness programs. Others must be acutely aware so that their activities in the laboratory do not promote such accumulation. For example, chromic acid, a solution of potassium dichromate ($K_2CrO_4$) and sulfuric acid, has been extensively used in laboratories as a glassware cleaning agent. Spent solutions of this cleaning solution cannot be disposed of down the drain because of chromium's toxicity.

*For Homework: Select two of the heavy metals listed above and write a report on the toxicity, reactivity and incompatibility, storage and handling, and disposal of these metals and their compounds. The following reference is suggested:* Prudent Practices in the Laboratory, Handling and Disposal of Chemicals, *National Research Council, National Academy Press (1995).*

---

The $n = 2$ principal level has a $p$ sublevel as well as an $s$ sublevel, so beryllium does not represent a filled principal level. When we proceed to boron, the fifth electron must be shown in the $2p$ level. Remember that $p$ sublevels have three orbitals. In this boxes and arrows symbolism, if a $p$ sublevel is used, all three $p$ orbitals are shown whether or not they have electrons:

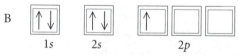

As stated previously, the three $p$ orbitals within a principal level are all equal in energy, therefore, since the assignment of an electron to a $p$ orbital is arbitrary, the individual $p$ orbitals are not given individual labels. The only distinction we have for these orbitals is that each is said to lie along an axis ($x$, $y$, or $z$) and they are, thus, perpendicular to each other (Fig. 3.9). With boxes and arrows, then, there is no distinguishing feature noted.

In the diagram for carbon, Hund's Rule must be invoked. Thus, two $p$ orbitals each get an electron:

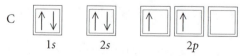

The correct diagram for carbon is actually a modification of this, but this will be addressed later in Chapter 6.

Following Hund's Rule and filling the $2p$ level before moving on to the $n = 3$ level results in the following diagrams for nitrogen, oxygen, fluorine, and neon:

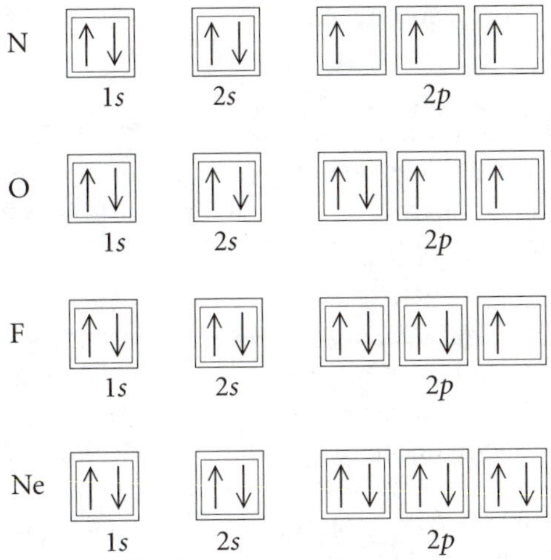

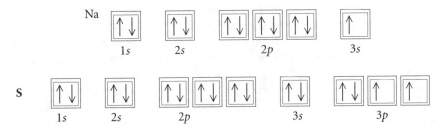

Elements beyond neon in the third period of the periodic table, involve the $n = 3$ principal level and, thus, the $3s$ and $3p$ sublevels. Examples include sodium and sulfur:

Notice that in all cases, the total number of arrows shown is the same as the total number of electrons in the neutral atom, which is the same as the atomic number of the element.

Elements beyond argon show some unexpected behavior, particularly in the order in which the sublevels are filled. The situation is, however, easily explained, as we shall see in the next section.

## 3.8 Electron Configuration

There is another technique that is commonly used to indicate electron structure. It is known as ***electron configuration***. Electron structure by electron configuration is not as comprehensive as the electron structure by orbital diagram method as discussed in the previous section. It does not designate orbital or spin, but only principal level, sublevel, and number of electrons in each sublevel. It utilizes the same symbolism, $1s$, $2s$, $2p$, etc., as used earlier, and indicates the number of electrons in each sublevel as a superscript following the sublevel symbol ($s$, $p$, etc.). A few examples will help illustrate this method.

hydrogen (H), atomic number $= 1$, $1s^1$

beryllium (Be), atomic number $= 4$, $1s^2 2s^2$

carbon (C), atomic number $= 6$, $1s^2 2s^2 2p^2$

sodium (Na), atomic number $= 11,$    $1s^2 2s^2 2p^6 3s^1$

sulfur (S), atomic number $= 16,$    $1s^2 2s^2 2p^6 3s^2 3p^4$

argon (Ar), atomic number $= 18,$    $1s^2 2s^2 2p^6 3s^2 3p^6$

Beyond argon, some unexpected things begin to occur. As stated previously, the $n = 3$ level has the $s$, $p$, and $d$ sublevels, and thus, with potassium, you would expect that the $3d$ sublevel would be the next sublevel to be utilized. The correct electron configuration for potassium (and also for calcium), however, shows that the $4s$ sublevel is utilized before the $3d$:

potassium (K), atomic number $= 19,$    $1s^2 2s^2 2p^6 3s^2 3p^6 4s^1$

calcium (Ca), atomic number $= 20,$    $1s^2 2s^2 2p^6 3s^2 3p^6 4s^2$

The explanation for this is simply that the $4s$ sublevel, while in a higher principal level ($n = 4$) than the $3d$ sublevel ($n = 3$), is actually lower in energy because the two principal levels overlap. The overlap is such that the $4s$ sublevel, the lowest energy sublevel in the $n = 4$ principal level, is lower in energy than the highest energy sublevel in the $n = 3$ principal level, the $3d$ level. Figure 3.10 should clarify this concept. Overlaps that affect the order of filling are prevalent into the $n = 4$ level and beyond. It is helpful to use the mnemonic device illustrated in Fig. 3.11 to remember the order of electron filling in the various sublevels. This method involves following the arrows starting with the topmost row ($1s$) and continuing downward to the lowest. The complete order of filling then would be: **$1s \rightarrow 2s \rightarrow 2p \rightarrow 3s \rightarrow 3p \rightarrow 4s \rightarrow 3d \rightarrow 4p \rightarrow 5s \rightarrow 4d \rightarrow 5p \rightarrow 6s \rightarrow 4f \rightarrow 5d \rightarrow 6p \rightarrow 7s \rightarrow 5f \rightarrow 6d$**. With this sequence, we can draw the complete electron configuration of all elements in the periodic table.

The periodic table itself can be used to determine electron configurations. Please refer to Fig. 3.12 for this discussion. Starting in the upper left corner (with hydrogen, period 1), proceed across to the right, then move to period 2, proceed to the right, etc. After filling the $6s$, proceed to the $4f$ as indicated, then back to the $5d$, and similarly from the $7s$ to the $5f$, then the $6d$. The order of filling determined in this way is the same as depicted in Fig. 3.11: $1s \rightarrow 2s \rightarrow 2p \rightarrow 3s \rightarrow 3p \rightarrow 4s \rightarrow 3d \rightarrow 4p \rightarrow 5s \rightarrow 4d \rightarrow 5p \rightarrow 6s \rightarrow 4f \rightarrow 5d \rightarrow 6p \rightarrow 7s \rightarrow 5f \rightarrow 6d$. We will expand on this in Chapter 4.

There are also some idiosyncrasies that occur *within* the $d$ and $f$ sublevels that warrant mentioning. The first of these occurs in the $3d$ sublevel. As you proceed across period #4, in the range of the transition elements, beginning with scandium, Sc, the electron configurations are those shown in Table 3.2. Notice

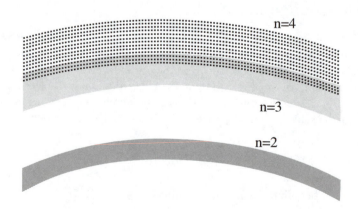

**FIGURE 3.10**    A drawing illustrating the overlap between the $n = 3$ and $n = 4$ principal levels.

what happens when proceeding from vanadium to chromium and also from nickel to copper. For chromium, the *d* sublevel appears to "borrow" an electron from the *s* sublevel, so as to leave only one electron in the *s* sublevel and to have five electrons (rather than the expected four) in the *d* sublevel. The "borrowing" when proceeding from nickel to copper also leaves one electron in the *s* sublevel while filling the *d* sublevel with ten.

The 4*s* and the 3*d* sublevels are extremely close in energy. It takes only a minor event to move an electron from one to the other. The minor events in the examples cited above are the filling or half-filling of the *d* sublevel. If borrowing an electron from the *s* sublevel will result in either half-filling the *d* sublevel with five electrons (in the case of chromium), or filling the *d* sublevel with ten electrons (in the case of copper), then

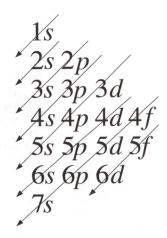

**FIGURE 3.11** A mnemonic device for assisting with the writing of electron configurations.

the borrowing will occur as seen in Table 3.2. This kind of unexpected behavior also occurs in the other *d* sublevels and in the *f* sublevels.

One result of this latter phenomenon is the variable oxidation numbers (Chapter 2) observed for the transition and inner-transition elements. A good example is iron. In Chapter 2, we mentioned that iron

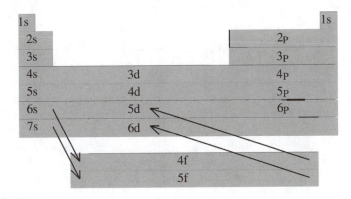

**FIGURE 3.12** The order with which the energy levels fill with electrons can be deduced from the periodic table. See text for discussion.

**TABLE 3.2** Electron Configurations of the 10 Transition Elements in the 4th Period

| Element (symbol), Atomic # | Electron Configuration |
| --- | --- |
| Scandium (Sc), 21 | $1s^2 2s^2 2p^6 3s^2 3p^6 4s^2 3d^1$ |
| Titanium (Ti), 22 | $1s^2 2s^2 2p^6 3s^2 3p^6 4s^2 3d^2$ |
| Vanadium (V), 23 | $1s^2 2s^2 2p^6 3s^2 3p^6 4s^2 3d^3$ |
| Chromium (Cr), 24 | $1s^2 2s^2 2p^6 3s^2 3p^6 4s^1 3d^5$ |
| Manganese (Mn), 25 | $1s^2 2s^2 2p^6 3s^2 3p^6 4s^2 3d^5$ |
| Iron (Fe), 26 | $1s^2 2s^2 2p^6 3s^2 3p^6 4s^2 3d^6$ |
| Cobalt (Co), 27 | $1s^2 2s^2 2p^6 3s^2 3p^6 4s^2 3d^7$ |
| Nickel (Ni), 28 | $1s^2 2s^2 2p^6 3s^2 3p^6 4s^2 3d^8$ |
| Copper (Cu), 29 | $1s^2 2s^2 2p^6 3s^2 3p^6 4s^1 3d^{10}$ |
| Zinc (Zn), 30 | $1s^2 2s^2 2p^6 3s^2 3p^6 4s^2 3d^{10}$ |

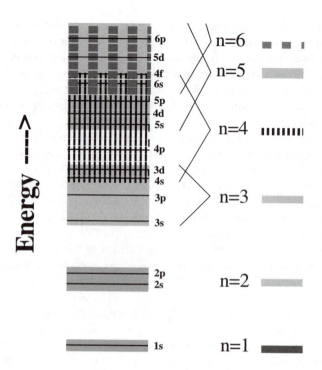

**FIGURE 3.13** A diagram of principal levels and sublevels illustrating the overlaps of principal levels and the closeness of sublevels.

sometimes gives up two electrons, forming the ferrous or iron(II) ion, $Fe^{+2}$, while other times it gives up three electrons, forming the ferric or Iron(III) ion, $Fe^{+3}$. Iron's electron configuration in Table 3.2 gives one possible explanation. When iron forms the $Fe^{+2}$ ion, one can imagine that it is the two $4s$ electrons that are lost, while when it forms the $Fe^{+3}$ ion, which is a more stable ion, it's reasonable to suggest that the two $4s$ electrons are lost along with one $3d$ electron because the loss of the one $3d$ electron results in this $d$ sublevel being half-filled.

Figure 3.13 illustrates the principal levels and sublevels through the $6p$ sublevel and represents a summary of these concepts. Notice that the overlap of principal levels is apparent, as well as the fact that some sublevels are so close in energy that movement up or down the energy scale even by the smallest increment can result in the reordering of the levels and the idiosyncrasies that we discussed.

## 3.9 Atomic Spectra Revisited

In Section 3.4, we briefly discussed the phenomenon of light emission by atoms, saying that under certain conditions, atoms of elements can be energized so that certain specific energy emissions in the form of light can be observed emanating from samples of these elements (see also Fig. 3.6). We said that Bohr explained this phenomenon by saying that it is possible for electrons to move from a Bohr orbit of high energy to another of lower energy. The resultant loss of energy is found in the form of light coming from the atoms—light that has a specific energy represented by the energy difference between the orbits. This is represented by the lines of energy—the atomic spectra, or line spectra, that were shown in Fig. 3.6.

The modern theory of the atom discussed in this chapter is compatible with Bohr's explanation, despite the fact that the quantum mechanical model has replaced the Bohr model. This is because the concept of electron energy is retained in the quantum mechanical model. Rather than referring to electrons as moving from one *orbit* to another, however, the modern theory refers to electrons moving from one *energy level* to another and thus explains the emission phenomenon in much the same terms.

The atom must be *energized* (i.e., it must absorb or gain energy) before it can emit light. This means that the atom must be hit with a burst of energy equal to that of the light emitted before the energy can be lost and the emission observed. The nature and source of this burst is quite variable. It can be in the form of light. It can be provided by a flame. It can be provided by a hot plasma source. It can also be provided by electricity. Thus, heat (thermal) energy, light energy, and electrical energy can all be sources of the required energy. In real-world laboratories, each of these is used in different situations for different reasons (see Section 3.10 for more information).

However, the fact that energy is needed to cause the phenomenon in the first place forces us to go back to our modern theory to look for the reason. The answer is that if an electron is to fall back from a level of high energy to one of lower energy, it must first be elevated to the higher level so that it has a vacant lower level to which it can fall back. Without this elevation, there would be no vacant orbital for it to fall back to, since other electrons have filled the orbitals in the lower levels. The result of this is that we focus on the outermost electrons, or those farthest from the nucleus. The outermost electrons are most easily raised to higher levels because it take less energy to do it. Elevation of an outermost electron from an *s* sublevel in the *n* = 3 level to a *p* sublevel in the *n* = 3 level requires less energy than the elevation of a *p* electron in the *n* = 2 level to that *p* sublevel in the *n* = 3 level. Thus, the outermost electrons are usually responsible for the emissions that are observed. Figure 3.14 shows examples of pathways that an outermost electron (in this case, a 3*s* electron) can take in the process. The "up arrows" symbolize the energy absorbed from the thermal, electrical, or light sources, or the elevation of the electron to the higher level. The "down arrows" symbolize the energy lost as the electron goes back to the lower level causing the observed emission of light.

In Section 3.4, we saw that Bohr explained the specificity of the emitted light (the fact that narrow energy "lines" are emitted, as shown in Fig. 3.6) by stating that the electron must be in one orbit or the other—it cannot exist in between. Thus, the emitted light is of very specific energies—the energy differences between the higher orbits and the lower orbits. The quantum mechanical model explains it in much the same way. The electron cannot come to a state between levels (as when climbing a ladder, you cannot stand between rungs). Therefore the light emitted is only the light equal in energy to the energy differences between any of the two levels involved (the down arrows in Fig. 3.14). No other light can be emitted.

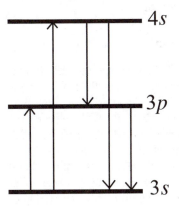

**FIGURE 3.14**   Some energy level transitions that an outermost electron in the 3*s* level can undergo. See text for discussion.

## 3.10   Spectroscopy

In Section 3.9 and in several of the "Chemistry Professionals at Work" boxes in this chapter, we alluded to the fact that some important laboratory analysis techniques are based on the absorption or emission of light. These techniques have been referred to in these boxes as both spectroscopy and spectrophotometry. In this section, we briefly summarize these important and well-known analytical methods, the facts about atomic and molecular structure that explain how they work, and some practical matters relating to the laboratory work.

### 3.10.1   Atomic Spectroscopy

*Atomic spectroscopy* refers to techniques that involve the absorption and emission of light by atoms, as opposed to molecules or ions. Those methods that are in this category and that have already been mentioned in this chapter include flame emission spectrscopy (FES) and atomic absorption spectroscopy

(AAS). Others to be discussed below include graphite furnace atomic absorption spectroscopy (GFAAS), and inductively coupled plasma emission spectroscopy (ICP).

FES relies on a flame to energize atoms and the resulting emission is measured. AAS utilizes a flame, but only for containing atoms and not for energizing them. Rather, a light beam is directed at the flame containing the atoms in order to energize them. The light is absorbed by the atoms and the absorption is measured. GFAAS also uses a light beam to energize the atoms, but a small quartz tube inside a very hot furnace is used to contain the atoms rather than a flame, and light absorption is measured. With ICP, an extremely hot plasma source is used to both contain and energize. The emission of light is measured.

Thus, some of these techniques measure the emission of light (FES and ICP) while some measure the absorption of light (AAS and GFAAS). The intensity of the emission and the degree of the absorption is related to the number of atoms involved, which in turn is related to the concentration of the element in the solution measured. The usual procedure is to prepare a number of solutions of the element being tested, each at a different concentration level, measure the emission intensity or the degree of absorption for each, and make a graph of the results. The graph would plot either (1) emission intensity on the *y*-axis and concentration level on the *x*-axis, or (2) degree of absorption (often called "absorbance") on the *y*-axis and the concentration level on the *x*-axis. Such a graph is often called the *standard curve* because the solutions are called *standard solutions*, or solutions which have known concentration levels. See Figs. 3.15 and 3.16. It is generally desirable for these curves to be linear, as shown in the figures. The concentration level is often expressed as parts per million, ppm (Chapter 10), although other units can be used. The concentrations of unknown solutions are then determined from the graph.

There are many more details concerning the use of these techniques but they are beyond the scope of this present discussion.

## 3.10.2   Molecular Spectroscopy

Molecules as well as atoms can absorb and emit light. The techniques for molecules and some ions come under the heading of molecular spectroscopy (also sometimes referred to as spectrophotometry or spectrometry). Visible light can be involved, but ultraviolet and infrared light are also used in certain situations. The energy transitions associated with ultraviolet and visible light involve electrons in the molecule gaining and losing energy. The energy transitions associated with infrared light represent much lower energy (infrared light is of much lower energy) and are not associated with electrons (electronic transitions). Since the ultraviolet and visible techniques are very different from the infrared techniques, we often speak separately of UV/Vis spectrophotometry and IR spectrometry. Infrared spectrometry will be discussed in Chapter 6.

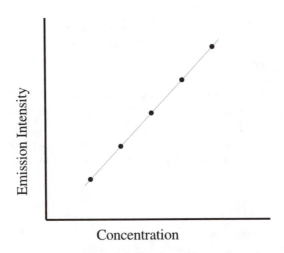

**FIGURE 3.15**   The standard curve for an analysis in which emission intensity is measured.

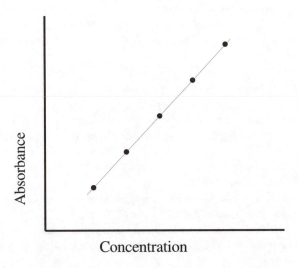

**FIGURE 3.16**    A standard curve for an analysis in which absorbance is measured.

For UV/Vis spectrophotometry, by far the most important techniques involve absorption rather than emission and for these, absorbance is plotted on the *y*-axis of the standard curve, as in Fig. 3.16. The molecules are always energized with light and never by using a flame. The samples are most often solutions and are held in a test tube-like container (called a "cuvette") inside the instrument, as described in CPW Box 3.3. A beam of light of a certain wavelength, the wavelength that the sample absorbs to the greatest extent, is passed through the cuvette and the absorbance recorded.

The procedure most often involves pouring the solution into the cuvette, making sure the cuvette is clean on the outside, placing it into the holder inside the instrument, closing the lid, and reading the absorbance. Again, there are many more important details of these and other techniques that are beyond the scope of this discussion.

## 3.11   Homework Exercises

1. In five words or less, tell what each of the following scientists is famous for as far as atomic theories are concerned: Dalton, Crookes, Thomson, Millikan, Rutherford, Chadwick, Bohr, and Schrodinger.
2. Match the item on the left with the name on the right by placing the appropriate letter on the line.

    _____ solar system model                (a) Schrodinger
    _____ nuclear model                     (b) Dalton
    _____ original atomic theory (early 1800s)   (c) Bohr
    _____ quantum mechanical model          (d) Thomson
    _____ plum pudding model                (e) Rutherford
    _____ neutron                           (f) Crookes
    _____ electron                          (g) Chadwick
    _____ electron charge                   (h) Millikan

3. (a) What scientist published a comprehensive theory of the atom in the early 1800s? What were the main points of this theory?
   (b) Briefly describe the atomic theory advanced by Neils Bohr.
   (c) What conclusion did Rutherford make when he observed that most of the alpha particles went through a gold foil as if it was not there?
4. What experimental evidence did Rutherford have that implied that the nucleus exists?
5. What experimental evidence did Rutherford have that implied that most of the atom is empty space?
6. What is an alpha particle?
7. If energy is the capacity to cause matter to move due to an applied force, how can a boulder at rest on the edge of a cliff be considered to have energy? What kind of energy is it?
8. Name six different forms of energy. Give an example scenario for each in which each is transformed into another form.
9. Light is said to have a "dual nature." Explain what that means.
10. What is different about light waves that allow them to travel through the void of space?
11. Define wavelength. Is a long wavelength high energy or low energy?
12. Which of the following are of sufficient energy to cause some kind of damage to our bodies: visible light, gamma rays, infrared light, radio waves, ultraviolet waves, microwaves, x-rays? Briefly give details.
13. Visible light consists of the rainbow colors. Arrange the following in the order of increasing energy: yellow, red, green, blue, orange, violet.
14. If a wavelength of visible light is 450 nm, what color is it? Answer this same question for the following wavelengths: 600 nm, 550 nm, 700 nm, 350 nm.
15. Explain how Eq. (3.4) is derived from Eqs. (3.2) and (3.3). Does a low frequency mean a long wavelength or a short wavelength? Does long wavelength mean high energy or low energy?
16. What is a "line spectrum?"
17. How are the different colors of light emitted by elements under certain conditions explained by the Bohr model?
18. What basic premise of the Bohr model was discounted by the modern theory?
19. What is wrong with this statement: "In the quantum mechanical model, the electron is pictured as a particle in an orbit around the nucleus, much like a planet is in an orbit around the sun."

20. Fill in the blanks:
    (a) What is the maximum number of electrons that can be found in the $n = 3$ principal level? _____
    (b) How many orbitals are there in the $n = 3$ principal level? _____
    (c) What is the maximum number of electrons that can be found in a $2p$ sublevel? _____
    (d) What is the maximum number of electrons that can be found in a single $d$ orbital? _____
21. How many orbitals are in the $n = 4$ principal level? How many are in the $n = 5$ level?
22. Identify the drawing in Fig. 3.17. How many different orientations within an energy sublevel can this drawing have and what are they? Where is the nucleus located?
23. What atomic orbital is spherical in shape? What is the lowest principal energy level in which this orbital is found? How many different orientations can it have?
24. Sublevels within principal levels are designated by the letters $s$, $p$, $d$, and $f$. Within a given principal level, which of these has the most energy? Which of these has an orbital that has a spherical shape? Which of these can hold 10 electrons? Which of these has only 3 orbitals?
25. What is the maximum number of electrons each of the following can hold: (a) a single $f$ orbital, (b) the $3p$ sublevel, (c) the $n = 4$ principal level, (d) any s sublevel.
26. Fill in the blanks with one of the following: principal level   sublevel   orbital
    (a) A region of space around the nucleus in which a maximum of two electrons can be found. _____
    (b) An energy level designated by $n$. _____
    (c) An energy level that is within a larger energy level. _____

**FIGURE 3.17**   Drawing for question 22.

27. Sketch a single s orbital and a single $p$ orbital and describe the orientation(s) each can have.
28. How many electrons are in each of the following: (a) the $n = 3$ level, (b) a single $d$ orbital, (c) a $p$ sublevel, (d) the $4s$ level?
29. Draw the orbital diagrams (boxes and arrows) for N, Mg, Al, Si, P, Cl, Ar, K, and Ca.
30. Draw the orbital diagrams (boxes and arrows) of the following: boron, phosphorus, sodium, palladium, zirconium, sulfur, and copper.
31. Write out the electron configuration of the following: Ni, As, O, V, Rb, Ba, Sn, Zn, Ag, and Nd.
32. What element is represented by each of the following electron configurations?
    (a) $1s^2 2s^2 2p^6 3s^2 3p^5$   (b) $1s^2 2s^2 2p^6 3s^2 3p^6 4s^2 3d^{10}$   (c) $1s^2 2s^2 2p^6$
33. In terms of energy level overlap, explain why the $4s$ level gets electrons before the $3d$ level when building electron configurations.
34. What are two consequences of the fact that principal levels overlap in energy?
35. What does it mean for an energy sublevel to be "half-filled." Explain what the half-filled condition has to do with the unexpected electron configuration of chromium.
36. What is the electron configuration of: (a) the potassium ion, $K^+$ (b) bromide ion, $Br^{-1}$?
37. What is meant by "outermost" electrons? Why do we pay particular attention to them?
38. Draw an energy level diagram showing energy transitions that can occur in an atom when an atom is energized.
39. Why must an atom be energized before it can emit light?

40. Why does the light emitted by atoms under certain conditions have such specific energy, or wavelength (line), rather than a wide range of wavelengths? What exactly does this energy represent in terms of energy levels within atoms?

41. For each of the following, tell how atoms get energized, whether it is absorption or emission that is measured, and what is plotted for the standard curve: FES, FAAS, GFAAS, ICP.

42. What is the basic difference between atomic spectroscopy and molecular spectroscopy in terms of what it is that is being energized, what contains the species that is energized, and what is plotted for the standard curve.

## For Class Discussion and Reports

43. It can be very important for chemistry professionals to characterize materials as being "very reactive" or "very unreactive." Elements that are reactive have electron configurations in which only one electron makes the difference between a filled energy level or an unfilled energy level. Discuss as a class what elements meet these criteria and hence are very reactive. Also discuss what elements do not meet these criteria and are thus very unreactive.

44. In your library, find a book of methods of chemical analysis. Examples include various volumes of the American Society for Testing and Materials (ASTM), *Standard Methods of AOAC International, Standard Methods for the Analysis of Water and Wastewater*, and methods from various government agencies, such as the Environmental Protection Agency (EPA) and the National Institute for Occupational Safety and Health (NIOSH). Find a spectroscopic method of analysis and report on what kind of sample is analyzed, what substance is being determined, what the spectroscopic method is, what the sample preparation involves, etc.

45. In the atomic spectroscopic methods, it is the outermost electrons in *atoms* that absorb or emit light. Yet, it is the *ions* that exist in the solution that are tested and these ions are atoms that have *lost* the outermost electrons. That is why they are ions. A process called atomization (making atoms out of ions) must occur before there can be absorption or emission. Consult available analytical chemistry and instrumental analysis references to discover how this atomiization takes place during the experiment. Write a report.

46. Ask your instructor for a reference that lists the vendors of atomic and molecular spectroscopy equipment. Then consult the Internet sites of these vendors to see what exactly is offered and how one vendor's product compares with another. Also compare price. Write a report.

4

# The Periodic Table

## 4.1 Introduction

The periodic table is an icon that people associate with chemists and chemistry, or the work that chemists do. Indeed, there is a wealth of information important to chemists that is available from the modern periodic table. Some of this information we have already discussed.

We introduced the periodic table in Chapter 1, mentioning first that the names and symbols of the pure substances known as elements are listed there. Such a list is important because all of matter is composed of these elements either in their pure state or in physical or chemical combination. In addition, we have observed that the periodic table presents a rather odd arrangement of the elements, but the manner of this arrangement provides us with additional useful information. For example, we have noted that the elements are subdivided into groups. These groups include metals, nonmetals, and metalloids; representative, transition, and inner-transition elements; alkali metals and alkaline earth metals; and halogens and noble gases (for a review of these groups, see Sections 1.3.2 and 1.7). These group names have since been used in Chapters 2 and 3 in our discussion of compound formation, the nomenclature of compounds, and atomic structure (review Sections 2.2 and 2.4, for example). One reason for assigning elements to groups is that all elements in a group tend to behave the same way. For instance, all alkali metals tend to form monatomic ions with a +1 charge.

# WHY STUDY THIS TOPIC?

The periodic table is a reference source by which a chemistry professional's understanding and appreciation for an element's properties becomes complete. For example, a chemistry technician assigned to analyze the air in your house for radon understands that the hazard with radon is its ionizing radiation. He/she also realizes, from the location of radon in the periodic table, that it is a noble gas and presents no hazard from chemical reactivity.

Similarly, a certain laboratory test for the determination of ozone content in air calls for the preparation of a series of iodine solutions. While the preparation of such solutions is routine for a laboratory analyst, the fact that iodine's position in the periodic table is that of a halogen tells this analyst that there are reactivity hazards to be considered.

When communicating with colleagues, it is important for a chemistry professional to present information accurately. For example, the calcium ion involved in water hardness has a +2 charge, a fact one can determine from calcium's position in the periodic table. If a chemist or chemistry technician were to erroneously indicate that this charge is +1, his/her credibility as a chemistry professional would be questioned. It is important to have full knowledge and appreciation for the use of the periodic table.

*For Homework: Research the problem of radon in the home and report on the source of the radon, the nature of the hazard, the concentration level that is hazardous, and the method by which a chemistry professional would sample and analyze the air for radon.*

In Chapter 1 we also defined two important properties of the elements, the atomic number and the atomic weight. Both of these properties are listed in the modern periodic table. The atomic weight is a property that will be especially important as we introduce some quantitative concepts of chemistry in later chapters.

Because of the importance of the periodic table to chemists, the general public's tendency to recognize the periodic table as the icon of chemists and their work is understandable.

## 4.2   Relationship to Electron Configuration

Let us now consider the matter of the odd arrangement of the elements in the periodic table. We would like to answer several questions about this arrangement (please refer to Fig. 4.1). For example, why is there such a large gap between hydrogen (the element with atomic number 1) and helium (the element with atomic number 2)? Why is there a gap, albeit a smaller gap, between beryllium (the element with atomic number 4) and boron (the element with atomic number 5)? Why is there a similar small gap between magnesium (atomic number 12) and aluminum (atomic number 13)? What dictates the sizes of these gaps? Why are there no other gaps. Why are the inner-transition elements listed apart from the others near the bottom of the table? What dictates when the end of a period is reached and when a new one must begin? The arrangement of elements in the periodic table is based on the electron

---

*Chemistry Professionals at Work* CPW Box 4.2

# THE "RARE EARTHS" ARE NOT RARE

T he lanthanide series of elements belong to a group that have come to be known as the "rare earth" elements. This name was given to them because it was originally thought that they represented only a small fraction of all the elements present in the Earth's crust. However, not only has this proven to be untrue, but at least two of them have been found to have important and common uses in analytical laboratories. The element lanthanum (La), for example, is added to the sample matrix when performing spectroscopic analysis of all types of materials for calcium content. In addition, solutions of cerium (Ce) are used in certain analysis schemes for oxidation/reduction reactions.

*For Homework: Research the use of lanthanum, as mentioned above, and determine the exact reason why lanthanum is added to the matrix when analyzing for calcium. Write a report on this.*

---

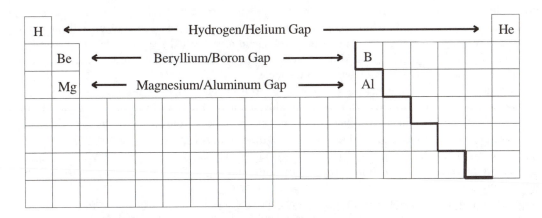

**FIGURE 4.1** An outline of the periodic table showing the gaps discussed in the text and also highlighting the group of inner-transition elements that is set apart from the others.

configuration of the elements, and, in particular, the electron configuration in the outermost sublevels. You will recall that in Section 3.8, the use of the periodic table was mentioned as a possible aid in the writing of electron configurations. Please refer to Fig. 4.2 for the following discussion. The first two families on the left in the periodic table constitute the "*s*-block" in which the *s* sublevels are being filled. The six families on the right constitute the "*p*-block" in which the *p* sublevels are being filled. The transition elements constitute the "*d*-block" and the inner-transition elements constitute the "*f*-block" for the same reasons.

Each element within a family in these blocks has the same number of electrons in the outermost sublevel identified by the block designation. For example, nitrogen, phosphorus, arsenic, and the other elements in group VA all have three electrons in their outermost *p* sublevel. The exception to this general statement

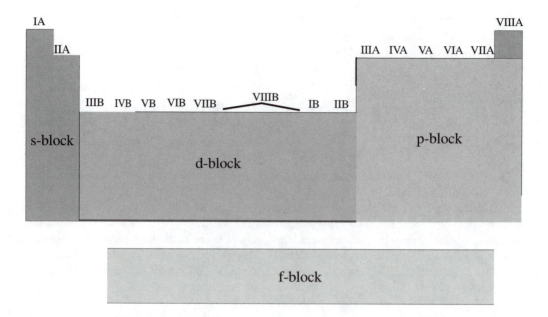

**FIGURE 4.2**   An outline of the periodic table showing the locations of the *s*-block, the *p*-block, the *d*-block, and the *f*-block. Note that helium, in the upper right corner, is an *s*-block element.

is the family of noble gases. Helium is an *s*-block element that is listed in the same family as some *p*-block elements, or the other noble gases. The noble gases are noted for their unusual stability or lack of tendency to combine with other elements. In all but helium, this unusual stability results from the fact that the outermost *p* sublevel is filled. Helium is the exception because it does not have a *p* sublevel in the outermost level that is being filled, the $n = 1$ principal level. Helium is unusually stable, however, because this $n = 1$ level is filled. This is why helium is included with the other noble gases and since it has no *p* electrons, a gap exists between hydrogen and helium in the table.

Each time we step from one element to the next and from left to right across a period, a proton and an electron are added to an atom, thus changing the identity of the element to the one next to it order of increasing atomic numbers. Hydrogen (1 proton, 1 electron) begins the process in the upper left corner. When enough electrons have been added to either fill the $n = 1$ principal level (helium) or fill a *p* sublevel (the other noble gases), the end of a period is reached and the next element in order begins the next period.

The beryllium/boron gap and the magnesium/aluminum gap are created, and the hydrogen/helium gap becomes even larger because as we proceed to elements in which the $n = 3$ level is involved, the number of elements that must be found between the first and last element in a period increases. A *d* sublevel is filled after an *s* sublevel and before a *p* sublevel. The end of a period does not occur until a *p* sublevel is filled. Thus in the fourth period from the top, the *d* sublevels begin to be involved and the spread between the *s*-block and the *p*-block opens, creating the gaps.

The above discussion indicates that the number of the period (1, 2, 3, etc., numbering from top to bottom in the table—see Fig. 4.3) coincides with the number of the principal energy level associated with the *s* and *p* sublevels that are being filled. The first *d* sublevel to be filled is in the $n = 3$ principal level, but the first line of *d*-block elements (scandium through zinc) doesn't appear until the fourth period of the table. The reason for this is that the first *d* sublevel (the 3d sublevel) doesn't fill until after the 4*s* sublevel and before the 4*p* sublevel (refer to Fig. 3.11). For this reason, it is logical to present the first line of *d*-block elements in the fourth period between calcium and gallium. The other lines of *d*-block elements are similarly placed in the fourth and fifth periods.

The inner-transition elements are *f*-block elements and represent a filling of *f* sublevels. The filling of the *f* sublevels begins to occur, according to Fig. 3.11, after the 6*s* level is filled and before the 5*d* level is filled. Referring to Fig. 1.2, this would place the first *f* line of elements in the 6th period between barium and

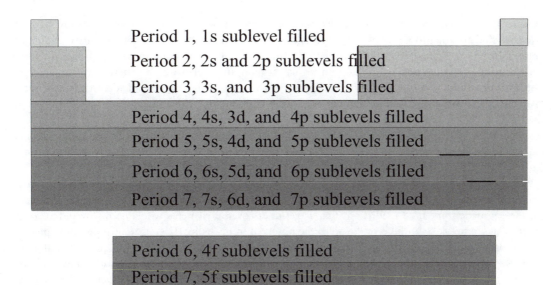

Period 1, 1s sublevel filled

Period 2, 2s and 2p sublevels filled

Period 3, 3s, and 3p sublevels filled

Period 4, 4s, 3d, and 4p sublevels filled

Period 5, 5s, 4d, and 5p sublevels filled

Period 6, 6s, 5d, and 6p sublevels filled

Period 7, 7s, 6d, and 7p sublevels filled

Period 6, 4f sublevels filled

Period 7, 5f sublevels filled

**FIGURE 4.3** The *s* and *p* sublevels being filled have the same principal level number as the period number. The *d* sublevels being filled have a principal level number one less than the period number. The *f* sublevels being filled have a principal level number two less than the period number

lutetium. A close look at Fig. 1.2 reveals that there is a gap of atomic numbers between barium and lutetium equal to the atomic numbers in the first row of the inner-transition elements from lanthanum and ytterbium. Similarly, there is a gap of atomic numbers in the seventh period between radium and lawrencium equal to the second row of elements in the inner-transition group, actinium to nobelium. The inner-transition elements thus represent the *f*-block of elements, those in which the *f* sublevels are being filled. The first row of this block is referred to as the **lanthanide series**, after the first element in the row, lanthanum. The second row in this block is referred to as the **actinide series,** after actinium. The reason these groups are set apart from the others is a matter of convenience. If they were to take their rightful place between barium and lutetium and between radium and lawrencium, an extraordinarily large gap would be created above these elements, and this would make for a needlessly wide periodic table with much empty space.

In summary, the arrangement of the elements in the modern periodic table of the elements is based on electron configuration. The first two families on the left (Groups IA and IIA) represent elements in which the outermost level being filled involves an *s* sublevel. This grouping of elements is thus referred to as the s-block. If the periods are numbered from top to bottom, the number of a period indicates in which principal level the *s* sublevel is located. The final six families on the right (Groups IIIA through VIIIA) represent elements in which the outermost level being filled involves a *p* sublevel. This grouping of elements is known as the *p*-block. Once again, the number of the period indicates in which principal level the *p* sublevel is located. The transition elements represent the grouping of elements in which *d* sublevels are being filled and this grouping of elements is known as the *d*-block. The number of the principal level involved is one less than the period number. The inner-transition elements represent the grouping of elements in which the *f* sublevels are being filled and this grouping of elements is called the *f*-block. The number of the principal level involved is two less than the period number.

## 4.3 Mendeleev and Meyer

A historical perspective on the development of the modern periodic table is both interesting and important. A Russian chemist by the name of **Dimitri Mendeleev** is generally credited with the development of the first periodic table, although a German chemist, **Lothar Meyer**, working independent of Mendeleev at approximately the same time, did much the same work. Their work was published in 1869, long before the modern theory of the atom and hence electron configuration, was developed. Their work involved an extensive study of the properties of the elements that were known at the time. They had observed that as one steps from one element to the next, increasing in atomic number and beginning with hydrogen, physical and chemical properties seem to repeat in a regular fashion. In other words, a pattern seems to emerge. The physical properties that they observed included such basic properties as color and density; the chemical properties they studied included the manner of reaction of an element with acid (no reaction, slow reaction, fast reaction, violent reaction, etc.) and the formula of the compound formed between each element and chlorine. Their published results included the grouping of elements into families, each containing elements that exhibited similar properties. The result was a periodic table with periods and families that were remarkably similar to the modern periodic table.

At that time, it was unknown to Mendeleev and Meyer that not all the elements had as yet been discovered. There were only 62 elements in Mendeleev's table. However, due to the regular pattern he observed, Mendeleev made some bold predictions that as yet undiscovered elements would be discovered at some point in the future so as to fill certain "holes" he observed in his table. He even went so far as to predict specific properties for these elements. For example, the element germanium was not yet known at that time. There was a space in Mendeleev's table betweem silicon and tin in Group IVA. Mendeleev not only predicted the discovery of an element to fill this space, but he even went on to predict specific properties this element would have. He named this as yet undiscovered element "eka-silicon." Germanium was later discovered and was found to have properties nearly identical to what Mendeleev had predicted.

The important lesson to be learned from this story is that the electron configurations of elements are closely related to the elements' properties. The periodic table based on electron configurations is remarkably similar to one based on properties.

# 4.4  Descriptive Chemistry of Selected Elements

What specifically were the properties that Mendeleev and Meyer observed leading up to the development of their periodic tables? What specific properties are we talking about when we say that elements in a given family have similar properties? What specific descriptions can be given for elements within a family that would demonstrate this similarity of properties? Specific descriptions of physical and chemical properties and other interesting facts about elements and compounds, is called *descriptive chemistry*. We now provide a summary of the descriptive chemistries of selected families and other elements and also present at least one important fact about each of these elements.[1]

## 4.4.1  Alkali Metals

Lithium, sodium, potassium, rubidium, and cesium are all highly reactive metals. They all react violently with water, forming hydrogen gas and the metal hydroxide with the formula MOH ("M" symbolizing the metal). They all react readily with oxygen in the air forming the metal oxide with the formula $M_2O$, but can also form *superoxides* of the formula $MO_2$. These latter compounds are used in self-contained breathing devices, such as those that deep-sea divers use to generate oxygen for breathing. Each alkali metal must be stored under mineral oil or kerosene in order to preserve them in the elemental state. Since they are so reactive, they are never found in nature in the free elemental state, only in compounds. Each is a silvery, soft, pliable metal that can be cut with a knife.

| Li |
| Na |
| K |
| Rb |
| Cs |
| Fr |

Lithium is used in lithium batteries because it is a light element and produces a higher battery voltage than other metals. Sodium forms a plethora of compounds used in consumer products important to our livelihood, including NaCl (table salt), $NaHCO_3$ (baking soda), and NaOH (drain cleaner, oven cleaner, soap making). Potassium is important for plant growth, and in the form of potassium nitrate ($KNO_3$) and other compounds, is an component of fertilizers. Rubidium is of little value. It is found as a trace impurity in ores

[1]See also Stwertka, Albert, *A Guide to the Elements,* revised edition, Oxford University Press, 1998.

of other alkali metals. The only naturally occurring isotope of cesium, cesium-133, is used as the world's official measure of time. Francium is radioactive (all isotopes) and unstable.

## 4.4.2  Alkaline Earth Metals

Beryllium, magnesium, calcium, strontium, barium, and radium are the alkaline earth metals. They are grayish white in color and vary in hardness from being able to scratch glass (beryllium) to being soft as lead (barium). They are chemically reactive, but less so than the alkali metals. They are, for the most part, not stable in air, reacting with oxygen to form the metal oxide with the formula MO. In fact, magnesium, strontium, and barium will actually burn in the presence of air, magnesium emitting an extremely bright light in the process. They also react with water to form hydrogen gas ($H_2$) and the metal hydroxide [$M(OH)_2$], but not in the violent manner by which the alkali metals are known to react. Like the alkali metals, however, their reactivity is such that they are never found in the free elemental state in nature.

| Be |
| Mg |
| Ca |
| Sr |
| Ba |
| Ra |

The precious gems of emerald and aquamarine are forms of the mineral beryl, a compound of beryllium, silicon, and oxygen. Compounds of magnesium are found in two consumer products. Magnesium hydroxide [$Mg(OH)_2$] is the active ingredient in milk of magnesia, and magnesium sulfate ($MgSO_4$) is the active ingredient in Epsom salt. Calcium is found in human teeth and bones, hard water, and limestone, which is an ingredient in cement. Strontium compounds are used in fireworks, emitting a red-colored light. A compound of barium, $BaSO_4$, is the white solid in the suspension we swallow so that x-rays can be taken of our intestinal tract. Radium has 25 known isotopes, the most common of which is radium-226, the form of radium discovered by Marie Curie (see CPW Box 3.2).

## 4.4.3  Halogens

The halogens, fluorine, chlorine, bromine, iodine, and astatine, are a group of highly reactive nonmetals. This reactivity is the primary way in which they are similar. Fluorine, $F_2$, and chlorine, $Cl_2$, are gases; bromine, $Br_2$, is a liquid; and iodine, $I_2$, is a solid. Astatine has 20 known isotopes and all are radioactive and highly unstable, which means that only a minute amount of this element exists. However, fluorine, chlorine, bromine, and iodine are quite available and the properties are well known.

| F |
| Cl |
| Br |
| I |
| At |

Fluorine is a yellowish gas and is the most reactive halogen. Its most recognized use is as sodium fluoride, NaF, which is used for tooth decay prevention in toothpastes and municipal water supplies. Chlorine is a yellow-green gas, is extremely poisonous, and has a stifling, pungent odor such as that found in common household bleaches. While it is indeed useful as a bleach, the primary ingredient in household bleach is sodium hypochlorite (NaClO), an important compound of chlorine.

Bromine is a dark red dense liquid with a strong peppery odor. It is highly toxic and can cause serious chemical burns. A compound of bromine, silver bromide (AgBr), is used in the manufacture of photographic film, as is silver iodide (AgI). Iodine is a solid existing as dark purple crystals. It is used as an antiseptic called tincture of iodine, which is a 50% solution of iodine in alcohol. A compound of iodine, potassium iodide (KI), is used as an additive to table salt to make iodized salt.

## 4.4.4  Hydrogen and Helium

Hydrogen and helium are the lightest of all elements. Both are gases. Hydrogen, $H_2$, is the most abundant element in the universe and helium is the second most abundant. The most common compound of hydrogen is water, but it is found in all organic compounds (see Chapter 6) and in acids and bases. Hydrogen gas is

| H | He |

extremely flammable and is used as a fuel for rockets and space shuttles. A very important use of hydrogen gas is in the manufacture of ammonia, $NH_3$ (see Chapter 11).

---

*Chemistry Professionals at Work*                                    *CPW Box 4.3*

# USING BARIUM TO HELP ANALYZE FOR RADIUM IN WATER

A substance that is sometimes added to the sample to eliminate inherent variables that can cause inaccurate results is called an **internal standard**. A standard method for analyzing for radium in water uses barium as an internal standard. It is because radium and barium are in the same family that makes this a useful procedure. The essence of the procedure is as follows.

The radium level in water is determined by adding sulfate to the water. The radium forms an insoluble solid (***precipitate***) with the sulfate. This precipitate, $RaSO_4$, can then be filtered, weighed, and the radium level calculated from this weight.

Because the radium level is typically very low, there is often not enough precipitate visibly formed. To solve the problem, a known amount of barium is added to the water. This barium will also form an insoluble precipitate, $BaSO_4$, when the sulfate is added, a process known as ***coprecipitation***. The precipitate weight that is then measured is a combination of the weight of the barium sulfate and the weight of the radium sulfate. Since the amount of barium added is known, it is possible to calculate the part of the precipitate that is barium sulfate. This will then lead to the amount of radium sulfate formed and ultimately the radium level in the water. Because the two elements are in the same family in the periodic table (alkaline earth metals), they exhibit similar behavior in terms of the percent of sulfate precipitate that is able to be recovered. As a result, this coprecipitation is a reliable method of analysis.

*For Homework: Look up this method for analyzing for radium in water in* Standard Methods for the Examination of Water Wastewater, *a manual published by the American Water Works Association. Write a report providing details of the procedure.*

---

Helium is a noble gas, the others being neon, argon, krypton, xenon, and radon (Family VIIIA). These gases are described as noble or inert because they only form compounds under extraordinary conditions. They are all monatomic, which means that the fundamental unit is the atom rather than the diatomic molecule found as the fundamental unit for other gaseous elements, such as $H_2$, $Cl_2$, etc. Helium is used whenever a very light unreactive gas is needed, such as in toy balloons, blimps, etc.

## 4.4.5   Carbon, Nitrogen, Oxygen, Phosphorus, and Sulfur

Carbon, nitrogen, oxygen, phosphorus, and sulfur are extremely important nonmetals. Carbon is important because it is the element upon which all life is based. All organic compounds, which encompass nearly all 13 million compounds that exist (see Chapter 6), contain carbon. Elemental carbon exists as several different ***allotropes***, or different forms of the same element.

| C | N | O |
|---|---|---|
| P | S |   |

These include diamond, graphite, and buckminsterfullerene, which is a form recently synthesized. Inorganic compounds of carbon are also important. These include carbon dioxide, $CO_2$, or dry ice, carbon monoxide, CO, and the ionic carbonates.

Of all chemicals produced in the U.S., nitrogen gas, $N_2$, ranks second in terms of quantity produced annually (see CPW Box 9.4), a fact that demonstrates its importance. It is a major component of air (78%). Liquid nitrogen is used to achieve extremely cold temperatures needed for some applications, such as in preserving biological systems. Compounds of nitrogen include organic compounds such as amino acids and proteins. Important inorganic compounds of nitrogen include ammonia, $NH_3$, nitric acid, $HNO_3$, and the gaseous oxides.

Oxygen is by far the most abundant element on earth and 46% of the Earth's crust is oxygen. Oxygen gas, $O_2$, supports all life on planet Earth. Oxygen is contained in a large number of organic compounds (see Chapter 6) but it is also contained in many inorganic compounds, including water and compounds of most polyatomic ions, such as nitrates, sulfates, etc. The most common of these is probably calcium carbonate, $CaCO_3$.

Phosphorus exists in two allotropic forms, white phosphorus and red phosphorus. Both are solids. The most important compound of phosphorus is phosphoric acid, $H_3PO_4$, which is important in fertilizer production. Phosphate compounds are used in fertilizers because compounds of phosphorus are key components of living plant cells. Phosphates have also been important ingredients of detergents because they soften hard water through the reaction of phosphate with the calcium and magnesium in the water, thus enhancing the cleaning action. However, this produces waste that has phosphates in it and such waste is not friendly to the environment, so many areas now ban the use of phosphate detergents.

Sulfur is a yellow solid that exists as several allotropes that form when sulfur is heated to different temperatures. Elemental sulfur is found in nature in underground deposits and is mined by the Frasch Process, which uses hot water to melt the sulfur so that is can be forced to the surface with compressed air. Most of the mined sulfur is used to make sulfur dioxide, $SO_2$. The most important compound of sulfur is sulfuric acid. This compound is made by converting $SO_2$ to $SO_3$ and then reacting the $SO_3$ with water. Sulfuric acid is the largest volume chemical produced in this country (see CPW Box 12.2).

### 4.4.6   Aluminum and Silicon

Silicon is the second most abundant element in the Earth's crust, second only to oxygen. Silicon and oxygen are found together in nature, chemically combined in compounds such as in silicon dioxide ($SiO_2$) in rock, sand, and soil. Elemental silicon is structurally similar to diamond, a form of carbon and an element that is in the same family as silicon, so similarity is expected. Silicon is used in computer chips, which is the reason the Silicon Valley is so named. A pure form of silicon dioxide is quartz, which has a modern use in clocks.

| Al | Si |
|----|----|

Aluminum is the third most abundant element in the Earth's crust. The aluminum industry is huge because aluminum metal has so many uses in consumer products. Examples are aluminum foil and beverage cans. It is manufactured by the Hall-Heroult Process by electroplating it from a molten mixture of two aluminum ores, bauxite and cryolite.

### 4.4.7   Selected First-Row Transition Elements: Chromium, Manganese, Iron, Cobalt, Nickel, Copper, and Zinc

While they are not in the same family, these elements display similarities because the $3d$ sublevel is being filled and they all have the

| Fe | Co | Ni | Cu | Zn |
|----|----|----|----|----|

---

*Chemistry Professionals at Work*                                    *CPW Box 4.4*

# THE BATTERY INDUSTRY

Nickel-cadmium batteries, lithium ion batteries, and dry cell batteries were mentioned in Section 4.4. Besides these we also have the lead-acid batteries used in automobiles and the batteries most of us use for long life in flashlights, cameras, etc., called alkaline batteries. The battery industry is obviously huge.

The key features of a battery—what the voltage is, whether or not it can be recharged, how long it will last, etc.—depends on what chemicals are used for the battery components, namely the anode material, the cathode material, the electrolyte used, and the basic design. Much research continues to occur in the battery industry in order to come up with a better product. For example, what materials, chemicals, and designs work best for the batteries used in battery-powered vehicles? In this case, the batteries should be lightweight, rechargeable using household voltages or gasoline-powered engines, and able to be used for relatively long distances before requiring recharging. Consumer demands are high, especially as long as gasoline-powered automobiles remain less expensive. The challenge is obviously a formidable one as chemists and chemistry technicians working as researchers in this industry continue their exciting work.

*For Homework: Check out the website below for a basic discussion of how batteries work. Then select one of the types of batteries mentioned above and find other references which give more details. Write a report.*

http://www.howstuffworks.com/battery.htm (as of 6/4/00)

---

same number of electrons in the $4s$ sublevel, which can be considered the outermost level. They are all fairly unreactive metals and their compounds are known for their striking colors.

Iron is a common, inexpensive metal and has the structural integrity needed to build buildings and bridges and a host of other manufactured products. It is probably the most recognizable metal we have in terms of uses and availability. It does have undesirable corrosion properties; it rusts in the presence of air and moisture. This problem is overcome, if necessary, by combining it with other metals in a homogeneous mixture to make various steels, including stainless steel, which is typically a combination of iron, nickel, and chromium (see CPW Box 1.3). Compounds of iron are usually red (like rust), orange or yellow. Iron (III) chloride, for example, is yellow.

Cobalt shares the property of magnetism with iron. Its compounds are usually either red or blue in color. Glass made with trace amounts of cobalt is blue, often called cobalt blue. It is much more rare than iron and that is why it is not as available or usable as iron.

Nickel is also rather rare. The common U.S. coin called the "nickel" is mostly composed of copper. It is an alloy that is about 25% nickel. An alloy of chromium and nickel, called **nichrome**, is a nickel/chromium alloy and is used to make heating elements and wires. Nickel is also used in nickel-cadmium (Ni-Cad) batteries, rechargeable batteries that have been used in laptop computers and other devices. Nickel compounds also display colors, typically green.

Copper is another metal recognizable to everyone. It is used to make pennies and other coins. An interesting fact about pennies is that before 1981, they were made of pure copper, but since 1981, they are composed of zinc with a copper layer deposited on the surface. Copper wire is used in electric power lines. It is also used in plumbing pipes. The popular alloy brass is a combination of copper and zinc, while bronze is a combination of copper and tin. Compounds of copper are typically blue or green.

Zinc is the most reactive of the transition metals, but after exposure to air it forms a coating of zinc oxide (ZnO) that protects it and makes it much less reactive. Zinc is used to coat steel to make a product called **galvanized steel**, that is used for such materials as household heating ducts, roofing nails, and highway guard rails. Galvanized steel does not disintegrate due to rusting and is therefore useful to maintain the integrity of steel. Zinc is also used in the ordinary dry cell battery used in flashlights and other devices.

## 4.5 Periodicity and Trends: Atomic Size, Ionization Energy, and Electron Affinity

From the discussion in Section 4.2, it is apparent that the elements in the periodic table are arranged according to a pattern based on an electron configuration. Beginning at the left side in a particular period, the outermost s sublevel acquires one electron (alkali metal) and then a second electron (alkaline earth metal). As the right side is approached, the outermost p sublevel acquires six electrons, one at a time, until the right side is reached and both of these s and p sublevels are filled. This pattern repeats in each and every period. In other words, the manner in which sublevels fill with electrons *recurs periodically* as atomic number increases.

In Section 4.3, we concluded that properties are closely related to electron configuration. Element properties also recur *periodically* as atomic number increases. Indeed, this plethora of periodic behavior is the reason that the table is called the *periodic table* and it is also the basis for the **Periodic Law**, which states that the properties of the elements are periodic functions of their atomic numbers. Let us examine three specific properties of elements in order to observe this periodicity.

### 4.5.1 Atomic Size

In Chapter 1 (Section 1.4.1), we defined the atom and mentioned how utterly small the atom is, indicating that it would take hundreds of thousands of even the largest of atoms lined side by side to make a "spot" visible to the naked eye. While atoms are extremely small particles, we would observe a significant variation in their sizes if we were to compare atoms of one element with atoms of another. In Section 1.4.1, for example, we indicated that comparing the size of a hydrogen atom (the smallest) with the size of a francium atom (the largest) would be like comparing a ping-pong ball to a basketball. A francium atom is nine times the size of a hydrogen atom. Let us now examine the trends and periodicity of the sizes of the atoms of the elements.

The size of an atom is directly related to two things: (1) how many principal energy levels it has, and (2) how tightly the electrons are held by the nucleus. The more principal energy levels an atom has, the more space is required to hold them and the larger the atom will be. The tighter the electrons are held, the more compact the atom becomes and the smaller the atom will be. As we move across a period, left to right, item (2) is more important because no new principal level is required and the electrons are held more tightly because of the increased charge (more protons) in the nucleus. Thus, the trend is for the atoms to decrease in size. As a new (higher) principal energy level is utilized at the beginning of the next period, and the size becomes larger again. Thus, atomic size exhibits periodic behavior: large atoms on the left side, smaller atoms as we proceed left to right, and then larger atoms again when a new period begins.

As we move vertically down a family of elements, item 1 becomes more important because a new principal energy level of electrons is present in each period. While the charge in the nucleus is substantially increased as we move down a family, and one may expect smaller atoms because of that, the effect of adding higher principal levels is more important.

To summarize these comments, we can draw arrows in the direction of an increase to show both the horizontal and vertical trends. A horizontal arrow pointing to the right indicates that the particular property increases as we go from left to right. If it points to the left, the property increases right to left. A vertical arrow pointing upward indicates that the property increases from bottom to top, while a vertical arrow point-ing downward indicates that the property increases top to bottom. Atomic size thus exhib-

# Atomic Size

**FIGURE 4.4**   Atomic size increases from right to left across a period, and from top to bottom within a family.

its the trends indicated in Fig. 4.4. A periodic table picturing the relative sizes of atoms is shown in Fig. 4.5.

It is apparent from Fig. 4.5 that there are exceptions to the trend. For example, the size of an Indium (In) atom is greater than the size of a Cadmium (Cd) atom even though the general trend is in the opposite direction. Such idiosyncrasies can sometimes be traced to sublevels being filled or half-filled with electrons. For example, the cadmium-to-indium size increase is likely due to the fact that cadmium represents a filled *d* sublevel, while indium represents the beginning of the *p* sublevel and is thus a slightly larger atom.

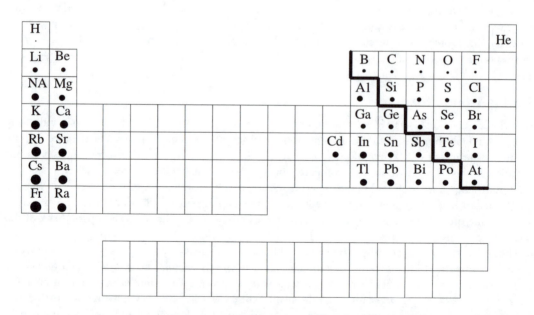

**FIGURE 4.5**   An outline of the periodic table showing the relative sizes of the atoms of elements in the *s*- and *p*-blocks. Notice the trend to smaller atoms as one goes left to right and top to bottom. The size of the cadmium (Cd) atom is also indicated in order to compare it with the indium (In) atom, as discussed in the text.

## 4.5.2 Ionization Energy and Electron Affinity

Aside from having relatively large atoms, elements that have one or two electrons in their outermost level (left side of the periodic table) also lose these electrons readily. It is almost as if these electrons, which are located on the outer fringes of these large atoms and far from the nucleus, are inviting substances that attract electrons to come by and take them away. Such atoms also have very little attraction for more electrons. Similarly, elements that have relatively small atoms and have their outermost level nearly filled (right side of the periodic table) readily take on more electrons. The outermost electrons in these atoms are also located on the outer fringes of these atoms, but, since the atoms are smaller, these electrons are located much closer to the nucleus. These atoms thus have very little tendency to give up electrons.

Elements whose atoms give up electrons readily and have a weak attraction for more electrons are located on the left side of the periodic table, while elements whose atoms do not give up electrons readily and have a strong attraction for more electrons are located on the right side of the periodic table. The energy required to remove an electron from an atom is called the ***ionization energy***. The attraction an atom has for more electrons is called the ***electron affinity***. Both increase as one steps left to right across a period, a trend opposite to atomic size. A strong attraction for electrons corresponds to a high ionization energy and vice versa. When the end of a period is reached and a new one begins, both ionization energy and electron affinity sharply decrease and the increase occurring from left to right begins again. Both ionization energy and electron affinity are thus periodic properties just like atomic size. Figure 4.6, which is a graph of ionization energy vs. atomic number, presents this in graphic form where the periodic behavior is indicated by the up-and-down changes occurring as atomic number increases.

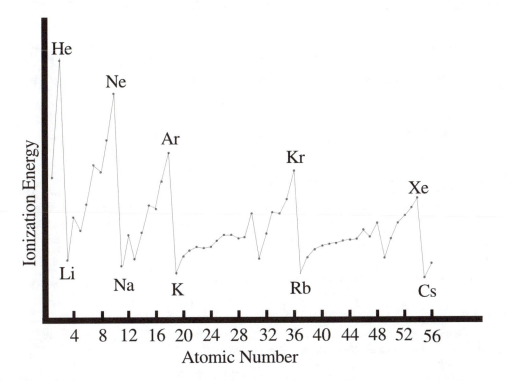

**FIGURE 4.6** A graph of ionization energy vs. atomic number demonstrating that ionization energy is a periodic property.

The behavior of these properties within a family is opposite to that observed for atomic size. Both ionization energy and electron affinity increase as we proceed from the bottom to the top within a family. Figure 4.7 summarizes both the left/right and the top/bottom trends. The top/bottom behavior is affected by what may be called the *screening effect*. Atoms that have a high atomic number (near the bottom of a family) have a large number of electrons between the nucleus and the outermost level. These electrons "screen" the outer electrons from the pull of the nucleus, causing them to be loosely held and causing the atom in general to have a low ionization energy. Thus, the ionization energy is high where there are fewer inner-level electrons to screen the outer-level electrons from the nucleus and low where there are more such electrons. We can make a similar statement about electron affinity. Electrons are less likely to be captured by an atom when the pull of the nucleus for such electrons is lessened by the opposing force of the inner-level electrons. They are more likely to be captured when there are fewer inner-level electrons.

# Ionization Energy and Electron Affinity

**FIGURE 4.7**    Ionization energy and electron affinity both increase from left to right across a period and from bottom to top within a family.

## Example 4.1

Which of the following has the largest atom according to periodic trends, Ca or Cl?

## Solution 4.1

Atomic size increases from right to left and top to bottom in the periodic table (Fig. 4.4). Calcium, Ca, is to the left and down from chlorine, Cl, so calcium has a larger atom.

## Example 4.2

Which of the following has the lower electron affinity according to periodic trends, Al or Zr?

## Solution 4.2

Electron affinity increases left to right and bottom to top in the periodic table (see Fig. 4.7). Thus, the further we go in this table to the left and down, the smaller the electron affinity will be. Zirconium (Zr) lies further to the left and is down from aluminum (Al), so it has the lower electron affinity.

## Example 4.3

Arrange the following in order of decreasing ionization energy: Mn, Ba, S, Se.

## Solution 4.3

Ionization energy increases from left to right and from bottom to top in the periodic table (Fig. 4.7). The correct order of decreasing ionization energy would be S, Se, Mn, and Ba.

# 4.6   Variability of Oxidation Number

Another characteristic of elements that must be addressed in terms of trends and periodicity is oxidation number. The concept of oxidation number was introduced in Chapter 2 (Section 2.4). Since there is an apparent relationship between the number of electrons that can be lost or gained and an element's

**TABLE 4.1** The Trend in Oxidation Numbers of Elements that One Might Predict Based on the Need to Lose and Gain Electrons in Order to Match the Electron Configuration of the Noble Gases

| Group IA | Group IIA | Group IIIA | Group IVA | Group VA | Group VIA | Group VIIA | Group VIIIA |
|---|---|---|---|---|---|---|---|
| +1 | +2 | +3 | +4, −4 | −3 | −2 | −1 | 0 |

position in the periodic table relative to the noble gases, it would seem that a trend in oxidation numbers might also be present. This is shown in Table 4.1.

On the left side of the periodic table, elements lose the electrons in their outermost level by creating ions with positive charges equal in magnitude to the number of electrons that are lost. On the right side of the table, elements gain electrons in their outermost level creating ions with negative charges equal in magnitude to the number of electrons gained. In the middle of the table, there would be a transition from a positive oxidation number to a negative one as the tendency switches from one in which electrons are lost to one in which electrons are gained. It would appear reasonable that this same trend would exist in all periods, resulting in periodic behavior similar to what we saw for atomic size, ionization energy, and electron affinity. However, in reality, we observe that oxidation numbers are much more variable and unpredictable. The only solidly predictable oxidation numbers are those designated in the rules for assigning oxidation numbers in Table 2.2 of Chapter 2, for alkali metals (+1), alkaline earth metals (+2) and aluminum (+3). In contrast, the oxidation numbers of elements in Groups IVA–VIIA are quite variable. For example, nitrogen exhibits seven different oxidation numbers and chlorine, bromine, and iodine each exhibit five different oxidation numbers.

This variability in oxidation number gives rise to a large number of compounds in which these elements may be combined and also complicates the nomenclature picture. We had a glimpse of this in Chapter 2 (Section 2.6) when seven different oxides of nitrogen were listed. We address many of these complications in the next section.

# 4.7 More Nomenclature

In this section, we address the naming of compounds that result from the variability of oxidation numbers, particularly occurring with elements on the right side of the periodic table.

## 4.7.1 Additional Polyatomic Ions and Oxyacids of Nitrogen, Phosphorus, and Sulfur

In Chapter 2 (Sections 2.3 and 2.7), we discussed the nomenclature schemes on which the names of ionic compounds of four of the "Big Six" polyatomic ions (carbonate, $CO_3^{2-}$, sulfate, $SO_4^{2-}$, phosphate, $PO_4^{3-}$, and nitrate, $NO_3^-$) are based. We also examined how the names of oxyacids corresponding to these ions are given, and we learned how to name ionic compounds such as $NaNO_3$ (sodium nitrate), $CaCO_3$ (calcium carbonate), and $K_2SO_4$ (potassium sulfate). This scheme is to simply write the name of the metal (or positive ion) and then to write the name of the polyatomic ion. We also learned how to name the corresponding oxyacids, such as $HNO_3$ (nitric acid), $H_2CO_3$ (carbonic acid), and $H_2SO_4$ (sulfuric acid). This scheme utilizes the stem of the nonmetal (other than hydrogen or oxygen), an *-ic* ending on the stem, and the word "acid."

The sulfur, the nitrogen, and the phosphorus in these polyatomic ions do not follow the trend in oxidation numbers that is indicated in Table 4.1. Only carbon follows the trend—its oxidation number is +4 in carbonate. Sulfur's oxidation number is +6 in sulfate; nitrogen's is +5 in nitrate; and phosphorus' is +5 in phosphate.

**TABLE 4.2**   Summary of Three Polyatomic Ions Derived from Several "Big Six" Ions Studied in Chapter 1[a]

| Polyatomic Ion with One Less Oxygen Than Related "Big Six" Ion | Related "Big Six" Ion | Acid and Example Salt |
|---|---|---|
| Sulfite, $SO_3^{2-}$ | Sulfate, $SO_4^{2-}$ | Sulfurous Acid, $H_2SO_3$ |
| | | Sodium Sulfite, $Na_2SO_3$ |
| Nitrite, $NO_2^-$ | Nitrate, $NO_3^-$ | Nitrous Acid, $HNO_2$ |
| | | Sodium Nitrite, $NaNO_2$ |
| Phosphite, $PO_3^{3-}$ | Phosphate, $PO_4^{3-}$ | Phosphorus Acid $H_3PO_3$ |
| | | Sodium Phosphite, $Na_3PO_3$ |

[a]These ions have one less oxygen in their formulas compared to their counterpart "Big Six" ion, and have an "ite" ending. The acids of these ions are characterized by an "ous" ending, rather than "ic."

In addition, sulfate, nitrate, and phosphate all have a twin polyatomic ion that has one less oxygen. These ions are named by replacing the "a" in the *-ate* ending with an "i." Thus we have sulfite, $SO_3^{2-}$, phosphite, $PO_3^{3-}$, and nitrite, $NO_2^-$. The charges on each of these ions are the same as the charges on those with the $-ate$ ending. The naming scheme for ionic compounds containing these ions is the same as the others, in which we use the name of the metal followed by the name of the ion. This gives us, for example, $KNO_2$ (potassium nitrite), $Na_3PO_4$ (sodium phosphite), and $MgSO_3$ (magnesium sulfite).

The oxyacids are named with an *-ous* ending rather than *-ic* on the first word of the name. Thus, we have $HNO_2$ (nitrous acid), $H_2SO_3$ (sulfurous acid), and $H_3PO_3$ (phosphorous acid). Table 4.2 summarizes these facts.

In sulfite, sulfur has an oxidation number of $+4$. In nitrite, nitrogen has an oxidation number of $+3$. In phosphite, phosphorus has an oxidation number of $+3$. Thus, in this series of polyatomic ions, the trend in oxidation numbers indicated in Table 4.1 is also not followed. This further demonstrates the variability of oxidation numbers that the elements in Groups IVA–VIA have.

## 4.7.2   Polyatomic Ions and Oxyacids of the Halogens

Let us now take a look at ions and oxyacids of Group VIIA elements, the halogens, and at chlorine (Cl), bromine (Br), and iodine (I), in particular. We indicated in Section 4.6 that these halogens, when chemically combined, exhibit five different oxidation numbers. Besides the $-1$ predicted from the periodic trend (Table 4.1), these halogens exhibit oxidation numbers of $+1$, $+3$, $+5$, and $+7$. This results in four different polyatomic ions that can form with oxygen and four different oxyacids for each halogen. These polyatomic ions have one, two, three, or four oxygen atoms associated with a single halogen atom, and each has a $-1$ charge. Thus, using chlorine as the halogen, we have the ions $ClO^-$, $ClO_2^-$, $ClO_3^-$, and $ClO_4^-$ representing oxidation numbers for chlorine of 1, $+2$, $+3$, and $+4$, respectively. The corresponding oxyacids would have the formulas $HClO$, $HClO_2$, $HClO_3$, and $HClO_4$.

The nomenclature scheme for these polyatomic ions and oxyacids utilizes the *-ous/-ite* and the *-ic/-ate* endings indicated previously for sulfur, nitrogen, and phosphorus. The *-ous/-ite* endings are used when there are two oxygens, and the *-ic/-ate* endings are used when there are three oxygens. In order to cover the other two possibilities, two new prefixes are introduced: *hypo-* and *per-*. *Hypo-* is used in conjunction with *-ous/-ite* when there is just one oxygen, and *per-* is used in conjunction with *-ic/-ate* when there are four oxygens. The same stems defined previously are used, *chlor-* for chlorine, *brom-* for bromine, and *iod-* for iodine. All twelve possibilities are summarized in Table 4.3.

---

*Chemistry Professionals at Work*          *CPW Box 4.5*

# PERCHLORIC ACID

Probably the most common compound containing a polyatomic ion of a halogen is perchloric acid, $HClO_4$. It is used in industrial laboratories to destroy or dissolve organic material in laboratory analysis samples. Chemistry professionals that use perchloric acid must take special safety precautions. For example, perchloric acid fumes can corrode the metals of which fume hood ducts are made. The resulting metal perchlorates that form on the interior of the fume hood ducts are very dangerous because they can explode. Thus, special fume hoods and special procedures must be used.

*For Homework: Do an Internet search on the topic of perchloric acid and discover as much safety information as you can. Write a report.*

---

**TABLE 4.3**    A Summary of the Polyatomic Ions of Halogens[a]

| Derived from Chlorine | Derived from Bromine | Derived from Iodine |
|---|---|---|
| $ClO^-$, Hypochlorite Ion | $BrO^-$, Hypobromite Ion | $IO^-$, Hypoiodite Ion |
| KClO, Potassium Hypochlorite | KBrO, Potassium Hypobromite | KIO, Potassium Hypoiodite |
| HClO, Hypochlorous Acid | HBrO, Hypobromous Acid | HIO, Hypoiodous Acid |
| $ClO_2^-$, Chlorite Ion | $BrO_2^-$, Bromite Ion | $IO_2^-$, Iodite Ion |
| $KClO_2$, Potassium Chlorite | $KBrO_2$, Potassium Bromite | $KIO_2$, Potassium Iodite |
| $HClO_2$, Chlorous Acid | $HBrO_2$, Bromous Acid | $HIO_2$, Iodous Acid |
| $ClO_3^-$, Chlorate Ion | $BrO_3^-$, Bromate Ion | $IO_3^-$, Iodate Ion |
| $KClO_3$, Potassium Chlorate | $KBrO_3$, Potassium Bromate | $KIO_3$, Potassium Iodate |
| $HClO_3$, Chloric Acid | $HBrO_3$, Bromic Acid | $HIO_3$, Iodic Acid |
| $ClO_4^-$, Perchlorate Ion | $BrO_4^-$, Perbromate Ion | $IO_4^-$, Periodate Ion |
| $KClO_4$, Potassium Perchlorate | $KBrO_4$, Potassium Perbromate | $KIO_4$, Potassium Periodate |
| $HClO_4$, Perchloric Acid | $HBrO_4$, Perbromic Acid | $HIO_4$, Periodic Acid |

[a]Potassium compounds are given only as illustrations of the formulas and names for ionic compounds that are formed from these ions. The formulas and names of the corresponding acids are also presented.

## 4.7.3 Partially Neutralized Salts of the Oxyacids of Carbon, Sulfur, and Phosphorus

In Chapter 2 (Sections 2.7 and 2.8) we defined the terms acid, base, and salt. An acid reacts with a base to form a salt and water in a reaction that is called neutralization. The acid neutralizes the properties of the base and the base neutralizes the properties of the acid. In this process, the hydrogen(s) of the acid is (are) replaced by the positive ion(s) of the base, forming a salt, which results from the combination of a positive ion from the base with a negative ion from the acid. Thus, we have such salts as NaCl (from HCl and NaOH), $K_2SO_4$ (from $H_2SO_4$ and KOH), and $NH_4NO_3$ (from $HNO_3$ and $NH_4OH$), as indicated in Section 2.8.

---

*Chemistry Professionals at Work*                                    *CPW Box 4.6*

# FACTS ABOUT SALTS THAT ARE
# PARTIALLY NEUTRALIZED ACIDS

A salt that can be described as partially neutralized carbonic acid, sodium bicarbonate ($NaHCO_3$), is familiar to most of us. It is the white solid commonly known as "baking soda." It is advertised to have a variety of real-world uses, such as baking powder, a deodorizer for refrigerators and freezers, and as an ingredient in personal care items like deodorants and toothpastes. Chemical process industries are responsible for the manufacture of this chemical and also for the formulations in which it is used. Chemistry technicians are needed in these industries for plant operations and for quality assurance analysis.

The monohydrogen and dihydrogen phosphates have very important uses in real-world analytical laboratories. Mostly, these uses center around the control of the acidity of solutions. Solutions of these partially neutralized acids resist changes in acidity level even when strong acids or bases are added. For this reason they are called "buffer solutions." For example, dilute solutions of potassium dihydrogen phosphate, $KH_2PO_4$, will maintain an acidity level near the neutral point between acids and bases. Solutions of sodium monohydrogen, $Na_2HPO_4$, will maintain an acidity level that is considerably below the neutral point, which means that base-like properties predominate.

*For Homework: Look in the* Concise Encyclopedia of Chemical Technology *to discover the details of the manufacture of sodium bicarbonate. Write a report.*

---

Acids having two or more hydrogens in their formula can be *partially neutralized*. This means that in the case of an acid that has two hydrogens, only one of the two may be replaced by the positive ion rather than both. Or in the case of an acid that has three hydrogens, only one or two hydrogens may be replaced rather than all three. Such a phenomenon results in an array of ions and salts that have not been discussed previously. Some acids that have two or three hydrogens in their formulas are carbonic acid ($H_2CO_3$), sulfuric acid ($H_2SO_4$), sulfurous acid ($H_2SO_3$), phosphoric acid ($H_3PO_4$), and phosphorous acid ($H_3PO_3$). The nomenclature schemes for the ions and salts obtained by partially neutralizing these acids are summarized in Table 4.4.

## 4.7.4  Other Polyatomic Ions

It should come as no surprise that in addition to the polyatomic ions addressed so far in this chapter and in previous chapters, there are quite a number of others that are of some importance. Some of these demonstrate the fact that oxidation numbers can vary in ways that can be explained by an element's position in the periodic table. What may be a surprise, however, is that metals can also be involved. Table 4.5 presents a list of additional polyatomic ions that have not yet been mentioned in this or previous chapters. Notice, in particular, that the chromate ion ($CrO_4^{2-}$), the dichromate ion ($Cr_2O_7^{2-}$), and the

**TABLE 4.4**  Summary of Polyatomic Ions with $-2$ or $-3$ Charge[a]

| Polyatomic Ion (Corresponding Acid) | Ion and Example Salt with Two Hydrogens Remaining | Ion and Example Salt with One Hydrogen Remaining |
|---|---|---|
| **Carbonate, $CO_3^{2-}$** (Carbonic Acid, $H_2CO_3$) | N/A | Bicarbonate Ion, $HCO_3^{1-}$ Sodium Bicarbonate, $NaHCO_3$ |
| **Sulfate $SO_4^{2-}$** (Sulfuric Acid, $H_2SO_4$) | N/A | Bisulfate Ion, $HSO_4^{1-}$ Sodium Bisulfate, $NaHSO_4$ |
| **Sulfite $SO_3^{2-}$** (Sulfurous Acid, $H_2SO_3$) | N/A | Bisulfite Ion, $HSO_3^{1-}$ Sodium Bisulfite, $NaHSO_3$ |
| **Phosphate $PO_4^{3-}$** (Phosphoric Acid $H_3PO_4$) | Dihydrogen Phosphate Ion, $H_2PO_4^{1-}$ Sodium Dihydrogen Phosphate, $NaH_2PO_4$ | Monohydrogen Phosphate Ion, $HPO_4^{2-}$ Sodium Monohydrogen Phosphate, $Na_2HPO_4$ |
| **Phosphite $PO_3^{3-}$** (Phosporous Acid, $H_3PO_3$) | Dihydrogen Phosphite Ion, $H_2PO_3^{1-}$ Sodium Dihydrogen Phosphite, $NaH_2PO_3$ | Monohydrogen Phosphite Ion, $HPO_3^{2-}$ Sodium Monohydrogen Phosphite, $Na_2HPO_4$ |

[a]Also includes the corresponding acids, and ions and salts derived from these acids that have one or two hydrogens remaining in their formulas.

**TABLE 4.5**  Formulas and Names of Other Common Polyatomic Ions

| Formula of Ion | Name of Ion |
|---|---|
| $CrO_4^{2-}$ | Chromate |
| $Cr_2O_7^{2-}$ | Dichromate |
| $MnO_4^-$ | Permanganate |
| $CN^-$ | Cyanide |
| $SCN^-$ | Thiocyanate |
| $C_2H_3O_2^-$ | Acetate |
| $C_2O_4^{2-}$ | Oxalate |
| $S_2O_3^{2-}$ | Thiosulfate |

permanganate ion ($MnO_4^{1-}$) all contain a metal. Chromium has a $+6$ oxidation number in both chromate and dichromate. Manganese has an oxidation number of $+7$ in permanganate. The oxidation numbers of the transition metals in these ions are related to the position of these metals in the periodic table. Chromium is in Group VIB, the sixth group from the left side (possibly indicating why the oxidation number can be $+6$), while manganese is in Group VIIA possibly indicating why its oxidation number can be $+7$). These are unstable oxidation numbers, however, and compounds containing these polyatomic ions are very reactive.

Naming salts containing the polyatomic ions in Table 4.5 is done by following exactly the same rules as for other polyatomic ions. First we write the name of the metal, followed by the name of the polyatomic ion.

# 4.8  Other Nomenclature Schemes

While official nomenclature schemes have been adopted, and traditional nomenclature schemes are in common use, naming schemes that reflect other norms such as detailed structural features, manufacturing processes, and trademarks, etc., are often encountered. Also, names that may not be based on any obvious property or formula relationship are sometimes found. For example, potassium carbonate, $K_2CO_3$, is often referred to as "potash," and sodium thiosulfate, $Na_2S_2O_3$, is often referred to as "hypo."

*Chemistry Professionals at Work*                                    *CPW Box 4.7*

# SUPPORT FOR SUBSIDIARIES
# IN THE ORIENT

My name is John Engelman. I am a chemistry professional employed by S.C. Johnson at their corporate research facility in Racine, Wisconsin. I am a Senior Research Technologist responsible for providing consumer products formulation and packing support to the S.C. Johnson subsidiaries in Japan, China, and Thailand. The support includes travel to the Orient to perform project updates and reviews as well as technical service at the subsidiaries. The consumer products I am responsible for cover the complete range of S.C. Johnson products including hair care, fabric care, insect repellants, bathroom cleaners, and products unique to the Orient.

*For Homework: Check out the Internet site for S.C. Johnson and write a report on this company.*

http://scjohnsonwax.com/welcome5.asp (as of 6/4/00)

The naming of phosphates, including the mono- and dihydrogen phosphates, as well as phosphoric acid, $H_3PO_4$, presents an interesting case in point. Table 4.6 presents some names of such compounds that may be encountered at various times. Each of these names represents a nomenclature method that has legitimacy because it is based on one of the following: an accepted formal naming scheme, a unique structural feature, a reaction mechanism, or certain assumptions that are made about the formula. The prefix "ortho" is used here to differentiate this one type of phosphate from metaphosphate, which has a unique ring-like structure that is based on the formula $(HPO_3)_n$.

**TABLE 4.6**  Phosphate Compounds and Their Various Names[a]

| Formula | Name |
|---------|------|
| $H_3PO_4$ | Phosphoric acid |
| | Orthophosphoric acid |
| $Na_3PO_4$ | Sodium phosphate |
| | Trisodium phosphate |
| | Sodium phosphate tribasic |
| | Sodium orthophosphate |
| $NaH_2PO_4$ | Sodium dihydrogen phosphate |
| | Monosodium phosphate |
| | Sodium phosphate monobasic |
| $Na_2HPO_4$ | Sodium monohydrogen phosphate |
| | Disodium phosphate |
| | Sodium phosphate dibasic |

[a]The first name listed in each case is the name that was given previously in this chapter.

The prefixes *tri-*, *di-*, and *mono-* can be used in front of the word sodium (or other metal) to indicate the number of sodiums in the formula. In these cases, it is assumed that the appropriate number of hydrogens are also present without specifically indicating this in the name. Finally, the words *tribasic, dibasic,* and *monobasic* can be used in these compounds to indicate how many more hydrogens can be added to the formula and replace the metal. In sodium phosphate tribasic, three hydrogens may be added to make $H_3PO_4$. In sodium phosphate dibasic, two hydrogens may be added, while in sodium phosphate monobasic, one hydrogen may be added.

## 4.9   Homework Exercises

1. Give the symbol of an element found in
   (a) the *d*-block of the periodic table.
   (b) the *s*-block of the periodic table.
   (c) the *p*-block of the periodic table.
   (d) the *f*-block of the periodic table.
2. Tell in which block of the periodic table, *s*-block, *p*-block, *d*-block, or *f*-block, each of the following elements are found.
   (a) Manganese   (b) Radon   (c) P   (d) Ca   (e) Yb   (f) Yttrium
   (g) Einsteinium   (h) Carbon   (i) Se   (j) Ru   (k) Rubidium
3. Describe what is meant by the "actinide series." Where does this occur in the periodic table?
4. Describe what is meant by the "lanthanide series." Where does this occur in the periodic table?
5. What is so remarkable about the work of the Russian chemist named Mendeleev?
6. Mendeleev did not know anything about electron configuration, yet he was able to create a table that looked remarkably similar to the modern periodic table, which is based on electron configuration. Explain.
7. On what did Mendeleev base the arrangement of the elements in his periodic table? On what is the arrangement of the elements in the modern periodic table based? Given that the two were based on two different things, how is it that Mendeleev's table looked so similar to the modern table?
8. What is descriptive chemistry?
9. (a)  What might mineral oil be used for in reference to the alkali metals? Why?
   (b)  Are any metals so soft you can cut them with a knife? If so, which?
   (c)  Write and balance the equation for the reaction of an alkali metal of your choice with water.

10. (a)  Give at least one common use of each of the alkaline earth metals.
    (b)  Write and balance the equation for the reaction of an alkaline earth metal of your choice with water.
11. Describe each of the halogens in terms of physical state and color.
12. Hydrogen and helium are both much lighter than air. Why is helium chosen over hydrogen for blimps?
13. What is an allotrope? Describe the different allotropes of carbon, phosphorus, and sulfur.
14. Give one important interesting or important fact about each of the following elements: silicon, aluminum, iron, cobalt, nickel, copper, and zinc.
15. When we say that a property is "periodic," what does that mean?
16. Why is the table of elements called the periodic table?
17. What is the "Periodic Law?" Explain why this is important in chemistry.
18. What is the trend that is observed in atomic size as you go from left to right and from top to bottom in the periodic table?
19. What is the explanation for the observed trend in atomic size in the periodic table?
20. What is the trend that is observed in ionization energy as you go from left to right and from top to bottom in the periodic table?
21. What is the explanation for the observed trend in ionization energy in the periodic table?
22. What is the trend that is observed in electron affinity as you go from left to right and from top to bottom in the periodic table?
23. What is the explanation for the observed trend in electron affinity in the periodic table?
24. What is the "screening effect?" How does this affect the properties that are observed for atoms?
25. One reason why an atom of fluorine has a higher electron affinity compared to an atom of bromine is the screening effect. Explain.
26. Based on trends, write the following in order of
    (a)  increasing atomic radius. Cr, S, Rb, K
    (b)  decreasing electron affinity. O, Ca, Cs, F
    (c)  decreasing ionization energy. Fe, Ge, Ba, F
    (d)  decreasing ionization energy. Hg, Fr, S, Te
    (e)  increasing electron affinity. Al, Tc, O, Ba
    (f)  decreasing ionization energy. Mo, Cs, Br, Sr
    (g)  increasing electron affinity. Au, S, Ba, Ir
    (h)  decreasing atomic size. Co, Pd, Pt, Ag
    (i)  increasing atomic size. Zn, O, Rb, Ag
    (j)  decreasing ionization energy. N, Mo, Ba
    (k)  increasing electron affinity. F, Sn, S, As
27. Based on trends,
    (a)  Which of the following has the smallest atom, Sr or Ca?
    (b)  Which of the following has the lowest electron affinity, O or S?
28. Does an element with a low ionization energy tend to form a cation or an anion? Explain.
29. In what part of the periodic table do you find elements that have relatively large atoms?
30. In what part of the periodic table do you find elements that have relatively small atoms?
31. Which of the following has the smallest radius, K or $K^{+1}$? Explain why.
32. Define ionization energy.
33. Name the following compounds.
    (a) $HNO_2$    (b) $CaSO_3$   (c) $HClO_4$   (d) $KIO_2$
    (e) $NaH_2PO_4$   (f) $Mg(HCO_3)_2$
34. Give the formula for each of the following.
    (a) Sulfurous acid   (b) Potassium phosphite   (c) Sodium bisulfate
    (d) Iodic acid   (e) Lithium monohydrogen phosphate
    (f) Calcium hypobromite

35. Name the following compounds.
    (a) $H_3PO_3$  (b) $Na_2HPO_4$  (c) $HIO_2$  (d) $Ca(NO_2)_2$  (e) $KHCO_3$
36. Write the formula of the following compounds.
    (a) Potassium sulfite  (b) Calcium chlorate  (c) Nitrous acid
    (d) Hypobromous acid  (e) Carbonic acid
37. Name the following compounds.
    (a) $HBrO_2$  (b) $KHCO_3$  (c) $Mg(IO_4)_2$  (d) $H_3PO_4$  (e) $Fe(NO_3)_3$  (f) $Na_3PO_3$
38. Give the formula of each of the following compounds.
    (a) Carbonic acid  (b) Potassium monohydrogen phosphate  (c) Iron (II) hypobromite
    (d) Chloric acid  (e) Nitrous acid  (f) Calcium sulfite
39. Name the following compounds.
    (a) $HNO_3$  (b) $HClO_3$  (c) $Mg(ClO_2)_2$  (d) $NaH_2PO_4$  (e) $NaHCO_3$
40. Write the formula of the following compounds.
    (a) Hypochlorous acid  (b) Sodium nitrite  (c) Potassium chlorate
41. Name the following compounds.
    (a) $K_2CrO_4$  (b) $BaCr_2O_7$  (c) $KMnO_4$  (d) $HCN$
    (e) $Mg(C_2H_3O_2)_2$  (f) $Na_2C_2O_4$  (g) $CaS_2O_3$
42. Write the formula of the following compounds.
    (a) Potassium thiocyanate  (b) Sodium acetate
    (c) Potassium dichromate  (d) Barium chromate
    (e) Calcium oxalate  (f) Sodium thiosulfate
    (g) Ammonium dichromate  (h) Sodium cyanide
    (i) Sodium permanganate
43. Write the formula of the following compounds.
    (a) Monocalcium phosphate  (b) Potassium phosphate dibasic
    (c) Magnesium phosphate monobasic  (d) Orthophosphoric acid
    (e) Aluminum orthophosphate  (f) Trilithium phosphate
    (g) Ammonium phosphate tribasic

## For Class Discussion and Reports

44. For Homework: Select your "favorite element" from the periodic table and write a 2–3 page paper telling why you think it is an interesting element. Describe what chemical and physical properties it has, what its history is (when discovered, by whom, how discovered, etc.), how it is commercially produced, what uses it has, where it is observed and used in everyday life, and what hazards it presents, if any. Use the following Website as the primary source of information, but also use other sources such as the source listed in the footnote in Section 4.4, and list these sources in a bibliography. http://www.webelements.com/index.html (as of 6/4/00)

45. For Homework: In Table 4.3, the oxyacids of the halogens are listed as well as the potassium salts of these acids. The total number of compounds listed is 24. Considering that we could potentially use sodium, ammonium, calcium, magnesium, or any other positive ion with any of the polyatomic ions listed, there would seem to be a huge number of compounds possible with these polyatomic ions. In reality, a very limited number of them actually exist or have any importance in the chemical industry or consumer products. Do some research and identify five compounds from the possible list and write a report giving some useful and interesting information about each. Also, indicate those compounds you researched but for which you found no information. Potential sources of information include a chemical catalog, the *CRC Handbook of Chemistry and Physics*, the *Merck Index*, the *Concise Encyclopedia of Chemical Technology*, and the Internet.

46. There are likely too many chemicals listed on the ingredients lists of consumer products (foods, personal care items, etc.) whose names you do *not* recognize from your chemistry studies thus far. Examine these lists. Find the name of one of these chemicals and do some investigating. What

is the chemical formula? What information is presented in the *Handbook of Chemistry and Physics,* or the *Merck Index*? Ask your instructor to help you locate a copy of the Material Safety Data Sheet (MSDS—see CPW Box 6.7). What safety issues do you discover on the MSDS? Ask your instructor for a copy of the catalog from the Fisher Chemical Company, or look up the chemical on the Fisher Chemical Company Internet site (www.fishersci.com) (as of 6/4/00). Is it available from the Fisher Chemical Company? If so, at what cost? Write a report that lists the consumer product and all of the above information.

# 5

# Chemical Combination

## 5.1   Introduction

With this chapter, we present the remaining details of the topic of chemical combination, a topic that has been a point of mystery and intrigue since Chapter 1. In Chapter 1, chemical combination was given as the explanation of why compounds, while consisting of more than one element, possess only some, or in many cases, none of the properties of those elements. We've indicated that there are forces that bind the atoms of these elements together and that these special forces, known as chemical bonding, dictate a complete transformation of properties. The result is that the compound in question possesses properties that are unique and that this new substance bears no resemblance to the elements of which it is composed. Our example in Chapter 1 was water, a liquid substance that extinguishes flames, yet it is composed of hydrogen, a highly flammable gaseous element, and oxygen, a gaseous element that supports combustion.

In Chapter 2, the formulas and names of compounds were introduced and, in order to establish naming schemes, a distinction was made between ionic compounds and molecular compounds. Ionic compounds were described as compounds, usually between metals and nonmetals, that are composed of formula units that include monatomic or polyatomic ions held together by electrostatic forces. Molecular, or covalent, compounds were described as compounds between two nonmetals that are not composed of ions, but of molecules—discrete units made up of atoms held together by other forces. As we have seen in Chapters 3 and 4, metals like to give up electrons to form positively charged ions, while nonmetals like to take on electrons producing negatively charged ions. Ionic compounds are often considered a result of the transfer of electrons from the metal to the nonmetal. The chemical bonding that occurs in ionic compounds, then, is called ***ionic bonding***, which is the electrostatic attraction between oppositely charged ions. Molecular compounds are composed of two or more nonmetals, both of which like to take on electrons. Thus, there can be no complete transfer of electrons

# WHY STUDY THIS TOPIC?

**M**any properties exhibited by chemical compounds are explained by citing the intimate details of the nature of the forces that bind atoms to one another in molecules and ionic units. Since a knowledge of these properties is important to all who handle these chemicals on a day-to-day basis, a knowledge of the intimate details can be very useful so that chemistry professionals can not only predict the physical behavior of compounds (and take appropriate precautions, if necessary), but also explain and/or predict observations that are made concerning these compounds in the laboratory. For example, being able to visualize molecules using the Lewis electron dot structure and the VSEPR theory allows a chemist to predict what compounds are soluble in what solvents. This could also help him to decide what stationary phases should be used when performing an analysis by popular chromatography methods. These and related concepts are presented in this chapter.

*For Homework: Look up "chromatography" in a basic analytical chemistry text or instrumental analysis text. Without going into great detail, report on what it is, what different types exist, and what is meant by "stationary phases," as mentioned above.*

from one element to another in these compounds. However, both nonmetals can achieve the same end result (i.e., a filled outermost *p* sublevel) by sharing two or more electrons. The chemical bonding that occurs in molecular, or covalent compounds is called ***covalent bonding***, which is the sharing of electrons between atoms.

## 5.2   Outermost Electrons

We have made numerous references to the outermost energy level and to outermost electrons (also called valence electrons) in the previous chapters. We have seen how filling or half-filling the outermost energy levels with electrons is a stable state sought by atoms. Since all the inner levels are filled with electrons already, it stands to reason that the outermost energy level of an atom, which in most cases is not filled when the atom is in a free, uncombined state (or elemental state), is the level that will be interacting with other atoms in order to satisfy the need to be filled. Because of this, a symbolism has been adopted that shows exactly what the situation is in the outermost level of atoms. This symbolism is known as the ***Lewis electron-dot structure*** of the atom, named after the chemist who proposed this idea, **G.N. Lewis** (1875–1946). His symbolism also applies to ions and molecules, as we will see in the following sections.

To draw a Lewis electron-dot structure of an atom, we only need to know the number of electrons in the outermost level of the atom and whether a given electron is alone in a given orbital or paired with another electron. The most convenient source of this information is the periodic table. As noted previously, in the first three periods of this table, the period number corresponds to the principal level being

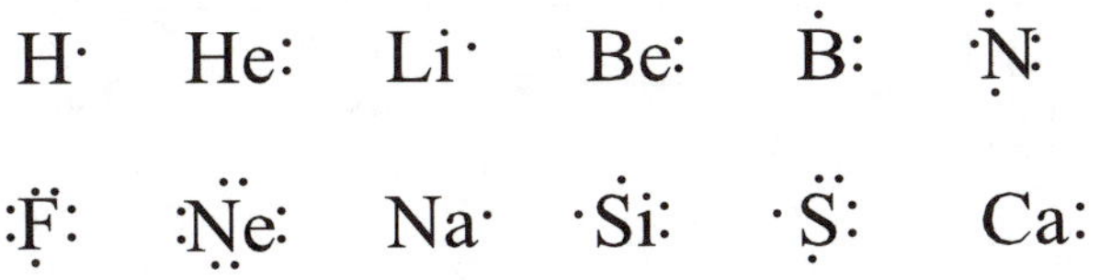

H·    He:    Li·    Be:    Ḃ:    ·Ṅ:

:F̈:    :N̈e:    Na·    ·Si:    ·S̈:    Ca:

**FIGURE 5.1**    Lewis electron-dot structures for some of the first 20 elements.

filled and the number of electrons in this level can be determined by simply counting elements starting from the left side of the table and proceeding to the right until we get to the element in question. Let us look at nitrogen as an example. Nitrogen is in period #2. Beginning with the first element in period #2, lithium, and counting to the right, we see that nitrogen is the fifth element. Thus, there are five electrons to show in the Lewis structure of nitrogen.

To determine whether electrons are paired or unpaired, it is useful to recall the previous orbital diagram discussion we had in Chapter 3. The first two electrons in a given principal level are always in the *s* sublevel and are together (paired) in the same orbital, the spherical *s* orbital. Recalling Hund's Rule from this same discussion, electrons in the *p* sublevel do not pair up until after all three orbitals get an electron.

Lewis structures utilize the element symbol with dots around the symbol to characterize the outermost electron level of that element. If you imagine the symbol enclosed in a square, each of the four sides of the square represents an orbital. Thus, one side of the square represents an *s* orbital and the other three sides represent *p* orbitals within the outermost level. Which side we use to represent each orbital is arbitrary. Figure 5.1 shows Lewis structures for some of the first 20 elements. It is important to note that elements within a given family or group (e.g., hydrogen, lithium, and sodium in Fig. 5.1) have the same Lewis structure. This should reinforce the notion that the elements are arranged according to their electron configurations.

## 5.3   Simple Ionic Compounds

Monatomic ions are easily represented in the format of Lewis electron-dot structures. Alkali metals, alkaline earth metals, and aluminum lose all the electrons in their outermost energy level when the corresponding monatomic ions form. Thus, the Lewis structures of monatomic ions of these metals simply show the positive charge that results from the loss of electrons but no dots. Monatomic ions of other metals may also be symbolized with Lewis structures as long as the charge on these ions is known. The Lewis structure of such metal ions also show no dots. Examples are shown in Fig. 5.2.

$$Na^+ \qquad Mg^{2+} \qquad Al^{3+}$$

**FIGURE 5.2**   Examples of Lewis electron-dot structures of metal ions.

Nonmetals gain electrons when they form monatomic ions, always filling the outermost *s* and *p* sublevels in the process. The result is a total of eight electrons in these sublevels, two in each of the four orbitals. The Lewis electron-dot structure in this case shows all eight electrons, two on each of the four sides of the invisible square, plus the charge on the ion. Examples are shown in Fig. 5.3.

$$:\overset{\cdot\cdot}{\underset{\cdot\cdot}{Cl}}:^- \qquad\qquad :\overset{\cdot\cdot}{\underset{\cdot\cdot}{O}}:^{2-}$$

**FIGURE 5.3**   Examples of Lewis electron-dot structures of monatomic anions.

Lewis structures of ionic compounds show the electron-dot structures of the positive ions together with the negative ions, as shown in Fig. 5.4.

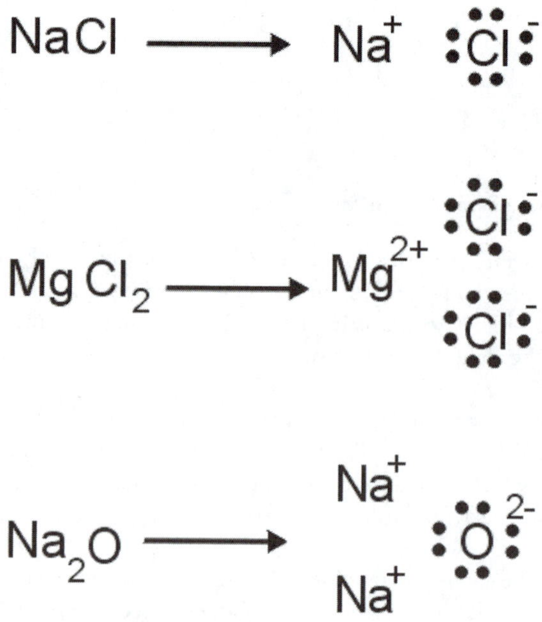

$$NaCl \longrightarrow Na^+ \quad :\overset{\cdot\cdot}{\underset{\cdot\cdot}{Cl}}:^-$$

$$MgCl_2 \longrightarrow Mg^{2+} \quad \begin{matrix} :\overset{\cdot\cdot}{\underset{\cdot\cdot}{Cl}}:^- \\ :\overset{\cdot\cdot}{\underset{\cdot\cdot}{Cl}}:^- \end{matrix}$$

$$Na_2O \longrightarrow \begin{matrix} Na^+ \\ \\ Na^+ \end{matrix} \quad :\overset{\cdot\cdot}{\underset{\cdot\cdot}{O}}:^{2-}$$

**FIGURE 5.4**    Examples of Lewis electron-dot structures of simple ionic compounds.

## 5.4   Simple Covalent Compounds—Slot Filling

The concept of sharing electrons in covalent bonding may also be represented by Lewis electron-dot structures. In simple cases we recognize that the nonmetal atoms have orbital slots to fill, which, when filled, would give these atoms a stable noble gas configuration. Thus, Lewis electron-dot structures of some covalent compounds can be determined by a slot-filling method. In these situations, an atom has at least one orbital with only one electron, the so-called **unpaired** electrons that were first defined in Section 3.5 of Chapter 3. In the process of sharing electrons, such an orbital is merged with a similar such orbital from the other atom. The result is a **molecular orbital** between the two atoms, and in which the two electrons reside. Lewis electron-dot structures show the two atoms in close proximity and horizontally adjacent to each other with two dots representing the two shared electrons in between. An example is molecular hydrogen, $H_2$ (Chapter 2, Section 2.4). Each individual hydrogen atom in $H_2$ has a single electron in the outermost $s$ sublevel. One additional electron for each would represent a filled $1s$ principal level, a stable situation that is equivalent to the stable electron configuration of helium. The Lewis electron-dot structure for $H_2$ shows two hydrogen atoms arranged with a pair of electrons shared between them. In the drawing below, two hydrogen atoms are shown coming together where each fills an orbital slot of the other. The completed structure is shown on the right.

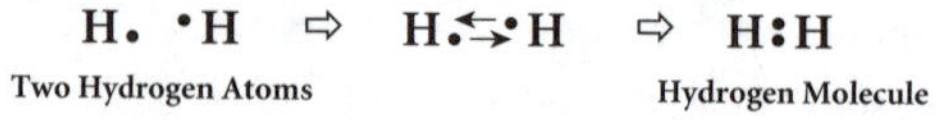

H·  ·H    ⇨    H·⇆·H    ⇨    H:H
Two Hydrogen Atoms             Hydrogen Molecule

Lewis electron-dot structures of molecular fluorine ($F_2$), chlorine ($Cl_2$), bromine ($Br_2$), and iodine ($I_2$) are drawn in a very similar way. In these cases, one $p$ orbital, rather than an $s$ orbital, has a slot to fill. All other orbitals in the outermost level are already filled. The merging of two $p$ orbitals thus occurs. The following shows this process for chlorine.

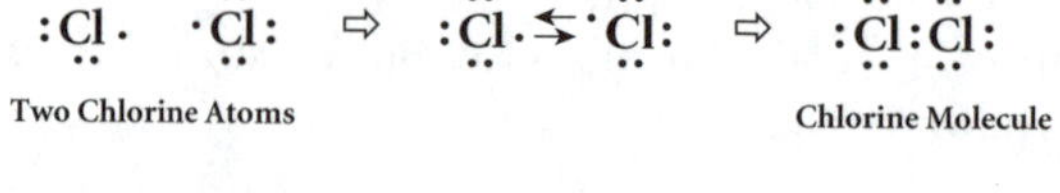

:Cl·  ·Cl:    ⇨    :Cl·⇆·Cl:    ⇨    :Cl:Cl:
Two Chlorine Atoms             Chlorine Molecule

The covalent bonds in molecular hydrogen, fluorine, chlorine, bromine, and iodine consist of a pair of shared electrons in what is called a ***single bond***.

Molecular oxygen ($O_2$) is slightly more complicated. Before bonding, each oxygen atom has two unpaired electrons and two slots to fill. There are two $p$ orbitals that have just one electron on each oxygen atom. In this case, both slots are filled by the lone electrons from the other atom such that *two* molecular orbitals form. This results in what is known as a ***double bond***. Both atoms in $O_2$ have the required eight electrons, four of which are shared in two covalent bonds, rather than two electrons in just one bond. The slot-filling process for molecular oxygen is shown below:[1]

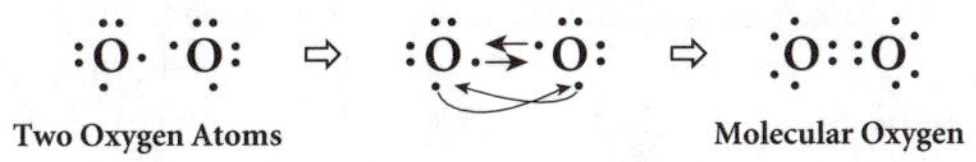

Two Oxygen Atoms          Molecular Oxygen

Molecular nitrogen ($N_2$) is a similar covalent bonding example. Before bonding, each nitrogen atom has three unpaired electrons and three slots to fill. The result is that the covalent bonding for molecular nitrogen consists of three bonds in which six electrons are being shared, a so-called ***triple bond***. Again, both atoms now have the required eight electrons: six are shared and one pair is unshared. This slot-filling process is shown below:

Two Nitrogen Atoms          Nitrogen Molecule

Simple formulas indicating combinations of unlike atoms, such as in HCl, $H_2O$, and $NH_3$, may also be represented by Lewis electron-dot structures determined by slot-filling. The slot-filling process for HCl is as follows.

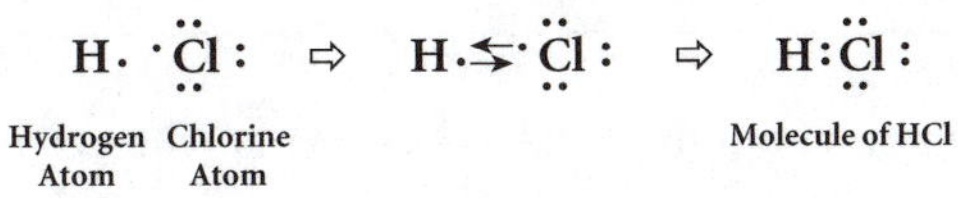

Hydrogen   Chlorine          Molecule of HCl
Atom        Atom

The process for $H_2O$ is shown below.

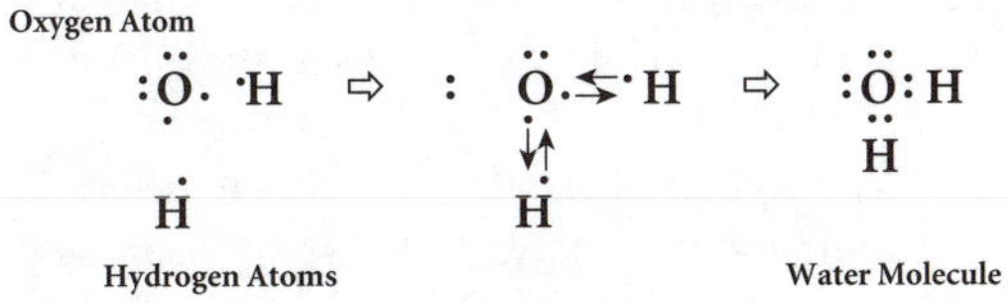

# 5.5   More Complicated Covalent Compounds

The slot-filling procedure works well for compounds containing elements that have oxidation numbers that can be predicted by the position of these elements in the periodic table. For example, a halogen can accept one electron to share in the slot-filling process just as it can accept one electron in becoming the chloride ion—a predicted oxidation number of $-1$. Oxygen can accept two electrons to share in the slot-filling process just as it can accept two electrons in becoming the oxide ion—a predicted oxidation number of $-2$.

---

[1]The actual structure of oxygen is even more complicated than can be predicted by slot filling. The fact that oxygen is "paramagnetic" indicates that the oxygen molecule must have unpaired electrons in its electron arrangement. Further discussion of this phenomenon is beyond the scope of this text.

**TABLE 5.1**    The Steps Required to Determine the Lewis Electron-Dot Structure of Covalent Compounds and of Polyatomic Ions that Are Too Complicated to Do by Normal Slot-Filling

| | |
|---|---|
| 1. | **Count the electrons.** Count the number of outermost electrons on each of the atoms in the formula before bonding and add to obtain the total number. This is the number of dots that must be shown in the Lewis electron-dot structure. If the formula is that of a polyatomic ion, also add to this number the extra electrons that are represented by the charge on the ion. |
| 2. | **Arrange the atoms.** Designate one of the atoms as the "central atom" and place the other atoms around it in some logical fashion. If there is more than one possibility for the central atom, choose the one that has the lowest electron affinity according to periodic trends. In addition, the following points should be considered. |
| (a) | Hydrogen cannot be a central atom. This is logical because hydrogen becomes stable by sharing just two electrons, which means it only forms one bond with another atom. One bond means just one other atom can be attached to hydrogen. Central atoms may have from one to six atoms attached, as we shall see later. In oxyacids, each hydrogen is bonded to an oxygen. |
| (b) | Oxygen never bonds to another oxygen except in oxygen gas ($O_2$), ozone gas ($O_3$), and peroxide ($O_2^{2-}$). Except for these, oxygen will never be the central atom. Normally, an oxygen atom bonds only twice, either with two single bonds or with one double bond. |
| 3. | **Distribute the electrons.** Without regard (yet) to the count in Step 1, place eight electrons (dots) around all atoms, one pair on each side of the invisible square (making only single bonds for now), except for any hydrogens, which get just two dots. |
| 4. | **Count the electrons** that were distributed in Step 3 and **compare this count** with the count that was obtained in Step 1. These counts must match. If the counts match at this point, the electron dot structure you have drawn is the correct one. If it is a polyatomic ion, place parentheses around it and write the charge to the upper right. |
| 5. | **If the counts in Steps 1 and 4 do not match**, make double or triple bonds between the central atom and other logical atoms. If the count in Step 4 is greater than the count in Step 1, the count in Step 4 must be reduced. This can be done by introducing double and/or triple bonding. Each double bond reduces the count in Step 4 by two. Each triple bond reduces the count in Step 4 by four. Of course, hydrogen cannot participate in a double or triple bond. |
| 6. | **If the central atom is phosphorus**, or atoms of elements to the right and/or below phosphorus in the periodic table, the central atom may accommodate more than eight electrons, if necessary, to make the counts in Steps 1 and 4 match. |

As we progress to covalent compounds that have elements with oxidation numbers that are not predicted by their position in the periodic table (such as the $+6$ or $+4$ oxidation numbers for sulfur in sulfate and sulfite, respectively, or the $+1$, $+3$, $+5$, and $+7$ for chlorine in hypochlorite, chlorite, chlorate, and perchlorate), we soon find the slot-filling procedure to be unworkable. In these examples, the atoms of interest (sulfur and chlorine) must have more shared electrons than would be determined by using just slot filling. In other words, the number of such electrons in these cases cannot be determined by the slot-filling process.

To solve this problem, a set of steps is given in Table 5.1 in order to determine the correct Lewis electron-dot structure for these types of compounds. We now present some examples on the use of these steps.

## Example 5.1
Draw the Lewis electron-dot structure of carbon tetrachloride, $CCl_4$.

## Solution 5.1
Step 1: Carbon has four electrons in the outermost level and each chlorine has seven. The total count is: $4 + 28 = 32$.

Step 2: Carbon is the logical choice for the central atom. It has a lower electron affinity than chlorine. Thus the arrangement would be:

Cl

Cl C Cl

Cl

Step 3: Giving each atom eight electrons would result in the following:

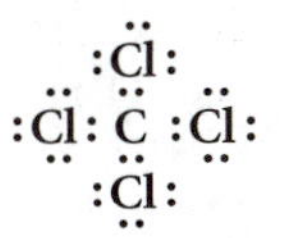

$$\overset{\displaystyle ..}{\underset{\displaystyle ..}{:\overset{..}{\underset{..}{Cl}}:}}$$

$$:\overset{..}{\underset{..}{Cl}}: \overset{..}{C} :\overset{..}{\underset{..}{Cl}}:$$

$$:\overset{..}{\underset{..}{Cl}}:$$

Step 4: The total count of electrons in Step 3 is: $4 \times 8 = 32$. This count matches the count in Step 1. Therefore the above drawing is the correct one.

## Example 5.2

Draw the Lewis electron-dot structure of chloric acid, $HClO_3$.

## Solution 5.2

Step 1: The count is tabulated as follows:

$$1 \times H = 1 \times 1 = \ \ 1$$
$$1 \times Cl = 1 \times 7 = \ \ 7$$
$$3 \times O = 3 \times 6 = 18$$
$$\overline{\qquad\qquad\qquad}$$
$$Total = 26$$

Step 2: Since hydrogen cannot be the central atom and since oxygen will not bond to another oxygen in this case, chlorine is the central atom. The oxygens are bonded to the chlorine and the hydrogen is bonded to an oxygen.

$$O$$
$$Cl\ O\ H$$
$$O$$

Step 3: Give the oxygens and the chlorine eight electrons and give the hydrogen two. This results in the following.

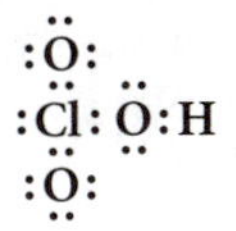

$$:\overset{..}{O}:$$
$$:\overset{..}{\underset{..}{Cl}}: \overset{..}{\underset{..}{O}}:H$$
$$:\overset{..}{O}:$$

Step 4: The total count from Step 3 is 26. This matches the count from Step 1, so the structure drawn in Step 3 is the correct one.

## Example 5.3

Draw the Lewis electron-dot structure of nitric acid, $HNO_3$.

## Solution 5.3

Step 1:

$$1 \times H = 1 \times 1 = \ \ 1$$
$$1 \times N = 1 \times 5 = \ \ 5$$
$$3 \times O = 3 \times 6 = 18$$
$$\overline{\qquad\qquad\qquad}$$
$$Total = 24$$

Step 2: Nitrogen is the only choice for the central atom, and the hydrogen bonds to an oxygen.

$$O$$
$$N\ O\ H$$
$$O$$

Step 3: Give hydrogen two electrons and each other atom eight.

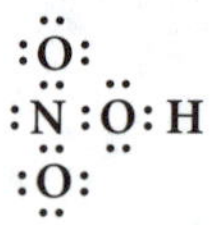

$$\ddot{:}\overset{..}{\underset{..}{O}}:$$
$$:\overset{..}{\underset{..}{N}}:\overset{..}{\underset{..}{O}}:H \quad \text{Incorrect structure due to too many electrons.}$$
$$:\overset{..}{\underset{..}{O}}:$$

Step 4: The total count here is 26. This does not match the count in Step 1, which is 24. As indicated above, this is, therefore, not the correct structure.

Step 5: Let us draw a double bond between the nitrogen and an oxygen:

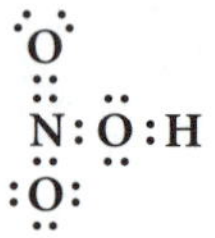

$$\overset{..}{\underset{..}{O}}$$
$$\overset{..}{\underset{..}{N}}:\overset{..}{\underset{..}{O}}:H \quad \text{Correct structure, 24 electrons.}$$
$$:\overset{..}{\underset{..}{O}}:$$

**Note:** Which oxygen gets the double bond is arbitrary. This bond may be shown between the nitrogen and any of the three oxygens.

## Example 5.4

Draw the Lewis electron-dot structure of the carbonate ion, $CO_3^{2-}$.

## Solution 5.4

**Note:** While the carbonate is an ion and is found in ionic compounds, the bonding between the carbon and the oxygens is covalent and the Lewis structure is determined by the steps in Table 5.1. This is true of all polyatomic ions. The Lewis electron-dot structures of ionic compounds that contain polyatomic ions are discussed in Section 5.7.

Step 1:
$$1 \times C = 1 \times 4 = \quad 4$$
$$3 \times O = 3 \times 6 = 18$$
$$\text{Extra electrons due to charge} = \quad 2$$
$$\overline{\hspace{6cm}}$$
$$\text{Total} = \ 24$$

Step 2: Carbon is the choice for the central atom:

$$O$$
$$C\ O$$
$$O$$

Step 3: Distribute the electrons so that each atom has eight. Note the parentheses and the charge in the following structure. These are necessary to show that the structure is a polyatomic ion.

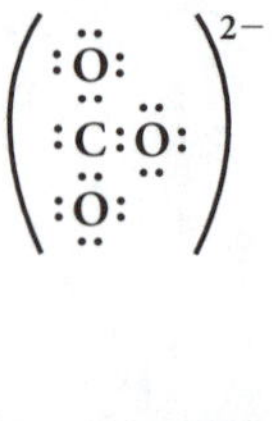

$$\left( \begin{array}{c} :\overset{..}{\underset{..}{O}}: \\ :\overset{..}{\underset{..}{C}}:\overset{..}{\underset{..}{O}}: \\ :\overset{..}{\underset{..}{O}}: \end{array} \right)^{2-} \quad \text{Incorrect structure—too many electrons.}$$

Step 4: The total count in the above structure is 26. Two electrons must by dropped in order for the count to equal that in Step 1 (24). Thus, this is not a correct structure.

Step 5: Placing a double bond between the carbon and an oxygen produces the following result.

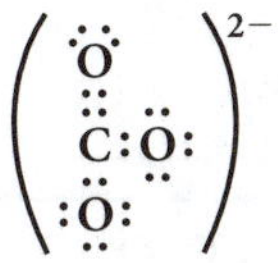

$$\left(\begin{array}{c} \ddot{\text{O}} \\ \vdots\vdots \\ \text{C}\!:\!\ddot{\text{O}}\!: \\ :\!\ddot{\text{O}}\!: \end{array}\right)^{2-} \quad \text{Correct structure.}$$

With this structure, the counts match at 24.

## *Example 5.5*

Draw the Lewis electron-dot structure of $PCl_5$.

## *Solution 5.5*

Step 1:

$$\begin{array}{rl} 1 \times P = & 5 \\ 5 \times Cl = & 35 \\ \hline \text{Total} = & 40 \end{array}$$

Step 2: Phosphorus is the central atom and has five chlorines around it.

$$\begin{array}{ccc} & \text{Cl} & \\ \text{Cl} & \text{P} & \text{Cl} \\ \text{Cl} & & \text{Cl} \end{array}$$

Step 3: Phosphorus can accommodate five chlorines by sharing five electron pairs, or ten electrons. Each chlorine will have eight electrons. This is an example of a central atom having more than eight electrons, so Step 6 is used.

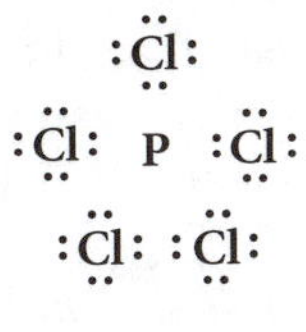

$$\begin{array}{ccc} & :\ddot{\text{Cl}}: & \\ :\ddot{\text{Cl}}: & \text{P} & :\ddot{\text{Cl}}: \\ :\ddot{\text{Cl}}: & & :\ddot{\text{Cl}}: \end{array}$$

Step 4: Forty electrons matches the count in Step 1.

Thus, this is the correct structure. It should be pointed out that all structures that can be drawn by slot-filling may also be drawn using the steps in Table 5.1.

## 5.6  Coordinate Covalent Bonds

Covalent bonds form as a result of an orbital on one atom merging with an orbital on another atom such that each shares a pair of electrons with the other. With the slot-filling process, each of these orbitals has one electron prior to bonding. Having one electron before bonding is not a requirement of these orbitals, however. It is possible for one orbital to have two electrons while the other has none. A bond formed in such a manner is called a *coordinate covalent bond.*

An obvious example of a coordinate covalent bond is one of the N–H bonds in the ammonium ion, $NH_4^+$. The ammonium ion forms when the ammonia molecule, $NH_3$, encounters a hydrogen ion, such as when a solution of ammonia is mixed with a solution of an acid. The hydrogen ion, which has no electrons, attaches itself to the nitrogen of the ammonia at the site of the nonbonded pair of electrons on the nitrogen. This creates a total of 4 N–H bonds and gives the ammonium ion a net positive charge.

# HYDROGEN—BE CAREFUL; NITROGEN—DON'T WORRY

G ases such as hydrogen, chlorine, nitrogen, and oxygen are used in a laboratory by having them contained in compressed gas cylinders. Such cylinders are made of high-strength steel and utilize a pressure regulator for dispensing and controlling the rate of release of the contained gas. For safety, the cylinders must be strapped to an immovable object, such as a table or wall, as pictured here.

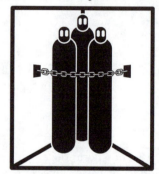

Many types of gases are required in analytical laboratory procedures that involve a technique known as gas chromatography (GC). Hydrogen gas ($H_2$), for example, is required for some GC methods. It is a very dangerous gas and chemistry professionals must take special precautions to assure that it stays contained. Nitrogen gas ($N_2$) is also a useful gas in the laboratory, but it is used when an inert environment is needed. It presents no danger due to its low reactivity.

The molecules of both $H_2$ and $N_2$ gases form from the need of their atoms to fill their outermost energy levels with electrons. It is easy to see how this happens based on the slot-filling process discussed here. It would appear that both gases would be quite stable, based on the fact that their outermost levels are filled following the slot-filling. However, hydrogen is not stable because the single bond between the atoms is much weaker than the triple bond between the two nitrogen atoms in nitrogen gas. Weaker covalent bonding means a greater tendency toward chemical reactions and, in the case of hydrogen, a greater hazard when this gas is used. See also CPW Box 9.3.

*For Homework: Visit several industrial laboratories in your area and inquire about the use of compressed gas cylinders. Ask how and where these are stored, how they are transported between the storage area and the laboratory, etc. Ask to view the areas where the cylinders are located. Write a report on this visit, commenting especially on how each laboratory addresses safety issues concerning location and the need to strap the cylinders to an immovable object.*

$$H^+ \; - \; - \; - \; \rightarrow \quad :\!\overset{\overset{\displaystyle H}{\cdot\cdot}}{\underset{\underset{\displaystyle H}{\cdot\cdot}}{N}}\!:\!H \quad \Rightarrow \quad \left( \overset{\overset{\displaystyle H}{\cdot\cdot}}{H\!:\!\underset{\underset{\displaystyle H}{\cdot\cdot}}{N}\!:\!H} \right)^+$$

**Hydrogen ion**          **Ammonia molecule**                **Ammonium ion**

Both electrons shared with the fourth hydrogen originated on the nitrogen and thus form a coordinate covalent bond.

## 5.7   More Complicated Ionic Compounds

We are now ready to draw Lewis electron-dot structures of ionic compounds that include at least one poly-atomic ion. The polyatomic ions in such structures are drawn according to the steps in Table 5.1. The total structure is then assembled as was done for the other ionic compounds that were discussed in Section 5.3.

### Example 5.6
Draw the Lewis electron-dot structure of sodium sulfate, $Na_2SO_4$.

### Solution 5.6
Two sodium ions, $Na^+$, will be shown with one sulfate ion. We draw the sulfate ion according to the procedure in Table 5.1.

Step 1:
$$1 \times S = 1 \times 6 = 6$$
$$4 \times O = 4 \times 6 = 24$$
$$\text{Extra electrons due to charge} = 2$$
$$\overline{\text{Total} = 32}$$

Step 2: The sodium ions are pictured apart from the sulfate ion. For the sulfate ion, the sulfur is the central atom with the oxygens around it.

$$
\begin{array}{cc}
Na^+ & O \\
& O \ S \ O \\
Na^+ & O
\end{array}
$$

Steps 3 and 4: If each oxygen is given eight electrons, the sulfur will have eight and the total count will be $4 \times 8 = 32$, matching the count in Step 1. With the parentheses and charge in place, we have the final structure.

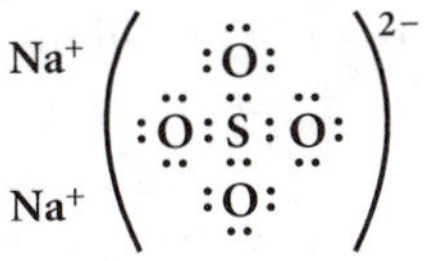

$$
\begin{array}{c}
Na^+ \\
\\
Na^+
\end{array}
\left(
\begin{array}{c}
:\ddot{O}: \\
:\ddot{O}:\ddot{S}:\ddot{O}: \\
:\ddot{O}:
\end{array}
\right)^{2-}
$$

### Example 5.7
Draw the Lewis electron-dot structure of ammonium phosphate, $(NH_4)_3PO_4$.

### Solution 5.7
The three ammonium ions are drawn as in Section 5.6 (see Steps 3 and 4 below). The phosphate ion must be drawn according to the procedure in Table 5.1.

Step 1:
$$1 \times P = 1 \times 5 = 5$$
$$4 \times O = 4 \times 6 = 24$$
$$\text{Extra electrons due to charge} = 3$$
$$\overline{\text{Total} = 32}$$

Step 2: The phosphorus is the central atom and the oxygens are around it:

$$
\begin{array}{c}
O \\
O \ P \ O \\
O
\end{array}
$$

---

*Chemistry Professionals at Work*                                     *CPW Box 5.3*

# COORDINATE COVALENT BONDING IN LABORATORY ANALYSIS

M etal ions are known to share pairs of electrons in the manner described in Section 5.6. For example, a single $Cu^{2+}$ ion will share the nonbonded pair of electrons on each of four ammonia molecules, forming a charged aggregate known as a ***complex ion***. The molecule donating the pair of electrons to such a bond can be large. For example, a single large molecule known by water chemists as EDTA is capable of forming six coordinate covalent bonds in the formation of a complex ion with a single metal ion. This reaction has become very useful in water analysis laboratories for determining the "hardness" of water samples. The hardness of water is due to the presence of calcium and magnesium ions in the water.

*For Homework: Ask a water softener salesman to test the water in your residence and provide you with a written analysis report. Write a paper describing the experience of the sales pitch. Include your personal evaluation of the analysis report and what you think of your water's quality.*

---

Steps 3 and 4: Giving each oxygen eight electrons will give the phosphorus eight and the total count will be 32, matching the count in Step 1. With the ammonium ions included, the final structure is:

$$\left( \begin{array}{c} H \\ H:N:H \\ H \end{array} \right)^{+}$$

$$\left( \begin{array}{c} H \\ H:N:H \\ H \end{array} \right)^{+} \qquad \left( \begin{array}{c} :O: \\ :O:P:O: \\ :O: \end{array} \right)^{3-}$$

$$\left( \begin{array}{c} H \\ H:N:H \\ H \end{array} \right)^{+}$$

## 5.8   Structure of Ionic Compounds

In Chapter 1 (Section 1.4.2), we made a distinction between the fundamental unit of an ionic compound and the fundamental unit of a covalent compound, saying that for an ionic compound it is the formula unit but for a covalent compound it is the molecule. Molecules are distinct individual units that exist independent of other molecules. Formula units, on the other hand, are collections of ions in the ratio indicated in the formula (and in their Lewis electron-dot structures). They do *not* exist as independent

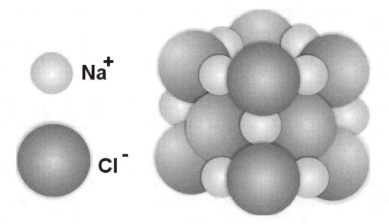

**FIGURE 5.5** The structure of NaCl. Note that no single sodium ion is associated with any single chloride ion—no molecule.

distinct units, however. The ions that make up ionic compounds, both anions and cations, exist in three-dimensional space like a stack of marbles, with one layer placed upon another. The cations and anions interact equally with all ions in their immediate vicinity, which is more than what is shown in the formula. It is this three-dimensional stack of cations and anions that makes up the structure of crystals of ionic compounds. Thus, while the formula indicates the whole number ratio of cations to anions, there is no distinct particle that can be called a molecule because there is no independent unit exemplified by the formula of the compound. The structure of NaCl is shown in Fig. 5.5 as an example.

## 5.9 Polar Bonds and Molecules

As mentioned in Section 5.4, the pair or pairs of electrons that are shared by atoms that are linked by a covalent bond occupy a space between the two atoms that is often called a ***molecular orbital***. A molecular orbital is the region of space in molecules that is analogous to atomic orbitals in atoms. The details of the formation, shape, orientation, and appearance of these orbitals will be presented in Section 5.11. They can be described as electron "clouds" that are located between the bonded atoms and in which the electrons reside.

### 5.9.1 Electronegativity and Bond Polarity

Each of the atoms taking part in a bond has an attraction for their shared electrons similar to the attraction nonbonded atoms have for electrons. This latter attraction was referred to as electron affinity in Chapter 4. The attraction that a bonded atom has for its shared electrons is called its ***electronegativity***. Linus Pauling, a prominent chemist of the 20th century, assigned numerical values to the electronegativities of elements for the purpose of quantitatively comparing one with another. Since covalent bonds are mostly between nonmetal atoms, it is the electronegativities of nonmetals that are important for the present discussion. Pauling's electronegativities of the nonmetals (excluding the noble gases) are listed in Table 5.2.

If the electronegativities of two bonded atoms are equal, the orbital between them is symmetrically shaped and the electron charge is balanced between the two atoms. Such a bond is said to be ***nonpolar***. If the electronegativities are different, the electrons in the bond are attracted more strongly toward the atom that is more electronegative. This means that the shared electrons are positioned more closely to that atom. This polarizes the orbital and makes the orbital become asymmetrical because the distribution of electron charge is not balanced between the two atoms. The bond then becomes like a small magnet that is slightly negatively charged on one end and slightly positively charged on the other end. Such a bond is said to be ***polar***, meaning that is has both a "negative pole" and a "positive pole" like the north

*Chemistry Professionals at Work*                                          *CPW Box 5.4*

# X-RAYS AND CRYSTAL STRUCTURES

T he manner in which positive and negative ions, molecules, or metal atoms stack together in the structure of a solid crystalline material varies because the "marbles" vary in size. Lithium ions are smaller than sodium ions and fluoride ions are smaller than chloride ions, for example. X-rays can be used to study crystal structures because when an x-ray beam is reflected from a crystal, it scatters according to a certain regular pattern that depends on the sizes of the "marbles" and the way they are stacked. The phenomenon is called **x-ray diffraction** and the technique **x-ray crystallography**. The pattern is so specific for a given solid that it can be thought of as a fingerprint of the material. The technique is then very useful for characterizing solids and detecting when a manufactured material is not within quality specifications. Chemistry professionals conduct the experiments, interpret the data, and report the results.

*For Homework: Investigate this phenomenon in more detail. Write a report detailing the principles of x-ray diffraction, its analytical uses, what different crystal structures exist, and the Bragg equation and how it is applied.*

**TABLE 5.2**    Pauling's Electronegativities of the Nonmetals, (Excluding the Noble Gases) in Decreasing Order

| Nonmetal | Electronegativity |
| --- | --- |
| Fluorine | 4.0 |
| Oxygen | 3.5 |
| Chlorine | 3.0 |
| Nitrogen | 3.0 |
| Bromine | 2.8 |
| Carbon | 2.5 |
| Iodine | 2.5 |
| Sulfur | 2.5 |
| Selenium | 2.4 |
| Astatine | 2.2 |
| Hydrogen | 2.1 |
| Phosphorus | 2.1 |
| Tellurium | 2.1 |
| Arsenic | 2.0 |
| Boron | 2.0 |
| Silicon | 1.8 |

and south poles of a magnet (see Fig. 5.6). This polarity is not to be confused with the total separation of charge that is seen in ionic bonds. Polar bonds are covalent bonds that have developed a charge imbalance due to the electronegativity difference between the bonded atoms.

Covalent bonds between two atoms of the same element are nonpolar. Both atoms, being the same element, have the same electronegativity. Thus, the H–H bond in $H_2$ is nonpolar, the Cl–Cl bond in $Cl_2$ is nonpolar, etc. Also, covalent bonds between atoms that have the same electronegativity value in Table 5.2 will also be nonpolar. For instance, a bond between hydrogen and phosphorus, or between chlorine and nitrogen would be nonpolar.

Covalent bonds between atoms that have different electronegativities are polar. Examples include the H–Cl bond in HCl, the H–O bonds in water, and each of the C–H bonds in methane, $CH_4$.

## 5.9.2 Polarity of Molecules

These latter examples warrant more discussion relating to the overall polarities of entire molecules. The water molecule is extremely polar while the methane molecule is nonpolar. This is true in spite of the fact that both of these mole-

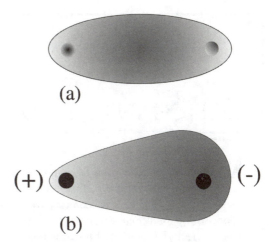

**FIGURE 5.6** (a) A nonpolar bond in which the electrons are distributed symmetrically in the space between the atoms. (b) A polar bond in which the electrons are not distributed symmetrically and slight charges appear on both ends.

cules contain polar bonds. The reason for this apparent contradiction has to do with the arrangement of the bonds within the two molecules. In methane, all bonds are C–H bonds and they are arranged in a perfectly symmetric fashion that is called a ***tetrahedral*** arrangement. Because of this symmetry, the individual bond polarities balance out, resulting in a nonpolar molecule. In water, the H–O bonds are not symmetrically arranged, and as a result, the *molecule is polar*. The molecular arrangements of the methane and water molecules are given in Fig. 5.7. These are "ball and stick" models in which the bonded atoms are represented as "balls" and the bonds are represented as "sticks" connecting the balls.

In Figure 5.7(b), the model for $CH_4$ shows the carbon atom as the larger ball in the center and the hydrogens as small balls connected to it. This is referred to as a tetrahedral arrangement because the four hydrogens connect to each other with imaginary lines to create a pyramid-like structure with four faces, one of which is the base of the structure. A four-faced three-dimensional shape of this nature is called a tetrahedron. The carbon atom is in the center of the tetrahedron (in a location where an Egyptian pharaoh's tomb may be found in a pyramid), and the hydrogens are located in the corners. As such, it

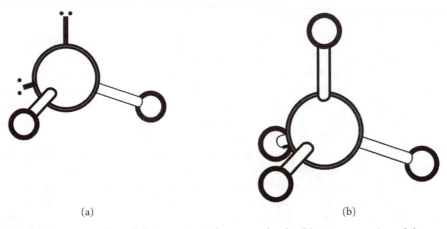

(a)                                               (b)

**FIGURE 5.7** (a) A representation of the asymmetrical water molecule; (b) a representation of the symmetrical methane molecule.

# DESICCANTS AND DESICCATORS

Chemists and chemistry laboratory technicians often have a need to dry chemicals, or, once they are dry, to keep them dry. This is done using small amounts of anhydrous, hygroscopic ionic solids that are held in a closed container. The environment in this closed container should be dry as long as the ionic solid continues to absorb water, or as long as its capacity for water has not been reached. The ionic solid used for this is called a *desiccant* and the container a *desiccator*. The container is often made of heavy glass and has a removable lid that is sealed with grease. Of course, in order to function properly, this container must be sealed from the laboratory air and the lid must be in place at all times.

*For Homework: Look up "desiccants" and "desiccators" in a chemical laboratory supply catalog, such as the Fisher Catalog. Report on the use of Drierite™ and on the different styles of desiccators that are available.*

is quite a symmetrical structure—so symmetrical that the individual C–H bond polarities exactly balance with the others, giving a nonpolar molecule.

The general structure of the water molecule [Figure 5.7(a)] is also a tetrahedron. In this structure, the oxygen atom is buried in the center and the two hydrogen atoms are located in two of the tetrahedral corners. The other two corners are occupied by the unshared pairs of electrons possessed by the oxygen atom in water. If one does not count the unshared electrons in a structure as part of the molecular geometry, the water molecule would not be referred to as tetrahedral, but *angular*. This is a highly unbalanced, asymmetric molecule. The individual H–O bond polarities *do not* balance, and thus the molecule is polar.

## 5.9.3   Water and the Hydrogen Bond

The polar/nonpolar nature of molecules dictates a host of important properties of compounds. Because water molecules are polar, they possess a strong attraction for each other. A container filled with water is like a container filled with small magnets. Relatively speaking, it takes a great deal of energy to separate these molecules from each other. For example, this means that water does not tend to evaporate as readily as its counterparts with nonpolar molecules, such as methane. Its boiling point is quite high compared with other substances with similar size and weight. The boiling point of water is 100°C, compared to −161.5°C for methane.

The intermolecular force that holds water molecules together is known as the **hydrogen bond**. This bond is technically defined as the intermolecular force in which the hydrogen that is bonded to a nitrogen, oxygen, or fluorine in one molecule is attracted to an unshared pair of electrons, usually on an oxygen, nitrogen, or fluorine, in another molecule. Such a bond is indeed an *intermolecular force*, and *not* a covalent or ionic bond. It is also much weaker than a covalent or ionic bond, although strong enough to dictate a host of unique properties. It occurs in water, but obviously from the above definition, is certainly not limited to water.

Besides the effect on the boiling point of water as mentioned above, the hydrogen bond in water creates other unique properties. One is the structure and properties of ice. In liquid water, the intermolecular attraction is a reality, but in spite of it, the water remains fluid, meaning that the molecules are still free to move and flow around one another. However, when water freezes, the molecules become immobilized and the hydrogen bonds become rigid. This creates a defined structure for ice that includes void spaces within. The presence of the void spaces mean that the ice must occupy a larger volume than it did when it was a liquid. This means that its density must be lower as it converts from liquid water to ice. These facts dictate properties with which we are all familiar, including the fact that ice floats on water (such as in a glass of iced tea), and the fact that freezing water can burst plumbing pipes and water hoses.

The hydrogen bond plays an important role in other systems too. An important example is the fact that hydrogen bonding is the force that holds the large DNA molecular strands together within the cells of all living things, an effect that we will study in more detail in Chapter 17.

## 5.9.4  Hydrates

There is one additional property of the polar water molecule that is important to consider. This is its tendency to become trapped within the marble-stack crystal structure of ionic compounds when these compounds crystallize from a water solution. Ionic compounds that have water molecules trapped in their crystal structure are called **hydrates** and are described as being **hydrated**. Hydrates are very common compounds. If we think of the cations and anions as having a spherical shape (as in Fig. 5.5), it is easy to visualize that there are, in fact, void spaces in which small particles, such as water molecules can fit. Due to the fact that water molecules are polar, they tend to cling to the cations and anions upon crystallization and then subsequently become trapped within the crystal structure. It is interesting that only a certain number of these molecules can fit there and that this number depends on the sizes of the cations and anions, which in turn depends on the identities of these ions and the way in which they fit together.

For example, for every formula unit of $CaCl_2$, six water molecules become trapped. For every formula unit of $BaCl_2$, two water molecules become trapped. For every formula unit of $MgSO_4$, seven water molecules become trapped, etc. This trapped water is not considered an impurity, but part of the structure of the crystal and thus it must be shown in the chemical formula of these compounds. The manner in which it is shown in the formula may be considered a bit unusual. Following the formula of the compound, a dot is written, followed by a number, indicating the number of water molecules per formula unit, and the water formula. The names of hydrates utilize the name of the compound that is represented by the ionic formula followed by a word indicating the number of water molecules that appear to the right of the dot. This word uses a prefix specifying the number of water molecules followed by the word "hydrate." For example, we have the following formulas and names for hydrates:

| | |
|---|---|
| $CaCl_2 \cdot 6H_2O$ | **Calcium chloride hexahydrate** |
| $BaCl_2 \cdot 2H_2O$ | **Barium chloride dihydrate** |
| $MgSO_4 \cdot 7H_2O$ | **Magnesium sulfate heptahydrate** |

The prefixes used in the names of these hydrates are mono (1), di (2), tri (3), tetra (4), penta (5), hexa (6), and hepta (7), etc.

The trapped water in hydrates can be removed. In some cases, the water comes out just by exposing the chemical to dry air. Such compounds are described as **efflorescent**. In other cases, heat is required. Substances that have had all the water removed, or that otherwise have no water within their structure are referred to as **anhydrous**. Some anhydrous substances, upon exposure to humid air, absorb water. Such substances are referred to as being **hygroscopic**. Sometimes these substances can absorb so much water that they dissolve in the absorbed water and form a solution. Such substances are referred to as **deliquescent**.

# 5.10   Valence Shell Electron-Pair Repulsion Theory

Molecules of different covalent substances possess a wide variety of geometric shapes and structures that dictate whether they are symmetric or asymmetric, whether they are polar or nonpolar, and thus, what properties they will have.

The theoretical basis for a molecule possessing a particular geometric shape is the concept that electron pairs, whether they are part of a covalent bond (as in a bonding pair) or not (as in a nonbonding pair), will repel each other. In methane, water, and all other molecules, this repulsion means that the electron pairs will get as far away from each other as they can get. The general name for this theory is the **Valence Shell Electron-Pair Repulsion theory**, or **VSEPR** theory.

The determination of a molecule's structure or geometry is accomplished by first looking at the Lewis electron-dot structure and counting the number of regions of high electron density around the central atom. A region of high electron density can mean a pair of nonbonding electrons, a single bond, a double bond, or a triple bond. The number of such regions dictates one of five possible general shapes. These are pictured in a vertical column along the left margin in Fig. 5.8. These shapes are known as linear, trigonal planar, tetrahedral, trigonal bipyramidal, and octahedral. Notice that for both water and methane, the general structure is tetrahedral.

$$:\overset{\cdot\cdot}{\underset{\cdot\cdot}{O}}:H \quad \text{Water, 4 regions of high electron density, 2 single bonds, 2 nonbonding pairs.}$$
$$\phantom{:}H$$

$$\phantom{:}H$$
$$H:\overset{\cdot\cdot}{\underset{\cdot\cdot}{C}}:H \quad \text{Methane, 4 regions of high electron density, 4 single bonds.}$$
$$\phantom{:}H$$

The actual molecular structure is determined by looking at the possibilities to the right of the general stucture in Figure 5.8. It is apparent from this that methane is tetrahedral and that water is angular. All molecular structures can be determined by this scheme.

## *Example 5.8*

What is the molecular structure of carbon tetrachloride, $CCl_4$? (See Example 5.1 for the Lewis electron-dot structure.)

## *Solution 5.8*

Referring back to the Lewis structure in Example 5.1, we see that there are four regions of high electron density around the central atom, Cl. Looking in Fig. 5.8, we see that this would dictate a tetrahedral general structure. Since all four regions are bonding regions, we see from Fig. 5.8 that the molecular structure is also tetrahedral.

## *Example 5.9*

What is the molecular structure of chloric acid, $HClO_3$? (See Example 5.2 for the Lewis electron-dot structure.)

## *Solution 5.9*

Referring back to the Lewis structure in Example 5.2, we see that there are four regions of high electron density around the central atom, Cl. From Fig. 5.8, we see that chloric acid would have a tetrahedral general structure. However, three of these regions are bonding regions and one region is a nonbonding region. Looking again at Fig. 5.8, we see that the molecular structure would be trigonal pyramidal.

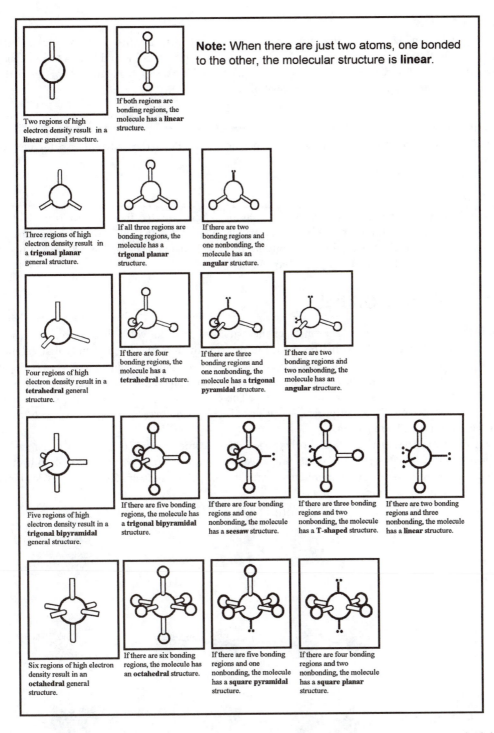

**FIGURE 5.8** The various structures that result from the VSEPR theory. General structures are on the left and possibilities for molecular structures are on the right in each row.

## Example 5.10

What is the molecular structure of nitric acid, $HNO_3$? (See Example 5.3 for the Lewis electron-dot structure.)

*Solution 5.10*

Referring back to the Lewis structure of nitric acid, we see that there are three regions of high electron density around the central atom, N. From Fig. 5.8, we see that nitric acid would have a trigonal planar general structure. All three regions are bonding regions. Looking again at Figure 5.8, we see that the molecular structure of nitric acid is also trigonal planar.

*Example 5.11*

What is the molecular structure of phosphorus pentachloride, $PCl_5$? (See Example 5.5 for the Lewis electron-dot structure.)

*Solution 5.11*

Referring back to the Lewis structure of $PCl_5$ in Example 5.5, we see that there are five regions of high electron density around the central atom, P. From Fig. 5.8, we see that $PCl_5$ would have a trigonal bipyramidal structure. All five regions are bonding regions. Looking again at Fig 5.8, we see that the molecular structure of $PCl_5$ is also trigonal bipyramidal.

## 5.11   Molecular Orbitals

What is the nature of the molecular orbitals, or the regions of space in which shared electrons reside in covalent bonding? One word that has been used to describe them is **overlap**, meaning that an atomic orbital on one atom occupies the same space as an atomic orbital on the other in a merging of two spatial regions. There are two kinds of overlap, the *end-to-end* overlap and *side-to-side* overlap. End-to-end overlap can occur between any two kinds of atomic orbitals. These orbitals can each be the same kind of orbital or they can be different. Fig. 5.9 gives representations of end-to-end overlap between two *s* orbitals, one *s* and one *p* orbital, and two *p* orbitals.

The pair of electrons shared when two hydrogen atoms come together to form hydrogen gas ($H_2$) occupy two *s* orbitals before the overlap (the unpaired outermost electrons occupy *s* orbitals in hydrogen atoms). The top example in Fig. 5.9 represents hydrogen. The pair of electrons shared when two chlorine

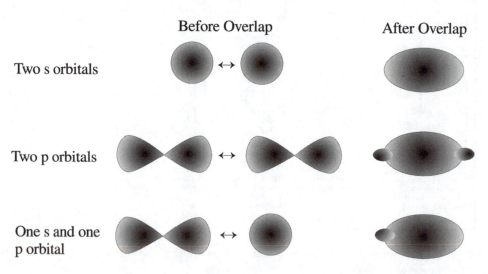

**FIGURE 5.9**   End-to-end overlapping of various orbitals. Top: two *s* orbitals as in hydrogen gas, $H_2$. Center: two *p* orbitals, as in chlorine gas, $Cl_2$. Bottom: one *s* and one *p* orbital as in hydrogen chloride, HCl. Each of these forms a sigma bond.

**TABLE 5.3**   A Summary of the Facts Concerning σ and π Bonds

| σ-bond | 1) | A σ-bond consists of the end-to-end overlap between any two kinds of orbitals. |
|---|---|---|
| | 2) | Whenever there is just one covalent bond between atoms, it is a σ-bond. |
| | 3) | A σ-bond is always one of the bonds in a double bond and one of the bonds in a triple bond. In these cases, it is an end-to-end overlap between two *p* orbitals. |
| π-bond | 1) | A π-bond consists of the side-to-side overlap between two *p* orbitals. No other kind of orbital can participate in a π-bond. |
| | 2) | A π-bond is always accompanied by a σ-bond. Thus, π-bonds only occur in double or triple bonds. |
| | 3) | In a double bond there is one π-bond. In a triple bond, there are two π-bonds. |

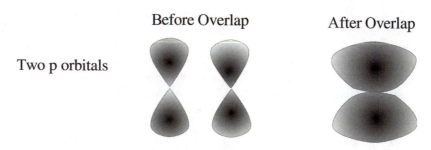

**Before Overlap**          **After Overlap**

Two p orbitals

**FIGURE 5.10**   The side-to-side overlap of *p* orbitals that constitutes a π-bond.

atoms come together to form chlorine gas ($Cl_2$) occupy two *p* orbitals before the overlap (the unpaired outermost electrons occupy *p* orbitals in chlorine atoms). The center example in Figure 5.9 represents chlorine. When one hydrogen atom and one chlorine atom come together to form hydrogen chloride (HCl), the unpaired electrons occupy one *s* orbital (the hydrogen) and one *p* orbital (the chlorine) before the overlap. The bottom example in Fig. 5.9 represents HCl. The *d* and *f* orbitals can also overlap in this manner, either with each other or with other kinds of orbitals. In addition, orbitals that are "hybrids" of the *s, p, d,* or *f* may also be involved. Hybridization will be mentioned again in Chapter 6.

The end-to-end overlap described here is called sigma bonding. A ***sigma bond*** (or, using the Greek letter σ for sigma, ***σ-bond***) is the end-to-end overlap *between any two orbitals.*

Side-to-side overlap can only occur between *p* orbitals. Figure 5.10 presents a drawing of two *p* orbitals coming together and overlapping side-to-side. The side-to-side overlap between *p* orbitals is called pi bonding. A ***pi bond*** (or using the Greek letter π for pi, a ***π-bond***), is the side-to-side overlap between *p* orbitals. Because the *p* orbitals in an atom are perpendicular to each other, a side-to-side overlap means that two other *p* orbitals also must come together end-to-end and form a σ-bond. Thus, a side-to-side overlap is always accompanied by an end-to-end overlap, meaning that two bonds, or a double bond, must form. This explains the formation of double bonds, such as in oxygen gas ($O_2$). One bond is a σ-bond between two *p* orbitals and the other is a π-bond between two *p* orbitals. In atomic oxygen, the two unpaired electrons occupy *p* orbitals so one σ-bond and one π-bond make sense.

A triple bond, such as that which occurs between the nitrogen atoms in nitrogen gas ($N_2$) consists of one σ-bond and two π-bonds, all between *p* orbitals. The three orbitals on atomic nitrogen that have unpaired electrons are *p* orbitals, each of which are perpendicular to each other. When two nitrogen atoms come together and form nitrogen gas, one bond is an end-to-end overlap (a σ-bond) and the others are side-to-side overlaps (π-bonds). All triple bonds consist of one σ-bond and two π-bonds. Table 5.3 summarizes these facts concerning σ and π bonds.

# ULTRAVIOLET ABSORPTION SPECTROSCOPY AND π-BONDING

A popular technique in the industrial analytical laboratory is a technique that utilizes a sophisticated electronic instrument known as an ultraviolet (UV) spectrophotometer. This instrument measures the degree to which a given quantity of a compound absorbs ultraviolet light. The amount of such light that is absorbed and the particular wavelength of UV light that is absorbed give both qualitative and quantitative information about the compound. Such information may be the object of the work. Curiously, it is the π-bonds in the structures of these compounds that generally give it the ability to absorb UV light in the first place. The qualitative information can thus include whether or not the compound has a π-bond in its structure, and sometimes also the position of a π-bond relative to other structural features.

*For Homework: Formaldehyde is an example of a compound that has a π-bond. Look up the EPA method for analyzing for formaldehyde in drinking water. It is a chromatography method with UV detection. Report on what UV wavelength is used for this detection. Also report on how the water sample is obtained and on any other aspect of the analysis that may be interesting to you.*

## 5.12 Homework Exercises

1. In terms of the behavior of outermost electrons, what is the basic difference between ionic and covalent compounds?
2. What is the difference between a covalent bond and an ionic bond? (To say that a covalent bond is a bond between two nonmetal atoms and that an ionic bond is a bond between a metal atom and a nonmetal atom is *not* sufficient.)
3. Is KBr an ionic compound or a molecular compound? Why? Is HBr an ionic compound or a covalent compound? Why?
4. What are the general rules for drawing a Lewis electron-dot structure for:
   (a) an element
   (b) a simple ionic compound
   (c) a simple covalent compound
   (d) a more complicated covalent compound
   (e) a more complicated ionic compound
5. Draw the Lewis electron-dot structure of the elements with atomic numbers between 1 and 20 that are missing in Fig. 5.1.
6. Draw the Lewis electron-dot structure of each of the following.
   (a) $K_2O$      (b) $H_2O$
   (c) Ca        (d) $H_2CO_3$
   (e) $NO_3^-$      (f) $AlCl_3$

(g) $Na_2SO_3$      (h) HCl

(i) S      (j) $O_2$

(k) CO      (l) $XeF_4$

7. Draw the Lewis electron-dot structure of each of the following.

(a) $IO_2^-$      (b) $Br_2$

(c) Al      (d) $K_2SO_3$

(e) $BaCl_2$      (f) $H_2S$

(g) $HNO_3$      (h) $IF_5$

(i) $CN^-$      (j) P

8. Draw the Lewis electron-dot structure of the following.

(a) $BrF_3$      (b) CO

(c) $PH_3$      (d) $SO_4^{2-}$

(e) $CaCl_2$      (f) $HClO_3$

(g) $KBrO_2$      (h) Al

(i) MgO      (j) $SO_3$

(k) $H_2PO_4^-$

9. Draw the Lewis electron-dot structure of the following.

(a) $Ca(NO_3)_2$      (b) $(NH_4)_2S$

(b) $Al_2(SO_4)_3$      (d) $(NH_4)_2CO_3$

10. Which of the following is the correct Lewis electron-dot structure of $MgCl_2$? Explain your answer.

(a) $:\ddot{Cl}:Mg:\ddot{Cl}:$

(b) $Mg^{2+}$ $\begin{matrix} :\ddot{Cl}\overline{:} \\ :\ddot{Cl}\overline{:} \end{matrix}$

11. Explain why the following Lewis electron-dot structures are incorrect.

(a) $\begin{matrix} \ddot{O} \\ S:\ddot{O}:H \\ :\ddot{O}: \\ \ddot{H} \end{matrix}$

(b) $\left( \begin{matrix} :\ddot{O}: \\ N:\ddot{O}: \\ :\ddot{O}: \end{matrix} \right)^-$

12. Why can hydrogen not be the central atom in a Lewis electron-dot structure?

13. Ethane ($C_2H_6$), ethylene ($C_2H_4$), and acetylene ($C_2H_2$) are organic compounds that we will study in Chapter 6. Draw and compare the Lewis electron-dot structures of these compounds.

14. One of the N–H bonds in the ammonium ion, $NH_4^{1+}$, is a coordinate covalent bond. Explain why.

15. What is a coordinate covalent bond? Give an example.

16. Explain why there is no such thing as a molecule for an ionic compound.

17. Differentiate between electron affinity and electronegativity.

18. What does electronegativity have to do with bond polarity?

19. What is a polar bond? Give an example.

20. What is a polar molecule? Give an example.

21. Consider the following. Both compound "A" and compound "B" consist of molecules that have polar bonds. However, compound A is polar while compound B is not. Explain this.

22. Tell whether the bond indicated is an ionic bond, a nonpolar covalent bond, or a polar covalent bond and explain your answer.

(a) H–O     (b) Br–Br     (c) Ca–O     (d) H–P

23. Carbon tetrachloride, $CCl_4$, does not mix with water, which indicates that it is nonpolar, and yet it has polar bonds in its structure. Explain this.

24. Consider a molecule that has polar covalent bonds in it. What is necessary in order for the molecule itself to be nonpolar?

25. What is meant by an "intermolecular force?" Give one specific example of such a force.

26. What is a hydrogen bond?

27. Identify at least two properties of water that are a direct result of the presence of the hydrogen bond.

28. Which compound would have a higher boiling point, chloroform, $CHCl_3$, or $CCl_4$, carbon tetrachloride? What factors enter into your answer?

29. What is a hydrate? Write the formulas of at least two examples.

30. Explain the use of the dot in formulas of hydrates.

31. Define: efflorescent, anhydrous, hygroscopic, and deliquescent.

32. What is the VSEPR theory?

33. How is the VSEPR theory used to help us understand the behavior and structure of molecules and ions?

34. Draw a ball-and-stick model for water and for methane based on the VSEPR theory. How are these models the same and how are they different?

35. Draw the general structures that are predicted by the VSEPR theory for compounds that have centers with two, three, four, five, or six regions of high electron density about them.

36. What is the molecular structure of each of the following? Possible answers include linear, trigonal planar, angular, tetrahedral, trigonal pyramidal, trigonal bipyramidal, seesaw, T-shaped, octahedral, square pyramidal, and square planar.

37. Ask your instructor to select certain Lewis electron-dot structures from Problems 1-4. Determine the molecular structure of each.

38. Look again at the molecular structures discovered for the compounds $XeF_4$ and $ICl_3$ from Problem 36, (B) and (E). Are these compounds polar? Explain.

39. What is meant by a "molecular orbital?"

40. What is the difference between a σ (sigma) bond and a π (pi) bond? Give an example of an element, compound, or ion that has a pi bond. Give an example of an element, compound, or ion that has two pi bonds.

41. Draw the shapes of the molecular orbitals that will be produced by each of the following combinations of orbitals.
    (a) Two *s* orbitals combined end-to-end
    (b) Two *p* orbitals combined end-to-end
    (c) One *s* and one *p* orbital combined end-to-end
    (d) Two *p* orbitals combined side-to-side

42. Which of the situations (a–d) in Problem 41 produces sigma bonds? Which of the situations in Problem 41 produces pi bonds?

43. Consider the following: (a) Cl–Cl   (b) O = O   (c) N ≡ N
    Fill in the blank with the word "all", the word "none", or one of the letters "a", "b", or "c" to correctly answer the question.
    (a) Which of these has only one sigma bond? _____
    (b) Which of these has only one pi bond? _____
    (c) Which of these has two pi bonds? _____
    (d) Which of these has two sigma bonds? _____

(e) Which of these has no sigma bond? _____

(f) Which of these has no pi bonds? _____

(g) Which of these has three sigma bonds? _____

44. Define "pi bond" with the help of drawings of the orbitals before and after a bond forms.

45. How many $\sigma$-bonds and how many $\pi$-bonds are there in a triple bond?

$\sigma$-bonds _____    $\pi$-bonds _____

46. (a) What kind(s) of atomic orbital(s) can be involved in a $\sigma$-bond? Is the overlap in a $\sigma$–bond end-to-end, or side-to-side?

(b) What kind(s) of atomic orbital(s) can be involved in a $\pi$-bond? Is the overlap in a $\pi$-bond end-to-end or side-to-side?

47. Fill in the blanks with either "sigma" or "pi".

(a) The side-to-side overlap between two $p$ orbitals is called a _____ bond.

(b) The end-to-end overlap between any two kinds of orbitals is called a _____ bond.

(c) A molecule of $N_2$ has two _____ bonds and one _____ bond.

(d) A molecule of $Cl_2$ has one _____ bond.

48. Does a double bond consist of two sigma bonds, two pi bonds, or one sigma bond and one pi bond? Explain your answer.

## For Class Discussion and Reports

49. Select one of the high-volume inorganic compounds from the following list and write a 2–3 page paper telling why you think it is an interesting compound, what chemical and physical properties it has, what its history is (when discovered, by whom, how discovered), how it is manufactured, details of its industrial production, what uses it has, where it is observed and/or used in everyday life, and what hazards it presents, if any. Suggested references include *The Merck Index* and the *Concise Encyclopedia of Chemical Technology*.

List:   sulfuric acid, ammonia, lime, sodium hydroxide, phosphoric acid, sodium carbonate, nitric acid, and hydrochloric acid.

50. It was mentioned in this chapter that the technique of ultraviolet spectroscopy is often used in the detection of compounds that have pi bonds in their structure. Look up this technique in a book on instrumental methods of chemical analysis. Write a report that includes some basic information on how this method is performed and on the design of the instrument that is used to conduct these types of measurements.

51. The ozone in our atmosphere helps to protect us from ultraviolet light by absorbing some of this light. Look up the structure of ozone. Use your understanding of pi and sigma bonds to help explain why ozone can absorb ultraviolet light. Discuss your conclusions with the rest of the class.

# 6

# Organic Chemicals

## 6.1   Introduction

In the mid- to late 1970s an "oil crisis" occurred in the U. S. when the oil-rich Middle Eastern countries slowed their production and sale of crude oil (petroleum) in order to drive up prices. What the average American remembers most about the crisis is the extraordinarily long lines of automobiles at the limited numbers of gasoline stations that were able to obtain and sell gasoline. What the average American does not remember about it is the effect the crisis had on the manufacture, sales, and overall economics of the chemical process industries that utilize petroleum in the manufacture of their products. In this chapter, we will take a look at the structure, classification, and importance of the chemicals these industries manufacture and use. These chemicals are called *organic chemicals* and the division of chemistry that deals with them is called *organic chemistry.*

     The organic chemicals manufactured by the chemical process industries have come to represent an extraordinarily important aspect of the everyday lives of every American. Virtually every aspect of life

*Chemistry Professionals at Work*                    *CPW Box 6.1*

# WHY STUDY THIS TOPIC?

T he production and sale of organic chemicals is a huge industry worldwide. A massive number of organic chemicals are produced, not just as end products in themselves, but also to support other industries that manufacture and sell consumer products based on these chemicals. Examples of such products include plastics, synthetic fibers, pesticides and herbicides, pharmaceuticals, dyestuffs, paints and coatings, soaps and detergents, and perfumes. Chemists and chemistry technicians are employed in all facets of this industry as process technicians and operators in chemical plants, as laboratory analysts for the quality assurance and analysis of such compounds, and in research. In addition to these examples in the chemical/consumer products industries, chemists and chemistry technicians may also be called upon to perform environmental analyses for pollutants, many of which are organic chemicals. A basic familiarity with the structure, composition, properties, and nomenclature of these compounds is essential to the career of a chemistry professional.

*For Homework: Find a consumer product in your home that has an organic compound as an ingredient. Then search the Encyclopedia of Chemical Technology for this compound and report on the details of the source and production of this compound.*

as we know it is affected by chemicals that are derived from petroleum. You cannot look around the room in which you are sitting or take a tour of your home and neighborhood without identifying dozens of consumer products that have their origin in petroleum. The obvious examples are fuels and oils such as gasoline, kerosene, propane (also called liquefied petroleum gas, or lp gas), butane (as in butane cigarette lighters), and motor oils. Other examples include plastics of every type, foams, paints and varnishes, paint thinners and removers, waxes, asphalt, clothing and carpet fibers, pharmaceutical products, food additives, agricultural chemicals, and synthetic rubber, vinyls, and leathers. The list seems endless.

Why are these chemicals called "organic?" To answer this, we need to study what the black syrup known as petroleum is and how it came to exist. The petroleum buried deep under the surface of the Earth in pockets called oil fields is the result of many hundreds of thousands of years of decay of living systems. A Chevron Oil Company advertising campaign in the 1960s and 1970s featured a dinosaur jumping out of the gasoline tank of a car, filling the tank, and jumping back into the tank before sticking its head out as the car drives off while the narrator says "Remember, that dinosaur gave his or her all for that tank of gas."

While petroleum (and thus living systems), is indeed the source of gasoline and other fuels and the usual source of the organic chemicals needed to manufacture many consumer products, it cannot be accurately stated that it is the only source of these chemicals. The decay of living systems produces vast supplies of organic chemicals, but some laboratory and industrial processes that utilize only chemicals derived from *nonliving* systems *(inorganic chemicals)* can also produce these chemicals. However, the

---

# THE REFINING OF PETROLEUM

The initial step in the refining of petroleum involves its separation into various component mixtures by distillation in a ***fractionating tower***. ***Distillation*** is the evaporation of mixture components by boiling, followed by recondensation of the vapors on a cold surface or condenser. Different fractions boil and recondense at different temperatures and at different heights in the tower and are subsequently drawn off and collected independent of each other. Gasolines, kerosenes, and naphthas (mixtures used mostly as solvents and diluents) are some of the principal fractions obtained in this way.

Gasoline is a mixture and consists of several dozen organic chemicals. The highest quality gasolines are actually formulated from fractions produced by ***catalytic cracking***, which is a chemical reaction used to produce the more desirable components. In addition, various large-scale processes produce the purified organic chemicals needed for other industrial processes and consumer products. Chemical process technicians operate, maintain, and troubleshoot the large-scale equipment for distillation, catalytic cracking, and other processes.

*For Homework: Find a reference that presents a list of the components of gasoline. Is the "several dozen" description above correct? Are any of these components familiar to you? Select one and report on how it is manufactured and for what industrial processes it is used.*

---

name *organic*, a name that implies an origin in living systems, has stuck over the years and that is how we identify such chemicals today.

## 6.2 The Chemistry of Carbon and its Compounds

Chemicals that are identified as organic constitute the vast majority of all chemicals that exist. A conservative estimate would be 13 million such chemicals. Why are there so many? The main reason for this is the proliferation of the bonding capabilities of the element carbon. It is important to realize that carbon is the commonality—the link between all these 13 million compounds. In fact, ***organic chemistry*** has often (and correctly) been defined as the chemistry of the compounds of carbon. While there are a small number of forms of carbon and compounds of carbon that are classified as ***inorganic*** (e.g., the graphite in your pencils, carbon dioxide and carbon monoxide, the metal carbonates, etc.), all organic compounds contain carbon atoms in their molecular structure. In fact, most molecular structures of such compounds have as their backbone a long string, or chain, of carbon atoms bonded together. Many even have a network of chains of carbon atoms in their molecular structure. Bonded to these backbones or networks are atoms of other elements. Hydrogen atoms are found in abundance in this arrangement. Found to a lesser extent are oxygens, nitrogens, sulfurs, and other nonmetals. Atoms of metals also find their way into these structures, but to a much lesser degree.

Life on planet Earth is sometimes referred to as carbon-based life. The obvious reason for this is that the molecular structures that make up living systems contain these carbon chains in abundance. Modern studies of biotechnology, biochemistry, and biology have as their basis the study of the chemistry of carbon compounds. One can easily see the vast scope of organic chemistry. Not only is organic chemistry the study of the chemistry of compounds derived from living systems (from petroleum), but it is also the study of the chemistry of systems that are living, and this list includes all vegetation, all animals, and all human beings. A timely example of the latter is *DNA chemistry*—an extraordinarily important topic in the modern world. But, alas, we must slow down a bit and consider the prolific behavior of that small atom we call carbon before tackling these other more engrossing topics.

## 6.3   Hybridization of Orbitals

The proliferation of the bonding of carbon atoms resulting in these backbones and networks is facilitated by the hybridization of carbon's atomic orbitals mentioned in a previous chapter. Let us briefly review the previous discussion. With carbon's electron configuration,

Carbon    $1s^2\ 2s^2\ 2p^2$

one would expect the following orbital diagram

and the following Lewis electron-dot structure.

However, a phenomenon occurs with carbon that is relatively rare among other elements. The phenomenon is called **hybridization**. All four electrons become unpaired and the orbital picture is transformed. We will not be examining the orbital picture here. The important point for us is that all four electrons become unpaired and available for bonding, resulting in the following Lewis electron-dot structure.

This produces a capability for covalent bonding that is unique. In all organic compounds, from the very simple to those that possess the complex network structure, carbon atoms always bond four times. While there may be single bonds, double bonds, and triple bonds involved, the total number of bonds to any carbon atom is always four, no more and no less, because of the four unpaired electrons resulting from hybridization.

## 6.4   Geometry Around Bonded Carbon Atoms

Let us apply the attributes of the VSEPR theory discussed in Chapter 5 to the structure of carbon compounds. When all four bonds to a carbon are single bonds, there are four regions of high electron density, and thus, the geometry around this carbon is tetrahedral. The molecular geometry of compounds containing just one carbon (as the central atom) is therefore tetrahedral. Examples of this are methane ($CH_4$) and carbon tetrachloride ($CCl_4$). These molecules would appear as shown in Fig. 6.1, with the large ball in the center representing the carbon and the four smaller balls surrounding it representing the hydrogens in the case of methane, or chlorines in the case of carbon tetrachloride. If the compound has two or more carbon atoms bonded together with nothing but single bonds attached (such as in many of the backbones or networks referred to earlier), then the molecular structure would consist of a tetrahedral geometry around each individual carbon. This would have a carbon-carbon geometry that would be linear for two carbons, but then take on a "sawtooth" appearance for three or more if the carbon skeleton (the structure showing only the carbon atoms, and not what is bonded to them) is laid out so that the bonds get as far from each other as they can get (see Fig. 6.2). This carbon skeleton and the sawtooth pattern are useful in defining the basic arrangement of the carbons and will be mentioned again in Section 6.8.

When a double bond is attached to a carbon, there are three regions of high electron density, and thus, the geometry around this carbon is trigonal planar. When a triple bond is attached to the carbon, there are two regions of high electrons and the geometry around this carbon is linear. It is *important to remember* that the planar or linear geometry appears only around carbons to which the double or triple bonds are attached. The remainder of the structure, where there are only single bonds, will have the sawtooth appearance.

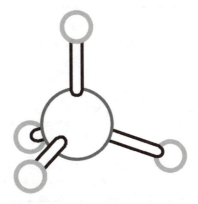

**FIGURE 6.1** The tetrahedral molecular geometry exemplified by such molecules as $CH_4$ and $CCl_4$.

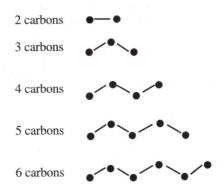

2 carbons

3 carbons

4 carbons

5 carbons

6 carbons

**FIGURE 6.2** The sawtooth pattern discussed in the text. The solid circles represent carbon atoms. Only carbon atoms are shown.

## 6.5 Physical Properties

Crude oil, as it comes from the oil well, is a black, odorous, syrupy liquid mixture. Gasoline, a mixture of organic compounds derived from crude oil, is a yellowish, odorous, volatile, and highly flammable liquid mixture. It should not be too surprising to realize that many organic compounds are liquids at ordinary temperatures and pressures, that many have an odor, that many are volatile, and that many are flammable. Compared to inorganic materials, this gives a new perspective on chemical compounds. Very few inorganic materials are liquids at ordinary temperatures and pressures and very few have an odor. Very few are volatile and few are flammable. The only common inorganic substance that is a liquid at ordinary temperatures and pressures is water. Of course, inorganic chemists commonly deal with water, water solutions, and other inorganic materials in their jobs. Thus, various physical properties such as solubility, boiling point, freezing point, density, specific gravity, color, and refractive index are properties that are commonly observed and/or measured. However, with organic compounds, a broad array of other physical properties are observed, if not for characterization of the materials themselves, at least for safety reasons. In addition to those physical properties listed earlier, others such as odor, flash point,

---

# FLUCTUATING VISCOSITIES

V iscosity, a liquid's fluidity or ability to flow (see Section 7.8), is sometimes measured as a test of the quality of a liquid pharmaceutical preparation. One pharmaceutical quality assurance laboratory relates the story that routine viscosity measurements made over a period of days by a chemistry technician on successive lots of a particular preparation were fluctuating. The technicians became part of a team organized to perform investigative procedures for identifying and rectifying the source of the problem. It was feared that, whatever the source of the problem, it may be threatening the integrity of the product.

Eventually the technicians isolated a rare, nonpathogenic bacteria that had adapted itself to the particular conditions of pH, temperature, and toxicity of the product. It was also discovered that pieces of equipment used in the formulation actually produced conditions favorable to the bacteria. The work led the company to rethink standard laboratory technique and procedure and the manufacturing process. The technicians were cited for their teamwork, persistence, and imagination in solving this problem.

*For Homework: Look up an ASTM (American Society for Testing and Materials) method for measuring viscosity. Briefly describe the method used, the material tested, and the reason for the test.*

---

flammability, viscosity, volatility, vapor pressure, miscibility with water, and optical rotation become important. Because of this, it is important for students of chemistry to gain first-hand experience with these compounds and with the observation and measurement of their physical properties in their laboratory work.

## 6.6  Hydrocarbons

The fact that there are millions of organic compounds dictates that we must find an alternative to studying them one by one. Chemical and physical properties of these compounds depend on the structural features that these compounds have. The traditional method of studying organic chemistry is to divide the huge number of compounds into classifications based on individual structural features and then to study these classifications, thus simplifying the study of both structure and properties. This will be our approach here. The structural features that place compounds into their particular classifications are called ***functional groups***. The first classification we will study is the group called the hydrocarbons.

The vast majority of organic compounds have hydrogen atoms bonded to the carbon atoms. Those that have only carbons and hydrogens (and no other elements) are called ***hydrocarbons***. Functional groups within the broader hydrocarbon classification include the carbon-carbon single bond, the carbon-carbon double bond, and the carbon-carbon triple bond. Compounds with these functional groups fall under the general heading of ***aliphatic hydrocarbons***. Aliphatic hydrocarbons that have nothing but single bonds between carbons are called ***alkanes***. They are also called ***saturated aliphatic hydrocarbons*** since they contain as many hydrogens as they can have. Aliphatic hydrocarbons that have at least one

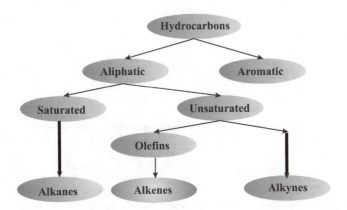

**FIGURE 6.3**   A flow diagram of the hydrocarbon classification.

double bond between carbons are called ***alkenes***, the carbon-carbon double bond being the functional group. They are also called ***olefins***. Aliphatic hydrocarbons that have at least one triple bond between carbons are called ***alkynes***, the carbon-carbon triple bond being the functional group. Compounds that are either alkenes or alkynes are **unsaturated aliphatic hydrocarbons**, since more hydrogens may be added by eliminating the multiple bonding between carbons. In addition to the aliphatic hydrocarbons, we also have the ***aromatic hydrocarbons***. Compounds that are aromatic hydrocarbons possess a structural feature such as the benzene ring, which is what we will define as the aromatic hydrocarbon functional group. Figure 6.3 summarizes these facts in the form of a flow chart.

Let us now look at some examples of compounds within each of these classifications, beginning with the alkanes, since the IUPAC nomenclature systems for other classifications is based on the alkane group.

## 6.6.1   Alkanes

The simplest alkane is one with just one carbon atom. The four bonds on the one carbon are all to hydrogen atoms. The name of this compound is ***methane***:

$$
\text{methane} \qquad\qquad CH_4 \qquad\qquad H-\underset{\displaystyle H}{\overset{\displaystyle H}{\underset{|}{\overset{|}{C}}}}-H
$$

Methane is the major component of natural gas.

The next simplest alkane is ***ethane***. It has two carbon atoms and six hydrogen atoms. Three of the four bonds on each carbon are to hydrogens and one is to the other carbon.

$$
\text{ethane} \qquad\qquad CH_3CH_3 \qquad\qquad H-\underset{\displaystyle H}{\overset{\displaystyle H}{\underset{|}{\overset{|}{C}}}}-\underset{\displaystyle H}{\overset{\displaystyle H}{\underset{|}{\overset{|}{C}}}}-H
$$

The structures shown in the center above for methane and ethane are called *condensed molecular structures*. In a condensed structure, bond lines are optional and are often not shown. Rather, the atoms to the right of each carbon are the atoms bonded to that carbon. In addition, the carbon atom or atoms adjacent to a given carbon are bonded to it. Comparing the dash structure on the right for ethane to the corresponding condensed structure just to its left should make this clear. Three hydrogens and the other carbon are bonded to the first carbon in the condensed structure. The other three hydrogens are bonded to the other carbon.

The alkane with three carbons is **propane**:

propane        $CH_3CH_2CH_3$

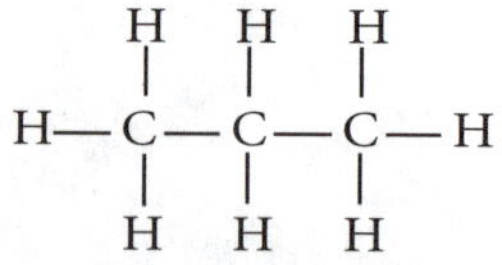

$$H-\overset{\overset{\displaystyle H}{|}}{C}-\overset{\overset{\displaystyle H}{|}}{\underset{\underset{\displaystyle H}{|}}{C}}-\overset{\overset{\displaystyle H}{|}}{\underset{\underset{\displaystyle H}{|}}{C}}-H$$

Like methane, propane is also a fuel used to heat homes, particularly in rural areas.

As we build on the carbon backbone and continue to write additional structures of alkanes, you may notice that we remove a hydrogen from the end of the chain and add a carbon with three hydrogens in its place, as we did in going from ethane to propane. A group consisting of a carbon and three hydrogens that we add to a chain (or to another site) is called a **methyl group**—a "$CH_3$–". You may wonder why we remove the hydrogen to the extreme right, as in proceeding from ethane to propane, rather than some other hydrogen, such as the one on the top of the chain. Removing a hydrogen from the top of the chain and replacing it with a methyl group would give, for example, the following structure of propane.

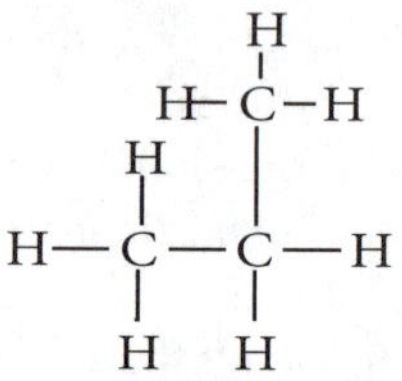

$$H-\overset{\overset{\displaystyle H}{|}}{C}-H$$

This structure is no different than the previous structure of propane. The chain of carbons is still a three carbon chain where there are three hydrogens bonded to the first and third carbons and two hydrogens bonded to the second, just as in the first structure. Thus, which hydrogen is removed and replaced with a methyl group in progressing from ethane to propane is not important—the result is the same either way.

Adding a methyl group, a group containing just one carbon, to a carbon chain is one method by which we can draw additional structures for a given classification. We can also add groups of two, three, or more carbons. A group containing two carbons is called an **ethyl** group, a group of three carbons is called a **propyl** group, and so on. For three carbons, there are actually two different groups that can be added. These are groups in which the bond to the carbon chain is to the middle carbon of the three-carbon chain giving an **isopropyl** group, or a group in which the bond is to one of the end carbons giving an **n-propyl** or **normal propyl** group.

ethyl $CH_3CH_2-$        n-propyl $CH_3CH_2CH_2-$        isopropyl $CH_3\overset{|}{C}HCH_3$

As we proceed to four or more carbons, the number of group arrangements like these increases dramatically. The listing of specific groups with four or more carbons is beyond our scope. However, one generalization can be made. In naming these groups, the "-ane" ending on the alkane with the same number of carbons is replaced with the "-yl" in each case. A general name for such a group is the **alkyl group**, as derived from the word alkane.

Now let us consider the alkane with four carbons, **butane**, which is a fuel used in cigarette lighters.

butane                              $CH_3CH_2CH_2CH_3$

With butane, it *is* possible to write a different structure depending on which hydrogen in propane is replaced with a methyl group. For instance, it is possible to remove one of the two hydrogens (it does not matter which) from the center carbon in the three carbon propane chain and replace this hydrogen with a methyl group. This results in the following butane structure known as ***isobutane***.

isobutane                     $CH_3CHCH_3$
                                                |
                                              $CH_3$

It does not matter whether the methyl group in isobutane is drawn above or below the chain—the compound would be the same.

Isobutane is an example of a *branched* chain structure—the methyl group attached to the center carbon of a three carbon chain is referred to as a **branch**. The possibility of branching boggles the mind as we proceed to longer chains of carbons. A very large number of different structures may be drawn as the number of carbons in an alkane increases. It quickly becomes easy to understand how there can be millions of compounds under the general heading of organic chemicals.

The straight chain structure that was shown for butane (more correctly referred to as ***n-butane***, or ***normal butane***) and the isobutane structure have the same molecular formula ($C_4H_{10}$). But these have different structures, which make them completely different compounds. Such compounds are called ***isomers***. Isomers are compounds that have the same molecular formula but different structures. We will see more examples of isomers as we proceed through the different classifications. Isobutane is actually named with propane as the base name under IUPAC rules, since the longest straight chain of carbons in isobutane is three. A detailed look at the IUPAC nomenclature rules is beyond the scope of this text. However, the base names of alkanes with carbon chains of one through nine are shown in Table 6.1. This table also lists the base names of the alkyl groups, as discussed earlier.

All open-chain alkanes, such as those discussed in this section, fit the general formula $C_nH_{2n+2}$, where *n* is the number of carbon atoms in the molecule.

## 6.6.2  Alkenes

Alkenes are aliphatic hydrocarbons that have at least one double bond between carbons in the structure. The simplest alkene has two carbons. Its IUPAC name is ***ethene*** and its common name, also endorsed by IUPAC, is ***ethylene***.

| Common IUPAC Name | Formal IUPAC Name | |
|---|---|---|
| Ethylene | Ethene | $H_2C=CH_2$ |

Notice that to accommodate the double bond, the carbons each have only two hydrogens. Remember that a carbon can have four bonds—no more and no less. This means that the carbons that are part of a double bond only have two remaining bonds each for use for other atoms such as hydrogen.

**TABLE 6.1**  The IUPAC Names for Carbon Chains and Alkyl Groups of Various Lengths

| Number of Carbons in Chain | Name of Alkyl Group | Name of Alkane |
|---|---|---|
| One | Methyl | Methane |
| Two | Ethyl | Ethane |
| Three | Propyl | Propane |
| Four | Butyl | Butane |
| Five | Pentyl | Pentane |
| Six | Hexyl | Hexane |
| Seven | Heptyl | Heptane |
| Eight | Octyl | Octane |
| Nine | Nonyl | Nonane |

*Chemistry Professionals at Work*                                   *CPW Box 6.4*

# ETHYLENE: A HIGH-VOLUME CHEMICAL

Ethylene, which is a gas at room temperature, is produced in chemical plants employing tubular reactor coils that are capable of very high temperatures. "Cracking" or "pyrolysis" (decomposition by heating) of hydrocarbons, especially ethane, propane, and butane, occurs in these coils and ethylene is formed and subsequently purified. It has been identified as the largest volume organic chemical produced today, meaning that no other organic chemical is produced in as large a quantity. Ethylene is used mostly in the production of polyethylene, of which many plastic materials and consumer products are composed. It is also used as one raw material (often called a *feedstock*) for many other organic chemicals and polymers produced by the chemical process industries. Chemistry professionals are employed in these plants as process operators and as laboratory workers for quality assurance.

*For Homework: Think of five organic chemicals and polymers (review Section 6.13.1 for help regarding polymers) that you think may use ethylene as a feedstock during their industrial production, then check the* Concise Encyclopedia of Chemical Technology *or other appropriate references to see if you are right. Write a report giving specific information for each.*

Notice, too, that the formal IUPAC name, ethene, is derived from the alkane with two carbons, ethane. The "a" in "-ane" changes to an "e," just as in the word "alkene," which is derived from the word "alkane."

The next simplest alkene is **propene**, or **propylene**, which has three carbons. Following the same pattern as ethene, the formal IUPAC name, propene, is derived from the name of the alkane with three carbons, propane, with the "a" in "-ane" changed to an "e."

| Common IUPAC Name | Formal IUPAC Name | |
|---|---|---|
| Propylene | Propene | $CH_2=CHCH_3$ |

The middle carbon of this three carbon chain and one of the carbons that has a double bond, has only one hydrogen attached. This is because the fourth bond is taken up by the methyl group on the right.

The double bond in this three carbon chain could also be placed between the second and third carbons. However, this would not be a different structure since the double bond would be between the first two carbons in a three carbon chain either way. It does not matter from which end one counts. There are no isomers of propene that are alkenes.

The alkene with four carbons, however, is a different story. Two structures can be drawn that are different because of the location of the double bond.

| | |
|---|---|
| 1-butene | $CH_2=CHCH_2CH_3$ |
| 2-butene | $CH_3CH=CHCH_3$ |

These two structures are isomers since they have the same molecular formula ($C_4H_8$), but a different structure. Remember when drawing organic structures that all carbons must have exactly four bonds. Notice that a number and a hyphen to the left of the name butene indicates the carbon in the chain on which the double bond first appears. The lowest number possible is always used. For example, if we had counted from the right, 1-butene would have been called 3-butene, which is incorrect.

As the chain gets longer, the double bond can appear between any two carbons, a fact that implies many possible isomers. That, coupled with the possibility of branching and the possibility of more than one double bond being present in a given structure, indicates that there are a large number of alkenes.

Replacing a single bond with a double bond reduces the number of hydrogens in the structure by two. Thus, all alkenes with just one double bond fit the general formula $C_nH_{2n}$, in which $n$ is the number of carbons in the chain.

## 6.6.3   Alkynes

The alkyne classification follows the same pattern as the alkene classification. The simplest alkyne has two carbons and is named ***ethyne*** (formal IUPAC name), or acetylene (common IUPAC name):

| Common IUPAC Name | Formal IUPAC Name | |
| --- | --- | --- |
| Acetylene | Ethyne | $HC \equiv CH$ |

The name acetylene may be familiar. This compound is the gas that welders use in their torches. The IUPAC name, ethyne, is derived from the name of the base alkane, ethane, as was done with alkene and ethene. In this case, the "a" changes to "y," as in the word "alkyne." This is true of the formal IUPAC names of all alkynes.

Note that the carbon atoms in ethyne each have just one bond available for bonding to other atoms. In the case of ethyne, this one bond is to a hydrogen on each carbon.

The next simplest alkyne is ***propyne***:

propyne            $CH_3C \equiv CH$

just as with propene, there are no isomers. The one bond remaining after the formation of the triple bond is taken up by a methyl group on one of the carbons and a hydrogen on the other. When there are four carbons, there is a pair of isomers.

1-butyne            $HC \equiv CCH_2CH_3$
2-butyne            $CH_3C \equiv CCH_3$

This is the same as we saw with butene. Once again, when more carbons are considered, and if there is branching and more than one triple bond possible in a structure, one can see that many alkynes can exist. Replacing a single bond in a structure with a triple bond reduces the number of hydrogens by four. Thus, all alkynes with just one triple bond fit the general formula $C_nH_{2n-2}$, in which $n$ is the number of carbons in the chain.

## 6.6.4   Ring Structures

As stated previously, most aromatic hydrocarbons possess a unique structural feature called the ***benzene ring***. The benzene ring is a structure that has six carbon atoms bonded together in a circle, or ring. It is not unusual to encounter ring structures in organic chemistry. There are many aliphatic hydrocarbons that have ring structures. An example is ***cyclohexane***, which is a six-membered ring with a molecular formula of $C_6H_{12}$.

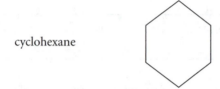

cyclohexane

Other aliphatic rings are common including cyclopentane (with five carbons) and cycloheptane (with seven carbons). Forming a cycloalkane ring from an open chain reduces the number of hydrogens by two. Thus, the general formula of a cycloalkane with just one ring and no multiple bonds is the same as that of alkenes, $C_nH_{2n}$. In addition, hydrogens can be removed and replaced by other substituents such as double bonds, branches, and chains, making many different structures possible.

The cyclohexane structure is sometimes drawn simply as a hexagon. It is understood that there is a carbon in every corner of the hexagon and that there are two hydrogens bonded to each carbon, even though the hydrogens are not shown. Such a structure is sometimes referred to as a ***bond-line structure***.

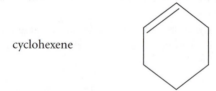

cyclohexane

It is possible to have a double bond in a ring. For example, cyclohexene exists.

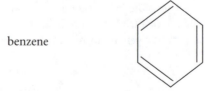

cyclohexene

Perhaps more interesting and important than these, however, is the benzene ring. Again, the benzene ring is the structural feature identifying a hydrocarbon as "aromatic"—the aromatic hydrocarbon functional group. The benzene ring is also a six-membered ring, but it is very different from cyclohexane and cyclohexene. The bonds between the carbons are not single bonds. The geometry around each carbon in the ring is identical to that around carbons to which a double bond is attached—planar. This means that each of the six carbons have a *p* orbital that overlaps side-to-side with the *p* orbital of the carbon next to it. Scientists originally thought that this resulted in alternating double and single bonds around the ring, or π-bonds on each carbon, which is a ***triene*** structure.

benzene

But properties of compounds with these ringed structures indicate that there are no double bonds present at all. For example, compounds with the benzene ring in the structure do not undergo the reactions that typically characterize alkenes. Studies indicate that the π-electrons are not localized between the carbons as in the triene structure but rather ***delocalized*** around the entire ring. The overlapping of the *p* orbitals occurs around the entire ring and creates donut-shaped molecular orbital regions both above and below the ring. These electron-rich "clouds" are responsible for the unique properties of aromatic compounds.

---

# BENZENE IS A KNOWN CARCINOGEN

**M**any organic chemicals are health hazards. Benzene is one that has been determined and confirmed to be a ***carcinogen***, which is a substance that is known to cause cancer. When process operators and laboratory technicians work with benzene, they must take special precautions to avoid contact and avoid breathing its vapors in order to protect their health. These precautions include wearing gloves, using fumes hoods, etc. The effects of over-exposure to carcinogens often do not appear until many years after the fact.

*For Homework: Look up benzene's Material Safety Data Sheet (MSDS) and also benzene's listing in the book* Prudent Practices in the Laboratory, Handling and Disposal of Chemicals *(National Academy Press, 1995). Give a full report on the hazards of benzene, the precautions that are necessary when working with benzene, and practices for the safe disposal of waste containing benzene.*

---

The most telling drawing for such a unique structure is a hexagon with a circle in the center. The following structure with no substituent, chain, or branch attached to any carbon is also an acceptable drawing for benzene, or the benzene ring, and is the one we will use in this text.

benzene

A single hydrogen is bonded to each carbon (at each corner of the hexagon). The molecular formula of benzene is therefore $C_6H_6$.

As with cyclohexane and other aliphatic hydrocarbons, the hydrogens attached to the ring of benzene can be removed and replaced with other substituents, branches, and chains. A common example of this has a hydrogen replaced by a methyl group. The IUPAC name for this compound is ***methylbenzene*** and the common name, endorsed by IUPAC, is ***toluene***.

| Common IUPAC Name | Formal IUPAC Name | |
|---|---|---|
| toluene | methylbenzene | 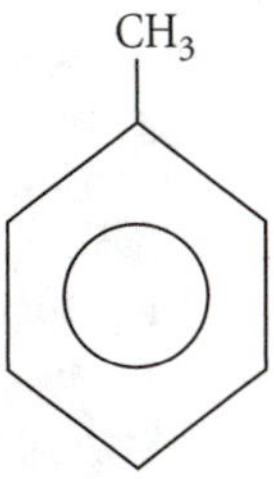 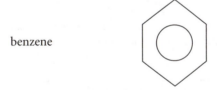 |

Once again we have a classification, when considering all the possibilities for substituting and branching, that enables many compounds from this group to be added to the large total number of organic compounds that exist.

# 6.7   Organic Compounds Containing Oxygen

As stated previously, oxygen, nitrogen, and other elements can be a part of the structure of organic compounds. The organic compounds containing oxygen is the largest group and warrants division into several classifications. These include functional groups and compounds that may be familiar to you. We will take a close look at the six groups shown in Fig. 6.4.

To facilitate an organized look at these classifications, we will utilize an abbreviation that will allow us to focus on the functional group itself and to help represent a general picture of each classification. This abbreviation is called an ***R-group***. An R-group represents a hydrocarbon grouping, or any organic grouping of elements. The functional group containing oxygen will be shown attached to it. The R-group may be a simple group, such as a methyl group, but it may also represent something much more complicated. In some cases it may represent a hydrogen. As we proceed through our study of the classifications, we will encounter some of the other relatively simple examples discussed in Section 6.6.1. These are the ethyl group, the n-propyl group, and the isopropyl group.

## 6.7.1   Alcohols

Let us begin with the alcohols. The functional group for alcohols is the " –OH" group. Using the R-group representation, the general alcohol structure would be:

$$\text{alcohol} \qquad\qquad \text{R—OH}$$

The common system for naming alcohols is

$$\text{name of R-group + "alcohol"}$$

and the formal IUPAC naming system is based on the name of the alkane with the same number of carbons where the "e" at the end of the name of the alkane is dropped and replaced with "ol." Here are the simplest alcohols and their names.

| Common IUPAC Name | Formal IUPAC Name | |
|---|---|---|
| Methyl alcohol | Methanol | $CH_3$—OH |
| Ethyl alcohol | Ethanol | $CH_3CH_2$—OH |
| n-propyl alcohol | 1-propanol | $CH_3CH_2CH_2$—OH |
| Isopropyl alcohol | 2-propanol | OH<br>\|<br>$CH_3CHCH_3$ |

Methanol is also known as "wood alcohol" because it can be distilled from wood. It is the major ingredient in automobile windshield washer fluid and is the automobile gasoline additive (under several different trade names) that helps car engines to start in cold weather. It has a low freezing point (and thus does not freeze while stored in the window washer fluid container under the hood of your car) and is completely miscible with water and gasoline, serving to dissolve water in gasoline tanks and thus assist in starting your car. It is extremely toxic if taken internally.

Ethanol is also known as "grain alcohol" because it can be distilled from grains such as corn. It is the alcohol that is present in alcoholic beverages. It is also the alcohol that is added to gasoline by oil companies to produce the gasoline that has been called "gasohol."

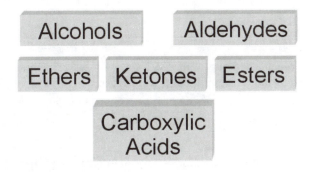

**FIGURE 6.4**    The classifications of organic compounds that contain oxygen.

Isopropyl alcohol is the alcohol found in rubbing alcohol. It is used as an antiseptic and is used in microbiology laboratories to kill bacteria on bench tops and other materials. It is toxic if taken internally. Notice that n-propyl alcohol and isopropyl are isomers.

As with the hydrocarbons, many more alcohols exist because of the large number of possibilities for branches and chains that are the R-group structures.

## 6.7.2   Ethers

The classification known as *ethers* have the following general structure, or functional group.

                 ethers                        R—O—R

The common naming system for ethers simply names the two R-groups, using the prefix "di-" if both R-groups are the same, with the word "ether" following. The formal IUPAC system is slightly more complex. The simpler of the two R-groups is named with the "-ane" replaced by "-oxy," as if this

R-O- is attached to a carbon of the other group; the name of the other group is then written as the base name. The three simplest examples are shown below.

| Common IUPAC Name | Formal IUPAC Name | |
|---|---|---|
| Dimethyl ether | Methoxymethane | $CH_3$—O—$CH_3$ |
| Ethyl methyl ether | Methoxyethane | $CH_3CH_2$—O—$CH_3$ |
| Diethyl ether | Ethoxyethane | $CH_3CH_2$—O—$CH_2CH_3$ |

*Diethyl ether* is the ether that has been used in hospitals as an anesthetic. It is also know as **ethyl ether** and as **ether**. It has some very noteworthy uses, and hazards, in the laboratory. It is highly flammable, and because it evaporates readily, a dangerous situation can be created in a laboratory because a room filled with its vapor presents an explosion hazard. In addition, it decomposes readily upon standing and turns into an explosive product. Ether must be stored in metal containers in refrigerators to slow its decomposition and must be used or discarded within nine months of its acquisition. The refrigerators in which it is stored should be explosion-proof.

Notice that dimethyl ether is an isomer of ethyl alcohol. Other ethers are also isomers of alcohols. A large number of ethers exist due to the large number of possible variations that can occur in the R-group.

### 6.7.3 Aldehydes

The four remaining classifications of oxygen-containing organic compounds have a common structural feature known as the carbonyl group.

$$\text{carbonyl group} \qquad \begin{array}{c} O \\ \parallel \\ -C- \end{array}$$

The general structure, or functional group, of the aldehyde classification has one bond from the carbonyl group carbon to an R-group and the other bond to a hydrogen:

$$\text{aldehydes} \qquad \begin{array}{c} O \\ \parallel \\ R-C-H \end{array}$$

In the simplest aldehyde, the R-group is a hydrogen. The common name of an aldehyde utilizes the prefix "form-" to designate one carbon (the carbonyl carbon) and "acet-" to designate 2 carbons. This indicates that there is some systematic nature to the common naming system, which we will see is utilized with the other three remaining oxygen-containing classifications as well. The formal IUPAC naming system for aldehydes changes the "e" on the base alkane to an "al." Thus, we have the following.

| Common IUPAC Name | Formal IUPAC Name | |
|---|---|---|
| Formaldehyde | Methanal | $\begin{array}{c} O \\ \parallel \\ H-C-H \end{array}$ |
| Acetaldehyde | Ethanal | $\begin{array}{c} O \\ \parallel \\ CH_3-C-H \end{array}$ |

Formaldehyde has been widely used as a preservative for biological specimens, although that use has been largely discontinued because it is a suspected carcinogen. It is, however, used in embalming fluids.

Again, with branches and chains creating a large number of possible variations in the R-group, there are a large number of aldehydes that exist.

## 6.7.4 Ketones

*Ketones* are compounds that have two R-groups, which may or may not be the same group, attached to the carbonyl group carbon as their functional group:

R′ is used here to differentiate one R-group from another. The simplest ketone has both R-groups as methyl groups. The common name of this compound is ***acetone***. Previously, we indicated that the "acet-" prefix designates two carbons. Acetone represents an exception to that rule, since it has three carbons. This compound may also be named like the ethers: dimethyl ketone. The IUPAC name changes the "e" of the base alkane to "one." Thus, the IUPAC name for acetone is propanone.

| Common IUPAC Names | Formal IUPAC Name | |
|---|---|---|
| Acetone, dimethyl ketone | Propanone | $\overset{\displaystyle O}{\overset{\displaystyle \|}{CH_3—C—CH_3}}$ |

Acetone has been used as a primary ingredient in fingernail polish remover. The four-carbon ketone is also fairly common.

| Common IUPAC Names | Formal IUPAC Name | |
|---|---|---|
| Methyl ethyl ketone | Butanone | $\overset{\displaystyle O}{\overset{\displaystyle \|}{CH_3CH_2CCH_3}}$ |

This latter compound is commonly abbreviated as "MEK." Both MEK and acetone are common laboratory solvents because they have intermediate polarity and are capable of mixing with and dissolving both polar and nonpolar substances.

## 6.7.5 Carboxylic Acids

The final two classifications have a structural feature in common that is derived from the carbonyl group. It is the carboxyl group.

$$\text{carboxyl group} \qquad \overset{\displaystyle O}{\overset{\displaystyle \|}{—C—O—}}$$

The first of these two classifications is the carboxylic acid classification, in which the functional group consists of an R-group attached to the carbon of the carboxyl group and a hydrogen attached to the oxygen.

$$\text{carboxylic acid} \qquad \overset{\displaystyle O}{\overset{\displaystyle \|}{R—C—O—H}}$$

These compounds are acids because the hydrogen shown in their carboxylic acid group is "acidic." This means it can be lost easily in water solution or in reaction with a base. Since these compounds are true acids, they have the "H" presented first in their chemical formula like all other acids. The simplest carboxylic acid also has an "H" for the R-group. The "form-" and "acet-" prefixes are used in the common

---

*Chemistry Professionals at Work*                    *CPW Box 6.7*

# MATERIAL SAFETY DATA SHEETS

Chemistry professionals must deal with safety issues almost every minute of every day in their jobs. With organic compounds, such issues include contact with skin, flammability, compatibility with other chemicals, toxicity of vapors, and disposal. To aid them in this situation, they can consult the Material Safety Data Sheet (MSDS) for the chemical in question. The MSDS is often a 4–8 page document that provides the information needed to protect the laboratory worker from harm when working with the chemicals. Federal law requires that MSDS be provided by manufacturers and distributors of chemicals. They should be filed in a location that is readily accessible in the laboratory in case of an emergency.

*For Homework: We have mentioned in the text that acetone and ethyl acetate are both sold to consumers as fingernail polish remover products. Ask your instructor for a copy of the MSDS for acetone and ethyl acetate and look for safety information that has relevance for users of fingernail polish remover. Write a report that includes your recommendations for safe use.*

---

naming system, as discussed for the previous classifications. In the formal IUPAC system, the "e" of the base alkane changes to an "oic" and is followed by the word "acid."

| Common IUPAC Name | Formal IUPAC Name | |
|---|---|---|
| Formic acid | Methanoic acid | $$\overset{\displaystyle O}{\overset{\displaystyle \|}{H-C-O-H}}$$ |
| Acetic acid | Ethanoic acid | $$\overset{\displaystyle O}{\overset{\displaystyle \|}{CH_3-C-O-H}}$$ |

Vinegar is about 5% acetic acid. It is the ingredient that gives vinegar its sour taste.

The molecular formula for formic acid would be written $HCHO_2$, while that for acetic acid would be $HC_2H_3O_2$. Again, the R-group can be any simple or complex organic group, and thus the number of isomers is large.

## 6.7.6   Esters

The final group of oxygen-containing organic compounds that we will study is the esters. Esters have the following functional group or general structure.

$$\text{ester} \qquad \overset{\displaystyle O}{\overset{\displaystyle \|}{R'-C-O-R}}$$

Again notice that, as with the ketones, there are two R-groups in an ester and they need not be the same group. Thus, one R-group is designated as R and the other as R'. The simplest ester is one in which R' is a hydrogen and R is a methyl group. We also show below the ester in which both R and R' are methyl groups and the ester in which R′ is a methyl group and R is an ethyl group. The common naming system for esters uses the "form-" and "acet-" prefixes followed by "-ate," a system similar to inorganic salts in which the "-ic" ending of an acid changes to "-ate" when naming the salt. In the IUPAC system, the "-ic" of the acid simply changes to "-ate." Thus we have the following names.

| Common IUPAC Name | Formal IUPAC Name | |
|---|---|---|
| Methyl formate | Methyl methanoate | $\begin{array}{c} O \\ \parallel \\ H-C-O-CH_3 \end{array}$ |
| Methyl acetate | Methyl ethanoate | $\begin{array}{c} O \\ \parallel \\ CH_3-C-O-CH_3 \end{array}$ |

One ester is an ingredient in some fingernail polish removers. It is ethyl acetate.

| ethyl acetate | ethyl ethanoate | $\begin{array}{c} O \\ \parallel \\ CH_3-C-O-CH_2CH_3 \end{array}$ |
|---|---|---|

Here again, the R-groups can be simple, like these that are shown here, or complex organic groupings. This means that many compounds exist that can be classified as esters.

Esters are responsible for the pleasant odors and unique flavors of many fruits. For example, pentyl acetate is found in bananas, octyl acetate is found in oranges, and isobutyl formate is found in raspberries.

## 6.8   Summaries of Hydrocarbon and Oxygen Classifications

Table 6.2 summarizes the hydrocarbon and oxygen classifications and functional groups, giving one simple example for each.

**TABLE 6.2**   Summary of the Hydrocarbon and Oxygen Classifications Discussed in Text

| Name of Classification | Functional Group | Simple Example | IUPAC Name | Common Name |
|---|---|---|---|---|
| Alkane | C—C | $CH_3-CH_3$ | Ethane | Ethane |
| Alkene | C=C | $CH_2=CH_2$ | Ethene | Ethylene |
| Alkyne | CH≡CH | CH≡CH | Ethyne | Acetylene |
| Aromatics | 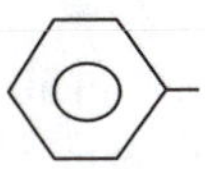 ⬡— | 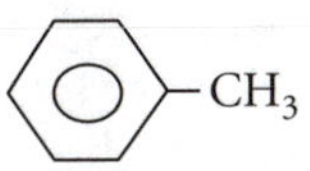 ⬡—CH_3 | Methylbenzene | Toluene |
| Alcohol | R—OH | $CH_3CH_2-OH$ | Ethanol | Ethyl alcohol |
| Ether | R—O—R′ | $CH_3-O-CH_3$ | Methoxymethane | Dimethyl ether |
| Aldehyde | $\begin{array}{c} O \\ \parallel \\ R-C-H \end{array}$ | $\begin{array}{c} O \\ \parallel \\ CH_3-C-H \end{array}$ | Ethanal | Acetaldehyde |
| Ketone | $\begin{array}{c} O \\ \parallel \\ R-C-R′ \end{array}$ | $\begin{array}{c} O \\ \parallel \\ CH_3-C-CH_3 \end{array}$ | Propanone | Acetone |
| Carboxylic Acid | $\begin{array}{c} O \\ \parallel \\ R-C-OH \end{array}$ | $\begin{array}{c} O \\ \parallel \\ CH_3-C-OH \end{array}$ | Ethanoic acid | Acetic acid |
| Ester | $\begin{array}{c} O \\ \parallel \\ R-C-O-R′ \end{array}$ | $\begin{array}{c} O \\ \parallel \\ H-C-O-CH_3 \end{array}$ | Methyl methanoate | Methyl formate |

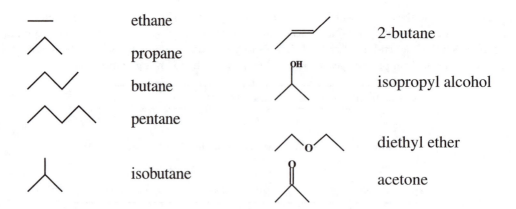

**FIGURE 6.5** Some examples of alkane structures drawn with the bond-line method.

**FIGURE 6.6** Some examples of structures of compounds containing various functional groups drawn using the bond-line method.

## 6.9   Alternate Structure-Drawing Methods

In Section 6.6.1, we noted that there are two acceptable methods of drawing organic structures—the dash method and the condensed method. In that same section, we presented two drawings of propane that at first glance may have appeared to be a pair of isomers. Upon closer inspection, it was observed that there was a three-carbon chain in each and that the hydrogens were bonded to the same atoms. The only apparent difference was that one of the structures showed one end carbon connected in a different direction, but this did not make them different structures. They were identical structures, just drawn slightly differently. In Section 6.6.4, we drew two structures for cyclohexane that were described as identical. One simply showed the carbons and hydrogens, while the other was a hexagon bond-line structure that did not show the carbons and hydrogens. Carbons were assumed to be at each corner of the hexagon while two hydrogens were assumed to be bonded at each corner. A drawing of a hexagon is a common way to show the structure of cyclohexane.

These are examples of the fact that there are several acceptable methods of drawing organic structures, all of which are used by chemists at different times for different reasons. Other texts and reference books may present structures in ways that have not yet been presented here. For this reason, we now discuss several other common methods and give examples.

One other way of drawing these structures is an extension of the bond-line method for cyclohexane that was mentioned above, but is now adapted for depicting non-ring structures. It is similar to the sawtooth picture described in Fig. 6.2, but without the solid circles. Carbon atoms are assumed to be at the corners and at the ends of the sawtooth with two hydrogens not shown, but assumed to be bonded to each corner carbon, and three hydrogens at each end carbon. Thus, we have structures such as those shown for the alkanes in Fig. 6.5. Compounds with functional groups can also be drawn as bond-line structures. Examples are shown in Fig. 6.6.

Finally, functional groups may be drawn in a condensed manner. Thus far in our discussion, all structures of compounds containing functional groups have shown the functional groups in a dash manner, while the rest of the structure was shown in a condensed manner (see the examples Section 6.7). The conventional ways to write entire structures as condensed structures, including functional groups, are shown in the following examples. Compare these with the structures for the same compounds in Section 6.7.

| | |
|---|---|
| $CH_3CH_2OHCH_3$ | isopropyl alcohol |
| $CH_3CH_2OCH_2CH_3$ | diethyl ether |
| $CH_3CHO$ | acetaldehyde |

$CH_3COCH_3$       acetone

$CH_3COOH$       acetic acid

$CH_3COOCH_3$       methyl acetate

The above structures for acetaldehyde, acetic acid, and methyl acetate are consistent with the standard *general* structures for compounds in the following list of classifications.

RCHO       aldehydes

RCOOH       carboxylic acids

RCOOR       esters

Notice that atoms bonded to a carbon are shown immediately to the right of that carbon, but to the left of the next carbon in the chain. In some cases, atoms between carbons, or between a carbon and an oxygen, are bonded to both. Double bonds to oxygens are known to be present in aldehydes, ketones, carboxylic acids, and esters, and thus do not need to be shown as such in these structures. Remember, too, that all carbons have four bonds, no more and no less, when writing or analyzing a formula of this type.

# 6.10 Organic Compounds Containing Nitrogen

Organic compounds containing nitrogen are of some importance. Here we list two sample classifications.

## 6.10.1 Amines

Amines are derived from ammonia. This means that they consist of the ammonia structure but with R-groups replacing one or more of the hydrogens. When one hydrogen is replaced by an R-group, it is a primary amine; when two hydrogens are replaced by R-groups, it is a secondary amine; and when all three hydrogens are replaced by R-groups, it is a tertiary amine.

primary amine      secondary amine      tertiary amine

Simple examples are those in which the R-groups are methyl groups and include methylamine $CH_3NH_2$, dimethylamine $(CH_3)_2NH$, and trimethylamine $(CH_3)_3N$.

Amines are bases because they have the power to neutralize acids. This is so because the nitrogen of an amine has a pair of electrons that is nonbonding and the hydrogen of an acid can attach itself to this pair (coordinate covalent bonding—Chapter 5) resulting in the acid being neutralized. An important such base (that also contains oxygens) is used routinely in analytical laboratories and is called tris (hydroxymethyl)aminomethane, $(HOCH_2)_3CNH_2$, commonly known as "tris" or "tham." A discussion of compounds that are bases because they bond to hydrogens is given in Chapter 12.

An important industrial use of amines is in the manufacture of "quaternary" ammonium salts, which in turn are used to produce fabric softeners. Specific amines have other uses too numerous to mention here.

## 6.10.2 Nitriles

Nitriles are organic compounds containing the cyano group, or a carbon triply bonded to a nitrogen.

$$-C \equiv N \qquad\qquad \text{cyano group}$$

The carbon is also bonded to an R-group. The simplest, and probably most important nitrile has a methyl group attached to the carbon and is called acetonitrile.

$$CH_3-C \equiv N \qquad\qquad \text{acetonitrile}$$

Acetonitrile is rather polar and, like acetone and methanol, is used in situations that require somewhat of a universal solvent, including the dissolving of polymers. It mixes freely with water just like acetone and methanol.

# 6.11 Organic Compounds Containing Both Oxygen and Nitrogen

We now mention some organic compounds that contain both oxygen and nitrogen in addition to the usual carbon and hydrogen.

## 6.11.1 Amides

Amides have structures in which a carbonyl group and the amine functional group are bonded to each other. The amine group can be primary, secondary, or tertiary.

$$
\begin{array}{ccc}
O & O & O \\
\| & \| & \| \\
R-C-NH_2 & R-C-NHR' & R-C-NR'R''
\end{array}
$$

The simplest example is acetamide.

$$
\begin{array}{cc}
 & O \\
\text{acetamide} & \| \\
 & CH_3-C-NH_2
\end{array}
$$

## 6.11.2 Isocyanates

Isocyanates have the following functional group.

$$R-N=C=O \qquad\qquad \text{isocyanates}$$

Because there are two double bonds adjacent to each other, isocyanates are very unstable and reactive. They are used in a great variety of reactions for synthesizing other useful organic compounds. An important example is the manufacture of urethanes, which are also called carbamates.

$$
\begin{array}{c}
O \\
\| \\
R-NH-C-O-R' \qquad\qquad \text{urethanes or carbamates}
\end{array}
$$

The simplest isocyanate, methyl isocyanate,

$$CH_3-N=C=O \qquad\qquad \text{methyl isocyanate}$$

is used in the production of the popular garden insecticide Sevin™, which is a urethane. Polyurethane, an important polymer, is derived from diisocyanates, as will be mentioned in Section 6.13.

## 6.12 Other Classifications

There are many additional classifications that are beyond the scope of the present chapter. These include anhydrides, halides (which contain halogens), thiols (which contain sulfur), nitro compounds, etc., and many biologically important compounds, such as carbohydrates, amino acids, nucleic acids, etc. The biologically important compounds are discussed in Chapter 17.

## 6.13 Polymers

When we introduced organic chemistry in this chapter, we spoke of the broad application of the subject to our everyday lives by referring to a number of examples of ordinary consumer products that are in fact organic compounds and mixtures of organic compounds (Section 6.1). A number of these examples can be classified as **polymers.** A *polymer* is an organic compound whose molecules consist of a very large number of carbon atoms, and other atoms that are bonded together. As a result, these molecules have very large formula weights. In addition, polymers have repeating structural units in the chain. While the examples of organic compounds cited in other sections of this chapter have 2–6 carbon atoms bonded together, polymer molecules can have hundreds and sometimes thousands of carbon atoms. The concept of the backbones or networks introduced and discussed in other sections of this chapter is especially applicable here for polymers, because this is indeed is their structure—extremely long chains and/or branches of carbons with hydrogens and functional groups bonded at regular intervals and making up a network of atoms that is the molecule. These compounds are not substances just gracing the laboratories of research scientists. They have extremely broad application in our everyday lives.

A number of examples of these applications were cited in Section 6.1 These included plastics, foams, paints, varnishes, clothing fibers, and synthetic rubbers and vinyls. In this section, we will give additional examples of ordinary consumer products that are polymers, discuss the structure of these molecules, and present some of the reactions by which polymers form. We will do this for two general classifications, addition polymers and condensation polymers.

### 6.13.1 Addition Polymers

Polymers classified as *addition polymers* are the most common as far as applications in our everyday lives. Molded plastic articles such as plastic bottles and other containers, children's toys, plumbing pipes—virtually anything you can think of that is a plastic—are addition polymers. We can also list floor tile, some textile fibers, ropes, rubber gaskets, saran wrap, and styrofoam cups among the materials that are composed of addition polymers. It is difficult to imagine what our lives would be like without addition polymers. Let us take a closer look at the structure of these substances.

The word "addition" is derived from the kind of chemical reaction that is used to manufacture these polymers, namely the addition reaction of alkenes. Addition reactions of alkenes involve the adding of a reacting molecule across the double bond of the alkene, as shown in Fig. 6.7. In this process, one of the two bonds in the double bond (the pi bond) is broken. The carbons that were doubly bonded become singly bonded, and the broken bond is replaced by bonds to the substituents from the reacting molecule, "A" and "B." The molecule represented by A-B can be H–H, Cl–Cl, H–Cl, etc., which results in organic

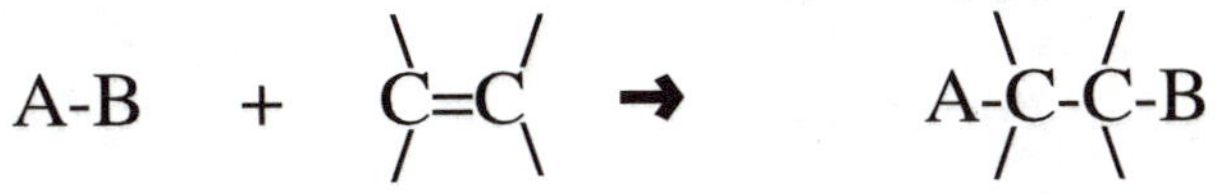

**FIGURE 6.7** The addition reaction of alkenes.

---

# GEL PERMEATION CHROMATOGRAPHY

Approximately 100 million metric tons of synthetic polymers are manufactured annually around the world. Special laboratory techniques have been developed for the analysis and research of polymers. One of these techniques is gel permeation chromatography. Chemists and chemistry laboratory technicians working at a polymer plant are likely familiar with this instrumental technique. It is specifically designed to separate different sized molecules. The very large molecules associated with polymers are separated from smaller ones through the use of a porous material. Small molecules fit in the pores while large molecules do not. Thus, a separation of these molecules and subsequent analysis can take place.

*For Homework and Class Discussion: Find a reference that defines and describes gel permeation chromatography, then visit with an industrial chemist or chemistry technician in your area to see how it is applied. Report back to the class.*

---

products that are not polymers. However, it has been discovered that under certain conditions, the broken bond can instead be replaced by bonds to carbons from other alkene molecules that have undergone the same change, as in the sequence in Fig. 6.8. This addition can continue in both directions, with alkene molecules adding to other alkene molecules until a very long molecule, a polymer molecule, is formed.

There can be many variations on the general description given above because the substituents bonded to the four carbons in the double bonds can be quite variable. For example, all four bonds may be to hydrogens ($CH_2=CH_2$, ethene, or ethylene), there may be one methyl group attached ($CH_2=CHCH_3$, propene, or propylene), there may be a chlorine attached ($CH_2=CHCl$, vinyl chloride) and so on. Polyethylene, polypropylene, polyvinyl chloride (PVC), and many other addition polymers have been manufactured in mass quantities by this approach and used for many consumer products. Table 6.3 lists some of the addition polymers that have been manufactured. Also listed are the individual alkene units (*monomers*) that are in these polymers and some of the uses of each.

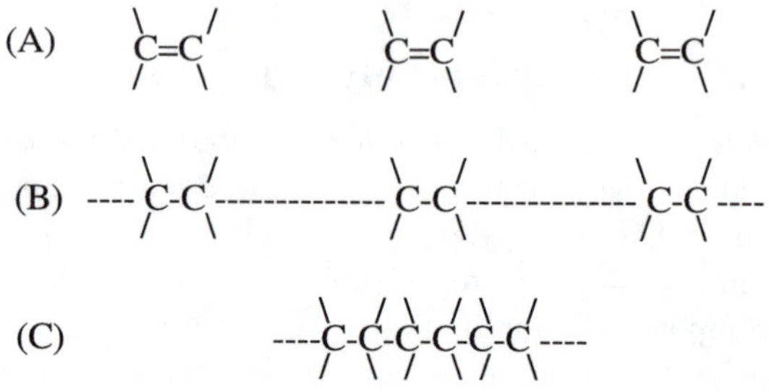

**FIGURE 6.8**    The formation of an addition polymer. (A) Separate alkene molecules. (B) Alkene molecules undergoing addition reactions with each other. (C) Addition polymer formed.

## 6.13.2  Condensation Polymers

*Condensation polymers* are also based on smaller organic units (monomers) that react with each other to form very large molecules. While in addition polymers, the functional group found in the monomer unit was a carbon-carbon double bond, the functional groups in the monomer for condensation polymers

**TABLE 6.3**   A Number of Addition Polymers with their Monomer Units and Uses

| Monomer | Polymer | Uses |
|---|---|---|
| $CH_2{=}CH_2$ ethylene | —$CH_2$-$CH_2$—$CH_2$-$CH_2$— polyethelene | Molded plastics, bottles, toys, packing material |
| $CH_2{=}CHCH_3$ propylene | —$CH_2$-$CHCH_3$-$CH_2$-$CHCH_3$— polypropylene | Molded articles, bottles, textile fibers, ropes |
| $CH_2{=}CHCl$ vinyl chloride | —$CH_2$-$CHCl$-$CH_2$-$CHCl$— polyvinyl chloride (PVC) | Garden hoses, plumbing pipes, floor tile, vinyls |
| $CH_2{=}CCl_2$ vinylidene chloride | —$CH_2$-$CCl_2$-$CH_2$-$CCl_2$— saran | Plastic wrap |
| $CF_2{=}CF_2$ tetrafluoroethylene | —$CF_2$-$CF_2$—$CF_2$-$CF_2$— Teflone® | Linings for pots and pans, gaskets, protective coatings |
| $CH_2{=}CH$ styrene | —$CH_2$-$CH$-$CH_2$-$CH$— polystyrene | Styrofoam cups and insulation, molded articles |
| $CH_2{=}CH$ &#124; $OC$-$CH_3$ ‖ $O$ vinyl acetate | —$CH_2$-$CH$ - $CH_2$-$CH$— &#124; &#124; $OC$-$CH_3$  $OC$-$CH_3$ ‖  ‖ $O$  $O$ polyvinyl acetate | Adhesives, paint and varnish |
| $CH_2{=}CHCN$ acrylonitrile | —$CH_2$-$CHCN$-$CH_2$-$CHCN$— Orlon®, Acrilan® | Textile fibers |

are variable. In addition, while with addition polymers there is only one kind of monomer unit building upon itself to form the polymer, there are typically two or three different units making up a monomer in condensation polymers which are found alternating throughout the chain. Typical functional groups involved in condensation polymers are alcohols, carboxylic acids, esters, amines, and isocyanates, and there are usually two such groups in one unit. A unit can be a diol, or glycol, a dicarboxylic acid, a diester, a diamine, or diisocyanate. (Note: A diol, for example, is a molecule with two alcohol functional groups present.) Thus we can have the following units as components of the monomers.

$$
\begin{array}{ccc}
CH_2C\ldots CCH_2 & HOCC\ldots CCOH & ROCC\ldots CCOR \\
\;|\qquad\quad | & \;\;\|\qquad\;\; \| & \;\;\|\qquad\;\; \| \\
OH\quad\;\; OH & O\qquad O & O\qquad O \\
\text{diol} & \text{dicarboxylic acid} & \text{diester}
\end{array}
$$

$$
\begin{array}{cc}
CH_2C\ldots CCH_2 & CH_2C\ldots CCH_2 \\
\;|\qquad\quad | & \;|\qquad\quad | \\
NH_2\quad\; NH_2 & N{=}C{=}O\;\; N{=}C{=}O \\
\text{diamine} & \text{diisocyanate}
\end{array}
$$

Just as the tendency of alkenes to undergo addition reactions was the driving force for addition polymerization, the tendency for one functional group to react with another is the driving force for condensation polymerization. For example, alcohols tend to react with carboxylic acids to form esters. Thus, each alcohol functional group in a diol will tend to react with each carboxylic acid group of a dicarboxylic

**TABLE 6.4**   Some Condensation Polymers

| Polymer | Other Names | Uses |
| --- | --- | --- |
| Polyester | Dacron®, mylar, kodel | Textile fiber, tire cord, fabrics, carpet fiber |
| Polyamide | Nylon | Textile fibers, plastics |
| Polyurethane | Foam, elastomers, thermoplastics | Cushioning in beds and furniture, insulation, packaging, weather stripping, paint and varnish |

acid forming linked units that are esters. This gives us polyesters, such as Dacron®, kodel, mylar, etc., which are condensation polymers.

Similarly, amines react with carboxylic acids forming amides. Dicarboxylic acids react with diamines to form polyamides, an example of which is nylon. Isocyanates react with alcohols to form urethanes, and diisocyanates react with diols to form polyurethanes, examples of which are foams found in insulation, in the stuffing of chairs and mattresses, and also paints and varnishes. Table 6.4 summarizes various condensation polymers and their uses.

## 6.14   Infrared Spectrometry

### 6.14.1   Introduction

Infrared light can be used to help identify and characterize material substances and is especially useful for organic chemicals and polymers. The experiment for this involves an instrument known as an *infrared spectrometer*, which measures the infrared light absorbed by a sample of the material. The sample to be tested is mounted in the path of a beam of the infrared light radiating from an infrared laser source located inside the instrument. Using a device known as an interferometer, which is also part of the instrument, the sample's absorption of each individual wavelength component of the infrared light can be recorded. A representation of the instrument is shown in Fig. 6.9.

The resulting output is called the *infrared spectrum* of the material. It is a plot of % transmittance vs. wavenumber. This spectrum is a "molecular fingerprint"—no two substances have exactly the same infrared spectrum. This is part of the usefulness of infrared spectroscopy as an identification and characterization tool. The infrared spectrometer is more fully explained in Section 6.14.4.

### 6.14.2   %Transmittance

The **% transmittance** is a measure of the amount of light that is absorbed by a material. If 100% of the light is transmitted by the material in the path of the light, then none is absorbed. If 50% of the light is transmitted, then 50% is absorbed. If 75% of the light is transmitted, then 25% is absorbed. The scale on the y-axis of the spectrum is in % and varies from 0 to 100%.

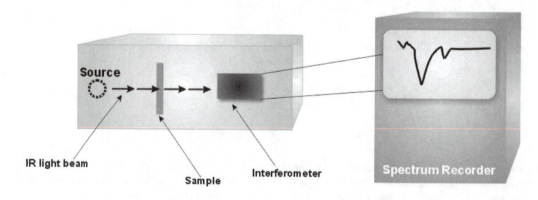

**FIGURE 6.9**   The essential components of an infrared spectrometer.

### 6.14.3   Wavenumber

The basic concepts of the wave theory of light, including wavelength, energy, the electromagnetic spectrum, etc., were discussed in Chapter 3, Section 3.3.2. Let us first briefly review the concept of wavelength. According to this theory, light consists of a continuum of repeating electromagnetic waves, one after another, that are emanating from the light source. The physical distance from the crest of one wave to the crest of the next wave is called the **wavelength**. Wavelength varies from lengths as short as atomic diameters (gamma rays) to as long as several miles (radio and TV waves). The various units for length in the metric system are reviewed in Chapter 7, Section 7.5.1. Wavelength is often expressed in nanometer (nm) units (1 nm = 1 billionth of a meter), although any unit of length could be used, including centimeters (cm). Visible light, which consists of the rainbow colors, and infrared light are approximately in the middle of this range. Visible light ranges from about 350 nm to about 750 nm. Infrared light ranges from about 750 nm to about 50,000 nm. Please refer back to Figs. 3.2 and 3.3 in Chapter 3.

**Wavenumber** is another way to express the wavelength of light. When wavelength is expressed in cm and inverted, it becomes the wavenumber. Therefore, wavenumber has the unit "inverse centimeters," or cm$^{-1}$. It can be calculated from the wavelength as shown in the following equation:

$$\bar{\nu}\,(\text{cm}^{-1}) = \frac{1}{\lambda(\text{cm})} \qquad (6.1)$$

In this equation, $\bar{\nu}$ ("nu bar") is the symbol for wavenumber and $\lambda$ is the symbol for wavelength. The infrared light wavenumbers typically utilized and displayed by infrared instruments range from 600 cm$^{-1}$ to 4000 cm$^{-1}$.

### 6.14.4   Infrared Spectrum

A plot of % transmittance vs. wavenumber, also known as the **infrared spectrum**, is displayed on the template shown in Fig. 6.10.

Infrared spectra are characterized by peaks of absorption, often called **bands**. The wavenumber at which a band occurs is important because it provides some key information about the bonding in the molecules of a substance. For example, the vast majority of organic molecules posses the C–H bond. There can be many IR absorption bands due to the presence of the C–H bond, but the characteristic band for those C–H bonds that are part of an alkyl structure (no double or triple

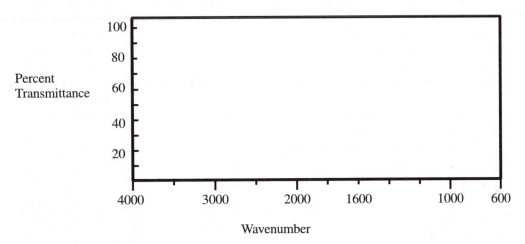

**FIGURE 6.10**   The % transmittance vs. wavenumber template for infrared spectra.

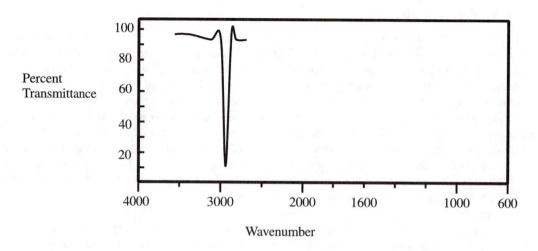

**FIGURE 6.11**   The location and typical appearance of an IR absorption band due to the presence of an alkyl C–H bond in a molecule.

bonds attached to the carbon) is located a little below 3000 cm$^{-1}$, often with two or more peaks, as shown in Fig. 6.11.

Alcohols possess the O–H functional group. The characteristic IR absorption band that is due to the O–H bond is a broad band centered around 3300 cm$^{-1}$. This is shown in Fig. 6.12. Since an alcohol would also have alkyl C–H bonds (as in ethanol, for example), the bands as indicated in Fig. 6.11 are also present in Fig. 6.12.

As we learned earlier, there are four classifications of organic compounds that possess the carbonyl group, a C=O bond. These are aldehydes, ketones, carboxylic acids, and esters. The characteristic absorption band for this bond is a strong and sharp absorption band that occurs around 1720 cm$^{-1}$. This is shown in Fig. 6.13. Since all compounds that possess the carbonyl group also possess alkyl C–H bonds, the C–H absorption band is also pictured in Fig. 6.13.

If the molecule has a benzene ring, the C–H bonds on the ring absorb a little above 3000 cm$^{-1}$, the C–C bonds will show two bands between 1500 and 1600 cm$^{-1}$, and there will usually be a series of weak bands between 1600 and 2000 cm$^{-1}$. These are shown in Fig. 6.14.

The region of the infrared spectrum between 600 and 1500 cm$^{-1}$ has not been mentioned thus far in this discussion. Unlike the region from 1500 to 4000 cm$^{-1}$, this region of the spectrum does not produce absorption bands that can be seen as specifically characteristic of any functional group. Almost every organic compound and polymer does exhibit significant absorption in this region, however. Its specific use, therefore, is not in identifying functional groups the molecule has, but rather in "fingerprinting." We mentioned earlier that the infrared spectra are fingerprints of compounds, and this is true of the entire range of the spectrum. But the region from 600 to 1500 cm$^{-1}$ is known as the ***fingerprint region*** because it is limited to that kind of use. Absorption patterns observed in the fingerprint region in an unknown's spectrum must be compared to that of knowns, as human fingerprints are, in order for this region to be useful in identification and characterization.

## Example 6.1

The infrared spectrum given in Fig. 6.15 is the spectrum for a certain unknown compound or polymer. Discuss what structural features this compound has and what structural feature it does not have. Based on your knowledge of infrared spectrometry and the structure of polymers, could this be the infrared spectrum of polystyrene?

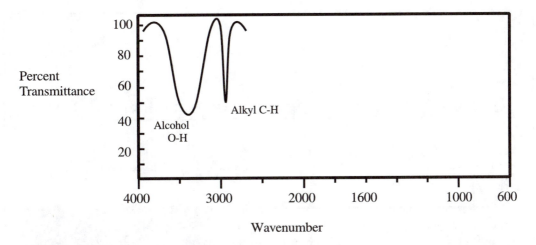

**FIGURE 6.12** The location and typical appearance of an IR absorption band due to the O–H bond. The C–H absorption described in Fig. 6.10 is also present.

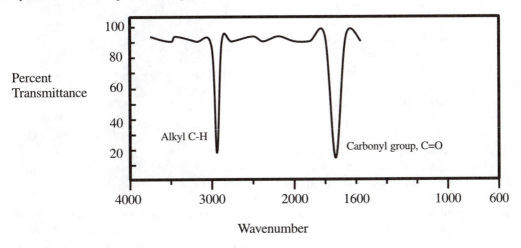

**FIGURE 6.13** Compounds that possess a carbonyl group have a strong, sharp absorption band at approximately 1720 cm$^{-1}$ as shown here.

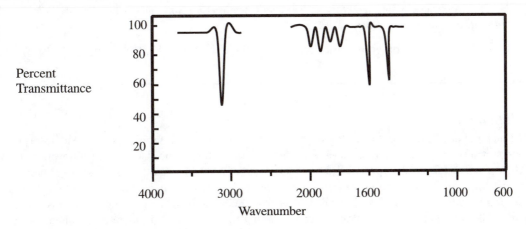

**FIGURE 6.14** The absorption bands for the benzene ring.

*Chemistry Professionals at Work*                                    *CPW Box 6.9*

# POLYMERIC MATERIALS AT THE DOW CHEMICAL COMPANY

My name is Connie Murphy. I work in Materials Research and Synthesis at The Dow Chemical Company in Midland, Michigan. I have an A.A.S. degree from Milwaukee Area Technical College. As a Sr. Research Technologist, I have many different responsibilities. My current project involves the preparation of new polymeric (plastic) systems targeted for packaging applications. I prepare these new materials either by synthesis from monomers and additives, or by the mechanical blending of commercially available polymers with other polymers or additives.

After preparing these materials, I fabricate them into articles that can be tested for a variety of properties. I do some of the testing myself and some is sent out to be done by our analytical lab. Properties important for these types of materials include glass transition temperature, tensile strength and modulus, flexural strength and modulus, gas barrier, color, haze, and impact strength. In addition to laboratory activities, I work up data with my computer, record experimental procedures and results in my data book, write technical reports and patent disclosures, and give oral presentations on my work. It is also my responsibility to train and supervise college co-op students assigned to work on our project.

*For Homework: Look at the internet site for The Dow Chemical Company and other sources, if available, and write a report on this company.*

http://www.dow.com/homepage/index.html (as of 6/4/00)

## Solution 6.1

The material does not have the alcohol functional group (–OH) because there is no broad absorption band at 3300 cm$^{-1}$. It does not have a carbonyl group because there is no sharp, strong absorption band 1700 cm$^{-1}$. There is an absorption band on the low side of 3000 cm$^{-1}$, so there are alkyl C–H bonds in the structure. All the absorption patterns for a benzene ring are present, including a band on the high side of 3000 cm$^{-1}$, a series of weak bands between 1600 cm$^{-1}$ and 2000 cm$^{-1}$, and two bands between about 1500 cm$^{-1}$ and 1600 cm$^{-1}$. We can conclude that the compound or polymer has a benzene ring along with alkyl structures. It could indeed be polystyrene (see structure in Table 6.3).

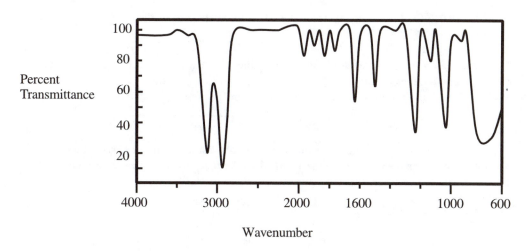

FIGURE 6.15   Infrared spectrum of the unknown compound in Example 6.1.

# 6.15   Homework Exercises

1. Give two different definitions of "organic chemistry." Where do most of the chemicals we call organic compounds come from?
2. To what does the punch-line "Remember, that dinosaur gave his or her all for that tank of gas" in the Chevron Oil Company commercial refer?
3. What is meant by a "backbone" or "network" of carbon atoms?
4. Define "hybridization."
5. Explain why hybridization is important in explaining the bonding of carbon to other atoms.
6. Sometimes the geometry around a bonded carbon atom is tetrahedral, sometimes it is trigonal planar, and sometimes it is linear. Under what conditions is the geometry tetrahedral? Under what conditions is the geometry trigonal planar? Under what conditions is the geometry linear?
7. List four physical properties that organic compounds generally have that inorganic compounds usually do not have.
8. Many organic compounds are liquids. List some physical properties of these liquids that are important for chemists to measure.
9. If you increase the chain length of a hydrocarbon, will the boiling point increase, decrease, or stay the same? Explain your answer.
10. What is a "functional group?"
11. How are aliphatic hydrocarbons different from aromatic hydrocarbons?
12. What is the difference between a saturated hydrocarbon and an unsaturated hydrocarbon?
13. Study the structures of methane, ethane, propane, n-butane, isobutane, and other alkanes and show that their formulas fit the general formula of alkanes, $C_nH_{2n+2}$.
14. Study the structures of ethene, propene, 1-butene, 2-butene, and other alkenes, as well as cyclopentane, cyclohexane, and other cycloalkanes and show that their formulas fit the general formula $C_nH_{2n}$.
15. Study the structures of ethyne, propyne, butyne, and other alkynes and show that their formulas fit the general formula for alkynes, $C_nH_{2n-2}$.
16. To what classification (chosen from the following: alkane, alkene, alkyne, cycloalkane, and aromatic) can each of the following compounds belong. Draw possible structures.
    (a) $C_3H_6$   (b) $C_4H_6$   (c) $C_6H_6$   (d) $C_8H_{16}$   (e) $C_8H_{10}$

17. Draw the structure of the simplest compounds that represent each of the following categories.
    (a) Alkane   (b) Alkene
    (c) Alkyne   (d) Aromatic
18. Draw condensed structures of all the isomers of the alkane with five carbons, pentane, $C_5H_{12}$. (Hint: there are three isomers.)
19. Which of the four structures below is not an isomer of this hexane structure?

$$CH_3CH_2CH_2CH_2CH_2CH_3$$

Explain why the suspected structure is not an isomer of hexane. (Hint: First write the molecular formula for each and then compare the structures of those that have the same formula. The formula for the hexane structure given is $C_6H_{14}$.)

(a) 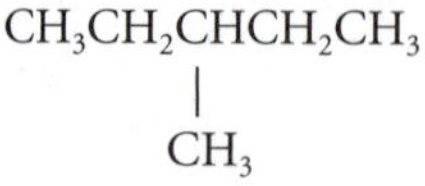   $CH_3CH_2CHCH_2CH_3$   (b) 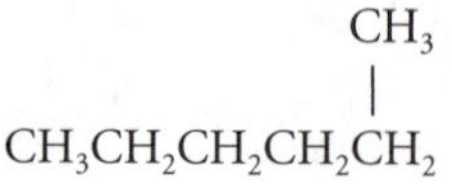                                    $CH_3$
          |                                                         |
          $CH_3$                          $CH_3CH_2CH_2CH_2CH_2$

                                          $CH_2CH_3$
(c) 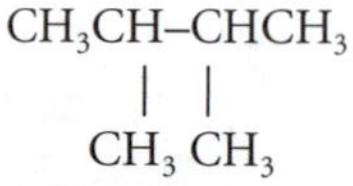                              (d) 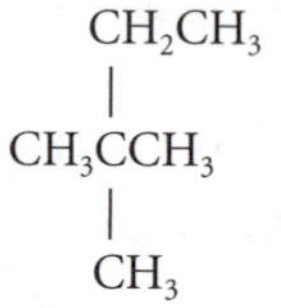        |
          $CH_3CH–CHCH_3$                  $CH_3CCH_3$
             |   |                          |
            $CH_3$ $CH_3$                   $CH_3$

20. What is the difference between a cyclohexane ring and a benzene ring? Explain the conventional way to draw the structures of these two compounds.
21. Pick out two structures below that are isomers of each other and tell why they are isomers. (See the hint in #19.)

(a)                     O            (b)
                       ||
          $CH_3CH_2CH_2C–H_2$              $CH_3CH_2–O–CH_2CH_3$

(c) 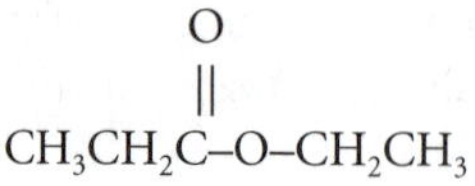                    O            (d) 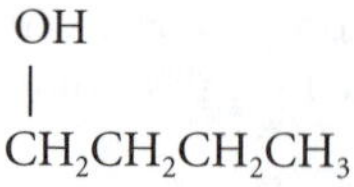        OH
                       ||                      |
          $CH_3CH_2C–O–CH_2CH_3$              $CH_2CH_2CH_2CH_3$

(e) 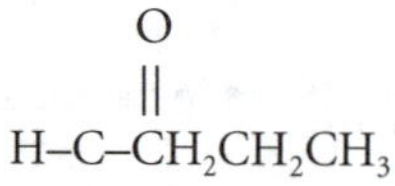                    O            (f) 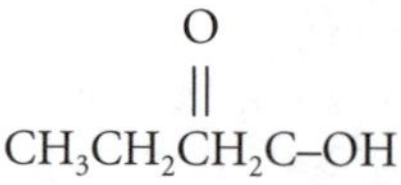                    O
                       ||                                  ||
          $H–C–CH_2CH_2CH_3$                   $CH_3CH_2CH_2C–OH$

22. Pick out the two structures below that are isomers of each other and tell why they are isomers. (See the hint in #19.)

(a) 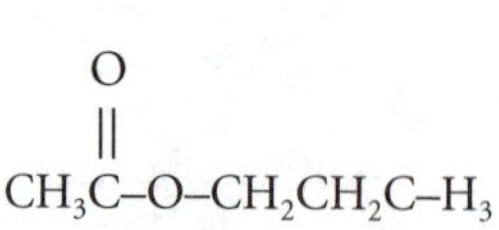           O                     (b) 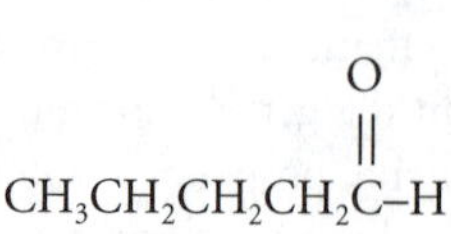                    O
              ||                                           ||
     $CH_3C–O–CH_2CH_2C–H_3$                   $CH_3CH_2CH_2CH_2C–H$

                       O
                       ||
(c) 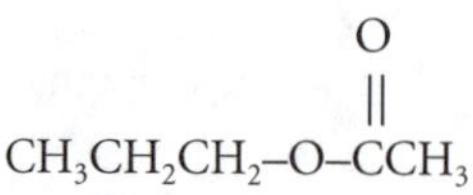     $CH_3CH_2CH_2–O–CCH_3$     (d)  $CH_3CH_2CH_2CH_2CH_2–OH$

                       O
                       ||
(e) 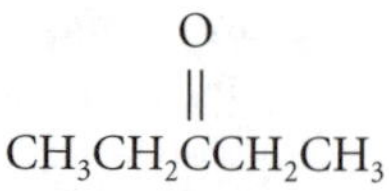     $CH_3CH_2CCH_2CH_3$        (f)     $CH_3CH_2–O–CH_2CH_3$

23. Draw the functional groups that are associated with each of the following oxygen-containing classifications.
    (a) Alcohol
    (b) Aldehyde
    (c) Carboxylic acid
    (d) Ester
    (e) Ether
    (f) Ketone

24. Classify each of the structures below as one of the following: alkane, alkene, alkyne, aromatic hydrocarbon, alcohol, ether, aldehyde, ketone, carboxylic acid, or ester.

(a) 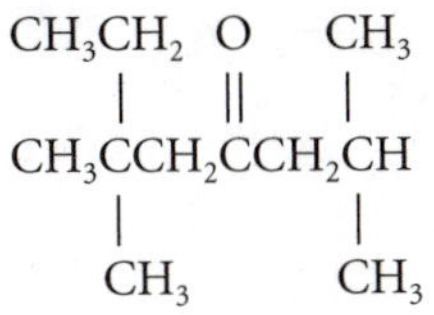
$$CH_3CH_2 \ O \ CH_3$$
$$| \ \ || \ \ |$$
$$CH_3CCH_2CCH_2CH$$
$$| \ \ \ \ \ \ \ \ \ \ |$$
$$CH_3 \ \ \ \ \ \ CH_3$$

(b) 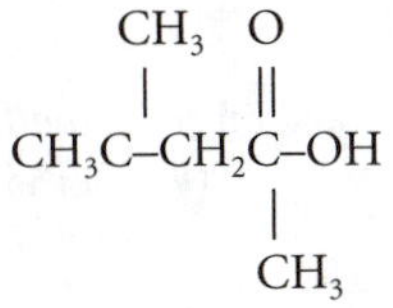
$$CH_3 \ O$$
$$| \ \ ||$$
$$CH_3C{-}CH_2C{-}OH$$
$$|$$
$$CH_3$$

(c) 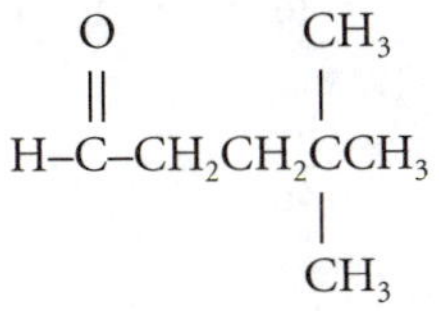
$$O \ \ \ \ \ \ \ \ \ CH_3$$
$$|| \ \ \ \ \ \ \ \ \ |$$
$$H{-}C{-}CH_2CH_2CCH_3$$
$$|$$
$$CH_3$$

(d) 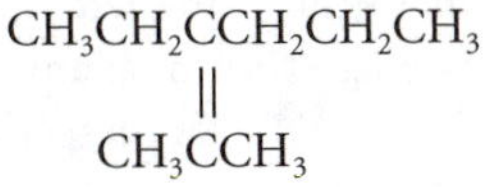
$$CH_3CH_2CCH_2CH_2CH_3$$
$$||$$
$$CH_3CCH_3$$

(e)

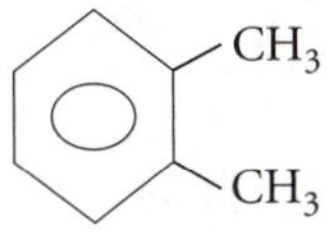

(f)

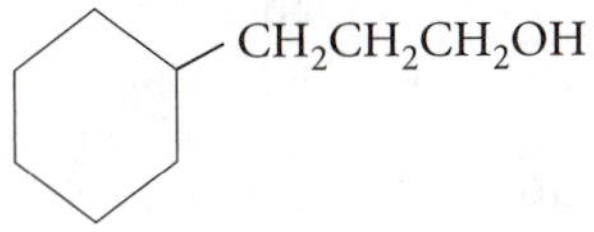

(g)
$$CH_3CH_2{-}C \equiv CH$$

(h)

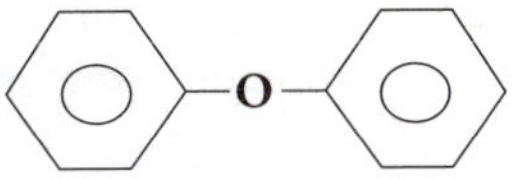

25. Name the following compounds.

(a) 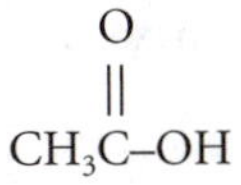
$$O$$
$$||$$
$$CH_3C{-}OH$$

(b) $CH_3CH_2{-}O{-}CH_2CH_3$

(c) $CH_3CH_2{-}OH$

(d) $CH_2{=}CH_2$

(e) 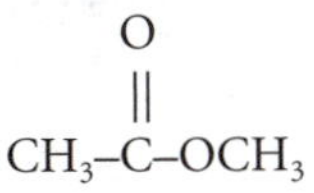
$$O$$
$$||$$
$$CH_3{-}C{-}OCH_3$$

(f) 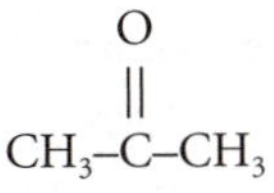
$$O$$
$$||$$
$$CH_3{-}C{-}CH_3$$

(g)

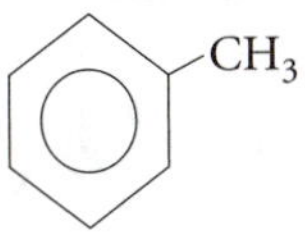

(h) 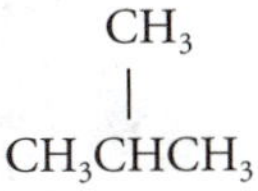
$$CH_3$$
$$|$$
$$CH_3CHCH_3$$

26. Name the following compounds.

(a) $CH_3$

(b) $CH_3CH_2CH_2CH_3$

(c) $CH_3CH{=}CH_2$

(d) 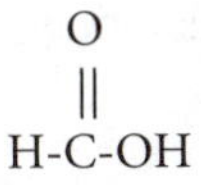
$$O$$
$$||$$
$$H{-}C{-}OH$$

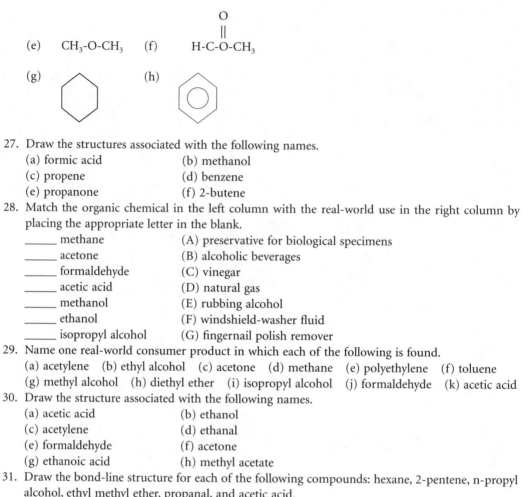

(e)    $CH_3-O-CH_3$    (f)    $H-\overset{\overset{O}{\|}}{C}-O-CH_3$

(g)    (h)

27. Draw the structures associated with the following names.
    (a) formic acid                    (b) methanol
    (c) propene                        (d) benzene
    (e) propanone                      (f) 2-butene

28. Match the organic chemical in the left column with the real-world use in the right column by placing the appropriate letter in the blank.
    _____ methane              (A) preservative for biological specimens
    _____ acetone              (B) alcoholic beverages
    _____ formaldehyde         (C) vinegar
    _____ acetic acid          (D) natural gas
    _____ methanol             (E) rubbing alcohol
    _____ ethanol              (F) windshield-washer fluid
    _____ isopropyl alcohol    (G) fingernail polish remover

29. Name one real-world consumer product in which each of the following is found.
    (a) acetylene   (b) ethyl alcohol   (c) acetone   (d) methane   (e) polyethylene   (f) toluene
    (g) methyl alcohol   (h) diethyl ether   (i) isopropyl alcohol   (j) formaldehyde   (k) acetic acid

30. Draw the structure associated with the following names.
    (a) acetic acid                    (b) ethanol
    (c) acetylene                      (d) ethanal
    (e) formaldehyde                   (f) acetone
    (g) ethanoic acid                  (h) methyl acetate

31. Draw the bond-line structure for each of the following compounds: hexane, 2-pentene, n-propyl alcohol, ethyl methyl ether, propanal, and acetic acid.

32. Name the following compounds.
    (a) $CH_3COOH$       (b) $HCHO$   (c) $HCOOCH_3$
    (d) $CH_3OCH_2CH_3$    (f) $CH_3CH_2OH$

33. Draw the functional groups that are associated with each of the following nitrogen-containing compounds and name a simple compound in each classification.
    (a)  Amide                         (b) Amine
    (c)  Isocyanate                    (d) Nitrile

34. Write the structure of the addition polymer that can be derived from the following alkene monomer.

$$CH_2=\overset{\overset{\displaystyle}{|}}{\underset{Cl}{CH}}$$

35. Name three addition polymers and list one corresponding consumer product that is associated with each.

36. How is the addition reaction of alkenes related to the formation of addition polymers?

37. What is a polymer? Differentiate between an addition polymer and a condensation polymer and give examples of each.

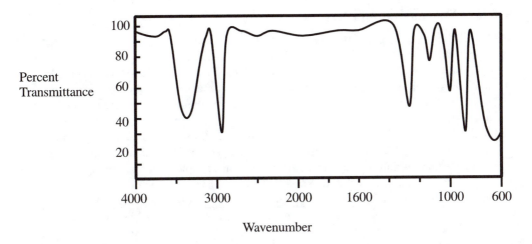

**FIGURE 6.16**  Infrared spectrum for Question #43.

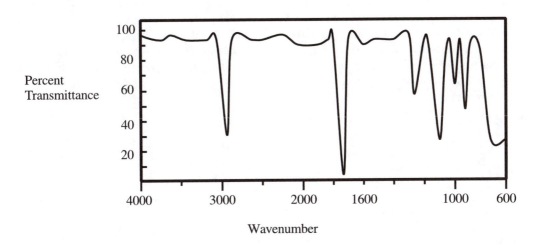

**FIGURE 6.17**  Infrared spectrum for Question #44.

38. Why must two functional groups be present in a molecule participating in a reaction that forms a condensation polymer?

39. Draw a simple diagram of an infrared spectrometer. Explain how it works.

40. Describe what is meant by the term "% transmittance." How is this determined? How is it used in infrared spectrometry?

41. Define what is meant by a "wavenumber." How is this related to wavelength? How are wavenumbers used in infrared spectrometry?

42. Draw a simple example of an infrared spectrum. Label the axes and discuss what types of information are provided by this spectrum about the compound that is being analyzed.

43. Consider the infrared spectrum in Fig. 6.16 and explain why it could be the spectrum of butyl alcohol, but not methyl ethyl ketone.

44. Consider the infrared spectrum in Fig. 6.17. Based on your knowledge of infrared spectra, into what classifications discussed in this chapter might the compound be placed. Explain.

## For Class Discussion and Reports

45. Look at the following Internet site and report on its usefulness in understanding polymers and their applications in everyday life as well as their structure and chemistry.
    http://www.homeworkcentral.com/Top8/files.htp?fileid=64345&use=hc (as of 6/4/00)

46. Many of the various shapes and sizes of consumer products made of polymers are created by a process called "extrusion." Find out as much as you can about extrusion and extruders and write a report.

47. In Section 6.12, we indicated that there are many classifications of organic compounds that have not been discussed in this text. Select one of these, perhaps found in an organic chemistry text, and report on the nature of the functional group, some simple/important examples, what specific uses compounds in the classification have, and how they are manufactured. If they are biologically important, or occur naturally in biological systems, report on the specifics.

48. Methyl isocyanate, an isocyanate that we mentioned in Section 6.11.2, is used in the production of the popular insecticide Sevin™ and is an extremely toxic gas. In 1984, a leak of this gas from a chemical plant in Bhopal, India, resulted in the death of over 1800 people. Find out as much as you can about this tragic accident, perhaps from the following Internet site, and write a report. http://www.bhopal.com/FactSheet.html (as of 6/4/00)

49. A common use of infrared spectroscopy is in identifying what types of functional groups might be present in an unknown compound. Some examples of this were given earlier where it was shown how specific bands are produced for such groups as alcohols, aliphatic hydrocarbon chains, and aromatic hydrocarbons. Look up more information on infrared spectroscopy and on where the bands for other types of functional groups typically appear. Report your findings to the class. Use this information to discuss why infrared spectroscopy is so useful in helping to identify organic compounds.

# 7

# Making and Using Measurements

## 7.1   Introduction

There are many, many occasions in our lives when we must make measurements on the numerous and varied systems that exist. For example, when you decide to buy a new carpet for your living room, you must measure the dimensions of the room so that you can know how much carpet to buy. When a cooking recipe calls for two cups of sugar, you must be able to make this measurement so that the cookies or pies you are making have the desired taste. When you buy an ice cream cone and it is guaranteed to have four ounces of ice cream, you would like the ice cream store clerk to have a scale so that you can be assured of having the promised amount. When you pump air into your bicycle tire, you measure the pressure so that you know if you have pumped in the correct amount.

Chemists, chemistry laboratory technicians, and chemical process operators also make measurements, and in fact, the accurate recording of measurements and the calculation of the results of their work and/or the conclusions they draw from these measurements can be exceedingly important. The importance of the knowledge and use of measurements in a chemistry laboratory is magnified by the fact that often some very critical decisions are made based on such measurements. Examples include the measurement of an illegal drug in the urine of a race horse, the measurement of nitrate in a drinking water supply, and the measurement of the active ingredient in a pharmaceutical preparation. Often our health, our economic well-being, our integrity, indeed our very lives depend on accurate measurements being made in a chemistry laboratory. In this chapter, the fundamentals of making and using laboratory measurements are discussed.

# Why Study This Topic?

The bulk of the work that chemistry professionals do can be described as "analytical." This means that chemistry professionals spend a rather significant amount of time analyzing samples that represent large bulk systems. Examples include analysis of environmental water for pollutants, analysis of pharmaceutical products for active ingredients, and analysis of manufactured chemicals for purity. Such analyses require making measurements on the samples. These measurements utilize laboratory measuring devices varying from the very simple, such as weighing devices, to the complex, such as computer-assisted measurements. In any case, a basic understanding of accurate measurement techniques is fundamental to a chemist's or chemistry technician's job and ultimately to his/her career.

*For Class Discussion: Consider the chemical process industries or government agency laboratories in your area and discuss examples of chemical analysis that might be performed there. What bulk systems must be analyzed and what must they be analyzed for?*

## 7.2   Reading a Measuring Device

The most fundamental rule regarding the use of any measuring device is that the device should always be read to its optimum capability with regard to precision. Precision is reflected in the number of digits in a measurement—the more digits this measurement has, the more precise it is. For example, the measurement 345.2 inches is more precise than the measurement 345 inches. If it is possible to obtain the first digit to the right of the decimal point from a device, then that digit should be recorded; if it is possible to obtain the second digit to the right of a decimal point, then that digit should be recorded, etc. Thus, a laboratory worker should endeavor to obtain the most digits possible from whatever device is being used whether it is digital or nondigital. This means recording all the digits shown if the device is digital and recording all digits that are known with certainty and then estimating the final digit if the device is nondigital. This statement is highlighted in Fig. 7.1.

> **Fundamental Rule of Measurement:**
>
> **When recording a reading from a digital measuring device, record all digits shown.  When recording a reading from a nondigital measuring device, write down all the digits that are known with certainty plus one digit that is estimated.**

**FIGURE 7.1**   A fundamental rule when using any measuring device.

**FIGURE 7.2**   Examples of readings on a digital device. The correct readings are (a) 23.4 units, and (b) 23.41 units.

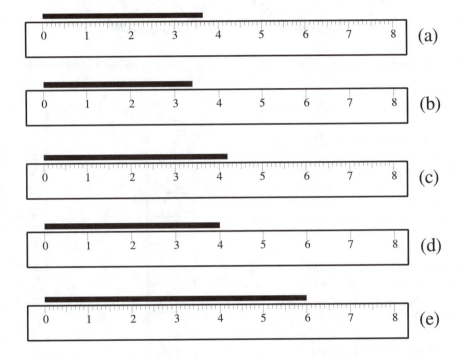

**FIGURE 7.3**   Several illustrations of the application of the basic rule of measurement for nondigital devices as stated in Fig. 7.1. The correct readings are (a) 3.67 units, (b) 3.3 units, (c) 4.20 units, (d) 4.0 units, and (e) 6.00 units.

This rule is not too difficult to apply if the device is digital. Examples are shown in Fig. 7.2. The rule for a nondigital device is illustrated in Fig. 7.3 for a simple length measurement, such as one you might make with an ordinary ruler. The length of the bold black line in Fig. 7.3(a) is between 3 and 4 units. It would be expressed as "three point something." The three is known with certainty. A closer look reveals that the length is in fact between 3.6 and 3.7, or "three point six something." Both the three and the six are known with certainty. Between the sixth and seventh graduation lines after the 3, there are no graduation lines. Thus, the second digit to the right of the decimal point would be the estimated digit and would perhaps be a seven. Thus, the measurement would correctly be recorded as 3.67 units.

If there were no graduation lines between the 3 and the 4, as in Fig. 7.3 (b), then the estimated digit would be the first digit to the right of the decimal point and the measured length would be 3.3 units. When the measurement falls exactly on a graduation line, the recorded reading must still reflect the precision of the measuring device. Thus, the correct reading in Fig. 7.3 (c) is 4.20 units, rather than 4.2 units. Similarly, the correct readings in Figs. 7.3(d) and (e) are 4.0 units and 6.00 units, respectively, rather than simply 4 units or 6 units. Examples of other measuring devices with their correct readings are shown in Figs. 7.4 and 7.5.

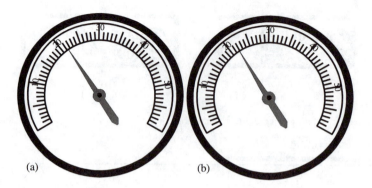

(a)                                    (b)

**FIGURE 7.4**  Additional illustrations of the application of the basic rule of measurement for nondigital devices as stated in Figure 7.1. The correct reading in (a) is 21.4 units. The correct reading in (b) 22.0 units, not 22 units.

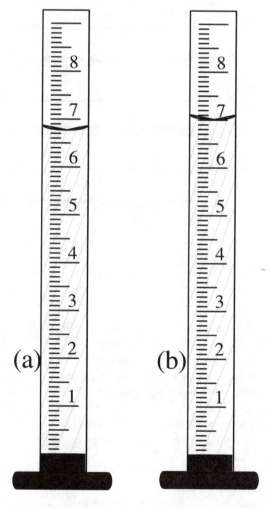

The measuring device illustrated in Fig. 7.5 merits additional comment. This device is known as a **graduated cylinder** and is used for measuring volumes of liquids, such as water and water solutions. The surface of the liquid, called the **meniscus**, is curved due to the fact that the glass molecules attract water molecules with greater force than do other water molecules. The question of exactly where in the vicinity of the meniscus to take the reading becomes important. The rule is that the position of the *bottom of the meniscus* is what is read and this is true of all volume measuring devices for liquids. Thus, the correct readings are those indicated.

## 7.3  Significant Figures

As stated in the last section, the number of digits in a given number indicates its precision, more digits implying greater precision. Unfortunately, there are occasions in which some of the digits in a given number are not included in the count of total digits. In other words, not all the digits found in a given number may be "significant." This problem involves only any zeros that may be in the number. Sometimes zeros are a part of the measurement, and therefore *are* significant, but sometimes zeros are present merely to locate the decimal point (called "placeholders") and are therefore *not* significant. For example, if the number 430,000 is given as the population of a city, it is not likely that all six digits reflect the true precision of the count. It is most likely that only the 4 and the 3 are significant and that the zeros are present only to locate the decimal point. The zeros in the number 0.0082 are likewise not significant because they are present only to show that the decimal

**FIGURE 7.5**  Illustration of measurement of liquid volume using a graduated cylinder. (a) 6.75 units, (b) 7.00 units.

---

*Chemistry Professionals at Work*             *CPW Box 7.2*

# SIZES OF GRADUATED CYLINDERS

The graduated cylinders drawn in Fig. 7.5 are similar to those in the laboratory that hold a maximum of 10 milliliters (mL) of liquid. Graduated cylinders that have a capacity for 25 mL, 50 mL, 100 mL, 250 mL, 500 mL, and 1000 mL are available. These cylinders are wider (have a larger diameter) as the capacity increases. A consequence of this is that the number of significant figures (precision) of the measurement decreases as the capacity increases. A general rule of thumb is that a laboratory worker should use a cylinder that provides the maximum precision for the measurement to be made. For example, if a laboratory procedure calls for a measurement of 150 mL, a laboratory worker should use the cylinder with a 250 mL capacity and not the 500 mL size or the 1000 mL size.

*For Homework: Ask your instructor to show you the graduated cylinders mentioned above. Then place water in the 100 mL cylinder so that the meniscus rests exactly on the 91 mL line. Record this reading with the correct number of significant figures. Now pour this water sequentially into the 250 mL, 500 mL, and 1000 mL cylinders and record the reading in each to the correct number of significant figures. Compare the precisions of the cylinders and write a brief report on your findings.*

---

point is located two places to the left of the 8. In some of these cases, the number can be expressed in a form known as *scientific notation* (described in Section 7.3.2) in order to show the significance, or lack of significance, of the placeholders.

In addition, when a number is used in a calculation and the calculator presents a long string of digits as the answer, it is important to realize that not all of these digits can be significant. A calculated answer cannot be more precise than the numbers that were used in the calculation. For example, when a calculator is used to multiply 7.51 by 149, the number found in the calculator display is 1118.99. If all the digits in this answer were counted as significant, then it would be a much more precise number than either number used in the calculation, and that is not possible. In cases such as this, a process known as rounding is used to reduce the count of digits to the proper number.

We now discuss in more detail, the concepts of rounding, scientific notation, and some special rules involving significant figures.

## 7.3.1 Rounding

*Rounding* is the removal of digits in a number when it is necessary to express the number with fewer digits. This process involves counting digits in the number beginning from the left. When the desired number of digits has been counted, any digits further to the right must be dropped. In this process, the final counted digit may increase by one. This occurs if the set of numbers immediately to the right would indicate that the number is closer to the next higher number. For example, if 8.264 is to be rounded to 3 digits, the last digit to be counted, beginning from the left, is the 6. The very next number is a 4. The number 8.264 is closer to 8.26 than to 8.27, and so the correctly rounded number is 8.26. If the number to be rounded to

3 digits is 9.1793, counting from the left, the last digit to be counted is the 7. Rounding to this digit gives us 9.17. However, 9.1793 is closer to 9.18 than to 9.17, and thus the correctly rounded number is 9.18.

The above discussion covers the vast majority of possibilities that may occur. The two instances that it does not cover are (1) the case in which there is just one number to the right of the last significant digit and that number is a 5, or (2) the case in which the first digit to the right is a 5 and the 5 is followed by any number of zeros. For example, if the number 2.65 is to be rounded to 2 digits, we find that 2.65 is just as close to 2.6 as it is to 2.7. Another example is when 7.21500 is to be rounded to 3 digits. The number 7.21500 is just as close to 7.21 as it is to 7.22.

The rule that is most often used to solve this last problem is the ***even/odd*** *rule*. If the last digit to be retained is odd, then we increase it by one. If the last digit to be retained is even, then we keep it the same. Applying this rule in the above two examples, 2.65 would round to 2.6 and 7.21500 would round to 7.22. There is no theoretical principle that defines the even/odd rule. It simply means that we recognize that the original number is just as close to the higher number as it is to the lower number, and thus in half the instances we increase the last digit to be retained by one while in the other half we do not.

## 7.3.2   Scientific Notation

In science, we often encounter extremely large numbers and extremely small numbers. For example, in Chapter 8 we will encounter a number that is so large that there are 20 zeros after the last nonzero digit before the decimal point is found. The number is

$$602200000000000000000000.$$

and is known as Avogadro's number.

In this type of situation, to make the number more manageable, we move the decimal point to the left such that it is immediately to the right of the first digit shown. Of course, each time the decimal point is moved by one digit, we are dividing the number by 10. In order to avoid altering the numerical value of the number, we must multiply the resulting number by 10. Thus, we have the following example.

$$723{,}000 = 72{,}300 \times 10$$

$$= 7{,}230 \times 100$$

$$= 723 \times 1000$$

Because $10 = 10^1$, $100 = 10^2$, $1000 = 10^3$, etc., the powers of 10 are used to shorten the expression of the number, as shown below.

$$723{,}000 = 72{,}300 \times 10^1$$

$$= 7{,}230 \times 10^2$$

$$= 723 \times 10^3$$

Moving the decimal to the right of the first digit, the correct way to express this number in scientific notation is as follows.

$$7.23 \times 10^5$$

The exponent of the 10 in this notation is the number of places we moved the decimal point. Expressing Avogadro's number in scientific notation, we would have the following number.

$$6.022 \times 10^{23}$$

---

*Chemistry Professionals at Work*                                    *CPW Box 7.3*

# MOISTURE IN SOIL SAMPLES

Chemistry professionals in analysis laboratories that routinely determine the moisture content of soil samples weigh the sample held in a suitable container both before and after a drying step. The difference in the two weights is the weight of the moisture. The weight of the sample is determined by subtracting the weight of the empty container from the weight of the container with the moist soil. The percent of moisture in this sample is then determined by dividing the weight of the moisture by the weight of the sample before drying and then multiplying by 100 to calculate percent. The calculation thus requires two subtraction steps, followed by a division, followed by a multiplication. In order to report the results of this experiment to the desired precision, the rules of significant figures in recording data and in calculating the results must be applied.

*For Homework: The following Website provides a procedure for the determination of moisture in soil. Study this procedure, and/or any other your instructor may suggest, and write a report describing the reasons for measuring moisture in soil, the methodology used, and the calculations needed.*

http://www.bsss.bangor.ac.uk/moisture4.htm (as of 6/4/00)

---

### 7.3.3  Rules for Significant Figures

It is important to be able to count the number of significant figures, or *sig figs*, in a given number. One reason that we express the answer to a calculation with the correct number of sig figs is to show how precise it is (see above example). Another is so that we can know how precise a measurement was when looking back some time after it was made without having to recall the nature of the measurement or the particular device that was used. Counting the sig figs in a number is not necessarily a trivial matter. The rules for counting sig figs is given in Table 7.1.

Determining the number of sig figs in the answer to a calculation requires its own set of rules. These are given in Table 7.2.

## 7.4  Dimensional Analysis

After making a measurement in a particular unit, it may be necessary to express it in some alternate unit. For example, consider a homeowner who is installing a new carpet in his living room. He measures the length of the room and finds it to be 15 feet. When he goes to the carpet store he discovers that the price of the carpet is given on a "per yard" basis. In order to know what the carpet costs, he needs to express the measured number, 15 feet, in yards. There are three feet in one yard. The question is "Do I multiply 15 by 3 or divide 15 by 3 in order to calculate the number of yards?"

While this is a very simple problem that most people may quickly respond to correctly by saying that 15 feet is 5 yards (i.e., divide 15 by 3), to some it may not be so simple. There are similar problems that chemistry technicians may encounter in which a more desirable method of making the decision of "Do

**TABLE 7.1**   Rules for Counting the Significant Figures in a Given Number

1. Any non-zero digit is significant.
   
   | | Example: | 916.3 | 4 sig figs |
   |---|---|---|---|

2. Any zero located between 2 significant figures is significant
   
   | | Example: | 1208.4 | 5 sig figs |
   |---|---|---|---|

3. Any zero to the left of nonzero digits is not significant unless it is also covered by Rule #2.
   
   | | Example: | 0.00345 | 3 sig figs |
   |---|---|---|---|

4. Any zero to the right of nonzero digits and also to the right of a decimal point is significant.
   
   | | Example: | 34.10 | 4 sig figs |
   |---|---|---|---|

5. Any zero to the right of nonzero digits and to the left of a decimal point and not covered by Rule #2 may or may not be significant, depending on whether the zero is a placeholder or was actually part of the measurement. Such a number should be expressed in scientific notation to avoid any confusion.
   
   | | Example: | 430 | don't know |
   |---|---|---|---|

   ($4.3 \times 10^2$ or $4.30 \times 10^2$ would be better ways to express this number, depending on whether there are 2 or 3 sig figs in the number.)

---

**TABLE 7.2**   Rules for Determining the Significant Figures (sig figs) in the Answers to Calculations

1. The answer to a multiplication or division has the same total number of sig figs as in the number with the least sig figs used in the calculation. Rounding to decrease the count of digits, or the addition of zeros to increase the count of digits, may be necessary.

   | Example 1: | $4.3 \times 0.882$ | = | 3.7926 (calculator answer) |
   |---|---|---|---|
   | | | = | 3.8 (answer with correct number of sig figs) |
   | Example 2: | $\dfrac{0.900}{0.2250}$ | = | 4 (calculator answer) |
   | | | = | 4.00 (answer with correct number of sig figs) |

2. The correct answer to an addition or subtraction has the same number of digits to the right of the decimal point as in the number with the least such digits that is used in the calculation. Once again, rounding to decrease the count of digits, or addition of zeros to increase the count of digits, may be necessary.

   | Example 1: | $24.992 + 3.2$ | = | 28.192 (calculator answer) |
   |---|---|---|---|
   | | | = | 28.2 (answer with correct number of sig figs) |
   | Example 2: | $772.2490 - 0.049$ | = | 772.2 (calculator answer) |
   | | | = | 772.200 (answer with correct number of sig fig) |

3. When several calculation steps are required, no rounding is done until the final answer is determined.

   | Example: | $\dfrac{3.026 \times 4.7}{7.23}$ | = | 1.9363762 (calculator answer with premature rounding) |
   |---|---|---|---|
   | | | = | 1.9 (incorrect answer due to premature rounding) |
   | | | = | 1.9671093 (calculator answer correctly determined) |
   | | | = | 2.0 (correct answer) |

4. When both Rules #1 and #2 apply in the same calculation, follow Rules #1 and #2 in the order they are needed while also keeping Rule #3 in mind.

   | Example: | $(3.22 - 3.034) \times 5.61 =$ | 1.0659 (calculator answer with premature rounding) |
   |---|---|---|
   | | = | 1.1 (incorrect answer due to premature rounding) |
   | | = | 1.04346 (calculator answer correctly determined) |
   | | = | 1.0 (correct answer) |

5. Conversion factors that are exact numbers have an infinite number of sig figs.

   | Example: | There are exactly 3 feet per yard. How many feet are there in 2.7 yards? |
   |---|---|
   | | $2.7 \times 3$ = 8.1 (two sig figs in correct answer, not one) |

---

I multiply or divide?" is important. For this reason, we use the concept of ***dimensional analysis*** in order to remove all guesswork from this kind of problem.

At the heart of the dimensional analysis method is a technique known as the *cancellation of units*. The basic rule is: When the same unit appears in both the numerator and denominator, these units "cancel."

This cancellation of units is analogous to the cancellation of numbers when the same number appears in both the numerator and denominator.

$$\frac{\cancel{4}}{5} \times \frac{21}{\cancel{4}} = \frac{21}{5}$$

For the conversion of feet to yards in the earlier example, we have two setup choices, (a) and (b), as follows.

$$(a)\ \frac{15\ \text{feet}}{} \times \frac{3\ \text{feet}}{1\ \text{yard}} = ? \qquad (b)\ \frac{15\ \text{feet}}{} \times \frac{1\ \text{yard}}{3\ \text{feet}} = ?$$

In (a) we would be multiplying by 3, while in (b) we would be dividing by 3. The correct setup is (b) because the unit "feet" appears in both the numerator and denominator (and thus cancels) and the unit "yard" remains in the numerator, indicating that the answer to the calculation will have the "yard" as its unit.

$$\frac{15\ \cancel{\text{feet}}}{} \times \frac{1\ \text{yard}}{3\ \cancel{\text{feet}}} = 5.0\ \text{yard}$$

The answer is thus 5.0 yards.

It is important to remember to follow the rules of significant figures in these unit conversion problems. Notice in this case that there are *exactly* three feet per yard (infinite number of significant figures—Rule 5, Table 7.1). Thus, the answer to the calculation should have the same number of significant figures as in the number 15, which is 2. The correct answer is therefore 5.0 yards and not 5 yards.

An equation that gives the relationship between one unit and another is called an ***equality***. The equality used in the feet/yard example above is shown below.

$$3\ \text{feet} = 1\ \text{yard} \tag{7.1}$$

A ***conversion factor*** is one of two ratios that are derived from an equality. The following are the two possible conversion factors derived from the feet/yard equality.

$$\frac{3\ \text{feet}}{1\ \text{yard}} \qquad \frac{1\ \text{yard}}{3\ \text{feet}}$$

A ***conversion*** is a calculation in which a conversion factor (or several conversion factors) is (are) used to convert from one unit to another, as in the above feet/yard example.

A useful approach to solving conversion problems is to write down the number, with units, that is to be converted followed by a times sign ($\times$). Leave a blank for the conversion factor(s) and follow this with an equal sign ($=$). To the right of the equals sign, leave a blank for the final numerical answer and follow this with the desired units. The decision of what conversion factor or factors to use will be obvious by the fact that you can see what units are to be cancelled (denominator) and what units are to be retained (numerator). The following example illustrates this procedure.

## Example 7.1
How many inches are there in 852 feet?

## Solution 7.1
The set–up for this problem would be:

$$\frac{852\ \text{feet}}{} \times \frac{\phantom{xxxx}}{\phantom{xxxx}} = \phantom{xxxx} \text{inches}$$

The equality that would apply is: 1 foot = 12 inches. The two conversion factors derived from this equality are:

$$\frac{1 \text{ foot}}{12 \text{ inches}} \quad \text{and} \quad \frac{12 \text{ inches}}{1 \text{ foot}}$$

The conversion factor that would cancel the feet unit and retain the inch unit is the one on the right. Thus, we have the following.

$$\frac{852 \text{ feet}}{} \times \frac{12 \text{ inches}}{\text{foot}} = 10,224 \text{ inches (calculator answer)}$$

$$= 1.02 \times 10^4 \text{inches (correctly expressed answer)}$$

## 7.5   The Metric System

The measurement system utilizing the familiar units of feet, yards, ounces, pounds, gallons, quarts, etc., is known as the **English system**. Another system of measurement, the **metric system**, will be emphasized in this chapter and subsequent chapters. The **metric system** of measurement is used almost exclusively in all chemistry laboratories worldwide.[1] The main reason for this can be summed up in one word: convenience. The entire metric system is constructed such that all units are related to other units by factors that are multiples of 10. For instance, we have the meter, which is 100 centimeters; the milligram which is one-thousandth of a gram; and the liter, which is 1,000,000 microliters. Each measurement domain (length, mass, volume, etc.) utilizes prefixes to indicate what multiple of 10 is involved. Some examples of these prefixes are shown in Table 7.3.

In this chapter, we are concerned with three measurement domains: length, mass, and volume. Let us now discuss each domain, giving the relationships with the English system, and presenting the most common units used in each domain.

### 7.5.1   Length

The base unit for length in the metric system is the meter. The *meter* (abbreviated "m") is slightly longer than the yard in the English system and equals 39.37 inches (in.). Other common metric system units of length are the centimeter (cm), the millimeter (mm), the micrometer (μm), which is also known as the micron (μ), and the kilometer (km). Please refer to Table 7.3 for the meaning of the prefixes. Equalities and conversion factors for length units compared to the meter are given in Table 7.4. One additional equality that is sometimes useful is the fact that one inch is equal to 2.54 cm.

### 7.5.2  Mass (Weight)

*Mass* is the measure of the amount of a substance based on its weight. In other words, the measurement of the amount of a substance is done by measuring the effect of the Earth's gravitational attraction for it. This means that, technically, mass and weight are not the same. If you were to take a 180 pound man to the moon and measure his weight there, it would not be 180 pounds, yet his mass has not changed. However, weight, or the effect of the Earth's gravitational attraction for an object, is a good measure of mass because it is very rare indeed that anyone would measure mass

---

[1]The actual system of measurement used worldwide is called the International System of Units (in French, Systeme International d'Unites, or the SI system). Besides including all units that are considered "metric," it has been extended to include modes of measurement that were not part of the original metric system.

*Chemistry Professionals at Work* *CPW Box 7.4*

# BALANCES

The measurement of mass or weight is probably the most important measurement made in any real-world laboratory. A laboratory device for measuring mass is called a **balance**. It is similar to a bathroom scale. The object to be weighed is simply placed on the pan of the balance and the weight is read.

Just as with graduated cylinders (see CPW Box 7.2), there are a wide variety of balances with varying degrees of precision available. A **top-loading balance** is one that is capable of mass readings accurate to the second decimal place [as in Figure 7.2 (b)]. An **analytical balance** is one that is accurate to the fourth or fifth decimal places. Analytical balances are obviously very precise measuring devices—so precise that a person's fingerprint makes a difference in the measurement. Analytical balances feature a pan that is completely enclosed in a transparent case with sliding doors in order to prevent air currents from affecting the measurement.

*For Homework: Ask your instructor to provide you with a scientific supply catalog. Look under "balance" to see what is available. Write a report on the various designs, capacities, precisions, etc. that are available and what the prices are.*

**TABLE 7.3** Some Common Prefixes Used in the Metric System and Their Meanings

| Prefix | Meaning |
|--------|---------|
| Deci | One-tenth of |
| Centi | One-hundredth of |
| Milli | One-thousandth of |
| Micro | One-millionth of |
| Kilo | One thousand |

**TABLE 7.4** Units of Length with Equalities and Conversion Factors Derived from the Equalities

| Meter Compared to | Equalities | Conversion Factors | |
|-------------------|------------|--------------------|--------------------|
| Inches | 39.37 in. = 1 m | $\dfrac{39.37 \text{ in.}}{1\,m}$ | $\dfrac{1\,m}{39.37 \text{ in.}}$ |
| Centimeter | 100 cm = 1 m | $\dfrac{100\,cm}{1\,m}$ | $\dfrac{1\,m}{100\,cm}$ |
| Millimeter | 1000 mm = 1 m | $\dfrac{1000 \text{ mm}}{1\,m}$ | $\dfrac{1\,m}{1000\,m}$ |
| Micrometer (micron) | 1,000,000 μm = 1m | $\dfrac{1{,}000{,}000\,\mu m}{1\,m}$ | $\dfrac{1\,m}{1{,}000{,}000\,\mu m}$ |
| Kilometer | 1000 m = 1 km | $\dfrac{1\,km}{1000\,m}$ | $\dfrac{1000\,m}{1\,km}$ |

**TABLE 7.5**  Units of Mass with Equalities and Conversion Factors Derived from the Equalities

| Gram Compared to | Equalities | Conversion Factors | |
|---|---|---|---|
| Ounce | 28.35 g = 1 oz | $\dfrac{28.35 \text{ g}}{1 \text{ oz}}$ | $\dfrac{1 \text{ g}}{28.35 \text{ g}}$ |
| Milligram | 1000 mg = 1 g | $\dfrac{1000 \text{ mg}}{1 \text{ g}}$ | $\dfrac{1 \text{ g}}{1000 \text{ mg}}$ |
| Microgram | 1,000,000 µg = 1 g | $\dfrac{1,000,000 \text{ µg}}{1 \text{ g}}$ | $\dfrac{1 \text{ g}}{1,000,000 \text{ µg}}$ |
| Kilogram | 1000 g = 1 kg | $\dfrac{1 \text{ kg}}{1000 \text{ g}}$ | $\dfrac{1000 \text{ g}}{1 \text{ kg}}$ |

**TABLE 7.6**  Units of Volume with Equalities and Conversion Factors Derived from the Equalities

| Liter Compared to | Equalities | Conversion Factors | |
|---|---|---|---|
| Quarts | 1.057 qt = 1 L | $\dfrac{1.057 \text{ qt}}{1 \text{ L}}$ | $\dfrac{1 \text{ L}}{1.057 \text{ qt}}$ |
| Milliliters | 1000 mL = 1 L | $\dfrac{1000 \text{ mg}}{1 \text{ L}}$ | $\dfrac{1 \text{ L}}{1000 \text{ mL}}$ |
| Microliters | 1,000,000 µL = 1 L | $\dfrac{1,000,000 \text{ µL}}{1 \text{ L}}$ | $\dfrac{1 \text{ L}}{1,000,000 \text{ µL}}$ |

anywhere except on the surface of the earth. Thus, we take mass and weight to be the same and they share the same units.

The base unit of mass in the metric system is the **gram** (g). One ounce (oz) is the same as 28.35 grams. One kilogram (kg), which is 1000 grams, is the same as 2.205 pounds (lb). Other common metric system units of mass are the milligram (mg) and the microgram (µg). Table 7.5 summarizes equalities and conversion factors for mass units compared to the gram.

### 7.5.3   Volume

**Volume** is the measure of the amount of space a quantity of matter occupies. The base unit for volume in the metric system is the liter. The *liter*, abbreviated "L," is the same as 1.057 quarts and 3.785 liters is one gallon. Other common metric system units of volume are the milliliter (mL) and the microliter (µL). Equalities and conversion factors for volume units compared to the liter are given in Table 7.6.

In addition to those units shown in Table 7.6, volumes are often designated with "cubic" length units. Thus we have cubic inches, cubic feet, cubic meters, and cubic centimeters, etc. These cubic units are based on the regularly shaped object we call a cube. A cube is a solid object that has regular and equal length, width, and height dimensions. A quantity of matter that has a volume of 1 cubic foot would have a volume equal to that of a cube that is one foot long, one foot wide, and one foot high. If an object is a cube, or otherwise has regular length, width, and height dimensions (such as an object that is a rectangle of a certain length and width that has been moved through space to create the height dimension—a rectangular prism) the volume of this object may be determined by multiplying together the length, width, and height of the object.

$$\text{volume of a cubic object } = \text{ length } \times \text{ width } \times \text{ height} \tag{7.2}$$

The units would then be cubic length units. The abbreviations for such units are in.³ for cubic inches, ft³ for cubic feet, m³ for cubic meters, and cm³ for cubic centimeters. The cubic centimeter unit is also

# MEASUREMENT OF VOLUME WITH A BURET

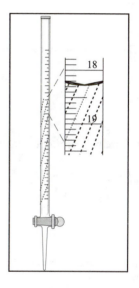

A special graduated cylinder that is often used in a lab is called a buret. A buret, pictured on the right, is different from a typical graduated cylinder in that it has a valve called a stopcock at the bottom. The liquid is dispensed by opening the stopcock rather than by inverting the cylinder. This means that the volume reading is made after the liquid has been dispensed, rather than before. Because of this, the graduation lines are inverted from those on the usual graduated cylinder. The zero-line is at the top and the graduations increase going down. This is demonstrated in the blow up on the right in the figure. The correct reading in the example shown is 18.28 mL, **not** 19.72 mL.

*For Homework: Obtain a scientific supply catalog from your instructor and look at the various designs of 50-mL burets that are available. Examine the differences that exist, such as stopcock design, cost, Class "A" vs. Class "B", etc. Write a report.*

abbreviated as "cc." In medical applications, the "cc" is the common unit for measuring volumes of liquid medications.

An additional convenience characteristic of the metric system is that the volume and length measurement domains are related in that one cubic centimeter, which is based on length measurement, is the same as one milliliter, which is strictly a volume unit.

$$1\,cm^3 = 1\,mL \tag{7.3}$$

This fact demonstrates that the volume domain of measurement was conceived based on the length domain, indicating that considerable thought and creativity were used in the process of setting up this system for the purpose of convenience. No similar such relationship exists within the English system because English system units were conceived in a completely arbitrary fashion. We will see examples of other metric system interdomain relationships later in this chapter.

## 7.5.4 Examples of Conversion Problems

Examples of English/metric and metric/metric unit conversions are given below.

*Example 7.2*

How many centimeters are there in 5.382 meters?

*Solution 7.2*

The setup is as shown below.

$$\frac{5.382 \text{ m}}{} \times \frac{}{} = \quad \text{cm}$$

The required conversion factor must have cm in the numerator and m in the denominator. Thus, the correct conversion factor, from Table 7.4, is as indicated below.

$$\frac{100 \text{ cm}}{1 \text{ m}}$$

This gives the following result.

$$\frac{5.382 \text{ m}}{} \times \frac{100 \text{ cm}}{1 \text{ m}} = 538.2 \text{ cm}$$

*Example 7.3*

How many grams does 492.7 milligrams represent?

*Solution 7.3*

The setup is as follows.

$$\frac{492.7 \text{ mg}}{} \times \frac{}{} = \quad \text{g}$$

The required conversion has g in the numerator and mg in the denominator. From Table 7.5, we find the following.

$$\frac{1 \text{ g}}{1000 \text{ mg}}$$

Thus, we have the answer given below.

$$\frac{492.7 \text{ mg}}{} \times \frac{1 \text{ g}}{1000 \text{ mg}} = 0.4927 \text{ g}$$

*Example 7.4*

How many microliters are there in 0.9174 liters?

*Solution 7.4*

The setup is as follows.

$$\frac{0.9174 \text{ L}}{} \times \frac{}{} = \quad \mu\text{L}$$

The conversion factor must have μL in the numerator and L in the denominator. This gives the following answer.

$$\frac{0.9174 \text{ L}}{} \times \frac{1,000,000 \text{ }\mu\text{L}}{1 \text{ L}} = 917400 \text{ }\mu\text{L} = 9.174 \times 10^5 \text{ }\mu\text{L}$$

## Example 7.5

How many kilometers are represented by 8274.5 inches ?

## Solution 7.5

Since we have not presented a conversion factor to transform inches directly to kilometers, let us first convert to meters and then to kilometers. This will require two conversion factors. The setup is as follows.

$$\frac{8274.5 \text{ in.}}{} \times \frac{\phantom{xxxx}}{\phantom{xxxx}} \times \frac{\phantom{xxxx}}{\phantom{xxxx}} = \phantom{xxx} \text{km}$$

The first conversion factor needed must convert inches to meters, while the second must convert meters to kilometers. The two conversion factors are as shown in the following.

$$\frac{1 \text{ m}}{39.37 \text{ in.}} \quad \text{and} \quad \frac{1 \text{ km}}{1000 \text{ m}}$$

Thus we have the answer given below.

$$\frac{8274.5 \text{ in}}{} \times \frac{1 \text{ m}}{39.37 \text{ in}} \times \frac{1 \text{ km}}{1000 \text{ m}} = 0.2101727 \text{ km} = 0.2102 \text{ km}$$

**Note:** The number 39.37 has only four sig figs in spite of the fact that it is a conversion factor (see Rule 5 in Table 7.2). No English/metric conversion factors are exact numbers and thus, all can limit the number of sig figs in the answer to the calculation in which they are used. In this case, the answer should have four sig figs. It may be possible to find a more precise number in a handbook in order to allow the measured number, rather than the conversion factor, to limit the sig figs in the answer.

## Example 7.6

How many cubic centimeters are there in 0.339 quart?

## Solution 7.6

If we are limited to the conversion factors in this text, this will require three conversion factors, one to convert from quart to liters, one to convert from liters to mL, and one to convert from mL to $cm^3$.

$$\frac{0.339 \text{ qt}}{} \times \frac{\phantom{xxxx}}{\phantom{xxxx}} \times \frac{\phantom{xxxx}}{\phantom{xxxx}} \times \frac{\phantom{xxxx}}{\phantom{xxxx}} = \phantom{xxx} cm^3$$

The conversion factors are as follows.

$$\frac{1 \text{ L}}{1.057 \text{ qt}} \quad \frac{1000 \text{ ml}}{1 \text{ L}} \quad \frac{1 cm^3}{1 \text{ mL}}$$

Thus we have the answer shown below.

$$\frac{0.339 \text{ qt}}{} \times \frac{1 \text{ L}}{1.057 \text{ qt}} \times \frac{1000 \text{ mL}}{1 \text{ L}} \times \frac{1 cm^3}{1 \text{ mL}} = 320.719 = 321 cm^3$$

# 7.6 Density and Specific Gravity

In Chapter 1, we discussed physical and chemical properties of substances. We mentioned several physical properties that have numerical values associated with them and indicated that these numerical values are found by making measurements. Two such properties are density and specific gravity. Both are helpful for determining the identity of an unknown substance because they are physical properties, meaning that only one substance is likely to have a density or specific gravity of a particular value. Thus, if that value for the unknown is determined by measurement, it can be matched to a particular known substance of that same value and identified.

## 7.6.1  Density

Density is a measure of how heavy a given volume of a substance is. For example, if you hold in your hands a lead ball and an aluminum ball of equal size, the lead ball appears to be much heavier than the aluminum ball. There are more grams of lead in the lead ball than there are grams of aluminum in the aluminum ball. We say that the lead is denser, or has a greater density, than the aluminum. This means that it has a greater mass packed into the same volume. Of course, when we compare densities, the two balls must be of equal volume. An aluminum ball the size of a baseball would obviously be heavier than one the size of a small marble. This does not mean that the larger ball has the greater density. When we speak of the density of a substance, we must not only speak of its mass, but also its volume. **Density** is therefore defined as the mass of an object per unit volume. A numerical value for density of a liquid or solid may be calculated by measuring the mass and the volume of a sample and then dividing the mass by the volume:

$$\text{Density} = \frac{\text{Mass}}{\text{Volume}} \qquad (7.4)$$

If "D" represents density, "M" represents mass, and "V" represents volume, this equation may also be written as follows.

$$D = \frac{M}{V} \qquad (7.5)$$

In the metric system, the density of a liquid or a solid is most often given as the number of grams in one milliliter, or grams per milliliter. When calculating density in this text, the mass will be expressed in grams and the volume in milliliters or cubic centimeters. The densities of some liquids and solids are given in Table 7.7.

The volumes of liquids depend on temperature. As the temperature increases, a liquid expands, meaning that the volume becomes larger even though the mass stays the same. This means that the density decreases as the temperature increases (i.e., the same number of grams are packed into a larger volume at the higher temperature). The reverse is true when the temperature decreases—the volume becomes smaller and the density increases. For this reason, the temperature must be specified when listing the densities of liquids. You will notice that the temperature is specified as 20°C (20 degrees Celsius) in Table 7.7. This is a temperature that is about the same as the temperature of a comfortable room, often called "room temperature." Concepts of temperature measurement are the subject of Section 7.7.

**TABLE 7.7**  Densities in g/ml of Some Common Liquids and Solids at 20°C.

| Substance Name | Physical State | Density (g/mL) |
|---|---|---|
| Lead | Solid | 11.34 |
| Aluminum | Solid | 2.70 |
| Iron | Solid | 7.86 |
| Gold | Solid | 19.3 |
| Water | Liquid | 0.998 |
| Ethyl alcohol | Liquid | 0.789 |
| Carbon tetrachloride | Liquid | 1.59 |

*Example 7.7*

What is the density of a solid object if it weighs 56.22 grams and occupies 6.15 milliliters?

*Solution 7.7*

$$D = \frac{M}{V}$$

$$= \frac{56.22\ g}{6.15\ mL} = 9.14\ g/mL$$

## 7.6.2 Measuring the Density of Solids Having Regular Dimensions

Measuring the mass of a solid substance is an easy task. We simply place the material on the laboratory device for measuring mass, a laboratory balance, and read the results either digital or nondigital. Measuring the volume, however, is more challenging. If the solid object is regularly shaped (i.e., if it has a shape that has regular dimensions that can be used to calculate volume), then it is a matter of measuring the dimensions and calculating the volume. Examples of shapes that have regular dimensions that can be used to calculate volume are rectangular prisms, cylinders, and spheres. Each of these has a mathematical formula for calculating the volume once the dimensions are measured. Drawings of these objects, along with the formulas for calculating their volumes, are given in Fig. 7.6.

*Example 7.8*

What is the density of a solid object in the shape of a rectangular prism if its mass is 78.28 grams and its length, width, and height are 2.11 cm, 1.86 cm, and 5.90 cm, respectively?

*Solution 7.8*

$$D = \frac{M}{V} = \frac{M}{length \times width \times height}$$

$$= \frac{78.28\,grams}{2.11\,cm \times 1.86\,cm \times 5.90\,cm}$$

$$= 3.38\,grams/cm^3$$

**FIGURE 7.6** Left to right, a rectangular prism (Volume = length $\times$ width $\times$ height), a cylinder (Volume = $\pi r^2 h$), and a sphere (Volume = $4/3\ \pi r^3$). ($\pi$ = 3.1416, r = radius, h = height).

### 7.6.3   Measuring the Density of Irregularly Shaped Solids

Irregularly shaped solids do not have regular dimensions and therefore cannot have their volumes calculated with a formula. Rather, we determine their volumes by water displacement (assuming that they do not dissolve in water). This means that the volume of water a solid displaces upon complete immersion is equal to the volume of the solid. In the laboratory, the process is as follows: a graduated cylinder in which the solid can fit is selected, a quantity of water is added (to a volume significantly less than the capacity of the cylinder, but enough so that the solid can be completely immersed), the volume is read, the solid is immersed (being careful not to splash any water out of the cylinder and not to trap any air under or within the solid), and the final volume is read. The difference in the two volumes is the volume of the solid.

*Example 7.9*

What is the density of an irregularly shaped piece of metal given the following data:
mass = 173.03 grams, water volume before immersion = 27.8 mL, water volume after immersion = 43.9 mL

*Solution 7.9*

$$D \ = \ \frac{M}{V} \ = \ \frac{173.03 \text{ grams}}{(43.9 - 27.8) \text{ mL}} \ = \ 10.7 \text{ grams/mL}$$

### 7.6.4   Measuring the Density of Liquids

Measuring the mass of solids is easy, but measuring the volume is a challenge requiring a calculation. With liquids, the opposite is true. It is a challenge to measure the mass, which usually requires a calculation, but it is easy to measure the volume. Volumes of liquids can conveniently be measured with a graduated cylinder or some similar volume-measuring device. The mass is more difficult because the liquid must be contained while on the balance pan, and thus extra mass is included in the measurement—the mass of the container. This procedure calls for the mass of the empty container to be determined before the liquid is introduced. After measuring the mass of the container and liquid together, the mass of the liquid is obtained by subtracting the mass of the empty container. It is convenient if the container is the same graduated cylinder in which the volume is read. Remember that volumes of liquids depend on temperature, so it is also prudent to measure the temperature of the liquid and report it along with the measured density.

*Example 7.10*

What is the density of a liquid if it is placed in a graduated cylinder weighing 81.42 grams to the 48.2 mL level, and the cylinder with the liquid in it weighs 133.89 grams?

*Solution 7.10*

$$D \ = \ \frac{M}{V} \ = \ \frac{(133.89 - 81.42) \text{ grams}}{48.2 \text{ mL}} \ = \ 10.9 \text{ grams/mL}$$

### 7.6.5   Use of Density as a Conversion Factor

The density of a substance is an equality from which conversion factors can be derived. For example, the density of gold (from Table 7.7) is 19.3 g/mL. In other words,

19.3 grams of gold (Au) = 1 mL of gold (Au).

---

*Chemistry Professionals at Work*            *CPW Box 7.6*

# CALIBRATION

When a chemistry professional makes a measurement, he/she depends on a measuring device to provide a correct, reliable number. This begs the question: "How can anyone be absolutely certain that a given measurement is correct?" The correctness of measurements is a constant concern to an analysis chemist. That is why a significant amount time and effort is spent on calibration.

*Calibration* is a process by which measuring devices are checked to see if they are providing accurate numbers. This is done by making measurements on known samples to see if they give the known result. An example is a balance. A set of known weights is purchased from a reliable vendor and periodically these weights are weighed on the balance. If the weight measured by the balance matches the weight the object is known to have, then the balance is said to be calibrated. If not, then the balance is taken out of service and repaired.

Some measuring devices can be electronically tweaked to bring them into calibration. Others require that a "calibration curve" be graphed so that an unknown sample can be compared to standard, or known substances.

*For Homework: How can a chemistry professional be certain that the calibration lines on a buret (see CPW Box 7.5) are positioned correctly? Devise a method of checking these calibration lines either on your own or by looking in another textbook. Write a report.*

---

This gives us the conversion factors that can be used to convert any number of milliliters of gold to grams, or any number of grams of gold to milliliters.

$$\frac{19.3 \text{ g Au}}{1 \text{ mL Au}} \qquad \frac{1 \text{ mL Au}}{19.3 \text{ g Au}}$$

In other words, density can be used to determine the volume that a certain number of grams of a substance occupies, or to determine the mass of a certain volume of a substance.

## Example 7.11
What is the mass of 7.84 mL (measured at 20°C) of ethyl alcohol?

## Solution 7.11
From Table 7.7, the density of ethyl alcohol at 20°C is 0.789 g/mL. The mass of 7.84 mL of ethyl alcohol is calculated as follows.

$$7.84 \text{ mL} \times \frac{0.789 \text{ g}}{1 \text{ mL}} = 6.18576 \text{ g} = 6.19 \text{ g}$$

*Example 7.12*

What volume will 83.224 grams of iron occupy?

*Solution 7.12*

From Table 7.7, the density of iron is 7.86 g/mL. We calculate the volume of 83.224 grams of iron as follows.

$$\frac{83.224 \text{ g}}{} \times \frac{1 \text{ mL}}{7.86 \text{ g}} = 10.588295 \text{ mL} = 10.6 \text{ mL}$$

*Sig fig note:* Density is not an exact number. The three sig figs in the density above do limit the number of sig figs in the answer. It is possible to measure the density more precisely so that it has more sig figs or to find a more precise density in the chemical literature. In that case, the measurement (83.224 g), and not the density, would limit the sig figs, in the answer. For reference, see Rules 1 and 5 in Table 7.2.

### 7.6.6  Density of Water

In Section 7.5.3 we discussed a link between the length domain of measurement and the volume domain, noting that 1 cm³ is the same as 1 mL. A link also exists between the mass and volume domains.

In Table 7.7, we see that the density of water at 20°C is equal to 0.998 g/mL. In the above discussion we noted that the density of liquids changes with temperature, decreasing with increasing temperature (as the volume expands) and increasing with decreasing temperature (as the volume contracts). The density of water is equal to 1.000 g/mL at 4°C, a temperature that is near the freezing point of water. Therefore, the link between the mass and volume domains is that 1 gram of water is the same as 1 mL of water at 4°C. No such link exists in the English system. Again, this shows the creativity and convenience that is exhibited by the metric system. Water has been used as a standard substance to create links and to set up entire domains of measurement in the metric system. We will see this again in the domains of heat and temperature measurement.

### 7.6.7  Specific Gravity

Density is useful for the identification of unknown substances or for the determination of the purity of known substances. If an impurity has a density different from the pure substance, the density of the impure substance will not match that of the pure substance. Another parameter that is also useful for this is *specific gravity*. If we divide the density of a liquid at a given temperature by the density of water at 4°C (1.000 g/mL), we get a number that is equal numerically to the density of the liquid, but with no units. This number is known as the specific gravity of the substance.

$$\text{sp.gr.} = \frac{\text{density at 20°C}}{\text{density of water at 4°C}} \tag{7.6}$$

*Example 7.13*

What is the specific gravity of ethyl alcohol given the density in Table 7.7?

*Solution 7.13*

$$\text{sp. gr.} = \frac{\text{density of ethyl alcohol at 20°C}}{\text{density of water at 4°C}}$$

$$= \frac{0.789 \text{ g/mL}}{1.000 \text{ g/mL}} = 0.789$$

Specific gravity (sp. gr.) has no units because all units cancel as shown.

The advantage of specific gravity is that it can be measured without having to make any volume measurements. However, the procedure does require the use of a device in which the volume is very accurately controlled and reproduced. A very good device for this is called a pycnometer. A *pycnometer* is a glass or metal container, holding up to 100 mL, that has a special glass stopper with a very small glass capillary tube in the center, through which the contained liquid can emerge when the cap is placed on the filled container. The contained liquid has a volume that is quite reproducible under these conditions. The specific gravity is determined by dividing the mass of the liquid required to fill the pycnometer by the mass of water (at 4°C) that is required to fill the pycnometer. The volumes of the two liquids are equal and hence do not have to be known. In order to determine the masses of these two liquids, the mass of the pycnometer filled with each liquid is determined, and the mass of the empty pycnometer is subtracted from each of these measurements.

## Example 7.14

The mass of a pycnometer filled with an unknown liquid is 93.99 grams. The mass of the same pycnometer filled with water at 4°C is 89.22 grams. The mass of the empty pycnometer is 14.39 grams. What is the specific gravity of the liquid?

## Solution 7.14

$$\text{sp. gr.} = \frac{(\text{mass of pycnometer with unknown} - \text{mass of empty pycnometer})}{(\text{mass of pycnometer with water at 4}°\text{C} - \text{mass of empty pycnometer})}$$

$$= \frac{(93.99 - 14.39)}{(89.22 - 14.39)} = 1.064 \text{ (no units)}$$

One may think it inconvenient to maintain the temperature of water at 4°C for specific gravity measurements. Actually, room temperature, or some other easily maintained temperature, can be used if desired, although the result would not be equal numerically to the density of the liquid or to the absolute specific gravity as it is defined in Eq. 7.6. Rather, this would be a specific gravity that would be measured under easily reproduced conditions. Also, some standard other than water may be used. Whatever temperature is used must be specified and maintained while making the measurement. Specific gravities are therefore often written as follows in order to specify the temperatures.

$$\text{sp. gr.}^{T_2}_{T_1}$$

In this notation, $T_2$ is the temperature of the measured liquid and $T_1$ is the temperature of the reference liquid, usually water. The specific gravity of ethyl alcohol at 20°C vs. water at 4°C would thus be expressed as follows:

$$\text{sp. gr.}^{20°\text{C}}_{4°\text{C}} \text{ of ethyl alcohol} = 0.789$$

For very precise specific gravity measurements, the temperature of the liquids held in the pycnometer (both the water and the test liquid) must be carefully controlled. This means that a device for maintaining a constant temperature must be employed. This usually consists of a water bath in which the temperature of the water in the bath is controlled with a thermostat. The pycnometers filled with the liquids are placed in the water bath before they are capped. Once the temperature is stabilized, the pycnometers are capped, and, in rapid succession, wiped off and weighed.

---

*Chemistry Professionals at Work*                    *CPW Box 7.7*

# SPECIFIC GRAVITY IN QUALITY ASSURANCE

One way to monitor the quality of liquid formulations in the pharmaceutical industry is to measure the specific gravity of samples of the product on a regular basis. The technician expects the specific gravity of such formulations to fall within a certain range. If it is found that it is outside the required range on a given day, contamination may be the cause. Contamination with substances of higher specific gravity increases the measurement, while contamination with substances of lower specific gravity decreases the measurement. Under the circumstances, the suspect product is then tested further with other techniques to determine if it is in fact contaminated and to identify what the contaminant is. The problem can thus be resolved.

*For Homework: Day-to-day quality checks such as this are often displayed graphically using what are called "control charts." Find basic information on control charting and write a report describing exactly what a control chart is, what warning limits and action limits are, what is meant by "statistical control," and examples of how control charts might indicate a problem.*

---

## 7.7 Temperature

### 7.7.1 Introduction

*Temperature* is the measure of the hotness or coldness of a body of matter. It is *not* a measure of heat, because heat is not a measure of the hotness, or coldness, but a form of energy. In this section we will discuss the English and metric system methods of measuring temperature. Both utilize a device called a thermometer. Although digital thermometers are in common use today, the traditional glass thermometer will be the type described here. This thermometer is typically a glass tube, closed on both ends and containing a liquid substance, which is either ethyl alcohol or mercury. The lower end is called the "bulb," which contains the bulk of the liquid. It is the temperature of the matter surrounding the bulb that actually determines the reading on the thermometer.

As we learned in the last section, the volume of a liquid increases as temperature increases and decreases as temperature decreases. This fact is utilized in the design of the glass thermometer. As temperature increases, the volume of the liquid inside increases and we observe it climbing in the tube. As temperature decreases, the volume decreases and we observe it lowering in the tube. A set of graduation lines on the tube allows us to monitor the level of the liquid in the tube, which then translates into the temperature of the matter surrounding the bulb (see Fig. 7.7).

### 7.7.2 Temperature Scales

The manner in which the positions of the graduation lines on a thermometer were determined for the English and metric systems was different. In the metric system, which measures temperature in Celsius degrees (°C), water is once again used as the standard substance. The temperature at which water freezes, or the temperature of a slushy mixture of crushed ice and water, was taken to be 0°C. The temperature at which water boils (at standard atmospheric pressure) was taken to be 100°C. These two temperatures define the magnitude of one Celsius degree, where each degree is one-hundredth of the length from 0 to 100 on a thermometer. This also defines the entire Celsius scale.

In the English system, we measure temperature in Fahrenheit degrees (°F). The Fahrenheit degree and the Fahrenheit scale was set up in a completely arbitrary fashion. There was no standard substance. The graduation lines were affixed on the tube with no particular standard involved. One Fahrenheit degree is very different from the Celsius degree. For example, the freezing point of water on the Fahrenheit scale is 32 degrees and the boiling point of water is 212 degrees. Thus the degree interval on the Fahrenheit scale is much smaller than the degree interval on the Celsius scale.

There is another metric system scale for temperature—the Kelvin scale. By convention of the SI system of units (see footnote under Section 7.5), the Kelvin system utilizes neither the degree symbol (°) nor the word "degree." Thus we designate a temperature as, for example, 50 K, and state it as "fifty Kelvin." The Kelvin scale is a metric system scale with one degree on the Celsius scale being equal to one unit on the Kelvin scale. The difference is that the Kelvin scale is an "absolute" scale. This means that zero on this scale is the lowest temperature that can theoretically be reached, or **absolute zero**. There is no "below zero" or negative temperature readings on this scale. Absolute zero occurs at 273.15 degrees below zero on the Celsius scale, or −273.15°C, and −459.67°F. The three scales, Celsius, Fahrenheit, and Kelvin, are shown in Fig. 7.8.

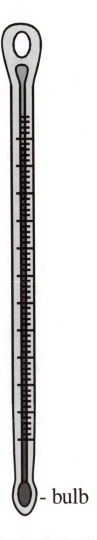

FIGURE 7.7   A drawing of a glass thermometer showing the bulb, the liquid inside, and the graduation lines.

Since one degree on the Celsius scale is the same as one unit on the Kelvin scale, the relationship between the two scales is a simple one.

$$K = °C + 273.15 \qquad (7.7)$$

or

$$°C = K − 273.15 \qquad (7.8)$$

On the Kelvin scale, the freezing point of water is 273.15 K while the boiling point of water is 373.15 K.

## Example 7.15
What is 45.2°C expressed in Kelvin?

## Solution 7.15

$$K = °C + 273.15 = 45.2 + 273.15 = 318.35 \text{ K} = 318.4 \text{ K}$$

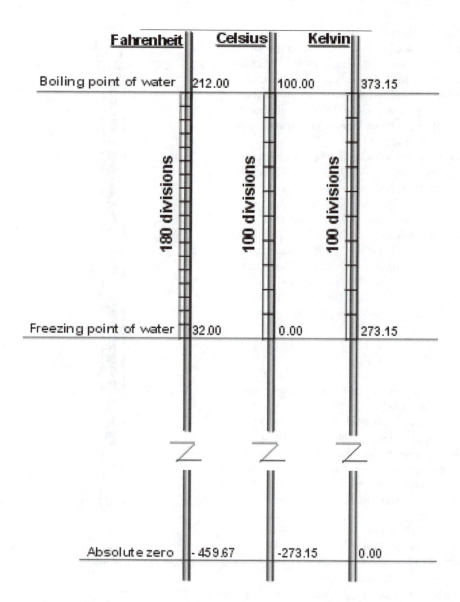

**FIGURE 7.8** A comparison of three common temperature scales, Celsius, Fahrenheit, and Kelvin.

The relationship between the Celsius scale and the Fahrenheit scale is more complicated, which is to be expected since the two scales are from two different systems of measurement. For converting from °F to °C, the following relationship is useful.

$$°C = \frac{(°F - 32)}{1.8} \tag{7.9}$$

For converting from °C to °F, Eq. (7.9) may be rearranged to give the equation shown below.

$$°F = 1.8 \times °C + 32 \tag{7.10}$$

All rules of significant figures discussed in Section 7.3 are in effect when performing calculations with these formulas.

## Example 7.16

A comfortable room temperature is 72°F. What is 72°F expressed in °C?

## Solution 7.16

Eq. (7.9) is selected for this calculation.

$$°C = \frac{(°F - 32)}{1.8} = \frac{(72 - 32)}{1.8} = 22.222222 = 22°C$$

**Sig fig note**: Both the number 32 and the number 1.8 have an infinite number of sig figs and thus never limit the number of sig figs in the answer. Rule #4 in Table 7.3 applies.

## Example 7.17

A cold day in the winter in the upper Midwest is one in which the temperature is −32°C. What is this temperature expressed in °F?

## Solution 7.17

Eq. (7.10) is selected for this calculation.

$$°F = 1.8 × °C + 32 = 1.8 × (−32) + 32 = −25.6 = −26°F$$

*Sig fig note*: The sig fig note in Example 7.16 applies here as well.

# 7.8   Viscosity

Imagine a large cylindrical container (like a test tube that is four inches in diameter) filled with honey. Now imagine dropping a heavy stainless steel ball the size of a ping-pong ball into the honey. The ball falls under force of gravity through the honey at a very slow rate, perhaps taking as long as 20 minutes to reach the bottom. Now, imagine filling the large test tube with water and doing the same experiment. The ball falls to the bottom in just a fraction of a second. The time it takes for the ball to fall is a measure of the *viscosity* of the liquid. Very thick liquids, like honey, are very viscous and have a high viscosity. Thin, watery liquids are not very viscous and have a low viscosity.

This "falling ball" method is one of many methods that can be used to measure viscosity. The most common methods that use commercially available equipment are (1) the *moving body method*, of which the "falling ball" method is one, (2) the *capillary method*, in which the time required for a given volume of the test fluid to pass through a capillary tube (a tube with an extremely narrow opening) is measured, and (3) the *rotational* method, in which a rotating metal spindle is immersed into the liquid and the resistance to the rotation is measured. Any of these devices, and any other device constructed to measure this thickness of a liquid, is called a *viscometer* or a *viscosimeter*.

The most common of the three general methods are the capillary method and the rotational method. There are three commercially available capillary viscometers. These are the Ostwald viscometer, the Cannon-Fenske viscometer, and the Ubbelohde viscometer. A Cannon-Fenske viscometer and a popular rotational viscometer are shown in Fig. 7.9.

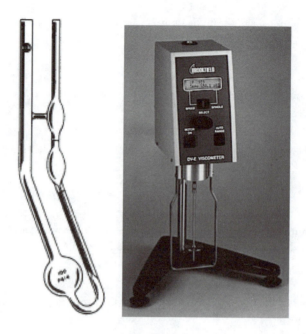

**FIGURE 7.9**   A Cannon-Fenske viscometer (left) and a popular rotational viscometer (right). (Rotational viscometer photo courtesy of Brookfield Engineering.)

In any case, it is important to recognize that viscosity depends on temperature. This means that a constant temperature water or oil bath must be used and the vessel containing the fluid being measured (in the case of the capillary viscometers, the viscometers themselves) must be immersed in it and the temperature stabilized before the measurement is made.

Absolute viscosity is measured in units of **poise** or **centipoise**. Kinematic viscosity is defined as the absolute viscosity divided by the density of the liquid. It is measured in units of **stokes** or **centistokes**. In the falling ball and capillary methods, viscosities of test liquids are determined by comparing the measured time for the test liquid to the time measured for a liquid whose viscosity is known. In the capillary methods, a **viscometer constant**, C, in centistokes (cS) per second (s), is determined. The equation for determining this constant is given below.

$$C = v/t \qquad\qquad (7.11)$$

In this equation, v is the kinematic viscosity of the known liquid in centistokes, and t is the time in seconds. This same equation, with the value of "C" calculated from the known viscosity, is used to calculate the "v" for the unknown. In the rotational method, the torque required to produce a given rotational velocity is a measure of the viscosity. Again, comparing this torque to that of a liquid with a known viscosity gives the viscosity of the test liquid.

*Example 7.18*

A given liquid has a kinematic viscosity of 15.61 centistokes. The time required for a volume of this liquid to pass through the capillary tube on a Cannon-Fenske viscometer is 14.98 seconds. The time required for the same volume of a test liquid to pass through the capillary is 12.35 seconds. What is the viscosity of the test liquid?

# THERMOGRAVIMETRIC ANALYSIS

In industrial laboratories, temperature is measured for a variety of reasons in a variety of applications. These not only include simple melting point and boiling point determinations, but any application in which temperature is a concern. We have already cited the importance of temperature when measuring density and specific gravity (see Section 7.6).

Another rather important application in which temperature is measured is in thermogravimetric analysis, or TGA. This is a technique in which the temperature and weight of a small sample of material are simultaneously monitored as the sample is being heated. A typical temperature range for this experiment would be from room temperature up to 1000 °C. When a temperature is reached at which a gaseous product of a physical or chemical change is released, a loss in weight is observed. The data usually consists of a graph of weight vs. temperature which would show one or more sharp drops in weight at key temperature values. If the moisture in the sample is being determined, for example, there would be a sharp drop in the sample weight at the boiling point of water. In addition to moisture, the method is useful for determining volatile matter, combustible material, and ash content of samples. Monitoring the temperature during the experiment is useful in order to know when the reaction of interest occurs.

*For Homework: The American Society for Testing and Materials (ASTM) publishes standard laboratory test methods for a variety of materials by a variety of techniques. Obtain a copy of the ASTM Standard Test Method for Compositional Analysis by Thermogravimetry (#E1131-98) and write a report on the use of TGA for compositional analysis.*

## Solution 7.18

First, calculating the viscometer constant, we get the following.

$$C = \frac{v}{t}$$

$$C = \frac{15.61\,cS}{14.98\ s} = 1.04206\ cS/s$$

Then, calculating v for the test liquid, we get the following.

$$v = Ct = 1.04206\ cS/s \times 12.35s = 12.87\,centistokes$$

---

*Chemistry Professionals at Work*                    *CPW Box 7.9*

# ACTIVITIES AT A STATE AGRICULTURE LABORATORY

**M**y name is Charlie Focht. I am a chemistry professional employed by the Nebraska Department of Agriculture Laboratory in Lincoln, Nebraska. I currently serve as Team Leader for the Feed, Fertilizer, and Agricultural Lime Laboratory. In addition to my supervisory duties in testing agricultural products for regulatory compliance, I am responsible for developing and monitoring the laboratory's Quality Assurance/Quality Control program. I also serve as the Department of Agriculture's Soil Survey Officer, conducting laboratory audits, and evaluating the proficiency performance of Nebraska's soil testing laboratories.

I am a member of the Oversight Committee for the North American Proficiency Testing Program, and I am the General Referee for Nutrients in Soils for the Association of Analytical Chemists International. I have taken an active role in the development of my department's data processing capabilities, and I chair its Computer Users' Group. In 1996 I was honored as Employee of the Year by the Nebraska Department of Agriculture. I have an A.A.S. degree in Environmental Laboratory Technology from Southeast Community College in Lincoln.

*For Homework: Identify a chemistry professional who works at a state government agency in your state. Interview this individual and write a report on his/her activities and on the mission of the agency.*

## 7.9   Homework Exercises

1. Discuss why measurements are important in chemistry. Give some specific examples of how measurements are used in chemistry.
2. What is the "fundamental rule of measurement"? Why is it important?
3. (a) What is the length of the bar that is being measured with the ruler in Fig. 7.10?
   (b) What is the length of the bar that is being measured with the ruler in Fig. 7.11?
4. What is the percent transmittance reading, read on the device in Fig. 7.12, to the correct number of significant figures? What volume will be delivered by the syringe shown in Fig. 7.13?
5. Summarize the rules that should be used when determining the number of significant figures that are present in a number.
6. How many significant figures (if known) are there in each of the following numbers?

| | | | |
|---|---|---|---|
| (a)  206 | (b)  210 | (c)  0.0540 | (d)  3510.0 |
| (e)  $7.30 \times 10^3$ | (f)  0.007 | (g)  1830 | (h)  4.320 |
| (i)  690.329 | (j)  50.00 | (k)  5023.4 | (l)  0.002380 |
| (m) 590 | (n)  0.0345 | (o)  $1.990 \times 10^2$ | |
| (p)  607 | (q)  607.0 | | |

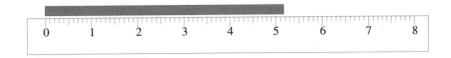

**FIGURE 7.10**   Drawing for Question 3(a).

**FIGURE 7.11**   Drawing for Question 3(b).

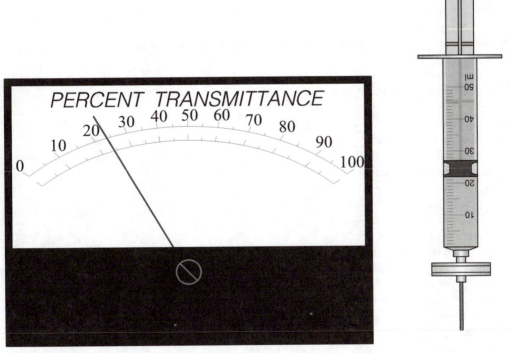

**FIGURE 7.12**   Drawing for Question 4, % transmittance reading.

**FIGURE 7.13**   Drawing of syringe for Question 4.

7. What is the difference between 4 grams and 4.00 grams?
8. What is meant by "rounding"?
9. What is the "even/odd" rule in rounding?
10. Summarize the rules that should be used when determining the number of significant figures that are to be given to the result of a calculation.
11. Perform the following calculations and express the answer in the correct number of significant figures.

   (a) $\dfrac{0.00576}{96}$

   (b) $723.5 \times 2.19$

   (c) $0.392 \times 15.934$

   (d) $98.04 + 5.6$

(e) 28.113 − 7.1130

(f) $\dfrac{(5.330 - 5.03)}{8.0023}$

(g) $6.20 \times 10^{-3} \times 2.001 \times 10^6$

(h) $3.229 \times 5.02$

(i) $3.2929 + 8.02$

(j) $5.229 \times 22$

(k) $9.20 \times 10^3 \times 8.333 \times 10^{-9}$

(l) $8.229 \times 2.84 - 0.3890$

(m) $\dfrac{5.309 \times 0.88}{9.34384}$

12. What is meant by "dimensional analysis?"
13. What is the basic rule of dimensional analysis?
14. Define each of the following terms, as related to chemical calculations.
    (a) Equality
    (b) Conversion factor
    (c) Conversion
15. State the equalities that pertain to each of the following comparisons.
    (a) inches vs. meters
    (b) micrometers vs. meters
    (c) milligrams vs. grams
    (d) quarts vs. liters
    (e) milliliters vs. liters
16. Give the conversion factors for each of the comparisons that are listed in Problem 15.
17. Derive conversion factors for each of the following changes from an English unit of measurement to a metric unit of measurement.
    (a) feet to meters
    (b) yards to kilometers
    (c) pounds to grams
    (d) quarts to milliliters
18. Derive conversion factors for each of the items listed in Problem 17, but now arranged so that the units are being converted from a metric unit of measurement to an English unit of measurement.
19. Provide the metric prefixes for each of the following items.
    (a) one-tenth of, or $10^{-1}$ times the base unit
    (b) one-millionth of, or $10^{-6}$ times the base unit
    (c) one thousand of, or $10^3$ times the base unit
    (d) one-hundredth of, or $10^{-2}$ times the base unit
20. What is the base unit of measurement for length in the metric system? What is the base unit of measurement for mass in this system? What is the base unit of measurement for volume?
21. How are the base units of volume and length related in the metric system?
22. Make the following conversions, expressing the answer with the correct number of significant figures. Show your setup for each conversion.
    (a) 2.27 kg to g        (b) 6224 mm to m
    (c) 0.482 L to $cm^3$      (d) 76.7 cm to m
23. Show your setup as well as the answers in all calculations that follow. Be sure to express the answers in the correct number of significant figures.
    (a) How many grams are represented by 0.338 kilograms?
    (b) How many L are represent by 0.23 mL?
    (c) How many cm are there in 5.4 m?
    (d) Convert 0.003380 kg to mg.
24. What is meant by "density?" How can this be measured?

25. Explain how density is different from mass?
26. How does density change with temperature?
27. Calculate the density of magnesium if 18.9 mL of magnesium weighs 32.85 grams.
28. What are three types of solids for which volumes can be easily calculated?
29. If the mass of a rectangular prism is 499.37 grams and its dimensions are length = 6.05 cm, width = 5.22 cm, and height = 18.93 cm, what is the object's density?
30. How can the volume of an irregularly shaped solid be determined?
31. Given the following data for an irregularly shaped solid, what is the object's density?
    mass of solid ..................................................................................... 54.30 g
    volume of water in graduated cylinder before immersion of solid ....... 43.6 mL
    volume of water in graduated cylinder after immersion of solid .......... 51.2 mL
32. A student measures the density of a liquid sample. The following measurements are made.
    (a) An empty graduated cylinder weighs 28.64 grams.
    (b) The liquid is added so that the volume reading is 35.4 mL.
    (c) The graduated cylinder with the liquid in it weighs 57.56 grams.
    What is the density of the liquid?
33. Convert 356 mL of a liquid to grams knowing that the density of the liquid is 1.27 g/mL.
34. What is the volume of a 33.60 gram piece of gold if the density of gold is 19.6 grams per mL?
35. If milk has a density of 1.030 g/cm³, and you need 21 grams of milk, how many milliliters of milk are needed?
36. How are the base units of mass and volume related in the metric system?
37. What is meant by "specific gravity?" How is this measured? How is it different from density?
38. What is the specific gravity of a liquid given the following data?
    Mass of pycnometer with unknown liquid:        102.45 grams
    Mass of pycnometer with water at 4°C:           101.24 grams
    Mass of empty pycnometer:                       15.62 grams
39. Write the equations that are needed for each of the following conversions.
    (a) Degrees Fahrenheit converted to degrees Celsius
    (b) Degrees Celsius converted to Kelvin
    (c) Degrees Celsius converted to Fahrenheit
    (d) Degrees Fahrenheit converted to Kelvin
40. (a) What is 45.3°C expressed in °F?
    (b) What is 134°F expressed in °C?
    (c) What is 23.18°C expressed in K?
    (d) What is 401.3 K expressed in °C?
41. (a) Convert 342 K to °C.
    (b) Convert 68°F to °C.
    (c) Convert 92°C to K.
    (d) Convert 14.3°C to °F.
42. Define what is meant by "viscosity." Describe how this is measured.
43. What is the viscosity constant for a viscometer if a liquid with a kinematic viscosity of 17.23 centistokes gives a time of 11.28 seconds?
44. What is the viscosity of a test liquid if the time measured for 21.70 mL is 13.82 and the viscosity constant is 1.768 cS/s?

## For Class Discussion and Reports

45. Ask your instructor to show you all the different styles of balances in your laboratory area. Ask him/her to give you some instructions for their use. Then, weigh an object such as your ball point pen on each one and record the weights, making sure that each has the proper number of significant figures. Compare the precisions of the balances and write a report on your findings.

46. In this chapter, we discussed the measurement and/or calculation of the following physical properties of materials: density, specific gravity, temperature (applicable to melting point, boiling point, etc.), and viscosity. Select two other physical properties from the following list and search for information on each. Write a report on each, defining what each is, what type of material each is applicable to, and how each is measured. Your instructor may be able to suggest references.
    List: optical rotation, refractive index, particle size, tensile strength, and hardness of metals and alloys

47. The "calibration curve" (mentioned in CPW Box 7.6) is the same as the "standard curve" discussed in Chapter 3 (Section 3.10). Citing an example of a spectroscopic method, explain how this can be a method of calibration as defined in CPW Box 7.6.

48. Obtain a copy of the ASTM standard method for measuring kinematic viscosity, #D 445. Summarize the method described there for calibration of a capillary viscometer and the general method for determining the kinematic viscosity of a liquid.

# 8 Moles and Stoichiometry

## 8.1   Introduction

We defined chemistry in Chapter 1 as the study of the composition, structure, and properties of matter and the changes that matter undergoes. As our study of chemistry has unfolded in Chapters 1–7, we have learned that a great deal of symbolism (element symbols, formulas of compounds, chemical equations, etc.) is used to describe chemical composition, structure, properties, and changes. We have also learned that some aspects of chemistry, particularly aspects of structure, can be very abstract, requiring us to think in terms of the extremely small invisible particles called atoms, molecules, and ions, and even the smaller subatomic particles called protons, neutrons, and electrons. The special language of chemists and chemistry technicians, however, is certainly not limited to the abstract or to very small quantities. In this chapter, we move from the realm of the abstract and unseen to the realm of observable and measurable realities. It is time to apply our symbolism and our knowledge to quantities of matter that are observable and measurable in the laboratory and to the very large quantities of matter that are generated and consumed every day in the chemical industry. The expression, measurement, and calculation of quantities of matter in compounds and in reactions is called stoichiometry.

*Chemistry Professionals at Work*                                    *CPW Box 8.1*

# WHY STUDY THIS TOPIC?

The mole is as basic to the language of the chemist and the chemistry technician as the dozen is to the language of the baker or the egg producer. It is the fundamental standard of quantity by which all substances are described, measured, compared, reacted, and produced. The mole is the quantity standard by which the composition of matter is expressed. It is the quantity standard by which the concentration levels of solutions are measured and expressed. It is the standard by which quantities of substances consumed and produced by chemical change are accurately represented. The study of the composition of matter, the measurement and preparation of solutions, and the measurement and calculation of the quantities of chemicals consumed and produced by chemical change are all typical tasks performed by chemists, chemistry laboratory technicians, and process operators on a daily basis. These professionals thus need to know the basis and uses of this most fundamental of units.

*For Class Discussion: Why is the dozen a useful and convenient quantity for egg producers and bakers? Why do you think a similar collection of items (the mole) is useful, convenient, and necessary for chemists?*

## 8.2   Atomic Weight Revisited

Let us begin with a brief review of the facts concerning atomic weight as discussed in Chapter 1. These facts are as follows:

1. A number, called the atomic weight, is listed in the periodic table. In most tables it is the number found near the bottom of each square in the periodic table below the symbol for the element (see the periodic table in Chapter 1).
2. The atomic weight of any element is the average weight of all the weights of all the atoms that exist naturally for that element. The weight of any one atom is the sum of the masses or weights of all the particles found in the atom, which are the protons, neutrons, and electrons.
3. The unit associated with these masses is the atomic mass unit (amu). One atomic mass unit is equal to approximately $1.66 \times 10^{-24}$ grams.
4. The mass of a single proton and the mass of a single neutron are each approximately 1 amu and the mass of an electron is approximately 0.00055 amu.
5. Weights of the atoms of a given element vary due to the existence of isotopes, which are atoms of a given element that differ due to differing numbers of neutrons in the nucleus.

A beginning chemistry student may conclude that an atomic weight should be approximately a whole number equal to the mass number, the sum of the protons and neutrons in an atom, since each of these weighs approximately 1 amu and since the weight of an electron is approximately zero amu. Points 2 and 5 above dispel this notion, however. A casual look at the periodic table reveals that atomic weights are anything but whole numbers.

We have used the word "approximately" in our references to the weights of protons and neutrons with respect to the amu. The question may arise as to what standard *is* used to define the amu.

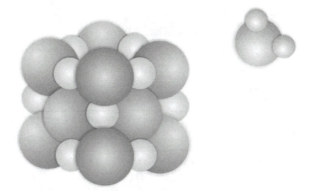

**FIGURE 8.1** Left, an example of a 3-dimensional array of ions (no molecule). Right, an example of a molecule, the fundamental unit of a molecular compound.

The answer is that the amu is defined as one-twelfth the mass of the isotope of carbon that has six protons and six neutrons, the carbon-12 isotope. From this it is easy to see why the masses of the proton and the neutron are *approximately* equal to 1 amu since it is carbon-12's six electrons that create the discrepancy.

## 8.3   Molecular Weight and Formula Weight

While we have dealt with the atomic weights of elements here and in Chapter 1, we have not dealt with the weights of molecules and formula units, which are the fundamental particles of compounds. First, by way of review, it is important to distinguish between molecular (or covalent) compounds and ionic compounds. Molecular compounds are those that consist of distinct individual particles called molecules. Molecules interact with one another in a sample of a molecular compound, but only in a rather mild way. Ionic compounds are those that consist of nondistinct particles that are most appropriately referred to as formula units, because a formula represents the simplest ratio between the cations and anions that constitute the compound (and not a distinct particle). Ionic compounds, on the atomic scale, consist of a three-dimensional array of cations and anions, all of which are strongly interacting with each other. Thus there is no such particle as a molecule in these cases. Figure 8.1 depicts the difference between these two kinds of compounds.

Molecular weight, or the weight of a single molecule in amu, is thus strictly applicable to molecular compounds. A more general term would be *formula weight*, or the weight of one formula unit. This latter term is descriptive of both molecular compounds and ionic compounds. In this text, formula weight will be the term used to describe the weight of the fundamental unit of all compounds.

The formula weight is determined by simply adding together the atomic weights of all the elements in the formula. Some examples are given below.

*Example 8.1*

What is the formula weight of ethyl alcohol, $C_2H_6O$?

*Solution 8.1*

$$2(C) = 2(12.011) = 24.022 \text{ amu}$$
$$6(H) = 6(1.008) = 6.048 \text{ amu}$$
$$1(O) = 1(15.999) = 15.999 \text{ amu}$$

Total:    Formula Weight = 46.069 amu

*Example 8.2*

What is the formula weight of calcium nitrate, $Ca(NO_3)_2$?

*Solution 8.2*

$$1(Ca) = 1(40.078) = 40.078 \text{ amu}$$
$$2(N) = 2(14.007) = 28.014 \text{ amu}$$
$$6(O) = 6(15.999) = 95.994 \text{ amu}$$

Total:    Formula Weight    = 164.086 amu

## 8.4   The Mole

How do we solve the problem of these fundamental units of matter being too small to measure in the laboratory, as suggested in Section 8.1? Scientists have defined a quantity of matter called the ***mole*** as a "substitute" entity for the atom, molecule, or formula unit. The mole *is* measurable and observable in the laboratory. It is simply the atomic or formula weight expressed in grams rather than amu. In other words, for elements the atomic weight as listed in the periodic table is not only the average number of amu per atom, it is also the average number of grams per mole of the element. For compounds, the formula weight as calculated from the atomic weights in the table is not only the average number of amu per molecule or formula unit, but it is also the average number of grams per mole of the compound. Examples of one mole of various elements and compounds are presented in Fig. 8.2.

### 8.4.1   Atomic and Formula Weight—The Number of Grams per Mole of an Element or a Compound

One result of this apparently simple concept is that the atomic weight can be used to interconvert grams and moles. Since the units of atomic weight, as defined above include both grams and moles, it can be used as a conversion factor to convert grams to moles and vice versa. The dimensional analysis would appear as follows for the conversion from grams to moles.

$$\text{grams} \times \frac{\text{mole}}{\text{grams}} = \text{moles} \tag{8.1}$$

**FIGURE 8.2**   A photograph showing one mole of, left to right: water, sodium chloride, copper, and sugar.

---

*Chemistry Professionals at Work* *CPW Box 8.2*

# MOLES AND INDUSTRIAL PROCESSES

T he concept of the mole translates tiny atoms, molecules, and formula units into quantities that are observable and measurable in the laboratory—quantities exemplified in the photograph in Fig. 8.2. But the concept of the mole is also very important to the thousands of industrial-scale chemical manufacturing processes that are in daily operation. Very large masses of elements and compounds are utilized and produced in these processes, and they are governed by the same rules as atoms and molecules, and by the same rules as laboratory quantities such as those pictured. For example, consider ammonia, $NH_3$. The fact that there is one atom of nitrogen and three atoms of hydrogen combined in one molecule of ammonia dictates the weight ratio of hydrogen and nitrogen that must come together to make ammonia, whether it is on the atomic scale, on the laboratory scale, or on the scale of the industrial process in which tons of nitrogen and hydrogen are needed and tons of ammonia are produced. The mole can be thought of as a "scaled-up" atom or molecule and it assists chemistry professionals in the lab, as well as process operators in the industrial plants to understand and measure the quantities involved.

*For Class Discussion: In large scale industrial processes, pounds rather than grams are often used to express weights of raw materials and chemicals. Discuss what this might mean in terms of atomic weights, formula weight, and the mole. What do you think a lb-mole is?*

---

and as follows for conversion from moles to grams.

$$\text{moles} \times \frac{\text{grams}}{\text{mole}} = \text{grams} \tag{8.2}$$

Some examples follow.

*Example 8.3*

How many moles does 17.3 grams of silver represent?

*Solution 8.3*

$$17.3 \text{ grams} \times \text{------} = \quad \text{moles}$$

From the periodic table, the atomic weight of silver is 107.87 grams per mole. So the problem is solved as below.

$$17.3 \text{ grams} \times \frac{\text{mole}}{107.87 \text{ grams}} = 0.160 \text{ moles}$$

*Example 8.4*

How many grams is 569.2 moles of water?

*Solution 8.4*

$$569.2 \text{ grams} \times \text{————} = \quad \text{grams}$$

The formula weight of water, $H_2O$, is needed.

$$2(H) = 2(1.008) = 2.016$$
$$1(O) = 1(15.999) = 15.999$$

Total:     Formula Weight = 18.015 grams per mole

Thus we have the following.

$$569.2 \text{ moles} \times \frac{18.015 \text{ grams}}{\text{mole}} = 1.025 \times 10^4 \text{grams}$$

## 8.4.2  Avogadro's Number—The Number of Particles per Mole

The mole is a concept of convenience similar to the dozen used for eggs and donuts. It is much more convenient for a baker to think in terms of a dozen donuts and for the egg producer to think in terms of a dozen eggs, than just one donut or one egg. The difference here is that the number of atoms, molecules, or formula units that make up one mole is an enormous number, not something easy to visualize like twelve. The number of atoms or molecules required for 107.9 grams of silver or 18.0 grams of water (the atomic or formula weights as in the above examples), or for one mole of anything, is $6.022 \times 10^{23}$. To illustrate the size of this number, one author[1] has described the mole in terms of garden peas. One mole of garden peas would be a pile so large that it would cover the entire U. S. to a depth of 6 km. This number has come to be known as **Avogadro's number**, named after **Amadeo Avogadro** (1776–1856), an Italian physicist. In the early 1800s, Avogadro hypothesized about the number of molecules of a gas in a certain volume of the gas (see Avogadro's Law, Chapter 9).

Avogadro's number can be used as a conversion factor to convert atoms, molecules, or formula units to moles or moles to atoms, molecules, or formula units. The dimensional analysis for converting moles to atoms is

$$\text{moles} \times \frac{6.022 \times 10^{23} \text{ atoms}}{\text{mole}} = \text{atoms} \qquad (8.3)$$

and for converting atoms to moles it is as follows.

$$\text{atoms} \times \frac{\text{mole}}{6.022 \times 10^{23} \text{atoms}} = \text{moles} \qquad (8.4)$$

---

[1]Ralph H. Petrucci, *General Chemistry Principles and Modern Applications*, 4th Edition, Macmillan Publishing Company (1985).

The dimensional analysis for converting moles to molecules or formula units and vice versa is similar. Some examples follow.

## Example 8.5
How many atoms are contained in 12.77 moles of titanium?

## Solution 8.5

$$\frac{12.77 \text{ moles}}{} \times \frac{6.022 \times 10^{23} \text{ atoms}}{1 \text{ mole}} = 7.690 \times 10^{24} \text{ atoms}$$

## Example 8.6
How many moles does 7.82 X $10^{26}$ molecules of formaldehyde represent?

## Solution 8.6

$$\frac{7.82 \times 10^{26} \text{ molecules}}{} \times \frac{1 \text{ mole}}{6.022 \times 10^{23} \text{ molecules}} = 1.30 \times 10^{3} \text{ moles}$$

Notice in Examples 8.5 and 8.6, while titanium and formaldehyde were specified, the identity of the element or compound did not enter into the solution of the problem. There are $6.022 \times 10^{23}$ atoms, molecules, or formula units in one mole of a chemical. Thus, this conversion factor is universal. It does not matter what the chemical is. Note that this is *not* the case with the use of atomic or formula weight as the conversion factor. These conversion factors (the atomic or formula weight) are specific to the chemical in question.

# 8.5   Percent Composition of Compounds

A certain percentage of the mass of a compound is attributable to the masses of the elements of which it is composed. For example, a certain percentage of the mass of a sample of ammonia ($NH_3$) is attributable to the mass of the nitrogen that it contains. A certain percentage of a sample of isopropyl alcohol ($C_3H_8O$) is attributable to the mass of the oxygen that it contains. We can also say that a certain percentage of the mass of a compound is attributable to the mass of a grouping of elements that are found in the formula, such as a polyatomic ion, or water of hydration. Thus, a certain percentage of the mass of a sample of ammonium chloride ($NH_4Cl$) is due to the ammonium ion, or a certain percentage of the mass of a sample of copper sulfate penthydrate ($CuSO_4 \cdot 5H_2O$) is due to water.

## 8.5.1   Percent Composition from Laboratory Data

Exactly what percentage of the mass of a sample of a compound is due to the individual elements, or groupings of elements, present can be determined experimentally in some cases. For example, the percentage of oxygen in $KClO_3$ can be determined by heating a weighed sample with a burner. A chemical reaction occurs in which the oxygen is freed and a weight loss occurs that is due to the oxygen being released. In much the same way, the percentage of water in a hydrate can be determined by heating a weighed sample of the hydrate with a burner. In this case, the water is released and there is a weight loss due to the loss of the water. The percentages can then be determined from the weight data and the mathematical definition of percent. Some examples follow.

## Example 8.7

What is the percent of oxygen in a sample of $KClO_3$, if the experiment described is performed and the sample before heating weighed 0.6692 grams while the sample after heating weighed 0.4109 grams?

## Solution 8.7

The calculation of the percent of an element in a compound is derived from the general equation for percent.

$$percent = \frac{part}{whole} \times 100$$

If the part is a weight and the whole is a weight, then we have the following.

$$percent = \frac{weight\ of\ part}{weight\ of\ whole} \times 100$$

Applying this to the percent of an element in a compound, we have the equation below.

$$percent = \frac{weight\ of\ element}{weight\ of\ compound} \times 100$$

In this case, the weight of the element oxygen is the weight loss of the sample,

$$percent = \frac{weight\ of\ oxygen}{weight\ of\ KCLO_3} \times 100 = \frac{weight\ loss\ of\ sample}{weight\ of\ sample} \times 100$$

and thus we get the percent of oxygen.

$$percent\ of\ oxygen = \frac{(0.6692 - 0.4109)}{0.6692} \times 100 = 38.60\%$$

## Example 8.8

What is the percent of water in a hydrate given the following data:

Weight of empty crucible ............................. 15.2294 grams

Weight of crucible containing hydrate ........ 16.1049 grams

Weight of crucible after heating .................. 15.8806 grams

## Solution 8.8

Following the same logic as in Example 7, we have the following.

$$percent\ of\ water = \frac{weight\ of\ water}{weight\ of\ sample} \times 100 = \frac{(16.1049 - 15.8806)}{(16.1049 - 15.2294)} \times 100$$

$$= 25.62\%$$

---

# CONCEPT OF PERCENT

Example 8.7 presents a basic discussion of the concept of percent. Despite the fact that this concept is covered in elementary school math, it sometimes causes confusion and frustration among chemistry students and chemistry professionals because it appears in so many different aspects of chemistry. It is most often used to quantitatively describe the composition of material substances (compounds, mixtures, laboratory samples, solutions, etc.). But it may also be used to describe the amount of light transmitted by a material sample (Section 7.10.2) or the yield of a product of a chemical reaction (Section 8.10). It may help to keep Example 8.7, or the discussion regarding solution composition (Section 10.6.1) in mind in order to assist with your understanding of percent whenever you encounter it.

*For Homework: Find as many references to percent as you can in this textbook and indicate what the "part" and the "whole" is for each.*

---

## 8.5.2 Percent Composition from Formulas and Atomic and Formula Weights

Percents of elements or groupings of elements in compounds can also be determined directly from the formulas of the compounds and the atomic and formula weights without doing any experimental work. This method, in fact, would have no experimental error associated with it, and thus, the results may be considered to be the theoretically correct percentages to which experimentally determined percentages may be compared. Of course, the formulas of the compounds must be known in order to use this method. This method is illustrated in the following examples.

### Example 8.9
What is the percent of oxygen in ethyl acetate, $C_4H_8O_2$?

### Solution 8.9
First the formula weight is needed.

$$4(C) = 4(12.011) = 48.044$$
$$8(H) = 8(1.008) = 8.064$$
$$2(O) = 2(15.999) = 31.998$$

Total: Formula Weight $= 88.106$ grams per mole

We then utilize the equation for percent and plug in the weights from the above calculation:

$$\text{Percent of oxygen} = \frac{\text{weight of oxygen}}{\text{weight of compound}} \times 100 = \frac{31.998}{88.106} = 36.318\%$$

## Example 8.10

What is the percent composition of sodium carbonate, $Na_2CO_3$?

## Solution 8.10

Percents of all three elements in the formula, Na, C, and O, are determined and reported just as the individual percent was in Example 8.9.

$$2(Na) = 2(22.990) = 45.980$$
$$1(C) = 1(12.011) = 12.011$$
$$3(O) = 3(15.999) = 47.997$$

Total:    Formula Weight $= 105.988$ grams per mole

$$\%Na = \frac{46.980}{105.988} \times 100 = 44.326\% \text{ Na}$$

$$\%C = \frac{12.011}{105.998} \times 100 = 11.332\% \text{ C}$$

$$\%O = \frac{47.997}{105.998} \times 100 = 45.285\% \text{ O}$$

Using the method from Example 8.9, the percentages of any and all elements in a compound may be determined and collectively reported. By definition, this collective report is defined as the ***percent composition*** of a compound.

The individual percents determined by the method of Example 8.10 should add to 100%. Occasionally, due to effects of the rounding of the atomic weights, they may not add to exactly 100%.

In a way similar to Example 8.10, the percent of a grouping of elements may be determined by this method, and once again, the result is a theoretically correct percent to which experimentally determined results may be compared.

## Example 8.11

What is the percent of water in $CuSO_4 \cdot 5H_2O$?

## Solution 8.11

First the weight of the five water units:

$$10(H) = 10(1.008) = 10.080$$
$$5(O) = 5(15.999) = 79.995$$

Total:    Weight of water units $= 90.075$

Then the formula weight of the compound is calculated.

$$1(Cu) = 1(63.546) = 63.546$$
$$1(S) = 1(32.066) = 32.066$$
$$4(O) = 4(15.999) = 63.996$$
$$5(H_2O) = 90.075 \text{ (from above)}$$

Total:    Formula Weight $= 249.683$ grams per mole

Finally, the percent of water is calculated.

$$\text{Percent of water} = \frac{90.075}{249.683} \times 100 = 36.076\% \ H_2O$$

# 8.6   Derivation of Empirical Formulas

While it is clear that percent composition can be derived from formulas, the reverse is also true. Empirical formulas, or formulas in which the subscripts are in their lowest terms (e.g., acetylene, $C_2H_2$, has an empirical formula of CH) can be derived from percent composition data. Empirical formulas may also be derived from laboratory data. Knowing what elements are in the compound and having some data about the relative amounts of these elements in the compound, we can determine the subscripts and the empirical formula.

As we learned in Chapter 1, the subscripts represent the number of atoms of each element in the formula. In the formula $H_2O$, for example, we know that the "2" subscript means that there are two atoms of hydrogen in a molecule of water. We also know that the lack of a subscript with the oxygen means that there is one atom of oxygen in one molecule of water. If we consider moles instead of atoms or molecules, we can say that the "2" subscript means that there are *two moles* of hydrogen in *one mole* of water, and the lack of the subscript means that there is *one mole* of oxygen in *one mole* of water. If we are able to translate the available information about a compound, whether it is percent composition or laboratory data, into *moles* of the elements involved, we can come up with the empirical formula.

## 8.6.1   Empirical Formulas from Percent Composition

Percent composition represents the number of grams of each element in every 100 grams of the compound. Grams can be converted to moles by dividing by the atomic weight. Subscripts can then be determined by comparing the number of moles of one element relative to another. If the number of moles of one element is twice the number of moles of another, then the subscript of the first element will be twice the other. The formal step-by-step procedure for converting percent composition data into subscripts is given in Table 8.1.

*Example 8.12*

What is the empirical formula of the compound that is 56.6% potassium, 8.7% carbon, and 34.7% oxygen?

*Solution 8.12*

**Step 1**: The percent figures translate directly to grams. So we have 56.6 grams of potassium, 8.7 grams of carbon, and 34.7 grams of oxygen.

**Step 2**: These grams are now converted to moles.

$$56.6 \ \text{grams K} \times \frac{\text{mole K}}{39.098 \ \text{grams K}} = 1.45 \ \text{moles K}$$

$$8.7 \ \text{grams C} \times \frac{\text{mole C}}{12.011 \ \text{grams C}} = 0.72 \ \text{moles C}$$

$$34.7 \ \text{grams O} \times \frac{\text{mole O}}{15.999 \ \text{grams O}} = 2.17 \ \text{moles O}$$

**Step 3:**

$$K: \frac{1.45}{0.73} = 1.99 = 2 \quad C: \frac{0.72}{0.73} = 0.99 = 1 \quad O: \frac{2.17}{0.73} = 2.97 = 3$$

The empirical formula is $K_2CO_3$.

## 8.6.2 Empirical Formulas from Laboratory Data

Chemical formulas may be determined by experiment. When a chemist or chemistry technician synthesizes (creates by chemical reaction) a new compound and does not know the empirical formula of the compound in advance, the formula may be determined from the weights of the individual elements of which the new compound is composed as measured in the various synthesis steps used in the laboratory procedure. The process involves the same calculation process as in Table 8.1, but without Step 1. As in Examples 8.7 and 8.8, one or more element weights may need to be determined by difference. Let us look at a simple example.

*Example 8.13*

A technician synthesized chromium chloride in the laboratory by reaction of $Cr_2O_3$ with $CCl_4$ vapors at an elevated temperature. If an amount of $Cr_2O_3$ equivalent to 0.935 grams of chromium metal were used and 2.848 grams of the chromium chloride formed, what is the empirical formula of the chromium chloride?

*Solution 8.13*

The weight of the chromium is known (0.935 grams) and the weight of the chlorine can be calculated by subtracting the weight of the chromium from the weight of the chromium chloride.

$$\begin{aligned} \text{weight of chlorine} \; &= \; \text{weight of compound} - \text{weight of chromium} \\ &= \; 2.848 - 0.935 \\ &= \; 1.913 \text{ grams} \end{aligned}$$

**TABLE 8.1**   The Procedure for Determining the Formula of a Compound When Given Percent Composition Data

1. Assume that we have 100 grams of the compound. The % composition data then translates to grams of each element.
2. Convert the number of grams of each element to moles by dividing each by the atomic weight of the element.
3. Divide each number of moles by the one that is the smallest number. If the result for each is a whole number, then these whole numbers are the subscripts in the formula.
4. If the result for each in Step 3 is not a whole number, multiply each by 2. If the result for each is now a whole number, then these whole numbers are the subscripts in the formula.
5. If the result for each in Step 4 is not a whole number, multiply each result from Step 3 by 3, then by 4, then by 5, etc., until whole numbers are obtained for the subscripts.

Note: This is one time that the rules for significant figures are unimportant since the subscripts always consist of one digit regardless of the number of significant figures in the data. It is often acceptable, then, to judge a "near whole number" as a whole number to avoid ridiculously large numbers as subscripts.

# ELEMENTAL ANALYSIS BY COMBUSTION

$\mathbf{A}$s we learned in Chapter 6, all organic compounds contain carbon and many contain nitrogen, sulfur, and other nonmetals covalently bonded to the carbon chain. The percent of these elements in natural and synthetic organic materials, or samples containing organic materials can be important determinations in industrial and environmental laboratories. A method often used for this is **combustion analysis**. This technique involves combusting the sample in a high temperature furnace (up to 1000°C) so that inorganic oxides of these elements, e.g., $CO_2$, NO, $SO_2$, are formed. These gases are then flushed from the furnace with an inert gas, detected, and measured by any one of a variety of methods. These combustion/analysis units are called **total carbon analyzers** (TOC), nitrogen analyzers, sulfur analyzers, etc. The TOC determinations are important for the analysis of environmental water, especially municipal and industrial wastewater. The analysis for nitrogen in animal feeds or other protein-containing products and analysis for sulfur in petroleum products are primary applications.

*For Homework: Search the Internet and other sources for specific information regarding the detection of the gaseous oxides in combustion analysis and write a report.*

We can now proceed with Step 2 of Table 8.1

**Step 2:**

$$0.935 \text{ grams Cr} \times \frac{\text{mole Cr}}{51.996 \text{ grams Cr}} = 0.0180 \text{ moles Cr}$$

$$1.913 \text{ grams Cl} \times \frac{\text{mole Cl}}{35.453 \text{ grams Cl}} = 0.05396 \text{ moles Cl}$$

**Step 3:**

$$\text{Cr:} \frac{0.0180}{0.0180} = 1 \qquad \text{Cl:} \frac{0.05396}{0.0180} = 3.00 = 3$$

The empirical formula is $CrCl_3$.

## 8.6.3 Empirical Formulas vs. Molecular Formulas

The subscripts in a molecular formula are some whole number multiple of the subscripts in the empirical formula. For example, the empirical formula of benzene is CH while the molecular formula is $C_6H_6$. The whole number multiple in this case is 6. The formula weight is the same whole number multiple of the empirical weight (the formula weight calculated from the empirical formula). In the case of benzene, the

*Chemistry Professionals at Work* *CPW Box 8.5*

## Determining Percent Composition and Formulas: Mass Spectrometry

In the modern research laboratory, chemists and chemistry technicians are sometimes involved in the synthesis of compounds that never existed before. For example, pharmaceutical companies are often involved in developing new drugs to combat bacteria and bacterial infections* and other health maladies. Often these compounds have very complicated formulas. To determine or verify the identity of a new compound, they often use an instrument called a ***mass spectrometer***. The mass spectrometer is a sophisticated, computerized instrument that utilizes a high-powered stream of electrons to divide the molecules of the compound into charged fragments. The detection of the masses and charges of the individual fragments, using a special magnet, detector, and computer that comprise the instrument, allows these laboratory professionals to piece together the individual molecule and determine its percent composition and formula. The percent composition and chemical formula are characteristics that then aid in the identification of the compound.

*See, for example, Stinson, Stephen C., *Chemical and Engineering News,* September 23, 1996, page 75.

*For Homework: Ask your instructor to suggest a reference giving an elementary discussion of mass spectrometry. Write a report on this technique emphasizing the kind of data obtained and how it is interpreted.*

empirical weight is 13.0 grams per mole and the molecular formula is 6 × 13.0 or 78.0 grams per mole. If the empirical formula is given and the molecular formula is desired, the formula weight must be known. The whole number multiple can then be calculated and the subscripts in the molecular formula determined.

### *Example 8.14*

The empirical formula of "borazole" is $BNH_2$. If the formula weight of borazole is 80.4 grams per mole, what it the molecular formula?

### *Solution 8.14*

First, calculate the empirical weight.

$$\begin{aligned} 1(\text{B}) &= 1(10.811) = 10.811 \\ 1(\text{N}) &= 1(14.007) = 14.007 \\ 2(\text{H}) &= 2(1.008) = 2.016 \\ \hline \text{Total:} \quad \text{Empirical Weight} &= 26.834 \text{ grams per mole} \end{aligned}$$

The whole number multiple is then,

$$\frac{80.4}{26.834} = 3$$

and the molecular formula is $B_3N_3H_6$.

## 8.7 Chemical Equations Revisited

The use of balanced chemical equations to deal with quantities of chemicals involved in chemical reactions is a very important aspect of stoichiometry. Let us briefly review the basics of chemical equations.

Chemical equations were introduced in Chapter 2 as an expression of the details of what occurs in a chemical reaction. For example, in the equation

$$CaCO_3(s) + 2HCl\,(aq) \rightarrow CaCl_2(aq) + H_2O\ell + CO_2(\uparrow) \tag{8.5}$$

$CaCO_3$ and $HCl$, the chemicals to the left of the arrow, are the reactants or the chemicals that are brought together to undergo reaction; $CaCl_2$, $H_2O$, and $CO_2$, the chemicals to the right of the arrow, are the products or the chemicals formed in the reaction; and the optional symbols "(s)", "(aq)," and "($\uparrow$)" are examples of symbols that provide information about the physical state and appearance of these chemicals.

We also dealt with balancing chemical equations in Chapter 2 in order to demonstrate the compliance of all chemical reactions with the Law of Conservation of Mass. Mass can neither be created nor destroyed in a chemical reaction and that means that the number of atoms of any element found in the reactants before reaction must also be found in the products after reaction, and vice versa. These facts can be easily expressed in the chemical equation with the use of balancing coefficients. Hence, besides being an expression of *what* chemicals are involved, and their physical state, chemical equations can also be an expression of *how much* of each chemical is involved. For example, in the following equation

$$N_2 + 3H_2 \rightarrow 2NH_3 \tag{8.6}$$

the balancing coefficients tell us that one molecule of $N_2$ and three molecules of $H_2$ react to give two molecules of $NH_3$. Recognizing that the mole is a substitute expression for a quantity of atoms or molecules, and that the mole represents a quantity analogous to the atom or molecule that is observable and measurable in the laboratory (as discussed in Section 8.4), we can also now state that *one mole* of $N_2$ and *three moles* of $H_2$ react to give *two moles* of $NH_3$. Thus, the chemical equation becomes an instrument through which observable and measurable quantities of matter entering into chemical reactions can be expressed and calculated.

## 8.8 Calculations Based on Equations

How can these observable and measurable quantities of matter be calculated? The balancing coefficients play the central role. Let us look at a simple example to illustrate the point, comparing the moles of oxygen that will react to a given number of moles of hydrogen.

$$2H_2 + O_2 \rightarrow 2H_2O \tag{8.7}$$

This equation states that two moles of $H_2$ react with one mole of $O_2$. It is a two to one ratio. There are twice as many moles of hydrogen involved as there are moles of oxygen. If four moles of hydrogen were to react, two moles of oxygen would be involved. If ten moles of hydrogen were used, five moles of oxygen would react. If 0.924 moles of hydrogen were involved, 0.462 moles of oxygen would be needed. It is always a 2:1 ratio of moles of hydrogen to moles of oxygen. Thus, the ratio of moles of one chemical to the moles of another is involved, and this ratio is determined from the coefficients in the balanced equation. Grams may also be involved, as we will see below.

### 8.8.1  Mole-to-Mole Conversions

The concepts of dimensional analysis may be applied to the above calculation exercise and is very useful in avoiding mistakes, particularly if the ratio is not as straightforward as 2:1. Using the above reaction as the example, the equality from which a conversion factor is derived is

$$1 \text{ mole of } O_2 = 2 \text{ moles of } H_2 \tag{8.8}$$

and the conversion factors are as shown below.

$$\frac{1 \text{ mole } O_2}{2 \text{ moles } H_2} \quad \text{and} \quad \frac{2 \text{ moles } H_2}{1 \text{ mole } O_2} \tag{8.9}$$

Such conversion factors can be used to convert moles of one chemical in an equation to moles of another.

### Example 8.15

How many moles of $Fe_3O_4$ can be made from 0.559 moles of Fe according to the following?

$$3\,Fe + 2O_2 \rightarrow Fe_3O_4$$

### Solution 8.15

There is one mole of $Fe_3O_4$ involved for every three moles of Fe. We want the moles of Fe to cancel and the units of the answer to be moles of $Fe_3O_4$.

$$0.559 \text{ moles Fe} \times \frac{1 \text{ mole } Fe_3O_4}{3 \text{ moles Fe}} = 0.186 \text{ moles } Fe_3O_4$$

Example 8.15 is an example of a "mole-to-mole" conversion—one in which moles of one chemical involved in a chemical reaction are converted to moles of another. The conversion factor is always the mole ratio, as determined from the balanced equation. A graphical picture of this is shown in Fig. 8.3.

It should also be pointed out that "A", the chemical for which an amount is given, can be a reactant or product, and "B", the chemical for which an amount is sought, can also be either a reactant or product. In the illustration of hydrogen, oxygen, and water, two reactants were involved, while in Example 8.15, one was a reactant and the other was a product. In any case, the mole ratio is always set up as

$$\frac{\text{sought}}{\text{given}}$$

so that the moles unit of the given chemical cancels and the moles unit of the chemical being calculated remains.

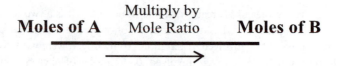

**FIGURE 8.3**   A graphical picture of the conversion of the moles of one chemical (A) to moles of another (B) involved in a chemical reaction with the use of the mole ratio.

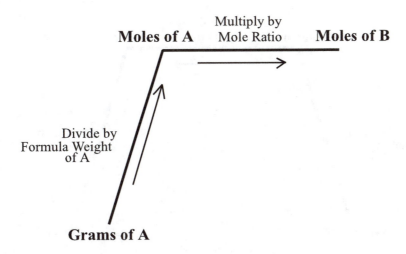

**FIGURE 8.4(a)**   A graphical representation of a gram-to-mole conversion.

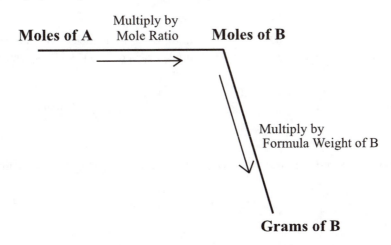

**FIGURE 8.4(b)**   A graphical representation of a mole-to-gram conversion.

## 8.8.2   Mole-to-Gram and Gram-to-Mole Conversions

We have learned that grams of an element or compound can be easily converted to moles and that moles can be easily converted to grams. The atomic or formula weight is the conversion factor in either case. Because of this, we can expand our discussion of calculations based on equations to include grams of the chemical sought or given. In other words, given grams of one chemical involved in a chemical reaction, we can calculate moles of another chemical, or given moles of one chemical involved in a chemical reaction, we can calculate grams of another. We thus have "gram-to-mole" conversions and "mole-to-gram" conversions, each requiring two conversions, or steps. Graphical representations of gram-to-mole and mole-to-gram conversions are presented in Fig. 8.4.

### Example 8.16
How many moles of oxygen, $O_2$, can be expected to form from 0.528 grams of $KClO_3$ according to the following?

$$2KClO_3 \rightarrow 2KCl + 3O_2$$

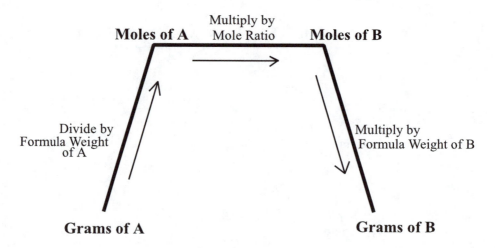

**FIGURE 8.5**    A graphical depiction of a gram-to-gram conversion.

*Solution 8.16*

This is a gram-to-mole conversion. Grams of $KClO_3$ must be converted to moles before we can use the mole ratio from the equation to get moles of oxygen. The formula weight of $KClO_3$ is 122.548 grams per mole.

$$0.528 \text{ grams } KClO_3 \times \frac{\text{mole } KClO_3}{122.548 \text{ grams } KClO_3} \times \frac{3 \text{ mole } O_2}{2 \text{ moles } KClO_3} = 0.00646 \text{ moles } O_2$$

*Example 8.17*

How many grams of Mg are needed to react with 1.392 moles of HCl according to the following?

$$Mg + 2HCl \rightarrow MgCl_2 + H_2$$

*Solution 8.17*

This is a mole-to-gram conversion. Moles of HCl are converted to moles of Mg using the mole ratio from the equation. The moles of Mg are then converted to grams using the atomic weight of magnesium.

$$1.392 \text{ moles HCl} \times \frac{1 \text{ mole Mg}}{2 \text{ mole HCl}} \times \frac{24.305 \text{ gram Mg}}{1 \text{ mole Mg}} = 16.91 \text{ grams Mg}$$

## 8.8.3  Gram-to-Gram Conversions

Finally, a calculation involving equations may be a gram-to-gram conversion. That is, given grams of one chemical involved in a chemical reaction, we can determine how many grams of another are involved. Such a calculation requires three steps: conversion of grams to moles, conversion of moles of one chemical to moles of the other, and conversion of these moles to grams. The graphical representation of this is given in Fig. 8.5.

*Example 8.18*

How many grams of $Al_2O_3$ can be obtained from 0.500 grams of Al according to the following reaction?

$$4Al + 3O_2 \rightarrow 2Al_2O_3$$

## Solution 8.18

This is a gram-to-gram conversion. Given grams of Al, how many grams of $Al_2O_3$ can form? Thus the three steps are needed. The formula weight of $Al_2O_3$ is 101.961 grams per mole.

$$0.500 \text{ g Al} \times \frac{\text{mole Al}}{26.982 \text{ g Al}} \times \frac{2 \text{ moles Al}_2O_3}{4 \text{ moles Al}} \times \frac{101.961 \text{ g Al}_2O_3}{\text{mole Al}_2O_3} = 0.945 \text{ g Al}_2O_3$$

## 8.9  Limiting Reactant

The quantity of product obtained in a chemical reaction is called the **yield.** As we have seen, the yield depends on what quantities of reactants are used. Using the methods discussed in Section 8.8 and given the moles or grams of a reactant, the moles or grams of a product (the yield) can be calculated. But what if there are two reactants and the moles or grams of both are given? What will be the yield in this case?

The yield that results when quantities of two reactants are given depends on which reactant is the limiting reactant. The **limiting reactant** is the reactant that limits the amount of product formed due to the fact that it is consumed first in the process. A simple analogy here will clarify the concept. Suppose you were preparing holiday greeting cards to mail to your family and friends. Each mailing requires an envelope (with a card) and a stamp. You begin with a book of stamps and a stack of envelopes with cards inside. The yield is the number of mailings that you are able to complete. You find that one book of stamps is not enough and you run out. The entire process stops until you are able to buy more stamps. The number of stamps limits the yield. The stamps are gone, but the envelopes are left over. The stamps represent the limiting reactant in a reaction. The completed mailings represent the product, the leftover envelopes represent the other reactant, which is present "in excess."

In a chemical reaction in which the amounts of two reactants are given, the yield will depend on the amount of the limiting reactant. The reactant that gives less product is the limiting reactant. To solve a limiting reactant problem (when the amounts of two reactants are given and you are asked to determine the limiting reactant and/or the yield), calculate the amounts of products possible from both reactants individually. The reactant that gives less product is the limiting reactant, the lesser amount of product is the yield, and the other reactant is said to be "present in excess."

## Example 8.19

What is the yield of $NH_4Cl$ (in grams) if 0.238 moles of $Cl_2$ and 1.627 grams of $NH_3$ are available for the following reaction? Which reactant is the limiting reactant?

$$8NH_3 + 3Cl_2 \rightarrow N_2 + 6NH_4Cl$$

## Solution 8.19

The amount of $NH_4Cl$ possible from the $NH_3$ given and the amount of $NH_4Cl$ possible from the amount of $Cl_2$ given must both be calculated. The reactant that gives less $NH_4Cl$ is the limiting reactant and the lesser amount of $NH_4Cl$ is the yield. The formula weight of $NH_3$ is 17.031 grams per mole, the formula weight of $Cl_2$ is 70.906 grams per mole, and the formula weight of $NH_4Cl$ is 53.492 grams per mole.

$$0.238 \text{ moles Cl}_2 \times \frac{6 \text{ moles NH}_4Cl}{3 \text{ moles Cl}_2} \times \frac{53.492 \text{ g NH}_4Cl}{\text{mole NH}_4Cl} = 25.5 \text{ g NH}_4Cl$$

$$.627 \text{ g NH}_3 \times \frac{\text{mole NH}_3}{17.031 \text{ g NH}_3} \times \frac{6 \text{ moles NH}_4Cl}{8 \text{ moles NH}_3} \times \frac{53.492 \text{ g NH}_4Cl}{\text{mole NH}_4Cl} = 3.833 \text{ g NH}_4Cl$$

# BATCH AND CONTINUOUS PROCESSES, REACTORS, RAW MATERIALS, AND METRIC TONS

In the chemical industry, chemicals are manufactured on a large scale by reaction of a large quantity of one reactant with a large quantity of another, the so-called "raw materials." Contact of one raw material with another occurs in a large chamber called a ***reactor***, the heart of the manufacturing plant. The reactor can be a *batch reactor*, or it can be a *continuous-flow reactor*. A batch reactor is one in which measured quantities of the raw materials are placed in the chamber, the reaction is allowed to occur, and the contents are removed at some later time. A continuous-flow reactor is one in which reactants and products are continuously added and withdrawn as the contents are stirred.

In either case, the quantities are quite large compared to laboratory operations, often measured in hundreds of thousands of grams. In fact, a large unit called the metric ton is often used. The *metric ton* is 1 million, or $10^6$, grams. The economics of the chemical industry are such that the quantities of chemicals used in the processes and purchased in bulk from suppliers must be carefully determined so that no excess of chemical needs to be stored for a long period, or discarded. Stoichiometry calculations such as those encountered in this chapter, are useful for such determinations in spite of the large quantities that are involved.

*For Homework: Find two manufacturing processes in the reference* Concise Encyclopedia of Chemical Technology, *one using a batch reactor and one using a continuous-flow reactor. Write a report describing the specifics of each process.*

The $NH_3$ is the limiting reactant because less product is obtained from it. The yield for the reaction is 3.833 grams because it is the lesser of the two amounts.

## 8.10   Theoretical Yield, Actual Yield, and Percent Yield

Often the yield realized from a given reaction is less than the yield calculated from the data. Some possible reasons for this include (1) sloppy lab technique, such that some reactant does not get into the reaction, or some product does not get measured, (2) side reactions, so that other products form that are not part of the reaction on which the calculation was based, and (3) reversible reactions, where the given reaction does not go to completion and some reactant material remains.

In these cases it may be customary to calculate a percent yield, which is a number that indicates how successful the process was. To calculate the percent yield, the ***actual yield*** (or the yield that was physically obtained in the reaction) and the ***theoretical yield*** (the yield that was expected from the calculations) must be known or calculated. The ***percent yield*** is calculated as follows.

# ORGANIC SYNTHESIS AND PERCENT YIELD

T he process of manufacturing an organic chemical by a chemical process industry likely began on a small scale in the laboratory with a chemistry laboratory technician who was assigned to research the reaction as part of a Research and Development effort. It is with these organic synthesis reactions that percent yield calculations are particularly applicable because of the possibility of side reactions exemplified by Example 8.20 in the text. Once the technician has established the conditions necessary to maximize the yield and determined all facets of safety associated with the process, the reaction may then be scaled up and become part of an overall manufacturing effort.

*For Homework: Visit with a chemistry professional who works in Research and Development at an industrial site in your local area. Ask him/her about current and past R&D efforts in the company. Ask him/her to give you an example of a laboratory process that became, or may become a full-scale industrial process. Write a report.*

$$\text{percent yield} = \frac{\text{actual yield}}{\text{theoretical yield}} \times 100 \qquad (8.10)$$

The actual yield and theoretical yield must have the same units (mole or grams). The percent yield should always be a number equal to 100% or less than 100%.

## Example 8.20

The following reaction can be used to synthesize t-butyl alcohol, $C_4H_9OH$,

$$C_4H_9Cl + H_2O \rightarrow C_4H_9OH + HCl$$

but the following is a side reaction that also occurs.

$$C_4H_9Cl \rightarrow C_4H_8 + HCl$$

If 2.19 grams of $C_4H_9Cl$ are used and 0.0211 moles of $C_4H_9OH$ actually form, what is the percent yield?

## Solution 8.20

We calculate the theoretical yield in moles, since the actual yield is in moles.

$$2.19 \text{ g } C_4H_9Cl \times \frac{\text{mole } C_4H_9Cl}{92.5 \text{ g } C_4H_9Cl} \times \frac{1 \text{ mole } C_4H_9OH}{1 \text{ mole } C_4H_9Cl} = 0.0237 \text{ moles } C_4H_9OH$$

Then the percent yield is as follows.

$$\text{percent yield} = \frac{\text{actual yield}}{\text{theoretical yield}} \times 100$$

$$= \frac{0.0211 \text{ moles}}{0.0237 \text{ moles}} \times 100 = 89.0\%$$

## 8.11   Stoichiometry with Molar Solutions

Chemists are often concerned with reactions in which one or more reactants are dissolved in water or some other solvent. In this case, it is convenient to express the quantity of reactants used in terms of the concentration of the reactant in the solution. *Concentration* is the quantity dissolved per unit volume of solution. Probably the most popular method of expressing concentration is molarity. *Molarity* is defined as the number of moles dissolved per liter of solution. The capital letter "M", is used to symbolize molarity. For example, a solution that has 0.25 moles of HCl dissolved per liter of solution would be referred to as a 0.25 M HCl solution. The word **molar** is often used. When describing this solution, we would say that this solution is a 0.25 molar solution of HCl. As you can tell from the definition of molarity given above, the units of molarity are moles per liter, or moles/L.

The molarity of a solution is the conversion factor necessary to convert from liters of a solution to moles. Knowing this, we can modify Fig. 8.5 to include calculations involving solution data. That is, liters (or milliliters) of a solution of a certain molarity may be given and the number of moles or grams of another chemical may be calculated using molarity as the conversion factor for the first step. The graphical representation for this would appear as in Fig. 8.6.

Multiplying the liters of solution by the solution molarity would give the moles of "A" needed for the second step:

$$\text{Liters} \times \frac{\text{moles of A}}{\text{Liter}} = \text{moles of A} \tag{8.11}$$

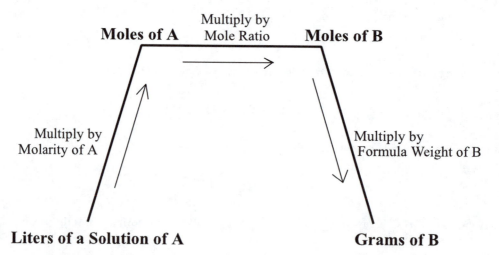

**FIGURE 8.6**   A representation of a liter-to-gram conversion when the liters or milliliters of a solution of a reactant of a certain molarity is to be converted to grams of another reactant or a product.

# RELEASE COATINGS FOR PRESSURE SENSITIVE ADHESIVES

My name is David Stickles. I am a Senior Technologist employed by Dow Corning Corporation at their Midland Development Labs. I have been employed by Dow Corning since December of 1973 after receiving an AAS in Industrial Chemistry Technology from Ferris State University in Big Rapids Michigan. I am currently working in the Paper Industry Group on silicone release coatings for organic pressure-sensitive adhesives. My main job activities include working with our customers introducing new products and formulations to them. My lab activities involve coating and curing our release coating onto various paper and film substrates and testing the release performance. I have had ten patents issued. I have received several awards from Dow Corning and several for my community service. I have been actively involved with the Mid-Michigan Technician Group and the American Chemical Society Midland Section.

*For Homework: Go to the Dow Corning Internet site and do a search for "adhesives." Explore the various Web pages accessible on this topic and write a report.*

http://www.dowcorning.com (as of 6/4/00)

## Example 8.21

How many grams of $(NH_4)_2SO_4$ will form from 50.0 mL of a 0.10 M solution of $H_2SO_4$ when it is mixed with an excess of $NH_3$, as in the following reaction?

$$H_2SO_4 + 2NH_3 \rightarrow (NH_4)_2SO_4$$

## Solution 8.21

$$0.050 \text{ L} \times \frac{0.10 \text{ mole } H_2SO_4}{\text{L}} \times \frac{1 \text{ mole}(NH_4)_2SO_4}{1 \text{ mole } H_2SO_4} \times \frac{132.140 \text{ g } (NH_4)_2SO_4}{\text{mole}}$$

$$= 0.66 \text{ g } (NH_4)_2SO_4$$

*Example 8.22*

How many grams of $Na_3PO_4$ will form if 75.0 mL of a 0.15 M solution of $H_3PO_4$ is mixed with 45.0 mL of a 0.25 M solution of NaOH, according to the following reaction?

$$H_3PO_4 + 3NaOH \rightarrow Na_3PO_4 + 3H_2O$$

*Solution 8.22*

This is a limiting reagent problem since the amounts of both reactants are given and no indication is given about either being in excess, or either being limiting. Thus, we must calculate the amounts of $Na_3PO_4$ to be obtained from the amounts of both reactants and determine which is limiting. The lesser amount of product is the amount that will be obtained.

$$0.075 \text{ L} \times \frac{0.15 \text{ mole } H_3PO_4}{\text{L}} \times \frac{1 \text{ mole } Na_3PO_4}{1 \text{ mole } H_3PO_4} \times \frac{163.940 \text{ grams } Na_3PO_4}{\text{mole}} = 1.8 \text{ grams}$$

$$0.045 \text{ L} \times \frac{0.25 \text{ mole } NaOH}{\text{L}} \times \frac{1 \text{ mole } Na_3PO_4}{3 \text{ mole } NaOH} \times \frac{163.040 \text{ grams } Na_3PO_4}{\text{mole}} = 0.61 \text{ grams}$$

Since 0.61 grams is less that 1.8 grams, 0.61 grams will be obtained in this example.

# 8.12 Homework Exercises

1. Define "stoichiometry."
2. Since the amu is defined as 1/12 the weight of the carbon-12 isotope, why are the weights of one proton and one neutron not equal to one atomic mass unit, since there are 6 protons and 6 neutrons in this isotope?
3. Define the following terms as they pertain to chemical substances.
   (a) Molecular weight
   (b) Formula weight
   (c) Mole
4. Why is it better to refer to the formula weight rather than the molecular weight, of compounds?
5. Calculate the formula weight of each of the following.
   (a) KCl   (b) $Al_2(SO_4)_3$   (c) $Mg(NO_2)_2$
6. Why was it useful for scientists to devise the mole unit?
7. Convert the moles of each to grams.
   (a) 82 moles of sulfur   (b) 7.3 moles of silver
   (c) 0.1883 moles of NaBr   (d) 34.1 moles of $BaCl_2 \cdot 2H_2O$
8. Convert the grams of each to moles.
   (a) 0.339 grams of phosphorus   (b) 0.04451 grams of antimony
   (c) 0.9338 grams of KCl   (d) 7.22 grams of $Ca(NO_3)_2$
9. How is the mole like the dozen? How is it different?
10. What is Avogadro's number? Why is this number important in chemistry?
11. Convert the moles of each to atoms.
    (a) 8.1 moles of gold   (b) 0.992 moles of lead
12. Convert the atoms of each to moles.
    (a) $9.21 \times 10^{25}$ atoms of antimony   (b) $4.29 \times 10^{27}$ atoms of tin
13. Convert the moles of each to molecules or formula units.
    (a) 0.338 moles of $CO_2$   (b) 43.2 moles of $Na_2SO_4$

14. Convert the molecules or formula units of each to moles.
    (a)  $9.227 \times 10^{24}$ molecules of $CH_4$   (b) $2.73 \times 10^{27}$ formula units of LiBr
15. Make the following conversions.
    (a)  1.2 grams of $K_3PO_4$ to formula units
    (b)  $7.183 \times 10^{25}$ atoms of Li to grams
    (c)  $6.912 \times 10^{24}$ formula units of $MgCl_2$ to grams
    (d)  0.889 grams of water to molecules
16. Which has more molecules,
    (a)  0.339 grams of $CO_2$ or 0.108 moles of $CO_2$? How many molecules is that?
    (b)  0.116 grams of $NH_3$ or 4.5 moles of $NH_3$? How many molecules is that?
17. Which weighs more, 0.416 moles of $CaCl_2$ or $5.11 \times 10^{22}$ formula units of $CaCl_2$? Show calculations and/or explain your reasoning.
18. What is the % oxygen in a compound if the compound weighed 0.9937 grams before removing the oxygen and 0.7803 grams after removing the oxygen?
19. What is the % water in a hydrate if a sample of the hydrate weighs 3.0028 grams and the same sample after driving off the water weighs 2.3744 grams?
20. What is the % water in a hydrate given the following data?
    Weight of empty crucible ...........................................17.038 grams
    Weight of crucible with hydrate in it.......................18.639 grams
    Weight of crucible with hydrate after heating ........18.403 grams
21. What is the % sulfur in $K_2SO_4$?
22. What is the % nitrogen in
    (a)  $KNO_3$   (b) $Ca(NO_3)_2$
23. Describe what is meant by the "percent composition" of a substance.
24. What is the % composition of
    (a)  $CCl_4$   (b) $(NH_4)_3N$
25. What is the % water in Epsom salt, $MgSO_4 \cdot 7H_2O$?
26. Two-thirds of the atoms in water are hydrogen atoms, yet only 11% of the weight of water is due to hydrogen. Explain.
27. What is meant by the "empirical formula" for a compound? How is this different from the molecular formula for that same compound?
28. How can the percent composition of a substance be used to determine the empirical formula for that same substance?
29. What is the empirical formula of the compounds in the following, given the % composition information listed.
    (a)  20.16% Al and 79.84% Cl.
    (b)  65.3% Sr, 10.6% Si, and 24.1% O.
30. What is the empirical formula of a copper oxide if 1.125 grams of this oxide formed from the combination of 1.000 grams of copper metal with oxygen?
31. A chemist reacts pure elemental phosphorus with oxygen to form a compound that has both phosphorus and oxygen in it. What is the formula of this compound if 0.559 grams of phosphorus resulted in 1.280 grams of the compound?
32. What is the molecular formula of a compound with an empirical formula of $C_2H_4O$ if the molecular weight is 132.0 grams per mole?
33. The % composition of a certain compound is 54.5% carbon, 9.1% hydrogen, and 36.4% oxygen. The formula weight of this compound is 88.0 grams per mole. Calculate both the empirical formula and molecular formula of this compound.
34. State the following equation in terms of atoms, molecules, and formula units and then restate the equation in terms of moles.

$$S + 3F_2 \rightarrow SF_6$$

35. Given the following reaction,

$$5Fe + 3O_2 \rightarrow Fe_3O_4 + 2FeO$$

(a) How many moles of $O_2$ are needed to react with 0.00382 moles of Fe?
(b) How many grams of FeO can be expected to form when 9.2 grams of Fe react?
(c) How many moles of $Fe_3O_4$ will form when 0.559 grams of FeO form?
(d) How many grams of FeO will form when 1.629 grams of Fe react?

36. Given the following reaction,

$$2Al + 6HCl \rightarrow 2AlCl_3 + 3H_2$$

(a) How many moles of Al will be used if 1.118 moles of HCl are used?
(b) How many moles of $H_2$ form if 938.2 grams of $AlCl_3$ form?
(c) How many grams of $AlCl_3$ will form if 0.338 moles of HCl are used?
(d) How many grams of Al are need to make 34.22 grams of $AlCl_3$?

37. If a chemist performs the following chemical reaction starting with 2.045 moles of FeS,

$$4FeS + 7O_2 \rightarrow 2Fe_2O_3 + 4SO_2$$

(a) how many moles of oxygen will react?
(b) how many grams of $Fe_2O_3$ should form?

38. If a chemist perfoms the following reaction and 5.29 grams of carbon are used,

$$Na_2SO_4 + 4C \rightarrow Na_2S + 2CO$$

(a) how many moles of $Na_2SO_4$ are needed?
(b) how many grams of $Na_2S$ can form?

39. Define what is meant by the term "limiting reagent."

40. Why is it important to know the limiting reagent when you are carrying out a chemical reaction?

41. Given the following reaction, if 0.229 moles of iron and 0.318 moles of $I_2$ are available, which would be the limiting reactant? Why?

$$2Fe + 3I_2 \rightarrow 2FeI_3$$

42. If 0.329 grams of $Fe_3O_4$ and 0.812 grams of Al are available for the following reaction, which would be the limiting reactant? How many grams of iron would be formed?

$$3Fe_3O_4 + 8Al \rightarrow 4Al_2O_3 + 9Fe$$

43. For the following reaction, 0.0728 grams of $Cu_2O$ and 0.822 moles of $O_2$ are available.

$$2Cu_2O + O_2 \rightarrow 4CuO$$

Which reactant would be the limiting reactant and how many grams of CuO would form?

44. Define yield, theoretical yield, actual yield, and percent yield.

45. If a chemist performs the following chemical reaction starting with 0.588 grams of antimony and actually obtains 0.643 grams of $Sb_4O_6$, what is the percent yield of the $Sb_4O_6$?

$$4Sb + 3O_2 \rightarrow Sb_4O_6$$

46. A laboratory technician performs the reaction below using 4.339 grams of carbon and obtains 1.993 grams of $CaC_2$. What is the % yield?

$$CaO + 3C \rightarrow CaC_2 + CO_2$$

47. Starting with 9.228 grams of Cr, a chemist is able to isolate 0.0803 moles of $Cr_2O_3$ according to the following reaction. What is the % yield?

$$4Cr + 3O_2 \rightarrow 2\,Cr_2O_3$$

48. How many grams of Zn are needed to quantitatively react with 200.0 mL of 0.50 M HCl according to the following reaction?

$$Zn + 2HCl \rightarrow ZnCl_2 + H_2$$

49. What is the limiting reactant if 50.0 mL of 0.34 M HCl are mixed with 25.0 mL of 0.46 M $Mg(OH)_2$ according to the following reaction?

$$Mg(OH)_2 + 2HCl \rightarrow MgCl_2 + 2H_2O$$

50. How many grams of NaCl will be obtained if 80.0 mL of 0.35 M $Na_2CO_3$ and 70.0 mL of 0.85 M HCl are mixed and the following reaction proceeds?

$$Na_2CO_3 + 2HCl \rightarrow 2NaCl + H_2O + CO_2$$

51. Imagine that a process development chemist is working on a reaction that will ultimately be a large-scale industrial process. The reaction is a new way to make sodium carbonate from carbon dioxide and it is according to the following reaction:

$$CO_2 + 2NaOH \rightarrow Na_2CO_3 + H_2O$$

She performs a preliminary experiment in which she uses 45.0 mL of a 0.15 M solution of NaOH. How many grams of $Na_2CO_3$ can she expect to get?

## For Class Discussion and Reports

52. In CPW Box 8.7, we mentioned how a small-scale reaction occurring in a laboratory often becomes a large-scale industrial process. Explore the history of Teflon® using the following Website and other sources and write a report on the history and uses of this extraordinarily useful product of the chemical process industry. http://www.dupont.com/teflon/index.html (accessed 6/4/00)
53. Phosphoric acid is manufactured by the reaction of phosphate rock, $Ca_3(PO_4)_2$, with sulfuric acid. Besides phosphoric acid, the reaction also yields calcium sulfate. Find as much information about this process as you can. Write the balanced equation. Write a report discussing how the stoichiometry topics discussed in this chapter are involved.
54. One method of preparing waste concentrated sulfuric acid (a solution that is 18 M) for disposal is to neutralize it with sodium carbonate. Write a brief report on this activity and include the balanced equation for the reaction (see Chapter 12, Section 12.6.1), all the safety details that must be involved and why, and how some of the details of stoichiometry discussed in this chapter would be involved.
55. One method of chemical analysis involves titrations or volumetric analyses in which a solution of one reactant is added to a solution of another until the latter reactant is completely and exactly consumed. The amount of the second reactant is then calculated using the concentration of the first reactant and the volume of it that was used. Locate an example of a real-world volumetric analysis, write the balanced equation involved, discuss the details of the method, and explain how the calculation involved utilizes concepts of stoichiometry.

9

# Gases and the Gas Laws

## 9.1    Introduction

The next topic for our study is that of gaseous substances, such as air, natural gas, and carbon dioxide. These substances require special attention because they display special and unique properties, properties different from liquids and solids. Gases play a significant role in many aspects of our lives. For example, air, a gaseous mixture, covers the surface of the Earth to a depth of several miles. We cannot, indeed, we do not desire to escape this safe gaseous envelope that sustains us. Life depends on air. The surface of the Earth and all of life have evolved here over time constantly exposed to air and its properties. The world is what it is because of this gaseous material substance. Thus, the physical and chemical properties of air and its components are very important to us.

Another example is natural gas, or methane. As we learned in Chapter 6, methane is an organic compound and is the simplest alkane hydrocarbon. It is found in underground pockets, much like petroleum. These pockets can be tapped and the methane piped to our cities and towns for the purpose of heating homes located in cold climates (in gas furnaces), for the purpose of cooking our food (with the use of gas stoves), for the purpose of heating our personal water supplies (in gas water heaters), and for the purpose of drying our laundry (in gas dryers). As a result, the physical and chemical properties of methane are very important.

The list of gases whose properties impact our lives in a major way continues. Carbon dioxide is a gas formed by the respiratory systems of humans and animals. We continuously exhale carbon dioxide, which makes it become a vital component of air near the Earth's surface. It is vital because carbon dioxide is

*Chemistry Professionals at Work* *CPW Box 9.1*

## Why Study This Topic?

In what may be a surprise to students in a beginning chemistry course, gases are very important substances in modern real-world laboratory applications and can also be important in certain specific chemical process industries. Let us briefly describe some examples of laboratory applications.

In modern analytical testing laboratories in which state-of-the-art laboratory instrumentation is used, metal cylinders filled with compressed gases such as helium, hydrogen, nitrogen, acetylene, nitrous oxide, and argon are found bound to walls and tables near sophisticated electronic instrumentation. These gases are required by these instruments in order for them to perform the work for which they are intended, such as in gas chromatography applications (separating and analyzing for components of mixtures of gases or volatile liquids) or in metals analysis using atomic spectroscopy (analyzing for elements in air, water, soil, food, etc.). While these and other applications are not the object of our current study, a familiarity with the properties of gases and of how to deal with gases provides us with insights that will become increasingly important as our understanding of chemistry broadens to include the above applications and a host of others.

*For Class Discussion: Section 9.1 mentions a number of gases, some of which should be quite familiar to most of us. Discuss as a class what personal knowledge you have or experience you may have had regarding any number of these or other gases.*

utilized by plants and trees in the process of photosynthesis. Carbon monoxide is a poisonous by-product of the burning of gasoline in automobile engines and is a component of smog in larger cities. It is an example of a gas that has properties that affect us in a negative way, as do all components of smog. Ozone is a gas found in an upper layer of the air envelope around our planet. Recent research has found that this ozone layer is becoming depleted due to the increasing concentration of freon (another gas that is present in the atmosphere), due to the use of air conditioners and aerosols by humans.

Ammonia is a gas utilized by farmers to fertilize their crops. Oxides of nitrogen are pollutant gases contaminating our atmosphere because of certain industrial processes. Their presence in our air envelope creates the problem of "acid rain," which destroys vegetation and metal statuary. (See Section 12.6.3 for more discussion of acid rain.)

The point is that gases and their physical and chemical properties impact our lives and must be understood in order for us to live the life we've come to expect.

## 9.2 Properties of Gases

As stated in the last section, gases, compared to liquids and solids, exhibit some very unique properties. Let us take a look at a few of these properties.

1. **Gases can be compressed.** Imagine a gas contained in a cylindrical container fitted with a movable piston, as shown on the left of Fig. 9.1. With some force, we can push the piston down and

compress the gas, or force it to occupy a smaller volume as shown on the right in this figure. If you imagine a liquid in the container, it may be possible to push the piston down a slight distance, but nothing like when it is filled with a gas. If a solid is in this container, it would not be possible to push the piston down at all. Gases can be compressed while liquids and solids cannot.

2. **Gases exert pressure in all directions.** In order to maintain the piston at the position indicated on the right in Fig. 9.1, the force used to push the piston down must be maintained. If we lessen this force, the piston will return to its original position. Thus, there is a pressure pushing up on the piston. The contained gas is exerting a pressure and this pressure is in all directions within the container. Liquids also exert pressure in all directions if they have been compressed as noted above, but solids (and liquids that have not been compressed) exert only a downward pressure, due to the force of gravity on them.

3. **Gases can expand.** Imagine again the gas contained in the cylindrical container that is fitted with the piston. This time, however, imagine that we pull up on the piston rather than push down, as shown in Fig. 9.2. With some force, we can pull the piston up. The gas would then fill the container and occupy a larger volume. With a liquid or solid filling the container, it would not be possible to move the piston upward. Thus, gases can expand in this situation while liquids and solids cannot.

4. **Gases diffuse through one another.** When natural gas (methane mixed with an additive to give it an odor) is released into a room, the occupants of the room, no matter where they are in the room, will soon detect the odor. The methane gas and the additive diffuse in all directions through air. We have all detected both pleasant and unpleasant

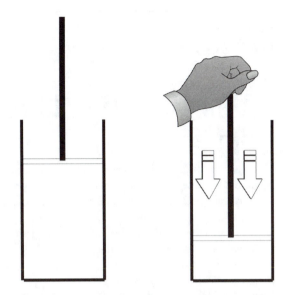

**FIGURE 9.1** Unlike liquids and solids, gases can be compressed, or forced to occupy a smaller volume.

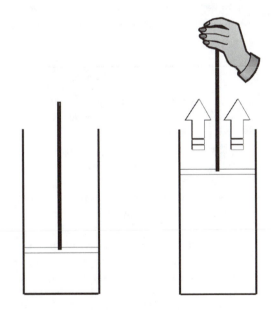

**FIGURE 9.2** Gases expand and fill a container when the container is made larger. Liquids and solids do not.

odors of various gaseous substances at different times. Examples include the odor of a skunk, the odor of a freshly baked apple pie, and the odor of perfume. It doesn't matter where we are standing relative to the source of the odor, we still smell it. We detect these odors because gaseous substances diffuse through air in all directions from the source. Indeed, all gaseous substances diffuse through all other gaseous substances regardless of the identity of the gases involved. Two liquids, such as alcohol and water, may diffuse through (or mix with) each other, but two other liquids such as oil and water, may not. Mixing of two solids never occurs regardless of the identities of the solids.

*Chemistry Professionals at Work*                    *CPW Box 9.2*

# THE BULGING DRUM STORY

T he pressure increase caused by more gas being added to a fixed volume can be a safety hazard because the container may not be able to withstand the additional pressure and may explode. This was the fear when an industrial company discovered that some 55 gallon drums of "ethoxylated" alcohol were "bulging," apparently under the strain of high internal gas pressure. Several theories as to what the gas might be and where it was coming from were tested before the chemistry technicians determined that the gas was hydrogen from a chemical reaction inside the drum. Thus, not only was there a hazard from the high gas pressure in a confined space, but the fact that the gas was hydrogen, a dangerously explosive gas, compounded the problem. Knowing what the gas was, enabled the company to safely relieve the pressure and dispose of the contents. The technicians' experience and ingenuity was very helpful in solving this problem.

*For Class Discussion: Suggest at least two ways to relieve the pressure in the drums in the scenario above based on what you learned in Section 9.2.*

5. **Gases expand when heated and contract when cooled.** If we were to place the cylinder with the movable piston on a hot plate as pictured in Fig. 9.3, and proceed to heat the contained gas without applying any upward or downward force on the piston, the piston moves upward as the gas expands and occupies a larger volume. Similarly, if we were to cool the gas by placing the cylinder in an ice bath, as pictured in Fig. 9.4, the gas contracts and occupies a smaller volume. The volume of a gas depends on its temperature. Liquids behave similarly, as was noted in Chapter 7 in the discussion of thermometers. The volume of a liquid depends on its temperature and that is why thermometers can be constructed with liquids inside—the liquid level depends on the volume of the liquid, which in turn depends on the temperature. There is only a very small effect of temperature on the volume of solids.

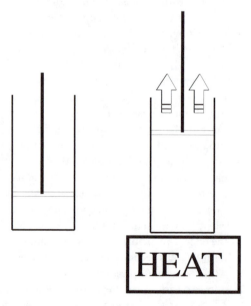

**FIGURE 9.3**   When a gas is heated, its volume increases.

6. **If the volume of a container filled with a gas is fixed, the pressure will increase if the gas is heated and decrease if the gas is cooled.** If the piston on the left in Fig. 9.3 were locked in

place, the gas volume would be fixed. When heated, the piston could not move, and as a result, the gas pressure would increase. When cooled, the piston could not move and the gas pressure would decrease. This would be similar to pushing down on the piston with increasing force (to prevent its upward movement) as the gas was heated, or pulling up on the piston with increasing force (to prevent its downward movement) as the gas was cooled. The increased downward force is due to the increased gas pressure and the increased upward force is due to the decreased gas pressure.

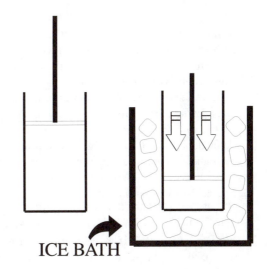

ICE BATH

**FIGURE 9.4** When a gas is cooled, its volume decreases.

7. **If more gas is added to a container of fixed volume, the pressure increases. If some gas is removed from a container of fixed volume, the pressure decreases.** When more air is pumped into a tire that is already filled with air, the pressure increases (as measured by the tire gauge). If some air is released from a tire filled with air, the pressure decreases.

## 9.3   The Kinetic-Molecular Theory of Gases

Scientists attempt to explain the various properties of gases, liquids, and solids in terms of the properties of the atoms, molecules, or formula units of which they are composed. The theory used to do this is called the "kinetic-molecular theory," or in the case of gases, the **Kinetic-Molecular Theory of Gases**. This theory has the following key points.

1. All gases are composed of atoms or molecules that are far apart from one another.
2. These atoms or molecules are in constant motion, colliding with each other and with the walls of the container.
3. Because these atoms or molecules are in constant motion, they possess the energy of motion, or **kinetic energy**. Kinetic energy (K.E.) depends on the mass of the moving atom or molecule and the velocity with which it moves. As discussed in Chapter 3, the mathematical definition of kinetic energy, K.E., is: K.E. $= \frac{1}{2}mv^2$. In this case, "m" is the mass of the atom or molecule, and "v" is the velocity with which it moves.
4. The collisions are *perfectly elastic*, meaning that the energy of the atoms or molecules is never diminished as a result of the collisions, and so the motion never stops.
5. The atoms or molecules of all gases have the same kinetic energy at a given temperature, regardless of the identity of the gas. Since the masses of the atoms or molecules of different gases (the atomic or molecular weight) vary according to their identity, the heavier the atom or molecule is, the slower it moves.
6. Kinetic energy changes with temperature. The heat energy added to increase the temperature causes the velocity of the atom or molecule, and its kinetic energy, to increase. Conversely, the heat energy removed to decrease the temperature causes the velocity, and thus the kinetic energy, to decrease.

Let us now proceed to explain the various unique properties that gases have by using this theory. (The numbered statements below correspond to the numbered statements in Section 9.2.)

1. Why can gases be *compressed*? The atoms or molecules are far apart from one another and thus can be made to come closer together.

2. Why do gases exert a *pressure in all directions*? The gas atoms or molecules are in constant motion and are constantly colliding with the walls of the container, so they exert a force on the walls. This force is observed as pressure.

3. Why can gases *expand*? Gas atoms or molecules are far apart from one another and can be forced even farther apart. Since the gas molecules are in constant motion, they immediately fill all parts of the container.

4. Why do gases *diffuse* through one another? Besides being far apart from one another, the atoms or molecules are in constant motion and thus move freely among other gas atoms or molecules regardless of their identities.

5. and 6. Why do gases *expand when heated* and *contract when cooled*, and why, if the volume is fixed, does the pressure increase when the gas is heated and decrease when the gas is cooled? The added heat energy is converted to kinetic energy, and thus the velocities of the atoms or molecules increase. If the velocities increase, they collide with the walls of the container more often. Since gas pressure is the result of these collisions, the gas pressure increases. If the volume is not fixed, however, it will increase as a result of the increased pressure. When the gas is cooled, heat energy is removed, causing the kinetic energy to decrease, which in turn means that the velocities decrease. Lower velocities mean that the collision rate with the walls of the container decreases, and thus the pressure is lowered. If the volume is not fixed, then the volume decreases.

7. Why does *pressure increase when more gas is added* to a container of fixed volume and decrease when gas is removed from a container of fixed volume? More gas atoms or molecules means more collisions with the walls of the container, and more collisions means increased pressure. Fewer gas atoms or molecules means fewer collisions and thus decreased pressure.

## 9.4   Gas Pressure

One of the properties of gases mentioned in our discussion here is that of ***gas pressure***. Gas pressure is something that is routinely measured in everyday life. For example, when you pump your bicycle tire or your automobile tire full of air, it is desirable to measure the pressure of the air inside the tire with a pressure gauge in order to determine whether you have added the correct amount of air. Also, when meteorologists measure *barometric pressure*, they measure the pressure exerted by the air in the vicinity of the weather station. Barometric pressure changes with changing weather conditions, so its measurement is often used to predict weather.

The pressure inside your bicycle tire is most often measured in *psi*, or *pounds per square inch*, the pounds of pressure exerted per square inch on the area of the inner wall of the tire. This would seem to imply that gas pressure is analogous to weight, which is also measured in pounds. The difference is that gas pressure is exerted in all directions, not just in the downward direction of gravity.

To illustrate the air pressure measured by a meteorologist, consider what happens when you fill a bottle with water by totally immersing it under the water in a sink, then pull the bottle up and out of the water bottom first, as illustrated in Fig. 9.5. Despite the fact that the water level inside the bottle is above the level of the water in the sink, the water does not flow out of the bottle. It isn't until the mouth of the bottle is pulled from the water that the water flows out. The water is held in the bottle because of air molecules exerting pressure on the surface of the water in the sink. This pressure is atmospheric pressure, the pressure exerted by the air in the Earth's atmosphere.

The air pressure measured by a meteorologist is most often reported in "inches of mercury." This unit of pressure is derived from the original mercury barometer invented by a 17th century scientist named **Torricelli**. Torricelli designed an apparatus in which a long glass tube, closed at one end and filled with mercury, is placed with its open end down in a container of mercury. Just as with the bottle

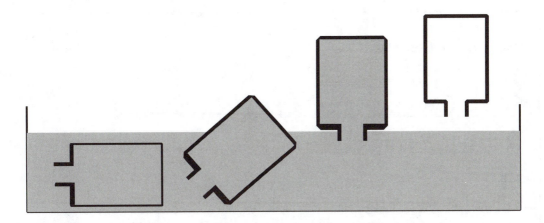

**FIGURE 9.5** The water contained in a bottle immersed in water does not flow out when lifted from the water bottom first until the mouth of the bottle is out of the water. See text for discussion.

and water experiment, the mercury does not flow from the tube. Instead, it drops to a level the height of which represents a measure of the atmospheric pressure. Torricelli measured the height of this mercury column with a meter stick as illustrated in Fig. 9.6 and this was the origin of a unit that has been widely used in chemistry applications, the mm Hg. Normal atmospheric pressure at sea level and 0 °C is 760 mm Hg. As the atmospheric pressure changes from one day to the next, the height of the mercury in the tube changes. The height of the mercury is thus an effective measure of atmospheric pressure. The height of the mercury can also be measured in inches (in. Hg); this is the source of the unit by which meteorologists express air pressure.

In addition to the units of psi, mm Hg, and in. Hg, there are other important pressure units. For example, for the International System of Units (see footnote in Chapter 7, Section 7.5.), the "bar" and the "pascal" are used. Additionally, the metric system utilizes the units of "atmosphere" and "torr." In this text, we shall only deal further with the units of psi, atmospheres, and torr.

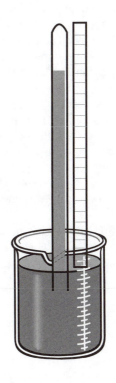

**FIGURE 9.6** The Torricelli barometer with meter stick.

The torr was defined to honor Torricelli. One torr is the same as 1 mm Hg. One atmosphere is defined as the pressure required to maintain a height of exactly 760 mm in the Torricelli barometer. In other words, 760 mm Hg, or 760 torr, is exactly one atmosphere (abbreviation: "atm"). Also, 14.696 psi is one atmosphere and 0.019337 psi is one torr. These are summarized in Table 9.1.

Normal atmospheric air pressure varies with altitude, but will usually be reported by meteorologists in the range of 27 to 31 inches of mercury. This corresponds to about 685 to 790 torr. Thus, 760 torr is within the range of normal atmosphere air pressure. Later in this chapter we will define 760 torr, or 1 atmosphere, as "standard pressure."

# Gas Cylinders and Pressure Regulators

The same heavy cylinder (pictured at right) filled with helium that is used to deliver helium gas to a balloon is also used to deliver helium to a popular analytical laboratory instrument known as a gas chromatograph. The helium is pumped into the cylinder by the manufacturer so that it is under a very high pressure. Because it is under very high pressure, when the helium is released to the balloon or to the gas chromatograph, its pressure must be regulated so that the flow of helium is controlled. To do this, a *pressure regulator*, a heavy pressure gauge/valve assembly, is attached to the valve at the top. The pressure regulator is not a flow controller but a pressure reducer, which allows a more manageable flow external to the cylinder. The pressure regulator can be interchanged from one cylinder to another, such as from an empty cylinder to a full one.

Gas cylinders under pressure, such as the helium cylinder described above, are safety hazards and must be handled properly. For example, if the pressure regulator is broken off, such as if the cylinder were to fall over, the cylinder will become like a moving rocket, similar to what happens with a filled balloon when it is released. For this reason, these cylinders are strapped to immovable objects such as walls and tables when in use, and are transported strapped on carts that are designed specifically for that purpose. (See also CPW Box 5.2.)

*For Homework: Ask your instructor for catalog information and references on pressure regulators for compressed gas cylinders. Find out if any special regulator is required for any particular gas, or if they all use the same regulator. Write a report.*

## Example 9.1
How many torr does 734 mm Hg represent?

## Solution 9.1
Since mm Hg and torr are the same unit, 734 mm Hg = 734 torr. Or, using dimensional analysis, we have the following.

$$734 \; \cancel{mm\,Hg} \times \frac{1 \; torr}{1 \; \cancel{mm\,Hg}} = 734 \; torr$$

**TABLE 9.1**   The Relationships between the Various Pressure Units Defined in the Text

1 atm = 760 torr
1 atm = 14.696 psi
1 torr = 0.019337 psi

## Example 9.2
How many psi are there in 1.05 atm?

## Solution 9.2
There are 14.696 psi in 1 atm.

$$1.05 \, \text{atm} \times \frac{14.696 \text{ psi}}{\text{atm}} = 15.4 \text{ psi}$$

## Example 9.3
How many atm does 778 torr represent?

## Solution 9.3
There are 760 torr in one atmosphere.

$$778 \text{ torr} \times \frac{1 \text{ atm}}{760 \text{ torr}} = 1.02 \text{ atm}$$

Gas pressures are measured using devices (pressure gauges) designed for that purpose. For example, the pressure of air inside a bicycle tire is measured with a tire gauge. The pressure in a compressed gas cylinder, such as the cylinder that is often used to fill balloons with helium, can be measured by a gauge that is part of the pressure regulator, the valve assembly by which the helium is delivered to the balloon. Such a gauge is used to indicate how much helium remains in the cylinder. The same kind of gauge is used on the acetylene cylinders that welders use, and on the large propane tanks that are situated near houses located in colder climates where propane is used to heat homes. We will not be concerned with the mechanism by which these gauges measure pressure. However, we will be concerned with the fact that such pressure can be measured and we will use this measured pressure in calculations.

## 9.5   Boyle's Law

One of the properties of gases mentioned in Section 9.2 was compressibility. We said that if we push down on the piston pictured in Fig. 9.1 with some force, the gas can be made to occupy a smaller volume. At any given point in this process, the force pushing down is equal to the force of the gas pushing up. In other words, the force of pushing down is equal to the pressure exerted by the gas. When the volume of the gas gets smaller by the act of compression, the force needed to push the piston down (the pressure of the gas) increases. The smaller the volume, the higher the pressure. This inverse relationship can be described mathematically. Multiplying the value of the pressure (expressed in torr, atm, or any pressure unit) by the value of the volume (expressed in liters, milliliters, or any volume unit) at any point along the way in the process of pushing down the piston always gives the same number. If we symbolize the pressure as "P" and the volume as "V," we have the following relationship.

$$PV = \text{constant} \tag{9.1}$$

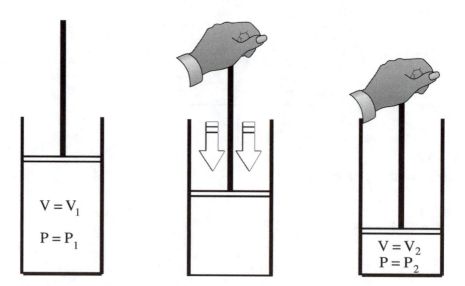

**FIGURE 9.7**   Pushing down the piston in the piston/cylinder arrangement changes the pressure from $P_1$ to $P_2$ and the volume from $V_1$ to $V_2$.

If we only push the piston down slightly, such that $P = P_1$ and $V = V_1$ as in Fig. 9.7, then we get the result shown below.

$$P_1V_1 = \text{constant} \tag{9.2}$$

If we push the piston down further, such that $P = P_2$ and $V = V_2$, then we get the following.

$$P_2V_2 = \text{constant} \tag{9.3}$$

The constants indicated in Eqs. (9.2) and (9.3) are the same number *if the pressures and volumes were measured in the same units each time*. Thus, we have the new relationship given in Eq. (9.4).

$$P_1V_1 = P_2V_2 \tag{9.4}$$

This is a statement of what has come to be known as **Boyle's Law**. It should be noted that Boyle's Law holds true regardless of whether the pressure increases (volume becomes smaller when the piston is pushed down) or decreases (volume becomes larger when the piston is pulled up as in Fig. 9.2). The units of $V_1$ and $V_2$ must be the same and the units of $P_1$ and $P_2$ must be the same. It should also be noted that the *temperature must not change* in the process of increasing or decreasing the pressure. If the temperature were to change, then the volume would be affected, as shown in Figs. 9.3 and 9.4, and Boyle's Law would not be valid.

## Example 9.4

A gas occupies a volume of 75.3 mL at a pressure of 708 torr. What volume will this gas occupy at 758 torr if the temperature does not change?

## Solution 9.4

The temperature has not changed, so Boyle's Law holds true. The initial pressure, $P_1$, is 708 torr, the final pressure, $P_2$, is 758 torr and the initial volume $V_1$ is 75.3 mL. We must calculate $V_2$. Solving for $V_2$

in Eq. (9.4), we have the following.

$$V_2 = V_1 \times \frac{P_1}{P_2}$$

The two pressure values have the same units, torr. The final volume will have the same units as the initial volume.

$$V_2 = 75.3 \text{ mL} \times \frac{708 \text{ torr}}{758 \text{ torr}}$$

$$V_2 = 70.3 \text{ mL}$$

### Example 9.5

What is the pressure of a sample of a gas held at a constant temperature when the volume is 6.02 liters, if the pressure was 0.893 atm when the volume was 3.41 liters?

### Solution 9.5

The temperature is constant, so Boyle's Law holds true. The final pressure is sought, so we must rearrange Eq. (9.4) to solve for $P_2$.

$$P_2 = P_1 \times \frac{V_1}{V_2}$$

$$P_2 = 0.893 \text{ atm} \times \frac{3.41 \text{ L}}{6.02 \text{ L}}$$

$$P_2 = 0.506 \text{ atm}$$

You may wonder why we rearrange Boyle's Law to show a pressure ratio ($P_1/P_2$ in Example 9.4) multiplying the initial volume, or a volume ratio ($V_1/V_2$ in Example 9.5) multiplying the initial pressure. Such ratios help us to see whether our answers make sense. For example, if the pressure is increased as in Example 9.4, the larger pressure will be in the denominator and the pressure ratio is a fraction less than 1. This would make the new volume smaller, as we would expect from an increase in pressure. If our calculation yields a larger volume, then we know we must have reversed the pressure values in the equation. Of course, if the pressure has decreased, the opposite is true. We would expect the new volume to be larger. Such ratios also help us to see that these calculations reflect the concepts of dimensional analysis. While we have not indicated that the units cancel, obviously if the units of pressure in the $P_1/P_2$ ratio are the same and the units in the $V_1/V_2$ ratio are the same, then they would indeed cancel.

## 9.6 Charles' Law

The fact that gas volumes change when the temperature changes can also be treated mathematically. In this case, it is a direct relationship rather than an inverse one. As temperature increases, the volume also increases. And as temperature decreases, the volume also decreases. This means that if we were to *divide* the volume by the temperature we would obtain a constant:

$$\frac{V}{T} = \text{constant} \tag{9.5}$$

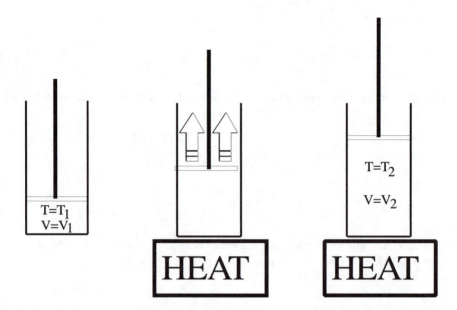

**FIGURE 9.8**  Heating the piston/cylinder arrangement causes the temperature to increase from $T_1$ to $T_2$ and the volume to increase from $V_1$ to $V_2$ at constant pressure.

Thus, at a particular temperature, $T_1$, a given sample of gas would occupy a volume, $V_1$, as shown on the left in Fig. 9.8. Dividing this volume by this temperature would then yield a constant.

$$\frac{V_1}{T_1} = \text{constant} \tag{9.6}$$

If the gas were heated to a new temperature, $T_2$, it would expand to a new volume, $V_2$, as shown on the right in Fig. 9.8. Dividing this volume by this temperature would yield the same constant as before.

$$\frac{V_2}{T_2} = \text{constant} \tag{9.7}$$

From this observation, we can obtain the following relationship.

$$\frac{V_1}{T_1} = \frac{V_2}{T_2} \tag{9.8}$$

This is a statement of what has come to be known as **Charles' Law**. Charles' Law holds true regardless of whether the temperature increases or decreases, meaning we could also have cooled the gas sample pictured in Fig. 9.8 with an ice bath rather than heated it with a hot plate. The smaller volume divided by the smaller temperature would yield the same constant.

Once again, the units of volume are irrelevant as long as both $V_1$ and $V_2$ have the same units. However, both temperatures must be in Kelvin units. Thus, if either temperature is given in Celsius or Fahrenheit units, they must be converted to Kelvin units before dividing. Also, Charles' Law only holds if the *pressure is held constant*. Obviously, if the pressure were to also change, then the volume would change for a reason other than a change in temperature and Charles' Law would not hold true.

---

*Chemistry Professionals at Work*                               *CPW Box 9.4*

# MANUFACTURING NITROGEN

**M**ore nitrogen is manufactured worldwide than any other gaseous substance. In fact, it is second only to sulfuric acid in total amount produced annually. The most popular method of producing nitrogen begins with a cryogenic cooling method for liquefying air, which is followed by a gradual heating to fractionally distill all components of air from the liquid. The initial steps of the cryogenic cooling involve the simultaneous compressing and cooling of air and both Boyle's Law and Charles' Law apply.

*For Homework: Look under cryogenics in the* Concise Encyclopedia of Chemical Technology *and other sources and learn as much as you can about its application to the production of gases. Write a report.*

---

## Example 9.6

A gas occupies a volume of 5.28 L at a temperature of 31°C. What volume will this gas occupy if the temperature is changed to 61°C while the pressure is held constant?

## Solution 9.6

Charles' Law applies because the temperature is changing at constant pressure. Solving Eq. (9.8) for $V_2$, we get the following.

$$V_2 = V_1 \times \frac{T_2}{T_1}$$

Plugging in the temperature and volume values and keeping in mind that the temperature units must be K, we get the answer shown below.

$$V_2 = 5.28 \text{ L} \times \frac{(61°C + 273.15) \text{ K}}{(31°C + 273.15) \text{ K}}$$

$$V_2 = 5.80 \text{ L}$$

## Example 9.7

What temperature would be required to change the volume of a gas from 935 mL at 316 K to 722 mL, assuming the pressure does not change?

## Solution 9.7

A change in temperature at constant pressure means that Charles' Law is applicable. Rearranging Eq. (9.8) for $T_2$, we get:

$$T_2 = T_1 \times \frac{V_2}{V_1}$$

and

$$T_2 = 316 \text{ K} \times \frac{722 \text{ mL}}{935 \text{ mL}} = 244 \text{ K}$$

Once again, the use of ratios ($T_2/T_1$ and $V_2/V_1$) helps us to see that the answer makes sense. For example, an increase in temperature should yield a larger volume and a decrease in temperature should give a smaller volume. Knowing this, we can see if the answer is appropriately larger or smaller and conclude that the ratio we used was the correct one. Also, we can see clearly that the units of temperature or volume would cancel appropriately when we use the ratio.

## 9.7   Combined Gas Law

What happens if both the pressure and temperature change? Boyle's Law and Charles' Law can be combined to produce a more general law that we can use to treat these situations. This law is called the **Combined Gas Law** and can be represented as follows:

$$\frac{P_1 V_1}{T_1} = \frac{P_2 V_2}{T_2} \tag{9.9}$$

*Example 9.8*
What volume will a gas sample occupy if initially the volume is 0.227 liters, and the pressure is changed from 716 torr to 734 torr while the temperature is changed from 31°C to 16°C?

*Solution 9.8*
Both the pressure and temperature are changing, so the combined gas law applies. Rearranging Eq. (9.9) to solve for $V_2$ yields the following.

$$V_2 = V_1 \times \frac{P_1}{P_2} \times \frac{T_2}{T_1}$$

Inserting the values for the initial volume and the initial and final pressures and temperatures, and remembering that the temperature units must be Kelvin gives the answer.

$$V_2 = 0.227 \text{ L} \times \frac{716 \text{ torr}}{734 \text{ torr}} \times \frac{(16°C + 273.15) \text{ K}}{(31°C + 273.15) \text{ K}}$$

$$V_2 = 0.211 \text{ L}$$

## 9.8   Standard Temperature and Pressure

In real-world analytical situations, pressure and temperature conditions can vary widely. Air samples for pollutant analysis, for example, can be taken on cold days or hot days, on a high mountain where the air is "thin" (i.e., lower pressure) or at sea level. The volume of the air sample depends on the pressure and temperature, as discovered in our discussion of Boyle's and Charles' Laws. This volume, in turn, is needed in the calculation of the concentration level of the pollutant in the air, which is the critical factor in knowing whether the polluted air is hazardous to our health. The bottom line is that the volume of air that we sample needs to be corrected to standard conditions of temperature and pressure so that we may accurately compare the results of the air analysis with accepted threshold safety limits.

Another example of this problem is where the measurement of a gas volume is needed to prove a point in a research experiment. Researchers in Denver, Colorado, (which is at low pressure) working at a laboratory temperature of 25°C will report one volume for their experiment while a researcher in Los Angeles, California, (at sea level) working at a laboratory temperature of 20°C will report a different volume, simply because conditions of temperature and pressure varied. However, if their volumes are corrected to an accepted standard for temperature and pressure, their answers should be the same, all other conditions being equal.

Scientists have therefore adopted specific pressure and temperature values that represent these standard conditions. Standard pressure is 760 torr, or 1 atmosphere. Standard temperature is 0°C, or 273.15 K. These values are referred to as **Standard Temperature and Pressure**, or **STP**.

## 9.9   Correction of a Gas Volume to STP

The correction of a gas volume to STP conditions, as alluded to in Section 9.8, utilizes the Combined Gas Law. The given conditions are the initial conditions ($P_1$, $T_1$), while the STP conditions are the final conditions ($P_2$, $T_2$).

*Example 9.9*
A gas occupies a volume of 27.8 liters at 744 torr and 288 K. What volume will it occupy at STP?

*Solution 9.9*
Rearranging the Combined Gas Law as in Example 9.8, we have:

$$V_2 = V_1 \times \frac{P_1}{P_2} \times \frac{T_2}{T_1}$$

which gives the following answer.

$$V_2 = 27.8 \text{ L} \times \frac{744 \text{ torr}}{760 \text{ torr}} \times \frac{273.15 \text{ K}}{288 \text{ K}}$$

$$= 25.8 \text{ L}$$

## 9.10   The Ideal Gas Law

If more gas is added to the container with the moveable piston and no additional force is applied to the piston to keep the volume fixed (and the temperature does not change), the piston will rise (the volume will increase) to accommodate the additional gas. There is a direct relationship between the volume of a gas and the amount of gas at constant temperature and pressure. This relationship can be expressed as follows.

$$\frac{V_1}{n_1} = \frac{V_2}{n_2} \tag{9.10}$$

where "n" is the number of moles of the gas. Combining Eqs. (9.9) and (9.10), we get:

$$\frac{P_1 V_1}{n_1 T_1} = \frac{P_2 V_2}{n_2 T_2} \tag{9.11}$$

---

*Chemistry Professionals at Work*                                    *CPW Box 9.5*

# SIX OUT OF TEN

In terms of quantity produced by the chemical industry, six of the top ten chemicals in the U.S. are gases. These six gases and their rank* are: nitrogen (2), oxygen (3), ethylene (4), ammonia (6), propylene (9), and chlorine (10). Nitrogen and oxygen are produced by separation from air. Nitrogen is used extensively as a purge gas to prevent reaction with oxygen. Oxygen is used in steel. Ethylene and propylene are produced from petroleum and are used mostly for manufacturing other chemicals and materials such as polymers. Ammonia is manufactured from the reaction of hydrogen with nitrogen (the Haber-Bosch Process) and is used in the fertilizer industry. Chlorine is produced by the electrolysis of brine solutions and is used mostly to make other chemicals.

*For Homework: Two more gases appear in the top twenty-five. These are vinyl chloride (18) and carbon dioxide (22). Look up one of these in the* Concise Encyclopedia of Chemical Technology *and other sources and write a report on the manufacturing methods and uses.*

*Source: Chemical and Engineering News, American Chemical Society, April 8, 1996.

---

or

$$\frac{PV}{nT} = \text{constant}. \tag{9.12}$$

Pressure, temperature, and the number of moles of a gas are the only parameters that affect gas volumes. If pressure, volume, temperature, or number of moles changes, one of the other parameters will also change so that the constant in Eq. (9.12) does not change. For example, if pressure increases, the volume decreases such that the constant remains the same. Thus, the constant in Eq. (9.12) can be assigned a numerical value and this value will be constant. Of course, this numerical value does depend on what units are used for the pressure and volume (the temperature units, as discussed previously, are always Kelvin). If the pressure units are atmospheres and the volume units are liters, then the value of this constant is as given below. This constant is often given the symbol "R" and is called the *universal gas constant*.

$$R = \frac{0.082057 \text{ L atm}}{\text{mole K}} \quad \text{or} \quad 0.082057 \text{ L atm mole}^{-1}\text{K}^{-1} \tag{9.13}$$

The units of R will be as shown unless different units for pressure and volume are used. Equation (9.12) can now be written as shown below.

$$R = \frac{PV}{nT} \tag{9.14}$$

Equation (9.14) can be rearranged to give:

$$PV = nRT \tag{9.15}$$

Equation (9.15) is known as the **Ideal Gas Law**. Given any three of the parameters—pressure, volume, number of moles or temperature—the fourth can be calculated from this equation.

*Example 9.10*

What volume will 0.449 moles of a gas occupy at a pressure of 1.29 atm, 293 K?

*Solution 9.10*

Solving Eq. (9.15) for volume, we get the following.

$$V = \frac{nRT}{P}$$

Inserting the values given and the value for R, we can then solve for V.

$$V = \frac{0.449 \text{ moles} \times 0.082057 \text{ L atm mole}^{-1}\text{K}^{-1} \times 293 \text{ K}}{1.29 \text{ atm}}$$

$$= 8.37 \text{ L}$$

## 9.11  Density and Formula Weight of a Gas

As we learned in Chapter 8, the atomic or formula weight of an element or compound (the number of grams per mole) can be used to convert grams to moles or moles to grams (Section 8.4.1). With this in mind, calculations using the Ideal Gas Law can be taken one step further in that, if we know the gas's identity, the formula weight is known and can be used to convert the number of moles calculated via the Ideal Gas Law to grams. Following this, the gas's density can be calculated, if desired, by dividing this number of grams by the volume of the gas. A mathematical relationship that includes density and formula weight can be derived via the following sequence beginning with Eq. (7.5) from Chapter 7.

$$D = \frac{M}{V} \tag{9.16}$$

in which "M" is the mass of the gas and "D" is its density. The mass is calculated by multiplying the number of moles, n, by the formula weight (FW).

$$D = \frac{n \times FW}{V} \tag{9.17}$$

The number of moles, n, is calculated using the ideal gas law.

$$D = \frac{\frac{PV}{RT} \times FW}{V} \tag{9.18}$$

Rearranging Eq. (9.18) gives Eq. (9.19) for calculating density.

$$D = \frac{P \times FW}{RT} \tag{9.19}$$

Another rearrangement gives Eq. (9.20) for calculating the formula weight.

$$FW = \frac{DRT}{P}$$ (9.20)

For the case in which density is not known or calculated in advance, we can substitute M/V for D and obtain the following.

$$FW = \frac{MRT}{PV}$$ (9.21)

*Example 9.11*

What is the density of carbon dioxide at 122°C and 0.944 atm?

*Solution 9.11*

The formula weight of $CO_2$ is 44.009 grams per mole. The density can be calculated using Eq. 9.19.

$$D = \frac{P \times FW}{RT}$$

$$D = \frac{0.944 \text{ atm} \times 44.009 \text{ grams / mole}}{0.082057 \text{ L atm / mole K} \times (122 + 273.15) \text{ K}}$$

$$= 1.28 \text{ grams / L}$$

*Example 9.12*

What is the formula weight of a gas if 3.264 grams of it occupies 1.82 L at 20°C and 1.08 atm?

*Solution 9.12*

Using Eq. (9.21), we have the following.

$$FW = \frac{MRT}{PV}$$

$$FW = \frac{3.264 \text{ grams} \times 0.08205 \text{ L atm / mole K} \times (20.0 + 273.15) \text{K}}{1.080 \text{ atm} \times 1.820 \text{ L}}$$

$$= 39.94 \text{ grams/mole}$$

## 9.12  Avogadro's Law

In all of the previous gas law discussions in this chapter, no particular gas has been specified. We have always made reference to "a gas" or an "ideal gas" without saying if the gas is air, oxygen, carbon dioxide, ammonia, helium, or what. One of the tenants of the Kinetic-Molecular Theory of Gases is that all gases behave the same way in terms of compressibility, expansion, diffusion, pressure, etc., and this "ideal" behavior extends to the gas laws. Thus, as long as a gas exhibits this ideal behavior,[1] the gas laws hold for that gas.

---

[1]Gases deviate from ideal behavior at very high pressures and very low temperatures. Under these conditions, the gas laws would not not necessarily be obeyed.

---

# CALCULATIONS FOR THE DETERMINATION OF GASEOUS AIR POLLUTANTS

The determination of the concentrations of gaseous pollutants in air is something that the U.S. Environmental Protection Agency (EPA) has comprehensively addressed in a set of official methods and procedures for laboratory personnel to follow. Many of these procedures culminate in a calculation to translate the laboratory result into a concentration level that can be compared to the published threshold limits that are deemed acceptable for human safety. This calculation often involves, in part, (1) the combined gas law to correct the sampled volume of air to STP, and (2) Avogadro's Law to convert the liters of gaseous pollutant to moles. Ultimately, what is usually calculated is the micrograms of pollutant per cubic meter of the air.

*For Homework: Find a method for determining the concentration of a gaseous pollutant in air and see if you can confirm the above statements. Write a report.*

---

The Ideal Gas Law affords us a good opportunity to summarize this in one easy statement. Using Example 9.10 as our basis, we can say that 0.449 moles of helium at 1.29 atm and 293 K, occupies 8.37 liters. We can also say that 0.449 moles of carbon dioxide at 1.29 atm and 293 K, also occupies 8.37 liters. Putting it another way, a sample of helium gas at 1.29 atm and 2.93 K that occupies 8.37 liters, will consist of 0.449 moles of helium, or, a sample of carbon dioxide gas at 1.29 atm and 293 K that occupies 8.37 liters will consist of 0.449 moles of carbon dioxide. If we were to convert the 0.449 moles to atoms or molecules, Avogadro's number ($6.022 \times 10^{23}$ atoms, or molecules, per mole) would be the conversion factor regardless of the type of gas. Each gas would therefore consist of the same number of particles, whether these were atoms or molecules. An easy statement for this is: Equal volumes of different gases at the same temperature and pressure consist of the same number of particles—either atoms or molecules. This is a statement of what has come to be known as **Avogadro's Law.**

Avogadro's Law may be expressed another way for the specific case of having exactly one mole of the gas held at STP. One mole of gas contains Avogadro's number of particles, so the next question is this: "What volume will this number of particles take up at STP?" If we were to calculate this volume using the Ideal Gas Law—Eq. (9.15)—we would discover that this volume would be 22.4 liters. In other words, *one mole of any ideal gas occupies 22.4 liters at STP*. Using this statement of Avogadro's Law, we can formulate an equality and two conversion factors:

$$\text{Equality:} \quad \text{1 mole of gas} = \text{22.4 liters (at STP)} \qquad (9.22)$$

$$\text{Conversion factors:} \quad \frac{22.4 \text{ L}}{1 \text{ mole}} \quad \frac{1 \text{ mole}}{22.4 \text{ L}} \qquad (9.23)$$

We make special note of the fact that the *22.4 L/mole conversion factor* is good **only at STP** and is good **only for gases**. We also point out that the number 22.4 is not an exact number and will limit the number of significant figures in the answer to a calculation if it has fewer such figures that the number being converted.

### Example 9.13

How many liters does 4.52 moles of argon gas occupy at STP?

### Solution 9.13

$$4.52 \text{ moles} \times \frac{22.4 \text{ L}}{1 \text{ mole}} = 101 \text{ L}$$

## 9.13  Dalton's Law

We often encounter mixtures of gases in our study of chemistry and in our everyday life. The one obvious gas mixture that comes to mind is air. Air is mainly a mixture of oxygen and nitrogen. However, other gases can be present, such as carbon dioxide, since humans and animals are constantly exhaling it into their surroundings. Other examples are carbon monoxide and unburned hydrocarbons, which are components of automobile exhaust. In addition, water vapor is a component of air in virtually all locations, giving rise to what we call **humidity**. Gaseous mixtures exhibit all the same properties and obey all the same laws that pure gases do.

However, there is a gas law that specifically addresses gaseous mixtures. **Dalton's Law** deals with the individual pressures that are exerted by the component gases, stating that the total pressure exerted by a gaseous mixture is the sum of the partial pressures exerted by each of the component gases. This can be summarized as follows.

$$P_{total} = p_1 + p_2 + p_3 + p_4 \ldots \ldots \tag{9.24}$$

in which $P_{total}$ is the total pressure exerted by the mixture, and $p_1$, $p_2$, $p_3$, etc., are the partial pressures of gases 1, 2, 3, etc. These various gas pressures can be expressed in any pressure units. However, the total pressure and partial pressures must all be expressed in the same units.

### Example 9.14

What is the partial pressure of hydrogen gas in a mixture of hydrogen, nitrogen, and ammonia gases if the total pressure exerted is 783 torr, the partial pressure of the nitrogen is 0.228 atm, and the partial pressure of the ammonia is 411 torr?

### Solution 9.14

All pressures must have the same units in order to use Dalton's Law. Let us first convert the partial pressure of the nitrogen to torr.

$$0.228 \text{ atm} \times \frac{760 \text{ torr}}{1 \text{ atm}} = 173.28 \text{ torr}$$

Next, we subtract the two partial pressures given for nitrogen and ammonia from the total to determine the partial pressure of the hydrogen.

$$
\begin{aligned}
P_{hydrogen} &= P_{total} - p_{nitrogen} - p_{ammonia} \\
&= 783 \text{ torr} - 173.28 \text{ torr} - 411 \text{ torr} \\
&= 199 \text{ torr}
\end{aligned}
$$

## 9.14   Vapor Pressure

A useful application of Dalton's Law is the case of a gas that is confined to a container in which there is also some liquid water. The water tends to evaporate, and thus, the confined gas is contaminated with water vapor. Perhaps the most obvious confined gas is the air that surrounds the Earth. This air is present in its "container" along with liquid water (the oceans, lakes, etc.). The result is air that is "contaminated" with water vapor due to the evaporation of water from these sources, creating the condition commonly known as **humid** air. There is a limit to how much water vapor the air can hold before precipitation, or rain, occurs. This limit is referred to as 100% humidity and at that point we say that the air is saturated with water vapor. Other humidity percentages, such as 45%, indicate the water vapor concentration level is at a given percentage of the saturation level (45% in this example). The actual concentration of water vapor in the air at the 100% humidity level varies with temperature.

Water (indeed, all liquids) tends to evaporate regardless of temperature. The boiling point of water does not need to be achieved in order for liquid water to escape the liquid phase and become a gas. We are all aware that a drop of water left on a table top would eventually evaporate to dryness even though we are not boiling it. Wet laundry hung on a clothes line dries because water tends to evaporate even though it is not boiling.

Let us examine this phenomenon in the laboratory setting. If we were to fill a beaker halfway with water and set it on the laboratory bench, it would slowly evaporate, as pictured on the left of Fig. 9.9. Eventually, perhaps over a period of days or weeks, the water would totally evaporate. If, however, we were to cover and seal the beaker so that none of the water vapor could escape into the atmosphere, eventually the water vapor would saturate the air space in the beaker and create the 100% humidity condition. At that point, the rate of evaporation would equal the rate of condensation, as pictured on the right of Fig. 9.9, and an equilibrium is achieved. This means that the concentration of the water vapor in the air space is constant and the liquid water would never totally evaporate.

The molecules of water in the air space exert a partial pressure like all gases in a mixture of gases. If the component of the mixture is present due to the evaporation of a liquid also present in the container, like the water vapor in this example, this partial pressure is called the **vapor pressure**. The formal definition of vapor pressure is the pressure exerted by the molecules of a gas in equilibrium with its liquid present in the same sealed container.

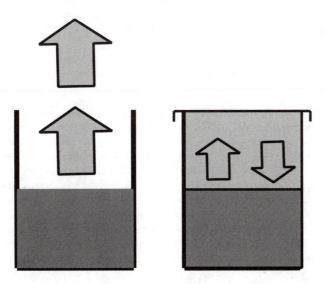

**FIGURE 9.9**   The water in the beaker on the left would eventually all evaporate. On the right, the water evaporates to the point of saturating the air space with water vapor. The partial pressure of the water vapor in the air space on the right is the "vapor pressure" of water at the given temperature.

# VAPOR PRESSURE AND GAS CHROMATOGRAPHY

**G**as chromatography is an instrumental technique in which the components of gas mixtures, or liquid mixtures that can easily be converted to gases, are separated and quantitated. The separation occurs in a small diameter tube called a **column**. The mixture is introduced in one end of the column and the individual gaseous components emerge from the other end. Inside this column, the mixture components move at different rates in and out of two phases, one liquid or solid and one gaseous. The gas phase (the **mobile phase**) is moving as it "blows" over the surface of the liquid or solid phase which is held stationary (the **stationary phase**).

The separation depends on two properties of the gaseous mixture components: (1) their relative affinity for the stationary phase, and (2) their relative vapor pressures, or tendency to evaporate and be in the gaseous mobile phase. Those components that have a low solubility and high vapor pressure tend to travel mostly with the mobile phase and emerge from the column quickly, while those with high solubility and low vapor pressure tend to be dissolved in the stationary phase and emerge from the column slowly. Thus a separation takes place. As the components emerge from the column, they are electronically detected and measured. This technique is very useful in measuring pollutants in air and water or in measuring volatile organic compounds in pharmaceutical formulations, petroleum products, beverages, and hygiene products.

*For Homework: Find a reference describing basic gas chromatography and discover exactly how a mixture of volatile liquids is introduced to the column and also exactly how they are electronically detected as they emerge from the other end as stated here. Write a report.*

---

Two important points about vapor pressure are (1) all liquids have a vapor pressure, and (2) vapor pressure varies with temperature. All liquids tend to evaporate, but some do so more than others. For example, liquid ether has a very strong tendency to evaporate, much more so than water, because ether has a very high vapor pressure compared to water. The 100% humidity level for ether would be a much higher concentration of ether vapor than water at its 100% humidity level. The concentration of the vapor, regardless of what the liquid is, depends on temperature. As temperature increases, the tendency to evaporate increases and, thus, more vapor from the liquid will result. The **normal boiling point** of a liquid is the temperature at which the vapor pressure equals standard atmospheric pressure, 760 torr. The dependence of the vapor pressure of water on temperature is obvious from Table 9.2. This table covers a limited range of temperatures, from 20°C to 30°C.

The vapor pressure of water is another example of a property of water that is influenced by hydrogen bonding (Chapter 5). The hydrogen bonds that bind water molecules to each other are relatively strong and so water molecules require more energy than most other liquids to break free of each other and enter the gas phase. Thus, the vapor pressure of water is typically much lower

**TABLE 9.2**  Vapor Pressures of Water at Various Temperatures

| Temperature, °C | Vapor Pressure, torr |
|---|---|
| 20 | 17.5 |
| 21 | 18.5 |
| 22 | 19.8 |
| 23 | 21.1 |
| 24 | 22.4 |
| 25 | 23.8 |
| 26 | 25.2 |
| 27 | 26.7 |
| 28 | 28.3 |
| 29 | 30.0 |
| 30 | 31.8 |

than other liquids at equivalent temperatures, reflecting the fact that there are fewer water molecules in the gas phase to be exerting pressure.

## 9.15   Correction of the Volume of a Wet Gas to STP

Sometimes it is necessary to collect a gas as it emerges from a tube, such as when it is the product of a chemical reaction and its volume or mass needs to be measured in order to study the reaction stoichiometry. A convenient way to collect such a gas is by using it to displace water from a bottle or tube filled with water. The usual procedure for this is to invert the filled bottle or tube and place its open mouth into a tub of water (similar to the process described by Fig. 9.5) and then deliver the gas via a tube inserted up into the bottle through the open mouth.

The problem with this procedure is that the gas is confined in a container in which there is also some liquid water. This means that the gas sample will be contaminated with water vapor. However, Dalton's Law can be used to examine this problem. Knowing the vapor pressure of water (from Table 9.2) at the temperature of the gas and knowing the total pressure of the system (by manipulating the apparatus, the total pressure can be made equal to the atmospheric pressure), the pressure of the dry gas can be determined by Dalton's Law. Following this, there can be a correction to STP to facilitate the study of the stoichiometry of the reaction.

*Example 9.15*

A gas is collected over water at 28°C and 731 torr. The volume was found to be 258 mL. What is the volume of the dry gas at STP?

*Solution 9.15*

The pressure of the dry gas is found by subtracting the vapor pressure of water at 28°C from the total pressure (Dalton's Law). The volume at STP is then found as in Example 9.9.

$$V_2 = V_1 \times \frac{(P_{total} - P_{H_2O})}{P_2} \times \frac{T_2}{T_1}$$

$$V = 258 \text{ mL} \times \frac{(731 - 28.3) \text{ torr}}{760 \text{ torr}} \times \frac{273.15 \text{ K}}{(28 + 273.15) \text{ K}} = 216 \text{ mL}$$

---

*Chemistry Professionals at Work*                    *CPW Box 9.8*

# A Challenging, Exciting Job in Texas

**M**y name is Gloria McPherson and I have worked for the E.I. Dupont DeNemours Company at La Porte, Texas, for 21 years. I worked as a Process Operator until seven years ago when an opportunity in the laboratories became available. I have worked in the Polymers Lab and the "Ag" Lab, but presently work in the Research and Development Lab. My current assignment as Chemist Assistant has given me many opportunities to learn more about the business as a whole. I have had training in Gas Chromatographs, Liquid Chromatographs, chemical reactions, organizing data, and operating process test facilities. I am currently working under the direction of a Ph.D. chemist.

In my job, I enjoy enthusiastic coworkers, challenging assignments, and a diverse environment. I feel that working as a Lab Technician for DuPont has made a tremendous impact on my family and me.

*For Homework: Visit the Internet site (address below) that provides an alphabetic listing of DuPont products and services. Select one product and navigate the site to find as much information as possible. Write a report.*

http://www.dupont.com/corp/products/prodlist.html (as of 6/4/00)

## 9.16   Stoichiometry of Reactions Involving Gases

In Chapter 8 reaction stoichiometry was a topic for our study. A summary of this might be the following question: "Given an amount of one substance, reactant, or product involved in a chemical reaction, how much of another substance is involved?" We saw five possible scenarios: a mole-to-mole conversion, a mole-to-gram conversion, a gram-to-mole conversion, a gram-to-gram conversion (Section 8.8), and a liter-to-gram conversion (Section 8.10). In a gram–gram conversion, for example, the amount of a reactant or product was given in grams and the amount of another reactant or product, in grams, was determined.

In a reaction in which gases are involved, the volume of a gas may be given and/or determined, since gases are most easily measured as volumes. If the conditions of the experiment are STP, Avogadro's Law may be used to convert liters to moles, or vice versa.

## Example 9.16
How many milliliters of $H_2$ can form at STP from 0.500 grams of aluminum according to the following reaction?

$$2Al + 6HCl \rightarrow 2AlCl_3 + 3H_2$$

## Solution 9.16
This is a gram–mL conversion and uses Avogadro's law (22.4 L/mole at STP) to convert moles of $H_2$ to liters.

$$0.500 \text{ g Al} \times \frac{1 \text{ mole Al}}{26.982 \text{ g Al}} \times \frac{3 \text{ mole } H_2}{2 \text{ moles Al}} \times \frac{22.4 \text{ L}}{1 \text{ mole } H_2}$$

$$= 0.623 \text{ L} = 623 \text{ mL}$$

# 9.17  Homework Exercises

1. List at least five properties that are characteristic of gases.
2. What is the kinetic-molecular theory of gases? What are the key points of this theory?
3. How does the kinetic-molecular theory of gases help explain each of the gas properties that you listed in Problem 1?
4. When you drive your car down the highway, the tires get hot. Explain, by the kinetic-molecular theory, why the air pressure inside the tires increases.
5. If a container of ether is opened in one part of a lab, its odor will soon be noticed by people in another part of the lab. Explain this using the kinetic-molecular theory of gases.
6. When you pump air into a bicycle tire, the air pressure inside the tire increases. Explain by the kinetic-molecular theory of gases (a) how it is that you are able to add more air to a tire that is already "filled" with air, and (b) why the pressure inside the tire increases.
7. Write the conversion factors for each of the following changes in pressure units.
   (a) torr to atm    (b) torr to psi    (c) atm to torr
   (d) atm to psi    (e) psi to torr    (f) psi to atm
8. What is 732 torr expressed (a) in psi, and (b) in atmospheres?
9. The pressure of a gas sample held in a closed container is 1.29 atmospheres. What is this pressure in torr? What is this pressure in psi?
10. What is 1.134 atm expressed in torr? What is it in psi?
11. The pressure of a sample of a gas held in a closed container is 743 torr. What is this pressure in atmospheres and in psi?
12. What is Boyle's Law? What is allowed to vary and what is assumed to be constant in this law?
13. The volume of a sample of a gas is 89.3 mL at 723 torr. What is the volume of this gas if the pressure is increased to 782 torr at constant temperature?
14. A sample of argon gas occupies 34.7 mL at 712 torr and 42°C. What volume will this gas occupy at the same temperature, but at 745 torr?
15. What is Charles' Law? What is allowed to vary and what is assumed to be constant in this law?
16. If the temperature of 234.3 mL of a gas is 56.6°C, what volume will this gas occupy at 78.4°C if the pressure doesn't change? What is the name of the gas law that applies to this situation?
17. The volume of a sample of a gas is 8.39 liters at 33°C. What would be the volume if the temperature is changed to 265 K, while the pressure is held constant?
18. What is the Combined Gas Law? What is allowed to vary and what is assumed to be constant in this law?
19. What will be the volume of a gas at 14°C and 0.892 atm if it has a volume of 3.29 liters at 21°C and 0.836 atm?

20. At 773 torr and 23.5°C the volume of a gas is 0.529 liters. What volume will it have if the pressure is decreased to 722 torr and the temperature increased to 30.4°C?
21. What is meant by the term "standard temperature and pressure?" What are the experimental conditions that are being used when you are working at a standard temperature and pressure?
22. A chemist has performed a chemical reaction and has captured a gas that forms, noticing that 68.2 mL of the gas formed at temperature of 25°C and a pressure of 753 torr. He wishes to publish his results in a chemistry journal but realizes that he must express this volume at STP. What is the volume at STP?
23. If a gas occupies 0.334 L at 563 torr and 94.2°C, what volume will it occupy at STP?
24. What is the Ideal Gas Law?
25. What is the value of the "R" term in the Ideal Gas Law?
26. How many moles of gas are there in a 56.3 liter container at 45.1°C and 0.933 atm?
27. What volume does 0.400 moles of oxygen gas occupy at 0.933 atm and 321 K?
28. Suppose a chemist performs a chemical reaction in which a gas is formed and finds 65.9 mL of the gas formed at 276 K and 0.987 atm. How many moles of the gas has this chemist produced?
29. What is the density of $N_2$ at 25°C and 0.855 atm?
30. What is the density of acetone ($C_3H_6O$) vapor at 801 torr and 81°C?
31. What is the formula weight of a gas if 1.08 grams of it occupy 0.800 L at 29°C and 17.6352 psi?
32. What is Avogadro's Law?
33. How many moles of $CO_2$ gas do you have if you have 0.733 liters of $CO_2$ at STP?
34. What is Dalton's Law?
35. Suppose a new chemistry technician in your lab is confused and wants to know exactly what vapor pressure is. What would you tell him?
36. What is meant by the "normal boiling point" for a compound?
37. How does the vapor pressure of a liquid usually change with temperature?
38. A gas is collected over water at 29°C and 745 torr. The volume collected is 873.2 mL. What volume would the dry gas occupy at STP? (Note: The vapor pressure of water at 29°C is 30.0 torr.)
39. The vapor pressure of water at 19°C is 16.5 torr. If a sample of oxygen measuring 2.31 liters is collected over water at 19°C and 729 torr, what volume will the oxygen occupy if it was dry and held at standard conditions of temperature and pressure?
40. A gas is collected over water at 24°C and 715 torr. If the volume of this "wet" gas is 9.88 L, what volume will the dry gas occupy at STP? (Note: The vapor pressure of water at 24°C is 22.4 torr.)
41. Consider the following reaction.

$$C_3H_8 + 5O_2 \rightarrow 3CO_2 + 4H_2O$$

How many milliliters of $CO_2$ gas can be produced at STP from 0.822 grams of $C_3H_8$?
42. How many milliliters of $O_2$ gas will be obtained at STP if 1.332 grams of $KClO_3$ are used in the following reaction.

$$2KClO_3 \rightarrow 2KCl + 3O_2$$

43. Consider the following reaction.

$$Zn + 2 HCl \rightarrow ZnCl_2 + H_2$$

How many liters of $H_2$ gas can be made from 0.500 grams of Zn at STP?

## For Class Discussion and Reports

44. A number of industrial companies produce and market specialty gases for consumer and industrial use. Visit any or all of the Internet sites (as of 6/4/00) listed below and write a report on one company and its activities in this area.

http://www.airproducts.com/corp/ctp/home.html
http://www.linweld.com/
http://www.scottgas.com/
http://www.airgas.com/

45. Ask your instructor for the name of an industry contact in your area whose work involves gas chromatography (see CPW Box 9.7) and/or atomic spectroscopy (see Chapter 3, Section 3.10.1). Visit him/her at the workplace and discover what specific gases are used, how they are contained, how they are piped to the gas chromatography or atomic spectroscopy systems, how they are stored, how they are transported from storage to the lab, and what precautions are taken in the event the gas is toxic or dangerous. Write a report.

46. Natural gas, or methane, is a vitally important resource. Visit the following Internet site (as of 6/4/00) and find out all you can about the exploration, extraction, production, transport, storage, and distribution of natural gas. Write a report.

http://www.naturalgas.org/

47. One important concern of the U.S. Environmental Protection Agency (EPA) is the safety and health of U.S. citizens with regard to air pollution. Visit the following Internet site (as of 6/4/00) and select a particular topic relating to air pollution and write a report.

http://www.epa.gov/students/air.htm

# 10

# Solutions

## 10.1   Introduction

An extraordinary amount of chemistry in our world occurs in water solution. A large percentage of the Earth's surface is covered by oceans, which are actually large bodies of water solutions. We are all very familiar with rain, which is a water solution. We are all familiar with lakes and rivers. These, too, are water solutions. All of us use water from wells and municipal water supplies for drinking, cooking, cleaning, and various and sundry other household chores; this water is also a water solution. Manufacturing industries use large amounts of water in their various processes for a variety of purposes, creating and using water solutions. Household waste as well as industrial waste is usually carried away by water, thus creating a water solution that can be hazardous and that must be properly treated so that we can all live safe and healthy lives. Chemists and laboratory technicians utilize water freely in their laboratory work, sometimes isolating and analyzing for chemicals dissolved in water while at other times creating water solutions in order to facilitate their work.

In addition to water solutions, other liquid chemicals can be the basis for solution formation and use. Crude oil (the source of organic chemicals), as well as manufactured oils, many petroleum distillates, and nonwater solvents used for cleaning are solutions. Gasoline, kerosene, and other fuels are solutions. In addition, organic solvents such as acetone, methyl alcohol, and acetonitrile (Chapter 6) are used as solvents in industrial laboratories. It follows that students of chemistry should devote much attention to this state of matter known as a solution.

---

*Chemistry Professionals at Work*          CPW Box 10.1

# WHY STUDY THIS TOPIC?

If there are two activities that all those working in the chemical process industries have in common, it is the preparation of laboratory solutions and the measurement of the qualitative and quantitative characteristics of unknown sample solutions. A quality assurance technician working in the pharmaceutical formulations industry may, for example, measure the concentration of the active ingredient in a cough syrup, which is a solution. A biotechnician may be assigned to measure the lead content in urine, which is a solution. An environmental laboratory technician may prepare a series of solutions of nitrate to obtain data crucial to the determination of the nitrate in a drinking water sample, which is also a solution.

A working knowledge of solution terminology, solution concentration, solution preparation, and solution properties is essential to these tasks.

*For Class Discussion: Each class member identifies a material that he/she thinks is a solution. Discuss each as a class and tell why each is or is not a solution.*

---

## 10.2 Terminology

In Chapter 1, we defined a ***solution*** as a homogeneous mixture which is a physical combination of two or more pure substances that are mixed as well as they can possibly be mixed so that we cannot tell, even with the most powerful microscope, that it is actually more than one substance. In order to tell that such a mixture is indeed a mixture and not a pure substance, one would need to be able to view the actual molecules, atoms, and/or ions of the pure substances involved, and in doing so, notice that there is more than one chemical species present.

Solutions, as defined above, do not have to be liquid mixtures. There are gaseous mixtures that are homogeneous and there are solid mixtures that are homogeneous. Any homogeneous physical mixture of two or more gases is a solution. Any homogeneous physical mixture of two or more solids is a solution. An example of a gaseous solution is air, although we would need to specify a particular altitude for the composition of this solution. Air gets thinner as one climbs a mountain, for example. We would also need to specify a particular locale. The composition of the air in a room full of people exhaling carbon dioxide is different from the air in an adjoining room where there are no people. An example of a solid solution is a sample of stainless steel alloy. Stainless steel is a homogeneous mixture of chromium, nickel, iron, and carbon. All alloys (brass, bronze, etc.) are also solid solutions.

The components of a solution are known as the solvent and the solute(s). There is only one solvent, but there can be more that one solute. The ***solvent*** is the component that is present in the greatest amount, or the component that has dissolved the other component(s) in itself. The ***solute(s)*** is (are) the component(s) present in the lesser amount or the substance(s) that has (have) dissolved in the solvent. For example, if you had a large bowl of pure, distilled water and you added a few crystals of salt plus a few crystals of sugar, and stirred this with a spoon, you would have a solution with water as the solvent and salt and sugar as the solutes.

Sugar and salt are examples of compounds that dissolve in water quite readily at ordinary room temperatures. There are, however, many substances that do not readily dissolve in water. Examples are sand and sawdust. There is some dependence on temperature for this ability to dissolve. However, at room temperature, substances that dissolve readily in a solvent are said to be **soluble** in that solvent. Substances that apparently do not dissolve readily in a solvent are said to be **insoluble** in that solvent. "Soluble" or "insoluble" are qualitative descriptors of the behavior of compounds in the presence of a solvent. This is simply a qualitative description of the ability of a given compound to dissolve in a solvent at a given temperature; either it does dissolve or it does not.

While a given solute may appear to dissolve in a given solvent as described above, and another may appear not to dissolve, solubility is actually a relative term that depends on both quantity and temperature. For example, even though sugar is soluble in water, it is not possible to dissolve a five-pound bag of sugar in a cup of water. All substances have an upper limit as to how much will dissolve in a given amount of solvent at a given temperature. By the same token, substances that appear to be insoluble do dissolve to a certain extent, however small that may be. In other words, all substances have a certain maximum *quantitative solubility*, such as a certain number of grams of solute per 100 mL of solvent at a given temperature, or a certain number of moles of solute per liter of solution at a given temperature. This quantitative **solubility** is defined as the maximum amount of a solute that will dissolve in a given amount of solvent at a given temperature. Table 10.1 presents the solubilities of some chemicals in water at 25°C.

It is important to note the solubility behavior of chemicals that are gases. Examples are oxygen, nitrogen, carbon dioxide, hydrogen chloride, and ammonia. Even though such chemicals are gases, they can be quite soluble in liquid solvents. It is common practice, for example, to measure the dissolved oxygen in treated and untreated wastewater at wastewater treatment plants. The fuming properties of concentrated HCl (hydrochloric acid, or hydrogen chloride) and of concentrated ammonia ($NH_3$) solutions are indicative of the fact that gases can sometimes be dissolved to a very great extent. Such solubility depends on temperature however, and the effect of temperature is the opposite of what we usually find for solids and liquids. As temperature increases, the solubility of gases decreases. As temperature increases, there is a greater tendency for dissolved gases to escape the liquid phase and be in the gas phase, a phenomenon which is identical to the concept of vapor pressure that was discussed in Chapter 9.

A solution that has the maximum amount of solute represented by the solubility dissolved in it is said to be **saturated** at that temperature. The solution is then saturated with the solute so that no more can be dissolved. Such a solution may be characterized by an amount of solute in the container that is undissolved and remains undissolved indefinitely, even with continuous shaking or stirring. A solution that does not have this maximum amount dissolved in it at the temperature specified by the solubility data is said to be **unsaturated**. Such a solution could be made saturated by simply dissolving more solute until the maximum amount is reached.

A solution may also be **supersaturated**. A *supersaturated* solution is one that has more than the maximum amount of solute specified by the solubility dissolved in it at the indicated temperature. It may seem unlikely that any solution would have more solute dissolved than is possible to be dissolved

**TABLE 10.1** Some Common Compounds and Their Solubilities in Water Expressed in grams/100mL at 25 °C

| Chemical | Solubility |
|---|---|
| NaCl | 35.7 |
| Sucrose (sugar) | 200 |
| $NaC_2H_3O_2 \cdot 3H_2O$ | 76.2 |
| $BaSO_4$ | 0.000222 |
| $Ca(OH)_2$ | 0.185 |

at a given temperature, yet it is possible, and such solutions are described as supersaturated. The question is, how do these solutions achieve the supersaturated state? The usual technique involves temperature.

In this discussion, we have stated repeatedly that temperature is a factor. The solubility of a solute in a given solvent does indeed depend on temperature. Usually, as the temperature increases, the solubility increases. It is usually possible to dissolve more solute if the temperature is increased. The exact behavior of solubility vs. temperature depends on what the solute and solvent are. For example, the solubility of sodium acetate in water increases markedly as temperature is increased. However, the solubility of sodium chloride in water increases only slightly, while the solubility of sodium sulfate in water actually decreases with a temperature increase. The sodium acetate/water behavior provides a good example of the process by which a solution may become supersaturated. If a small amount of water (approximately 5 mL) is added to a relatively large amount of hydrated sodium acetate crystals (approximately 50 grams of $NaC_2H_3O_2 \cdot 3H_2O$) in a beaker or flask, there will be very little evidence of dissolving at 25°C. However, if the container is placed on a hot plate and the temperature raised to near the boiling point of water, all of the sodium acetate will melt and dissolve. If this is followed by a cooling of the container and the sodium acetate crystals do not have a site to form and grow (something absent in a glass container with a smooth interior surface, like a beaker or flask), the solid will stay dissolved and the resultant solution will be supersaturated. The supersaturation can be easily proven by providing the required surface on which the crystals can form and grow. An excellent such surface is a *"seed crystal,"* a tiny solid crystal of the dissolved compound. In our example, if a tiny crystal of solid sodium acetate is added to the solution, the solution will turn completely to solid in a matter of seconds and the large amount of additional solute dissolved at the elevated temperature will precipitate or crystallize.

Two terms that describe the solubility behavior of two liquid substances in each other are miscible and immiscible. *Miscible* describes two liquid substances that completely mix with each other in all proportions. Examples are ethyl alcohol and water, acetone and water, and toluene and benzene. *Immiscible* describes two liquids that apparently do not mix when placed in the same container, even after vigorous shaking or stirring. Examples are oil and water, carbon tetrachloride and water, heptane and water, and gasoline and water. In these cases, two distinct liquid layers form in the container, the heavier (denser) liquid sinking to the bottom and the lighter (less dense) liquid floating on top. As with solids that are insoluble in water, however, there is likely to be at least a small amount of mixing that may not be detectable by simply looking at the mixture. Thus, it is also appropriate to quantify this amount by referring to a particular liquid's quantitative solubility in another, if it obviously is not miscible. For example, Table 10.2 lists the solubilities of a number of alcohols in water at 25°C.

Finally, there are two frequently used terms that rather loosely describe the relative amount of a solute that is dissolved in a solvent. The terms are dilute and concentrated. *Concentrated* describes a solution in which there is a relatively large amount of solute dissolved, while *dilute* refers to a lesser amount. These terms are used when describing solutions of common laboratory acids, such as sulfuric acid and hydrochloric acid. These acids are commonly purchased from a chemical supply company in the concentrated form and, as such, are known as "concentrated sulfuric acid" or "concentrated hydrochloric acid," etc.

**TABLE 10.2**  Solubilities of Common Alcohols in Water Expressed in grams/ 100mL at 25 °C[a]

| Alcohol | Solubility |
|---|---|
| 1-butanol | 8.0 |
| 2-butanol | 12.5 |
| 2-methyl-1-propanol | 10.0 |
| 1-pentanol | 2.4 |
| 1-hexanol | 0.6 |

[a]Alcohols with three or fewer carbons are totally miscible with water.

---

*Chemistry Professionals at Work*      *CPW Box 10.2*

# HIGH PERFORMANCE LIQUID CHROMATOGRAPHY

T he properties of solubility and miscibility are important in some rather sophisticated techniques in analytical laboratories. In CPW Box 9.7, we discussed gas chromatography and how a separation and quantitation of mixture components depends on (1) their relative affinities for the stationary phase, and (2) their relative vapor pressures. In another instrumental chromatography technique called **High Performance Liquid Chromatography** or **HPLC**, both the stationary phase and the mobile phase are liquids that are immiscible. The separation of the mixture components in the column depends on (1) their relative solubilities in the stationary phase, and (2) their relative solubilities in the mobile phase. Those components that have a high solubility in the mobile phase but a low solubility in the stationary phase will emerge from the column quickly, but those with a low solubility in the mobile phase and a high solubility in the stationary phase will emerge from the column slowly. Hence the separation. As with gas chromatography, mixture components are detected electronically as they emerge from the column. HPLC has broad application in many different laboratories.

*For Homework: Find a basic description of HPLC in a reference book and discover how the mixture is introduced into the column and how the electronic detection system works. Write a report.*

---

These concentrated solutions are nearly saturated with the solute. However, after a chemist mixes any of these with water, even if it dilutes the acid only slightly, such as in a 50-50 mixture, the resulting solution is often referred to as "dilute sulfuric acid" or "dilute hydrochloric acid."

It should be noted that when making solutions of these acids, one should always add the acid to the water and not the water to the acid. This has to do with the hazards that acids present to the lab worker. See Chapter 12, Section 12.3.3 for more details on this general rule.

## 10.3 Increasing Dissolution Rate

Sometimes, chemists and laboratory technicians attempt to dissolve a solid in water only to find out that it apparently will not dissolve, even if solubility data show that it will. Such a situation is more a problem with the rate of dissolving than with solubility. If the solute were allowed the time needed, it would dissolve. However, there are several methods for increasing the rate of dissolving.

### 10.3.1 Particle Size

The greater the surface area is for a solid that is exposed to a solvent, the faster the solid will dissolve in the solvent. The surface area of a solid can be dramatically increased by crushing the solid into very small pieces. Thus, large clumps of a solid will dissolve more slowly than powdered solid.

### 10.3.2 Agitation

Shaking a solution can also be an aid to dissolving and is probably the first technique that a novice would try. If the time to dissolve is on the order of hours, however, shaking can be monotonous. In this case, a mechanical shaker can be used. The solution can also be stirred, and stirring can be accomplished with a mechanical (magnetic) stirrer with a magnetic stirring bar. If one is available, an ultrasonic cleaner bath can be very effective, often requiring a matter of minutes when a stirrer or shaker may require hours.

### 10.3.3 Temperature

As mentioned in Section 10.2, higher temperatures can increase solubility. Higher temperatures can also increase dissolving rate. Thus, elevating the temperature and combining this with shaking or stirring can be very effective in getting a solute dissolved quickly.

# 10.4 Mechanism of Dissolving

A look at what happens to molecules and ions at the surface of a solid that has been introduced into a liquid solvent, or at the interface of the two liquids when a liquid is introduced to a liquid solvent, helps to explain why some things dissolve and others do not. In this section, we examine the interactions between the molecules and ions of a solute with the molecules of a solvent. We consider water as the solvent in Sections 10.4.1 and 10.4.2 and nonaqueous solvents in Section 10.4.3.

### 10.4.1 Ionic Solutes in Water

We first take a look at ionic solutes, such as sodium chloride, NaCl. As we stated in Section 5.8, such compounds do not exist as molecules, but rather as formula units in a three-dimensional array of ions. That is why we refer to formula units, rather that molecules, and to formula weights, rather than molecular weights.

When a crystal of an ionic compound is dropped into a container of water, there is an immediate attraction of the little "magnets" we call water molecules for the ionic charges in the three-dimensional ion array, the positive end of the water molecule being attracted to the negative ions and the negative end of the water molecule being attracted to the positive ions. As the water molecules gather around and align themselves, positive ends to negative ions and negative ends to positive ions, they begin to tug away at the ions, attempting to pull them out of the array. In most cases, they succeed in breaking the ionic bonds and in ripping the ions from the array. The result is a *solvated* ion—an ion that is carried off into the water as a separate and distinct positive or negative ion, totally separated from its partner in the array and surrounded by water molecules. At some point, the entire crystal disintegrates and disappears. We then say that the crystal has dissolved. Figure 10.1 is an illustration of this process.

If the attractive strength of the water molecules is not sufficient to break the ionic bonds, then the compound does not dissolve. While most ionic compounds dissolve in water, there are a number that do not dissolve in water. Precipitation formation in a chemical reaction is related to these concepts. If, after mixing two separate solutions together, there are two kinds of ions present that form a stronger ionic bond than the strength of the "solvation bonds" to water molecules, the ionic bonds will form and an undissolved compound (or precipitate) will result.

In general, we can make the following statements about common ionic compounds dissolving in water:

- All nitrates are soluble.
- Most acetates, sulfates, and chlorides are soluble.
- Most carbonates, hydroxides, and phosphates are insoluble.
- Most sodium, potassium, and ammonium salts are soluble.

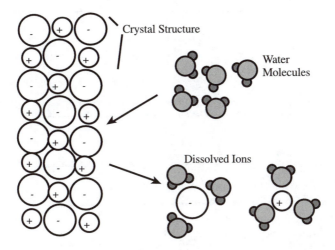

**FIGURE 10.1** An illustration of an ionic crystal dissolving in water. The water molecules pull the positive and negative ions out of the ionic array and the crystal dissolves.

## 10.4.2  Molecular Solutes in Water

Some compounds that consist of molecules (covalent compounds) are soluble in water and some are not. Some of those that are soluble form ions when they dissolve, but most do not. Let us now take a look at what happens at the molecular level in these situations.

Polar solute molecules interact with water molecules in a fashion similar to the way ions interact with water molecules. The negative end of a water molecule is attracted to the positive end of a polar solute molecule, and the positive end of a water molecule is attracted to the negative end of the polar solute molecule. The net effect is a breakdown of the cohesive forces holding the solute molecules together, which causes dissolving or mixing to occur. Thus, polar solutes such as sugars, low molecular weight alcohols, and acetone dissolve readily in water.

A few polar covalent solutes, namely acids, dissolve in water and also form ions in the process. The strength of the interaction with the water molecule is sufficient to break a covalent bond in these solutes and to form ions. Inorganic acids, such as HCl, $H_2SO_4$, and $HNO_3$, completely ionize and are called ***strong acids***. This ionization can be written as follows for hydrochloric acid, HCl, for example.

$$HCl \rightarrow H^+ + Cl^- \tag{10.1}$$

In this example, there are no undissociated HCl molecules remaining. After HCl has dissolved, all HCl molecules are present as the ionized acid, with $H^+$ ions and $Cl^-$ ions now being present in solution.[1]

Organic acids, such as the carboxylic acids (Chapter 6), can dissolve in water but only partially ionize. The usual example of such an acid is acetic acid, $HC_2H_3O_2$. Details of this partial ionization will be discussed in Chapters 11 and 12. Acids that partially ionize are called ***weak acids***.

If the molecular solute is nonpolar, there is no interaction with the polar water molecules, and so no mixing occurs. Thus, nonpolar compounds like alkanes, alkenes, aromatic hydrocarbons, and carbon tetrachloride are immiscible with water.

---

[1]Technically, the hydronium ion, $H_3O^+$, a combination of the hydrogen ion and a water molecule, forms instead of the hydrogen ion, $H^+$. The hydronium ion will be mentioned in Chapter 12, Section 12.4.2.

# LIQUID-LIQUID EXTRACTION

Chemical laboratory workers frequently encounter a technique called *liquid-liquid extraction* in their work. This is a technique in which two immiscible liquids are brought together in a container called a separatory funnel (pictured at right) for the purpose of transferring a solute from one liquid to the other while shaking. The solute transfers from the first liquid to the second because it is more soluble in the second. An example of the application of this technique is in the laboratory analysis of an environmental water for pesticide residue. The water containing the pesticide is placed in the funnel and a nonpolar solvent, such as hexane, is added. The water and the hexane are immiscible and form two separate liquid layers in the funnel as shown. When the funnel is shaken, the pesticide transfers from the water to the hexane. The hexane is then quantitatively analyzed for the pesticide by a technique that requires the solvent for the pesticide be hexane. Dozens of examples of this procedure exist in real-world analytical laboratories.

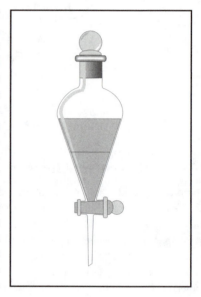

*For Homework: Interview a chemistry professional in your local area that analyzes environmental water samples for pesticide or herbicide residues and find out about his/her use of liquid-liquid extraction. Obtain detailed information about the analysis and write a report.*

## 10.4.3  Nonaqueous Solvents

If water is not the solvent, the interaction between solvent molecules and solute ions or polar solute molecules decreases substantially and can disappear completely. This means that, while ionic compounds and polar molecular solutes may partially dissolve in nonaqueous polar solvents (such as methyl alcohol), their solubility is not as high as in water.

Perhaps more interesting is the case of a nonpolar solvent with a nonpolar solute. Obviously there will be no polar–polar interaction between solvent molecules and solute molecules. One may conclude that such substances are insoluble because of this lack of interaction. However, since the molecules of both the solute and solvent are nonpolar, there are only very weak forces, called **London forces**, that attract solute molecules to other solute molecules and solvent molecules to other solvent molecules. Thus, these two mix and dissolve due to a lack of interactions.

The old adage "like dissolves like" appears to cover all of these cases. Polar (or ionic) substances dissolve in other polar substances, polar (or ionic) substances do not dissolve in nonpolar substances, and nonpolar substances dissolve in other nonpolar substances.

## 10.5  Electrolytes

We have seen that when ionic compounds dissolve in water, the cations dissociate from the anions and both diffuse into the bulk of the solvent as they are separated from each other. This results in free independent cations and free independent anions, each able to move freely through the solution. We have also seen that when inorganic and organic acids dissolve in water, hydrogen ions (hydronium ions) form along with the accompanying anion and these ions are also free independent ions that move freely through the solution. The presence of free, independent ions in solution gives the solution the unique property of electrical conductivity, which means that the solution is capable of conducting an electrical current.

*Electrical current* is the movement of electrical charge through a metal, or otherwise electrically conductive material. For example, when the poles of a battery are connected externally with a wire, electrical current flows through the wire. The charge movement in this case is the movement of electrons through the wire. This current can be used to run various devices that utilize electricity in order to function. These include our calculators, flashlights, automobile starters, etc. Electricity is brought into our homes by this method, the movement of electrons, or electrical current, through power lines. Thus, the various household appliances and convenience devices used in our homes that require household electricity utilize this flow of electrons in order to serve us in the way that they do.

Solutions that conduct an electrical current are able to do so because of the ability of free independent ions to move or diffuse through the solution. An apparatus used for demonstrating this is shown in Fig. 10.2. The apparatus consists of a beaker to hold the solution, a battery as the source of electrical current, a current-measuring device (ammeter), two electrodes consisting of either metal rods, wires, or strips, and wires that connect the battery, ammeter, and electrodes in a complete electrical circuit. The apparatus is often referred to as an *electrolytic cell*. Once all components are connected, electrical current will flow from the battery through the ammeter to one of the electrodes, through the solution (if ions are present) to the other electrode, and back to the battery. If ions are present in the solution, a current will register on the ammeter. If there are no ions present, no current will register.

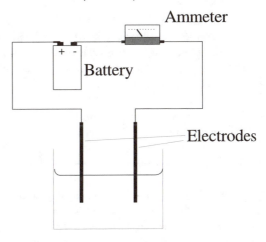

**FIGURE 10.2**  An apparatus for demonstrating the electrical conductivity of solutions.

Solutions that conduct electricity, as well as the chemicals that provide the ions for such conductivity (most ionic compounds and many acids), are called *electrolytes*. Compounds that completely ionize when dissolved in water (most ionic compounds and all strong acids) and their resulting solutions are called *strong electrolytes*. Weak acids (substances that only partially ionize when dissolved in water), water solutions of weak acids, and dilute solutions of strong electrolytes are called *weak electrolytes*. Compounds that dissolve in water but do not ionize, or solutions that have few or no dissolved ions, do not conduct electricity and are called *nonelectrolytes*.

One intriguing mystery surrounding the demonstration of electrical conductivity of solutions, as discussed thus far, is the fact that electrons flow through the wires connecting the battery to the other circuit components, but do not flow through the solution. Electrons do not, indeed cannot, flow through the solution. It is the ability of the ions present to diffuse through the solution that carries the current through the solution.

Therefore the questions are: "What happens to the electrons that flow from the battery to one of the electrodes immersed in the solution?" and "How are electrons generated at the other electrode, which allows electrons to flow back to the battery?" The answer is that oxidation and reduction processes (see Chapter 16) occur at the surfaces of the two electrodes. At the electrode where electrons are apparently

*Chemistry Professionals at Work*                                    *CPW Box 10.4*

# CHLORINE AND SODIUM HYDROXIDE

In 1995, sodium hydroxide NaOH (also called "caustic soda"), ranked eighth and chlorine, $Cl_2$, ranked tenth in quantity of chemicals produced in the United States. They are produced by the "Chlor-Alkali" industry by exactly the process described in the text, through the electrolysis of salt solutions, or "brine." The electrolytic cell is usually a large "diaphragm" cell in which a diaphragm is used to separate the chlorine that forms at the anode and the hydrogen and sodium hydroxide that form at the cathode. Chlorine is a gas that is collected and bottled. Sodium hydroxide is a white, water-absorbing, highly corrosive solid that is crystallized from the electrolysis cell and purified.

Chlorine is used as a raw material in a large variety of products including vinyl chloride, the monomer used in the manufacture of polyvinyl chloride, PVC (see Chapter 6). Sodium hydroxide also has a wide variety of uses. It is used mainly in the paper industry for the treatment of wood, in the manufacture of other chemicals (such as sodium hypochlorite), and in the production of various organic chemicals.

*For Homework: Look up the industrial process by which vinyl chloride is made from chlorine and give the full details in a written report.*

"consumed," **reduction** (or the gaining of electrons by some species in the solution) is taking place. This electrode is called the **cathode** and is connected to the negative pole of the battery so that positive ions diffuse toward it through the solution. At the electrode at which electrons are apparently being generated, **oxidation** (or the loss of electrons by some species in the solution) is taking place. This electrode is called the **anode** and is connected to the positive pole of the battery, thus attracting negatively charged ions in the solution. This process means that new chemical species, the products of the oxidation and reduction, are being formed at the respective electrodes. These new species are often formed in such a quantity that they are visible to the naked eye. The process of oxidation and reduction occurring at the anode and cathode in an apparatus such as the one described here is called **electrolysis**.

For example, a solution of NaCl conducts a current. The oxidation–reduction processes that occur at the electrode surfaces are as follows:

$$\text{Cathode:} \quad Na^+ + e^- \rightarrow Na \tag{10.2}$$

$$2Na + 2H_2O \rightarrow 2NaOH + H_2(g) \tag{10.3}$$

$$\text{Anode:} \quad 2Cl^- \rightarrow Cl_2(g) + 2e^- \tag{10.4}$$

The evidence that these reactions do in fact occur at the electrode surfaces is seen by the vigorous bubbling that occurs at both electrodes, indicating that gases have formed. Evidence that one of the gases is chlorine is the sharp odor of chlorine that can be detected near the beaker.

The industrial production of chlorine is accomplished by just this process. Large electrolytic cells, similar to the design described here, are used. The electrolyte solution is brine (salt water) pumped from the earth. The chlorine is collected and purified.

The electroplating industry also utilizes the concept of the electrolytic cell. An example is in the electroplating of copper. In this example, the process occurring at the cathode is the reduction of copper ions to copper metal.

$$Cu^{2+} + 2e^- \rightarrow Cu \ (s) \tag{10.5}$$

Under proper conditions of cathode preparation, copper ion concentration, acid concentration in the cell, etc., the copper metal that forms will deposit and build up on the surface of the cathode. The cathode in this case is often a decorative piece on which the operator intends to form a layer of fresh copper metal. One requirement of this decorative piece is that it be electrically conductive.

## 10.6  Expressing Concentration

The concentration of a solute in a solution (i.e., the amount of a solute dissolved in a particular amount of solvent or in a particular amount of solution) is a vitally important concept in a variety of situations in real-world chemistry. For example, nitrate in drinking water is unsafe for infants. The amount of nitrate dissolved in a given amount of drinking water, or the concentration of nitrate in drinking water, is important to know in order to determine whether that water is safe for infants. Thousands of chemists and chemistry technicians in the U.S. analyze solutions such as drinking water daily to determine the quantity of a solute dissolved per unit volume of solution. This concentration level of a solute in a solution is thus a very important topic for students of chemistry. In this section, various methods of expressing concentration are defined and calculations associated with the determination of solution concentration are studied.

### 10.6.1  Percent

The concept of percent is well known and was discussed in Chapter 8 in Example 8.7. The basic definition is as illustrated in Eq. (10.6).

$$Percent \ (\%) \ = \ \frac{part}{whole} \times 100 \tag{10.6}$$

For solutions, however, this concept can be confusing because the units used for the "part" and the "whole" are variable. Not only is it common for a solution concentration to be expressed as a "volume percent" (the numerator and denominator both have volume units), or a "weight percent" (the numerator and denominator both have weight units), but it is also common for the numerator units to be weight units and the denominator units to be volume units (weight/volume percent).

$$volume \ (v/v) \ percent \ = \ \frac{volume \ of \ solute}{volume \ of \ solution} \times 100 \tag{10.7}$$

$$weight \ (w/w) \ percent \ = \ \frac{weight \ of \ solute}{weight \ of \ solution} \times 100 \tag{10.8}$$

$$weight/volume \ (w/v) \ percent \ = \ \frac{weight \ of \ solute}{volume \ of \ solution} \times 100 \tag{10.9}$$

Note that the denominator in each case is the quantity of *solution* and not solvent. Also, the units in Eqs. (10.7) and (10.8) can be any unit of weight or volume as long as they are the same in both numerator and denominator. The units in Eq. (10.9) must be grams and milliliters, kilograms and liters, or milligrams and microliters.

### Example 10.1

If 7.3 mL of acetone are dissolved in a water solution such that the total volume of solution is 500.0 mL, what is the concentration of the acetone expressed as a volume percent?

### Solution 10.1

Eq. (10.7), the definition of volume percent, may be used.

$$\text{volume percent} = \frac{\text{volume of solute}}{\text{volume of solution}} \times 100$$

$$\text{volume percent} = \frac{7.3 \text{ mL}}{500.0 \text{ mL}} \times 100$$

$$\text{volume percent} = 1.5\%$$

### Example 10.2

What is the weight percent concentration of NaCl in water if 6.29 grams of NaCl and 250.0 grams of water are used to make the solution?

### Solution 10.2

Eq. (10.8) defines weight percent.

$$\text{weight percent} = \frac{\text{weight of solute}}{\text{weight of solution}} \times 100$$

$$\text{weight percent} = \frac{6.29 \text{ grams}}{(6.29 \text{ grams} + 250.0 \text{ grams})} \times 100$$

$$\text{weight percent} = 2.45\%$$

### Example 10.3

What is the weight/volume percent concentration of a solution if 2.48 grams of solute are dissolved in 600.0 mL of solution?

### Solution 10.3

Eq. (10.9) defines weight/volume percent.

$$\text{weight/volume \%} = \frac{\text{weight of solute}}{\text{volume of solution}} \times 100$$

$$= \frac{2.48 \text{ grams}}{600.0 \text{ mL}} \times 100$$

$$= 0.413\%$$

## 10.6.2 Molarity

We introduced molarity in Section 8.11. *Molarity* is defined as the number of moles of solute dissolved per liter of solution and may be calculated by dividing the number of moles dissolved by the number of liters of solution.

$$\text{molarity} = \frac{\text{moles of solute}}{\text{liters of solution}} \qquad (10.10)$$

As mentioned in Section 8.11, solutions are referred to as being, for example, 2.0 molar, or 2.0 M. The "M" refers to "molar" and the solution is said to have a molarity of 2.0, or 2.0 moles dissolved per liter of solution. It is important to recognize that, for molarity, it is the number of moles dissolved per liter of solution and not per liter of solvent.

### Example 10.4

What is the molarity of a solution that has 4.5 moles of solute dissolved in 300.0 mL of solution?

### Solution 10.4

Eq. (10.10) defines molarity. The number of mL given must first be converted to liters, and so 0.3000 L is used in the denominator.

$$\text{molarity} = \frac{4.5 \text{ moles}}{0.3000 \text{ L}} = 15 \text{ M}$$

### Example 10.5

What is the molarity of a solution of NaOH that has 0.491 grams dissolved in 400.0 mL of solution?

### Solution 10.5

In order to use Eq. (10.10), the grams of solute must be converted to moles by dividing the grams by the formula weight (FW) as we did often in Chapters 8 and 9 [see Eq. (8.1) for cancellation of units]. Once again, the mL are converted to L.

$$\text{molarity} = \frac{\text{grams/FW}}{\text{Liters of solution}} = \frac{(0.491/39.997) \text{ moles}}{0.4000 \text{ L}}$$
$$= 0.0307 \text{ M}$$

## 10.6.3 Molality

The *molality*, m, of a solution is the number of moles of solute dissolved per kilogram of solvent.

$$\text{molality} = \frac{\text{moles of solute}}{\text{kilograms of solvent}} \qquad (10.11)$$

Notice that the denominator here is an amount of solvent and not of solution.

### Example 10.6

What is the molality of a solution of KCl if 0.722 grams of KCl are dissolved in 500.0 grams of water?

---

*Chemistry Professionals at Work*                                    **CPW Box 10.5**

# NITRATE LEVELS IN DRINKING WATER

T
he U.S. Environmental Protection Agency has said that the nitrate level in drinking water should not exceed 10 ppm N in order to be safe. A possible point of confusion is that this threshold level is expressed as ppm nitrogen (N) and not as ppm nitrate ($NO_3^-$). The ppm N is a perfectly legitimate way of expressing the nitrate level because the weight of nitrate is easily converted to the weight of atomic nitrogen and vice versa using the weight fraction defined in Section 10.6.4.

*For Homework: Obtain information about nitrate levels in the drinking water in your state from a state environmental protection agency or health agency. Write a report.*

---

*Solution 10.6*

$$\text{molality} = \frac{\text{moles of KCl}}{\text{kilograms of water}}$$

To calculate moles of KCl, the number of grams is divided by the formula weight. Also, the grams of water need to be converted to kilograms.

$$\begin{aligned}
\text{molality} &= \frac{\text{grams of KCl/FW of KCl}}{\text{grams of water/1000 g per kg}} \\
&= \frac{(0.722/74.551) \text{ moles}}{(500.0/1000) \text{ kg}} \\
&= 0.0194 \text{ m}
\end{aligned}$$

## 10.6.4 Parts per Million

*Parts per million*, or **ppm**, is, as the name implies, the parts of solute per million parts of solution. A "part" can be a mass unit or a volume unit. Typically for solutions, the parts of solute is in milligrams of solute and the million parts of solution is in liters of solution.

$$\text{ppm} = \frac{\text{milligrams of solute}}{\text{liter of solution}} \tag{10.12}$$

Thus, if a solution of copper is labeled as 5 ppm, there are 5 milligrams of copper dissolved per liter of solution. The number of micrograms per mL, $\mu$g/mL, is also ppm.

   The calculation of ppm is more complicated if the weight of solute is given but the ppm of an ion derived from the solute is to be calculated. As an example, if the weight of NaF dissolved in a particular volume is given and the ppm F in the solution is to be calculated, the weight of F contained in the given weight of NaF must be determined. This can be done by multiplying the weight of NaF by the *weight fraction* of F in NaF. The weight fraction is calculated in a manner similar to the calculation of the percent of an element in a compound in Chapter 8 (Section 8.5.2), or by dividing the atomic weight of the element (multiplied by the number of its atoms appearing in the formula) by the formula weight of the compound.

$$\text{weight fraction} = \frac{\text{atomic weight of element} \times \text{number of element's atoms in compound's formula}}{\text{formula weight of compound}}$$

(10.13)

## Example 10.7

If there are 0.338 grams of a solute dissolved in 2000.0 mL of solution, what is the concentration in ppm?

## Solution 10.7

The concentration in ppm is calculated by dividing the mg of solute by the liters of solution to obtain the units seen in Eq. (10.12). We first multiply the 0.338 grams by 1000 to get milligrams and divide the 2000 by 1000 to get liters.

$$\text{ppm} = \frac{338 \text{ mg}}{2.000 \text{ L}} = 169 \text{ ppm}$$

## Example 10.8

What is the ppm F in a solution in which 0.0449 grams of NaF are dissolved in 750.0 mL of solution?

## Solution 10.8

The weight fraction of F in NaF, calculated using Eq. (10.13), is as follows.

$$\frac{\text{atomic weight of F} \times 1}{\text{formula weight of NaF}} = \frac{18.998}{41.988} = 0.452463$$

Then, calculating ppm F, we have the following.

$$\text{ppm F} = \frac{\text{mg F}}{\text{liters of solution}} = \frac{44.9 \text{ mg NaF} \times 0.452463}{0.7500 \text{ L}}$$
$$= 27.1 \text{ ppm F}$$

## 10.6.5  Normality

*Normality* is similar to molarity except that it utilizes a quantity of chemical called the *equivalent* rather than the mole, and is defined as the number of equivalents per liter rather than the number of moles per liter.

$$\text{normality} = \frac{\text{equivalents of solute}}{\text{liter of solution}}$$

(10.14)

If there are 2.0 equivalents dissolved per liter, for example, a solution would be referred to as being 2.0 normal, or 2.0 N. The equivalent is either the same as the mole or it is some fraction of the mole, depending on the reaction involved, and the *equivalent weight*, or *the weight of one equivalent*, is either the same as the formula weight or some fraction of the formula weight. Normality is either the same as molarity, or some multiple of molarity. Let us illustrate with acids in acid/base neutralization reactions.

The *equivalent weight of an acid in an acid/base neutralization reaction* is defined as the formula weight divided by the number of hydrogens lost per molecule in the reaction. Acids may lose one or more hydrogens (per molecule) when reacting with a base.

$$HCl + NaOH \rightarrow NaCl + H_2O \quad \text{(one hydrogen lost by HCl)} \tag{10.15}$$

$$H_2SO_4 + 2\ NaOH \rightarrow Na_2SO_4 + 2\ H_2O \quad \text{(two hydrogens lost by } H_2SO_4\text{)} \tag{10.16}$$

The equivalent weight of HCl is the same as its formula weight and the equivalent weight of $H_2SO_4$ is half the formula weight. The equivalent weight is analogous to the formula weight in the molarity discussions in Section 10.6.2. In the case of the HCl, there is one hydrogen lost per molecule, one equivalent is therefore the same as the mole, and the equivalent weight is the same as the formula weight. In the case of $H_2SO_4$ in the reaction in Eq. (10.16), there are two hydrogens lost per molecule, two equivalents per mole when used as in the above reaction, and the equivalent weight is half the formula weight.

## Example 10.9

What is the normality of a solution of sulfuric acid if it is used as in Eq. (10.16) and there are 2.48 moles dissolved in 250.0 mL of solution?

## Solution 10.9

$$
\begin{aligned}
\text{normality} &= \frac{\text{equivalents of solute}}{\text{liter of solution}} \\
&= \frac{\text{moles of solute} \times \text{equivalents per mole}}{\text{liters of solution}} \\
&= \frac{2.48\ \text{moles} \times 2\ \text{equivalents per mole}}{0.2500\ \text{L}} \\
&= 19.8\ \text{N}
\end{aligned}
$$

Normality applies mostly to acid/base neutralization, but the concentrations of other chemicals used in other kinds of reactions may also be expressed in normality. Treatment of these other chemicals and reactions is beyond the scope of the present discussion.

# 10.7  Preparing Solutions

A very important activity in real-world chemistry laboratories is the preparation of solutions. A variety of reasons exists to prepare a given volume of solution of a given solute at a given concentration. The given volume unit is almost always liters or milliliters and the given concentration unit can be any of those defined in Section 10.6. We now proceed to review the specific processes and calculations involved in this activity for each of the concentration units.

## 10.7.1  Dilution

One method of preparing solutions is by **dilution**. In other words, the solute is already in solution but at a higher concentration than desired. This more concentrated solution must be diluted, or more solvent must be added to it, to get the desired concentration of solute. An important example of this is when making solutions of acids from the concentrated acid solutions, such as 12 M HCl and 18 M $H_2SO_4$, sold by chemical supply companies. Adding a quantity of the solvent to a particular volume of the more concentrated solution decreases the concentration to the desired value. The questions to answer here are: "How much of the more concentrated solution is to be used?" and "How much solvent is to be added?" The volume of the more concentrated solution to be used must be calculated and the amount of solvent is the volume required to bring the solution volume to the desired total volume. Let us now derive a *dilution equation* that can be used for this calculation. Let us begin by assuming molarity is the concentration unit to be employed.

When diluting a solution of a given molarity to one of lesser molarity, solvent is added. The number of moles of the solute does not change. The total number of moles of solute is the same regardless of the amount of solvent present.

$$\text{moles of solute before dilution} = \text{moles of solute after dilution} \qquad (10.17)$$

or, using "B" to mean "before dilution," and "A" to mean "after dilution," we have the following true statement.

$$\text{moles}_B = \text{moles}_A \qquad (10.18)$$

The number of moles of a solute can be calculated by multiplying the molarity (moles per liter) by the volume of solution in liters.

$$\frac{\text{moles}}{\text{L}} \times \text{L} = \text{moles} \qquad (10.19)$$

Thus, Eq. (10.18) is transformed as follows.

$$\frac{\text{moles}}{\text{L}}_B \times \text{L}_B = \frac{\text{moles}}{\text{L}}_A \times \text{L}_A \qquad (10.20)$$

and we then have Eq. (10.21).

$$M_B \times L_B = M_A \times L_A \qquad (10.21)$$

While it appears that the "L" unit (liters) is required for proper dimensional analysis and cancellation with the "L" unit in the denominator of "M", the "mL" unit can also be used. This is acceptable if we include the use of the conversion factor required to convert "mL" to "L", 1000 mL per L,

$$\text{mL} \times \frac{1 \text{ L}}{1000 \text{ mL}} = \text{L} \qquad (10.22)$$

which is the same on both sides of the equation and thus cancels.

$$M_B \times \text{mL}_B \times \frac{1 \text{ L}}{1000 \text{ mL}} = M_A \times \text{mL}_A \times \frac{1 \text{ L}}{1000 \text{ mL}} \qquad (10.23)$$

$$M_B \times \text{mL}_B = M_A \times \text{mL}_A \qquad (10.24)$$

The same can be said about the molarities. Any concentration unit can be used because the conversion factor to convert to molarity is the same on both sides and also cancels. Thus, we can have a more general formula in which the units of concentration and volume are not specified.

$$C_B \times V_B = C_A \times V_A \qquad (10.25)$$

We will be referring to Eq. (10.25) in future discussions as the **dilution equation**. In this equation, "C" symbolizes concentration and "V" symbolizes volume. Any concentration unit (M, %, ppm, N, etc.) can be used and any volume unit (L, mL, etc.) can be used, as long as we recognize that whatever unit is used, it is the same on both sides of the equation. Thus, if the volume after dilution is expressed in mL units, then the volume before dilution will also be expressed in mL units. If the unit of the given concentration after dilution is different from that of the given concentration before dilution, one of the two units must be converted to the other before plugging it into the equation. The form of the dilution equation needed to calculate the volume of the more concentrated solution required to make the final solution is shown below.

$$V_B = \frac{C_A \times V_A}{C_B}$$

(10.26)

## Example 10.10

How many mL of concentrated HCl (12 M) are needed to prepare 250.0 mL of a 0.15 M solution of HCl?

## Solution 10.10

Using Eq. (10.26), we have the following.

$$V_B = \frac{C_A \times V_A}{C_B} = \frac{0.15 \text{ M} \times 250.0 \text{ mL}}{12 \text{ M}} = 3.1 \text{ mL}$$

Note: When diluting solutions of concentrated acids, a practical rule to follow is the "add acid to water rule," which means that some water should be placed in the container first as a safety precaution. See Chapter 12 (Section 12.3.3) for more details.

The process by which solutions are prepared by dilution is important to consider. The volume of the more concentrated solution, $V_B$, as calculated by the dilution equation (see Example 10.10), must be measured via a graduated cylinder, pipet, or similar volume transfer device and placed in a container with a graduation line corresponding to the volume of the more dilute solution being prepared, $V_A$. The choice of transfer device and container depends on the precision that is required. Some pipets, such as the volumetric pipet (see Fig. 10.3(a)) are high-precision transfer devices and are used for analytical work that requires more significant figures than a graduated cylinder can provide. Similarly, volumetric flasks (see Fig. 10.3(b)) are high-precision solution containers, while bottles, beakers and Erlenmeyer flasks are less precise.

Having placed the more concentrated solution in the chosen container, the solvent, usually water, is added in such a quantity that the bottom of the meniscus (the curved surface of the solvent) is resting on the calibration line. For this latter observation, it is necessary for the meniscus to be at eye level. Once the appropriate amount of solvent is added, the solution is shaken, stirred, or swirled so as to make the solution homogeneous. Figure 10.4 illustrates this process using an Erlenmeyer flask as the container.

## Example 10.11

How would you prepare 100 mL of a 5.0 ppm copper solution from a 1000 ppm copper solution?

## Solution 10.11

Using Eq. (10.26), we can calculate how many mL of the 1000 ppm solution are required.

$$V_B = \frac{C_A \times V_A}{C_B} = \frac{5 \text{ ppm} \times 100 \text{ mL}}{1000 \text{ ppm}} = 0.50 \text{ mL}$$

You would thus measure 0.50 mL of the 1000 ppm solution, place this volume in a 100-mL flask, dilute with water to 100 mL, and shake.

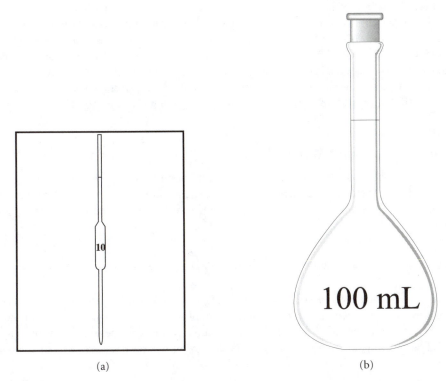

(a)                                                          (b)

**FIGURE 10.3** (a) A volumetric pipet—a high-precision device for measuring the volume of the solution to be diluted ($V_B$). Volumetric pipets have just one calibration line. This example is a 10-mL size. (b) The volumetric flask—a vessel for precisely measuring the volume of solution being prepared ($V_A$). It has just one calibration line. This example is a 100-mL size.

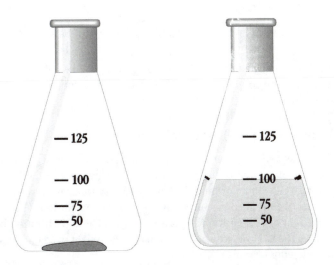

**FIGURE 10.4** An illustration of the preparation of 100 mL of a solution by dilution. The flask at left has the measured volume of the undiluted solution ($V_B$) inside. The flask on the right contains the volume of diluted solution ($V_A$) after swirling. Erlenmeyer flasks such as this one can be used if the accuracy of the concentration is not important to the procedure.

If high precision is important, it should be emphasized that a volumetric pipet or other high-precision device would be used for the transfer of $V_B$ to the container and a volumetric flask would be used for the container. Also, if too much water is added by mistake, such that the solution level is above the required calibration line, the solution must be discarded and the entire process repeated. It is not acceptable in the accurate preparation of solutions to remove solution from the container in order to correct such a situation. That is because the solution being removed likely contains some solute and thus the concentration of the solute would change in an indeterminable way if some were removed.

## 10.7.2   Solid Solute and Weight Percent

Preparation of solutions when the solute is a solid and the concentration is given as a weight percent is fairly common because the weight of a solid solute is easy to measure. Given the weight percent required and the total weight of solution required, the weight of solute is calculated via a rearrangement of Eq. (10.8):

$$\text{grams of solute } = \text{ (\% desired/100)} \times \text{solution weight desired} \qquad (10.27)$$

The actual steps in this preparation involve weighing the amount calculated from Eq. (10.27) and placing it in a container that will hold the required weight of solution, then, while keeping the container on the balance, adding solvent until the total weight required is reached. The solution is made homogeneous by shaking, stirring, or swirling, whichever is most convenient for the vessel chosen. Notice that the identity of the solute is not relevant to the calculation or the process of preparation.

### Example 10.12
How would you prepare 500.0 grams of a 13% (w/w) solution of NaCl in water?

### Solution 10.12
Using Eq. (10.27), we have the following:

$$\text{grams of NaCl } = \text{ (13\%/100)} \times 500.0 \text{ grams} = 65 \text{ grams}$$

Thus, 65 grams of NaCl are weighed and placed in a container capable of holding 500.0 grams of solution. Solvent is then added until the solution weighs 500.0 grams. The solution is made homogeneous by shaking, stirring, or swirling.

## 10.7.3   Liquid Solute and Volume Percent

Expressing concentration as a volume percent is appropriate and common when the solute is a liquid. This is because the volume of a liquid is easy to measure with a graduated cylinder or pipet (whereas the volume of a solid is not). The volume of the solution is also easy to measure with a graduated container or volumetric flask, as has already been discussed for preparation by dilution. The calculation and procedure can actually be thought of as identical to that for dilution. The concentration before dilution, $C_B$ in Eqs. (10.25) and (10.26) is 100%. However, it is also similar to the weight percent calculation and procedure discussed above. The following formula is useful:

$$\text{mL of solute } = \text{ (\% desired/100)} \times \text{volume desired} \qquad (10.28)$$

---

*Chemistry Professionals at Work*                    *CPW Box 10.6*

# COLLIGATIVE PROPERTIES OF SOLUTIONS

Colligative properties of solutions are properties that depend only on the number of particles present in the solution and not on the identity of these particles. Three such properties are boiling point elevation, freezing point depression, and osmotic pressure. For example, the more sugar that is dissolved in a solution, the higher the boiling point, the lower the freezing point, and the higher the osmotic pressure. By knowing how many grams of solute are dissolved and by measuring the degree to which a boiling point is elevated, the freezing point is depressed, or the osmotic pressure is increased, a chemistry professional can determine the formula weight of a molecular solute. This can be especially useful for the determination of the molecular weight of an unknown polymer dissolved in a solvent.

*For Homework: Find some references that provide more detailed information about colligative properties. Write a report giving more details including the definition of osmotic pressure, the calculations involved in the determination of a molecular weight, and the procedures that a chemistry professional might use to determine a molecular weight.*

---

The solute is measured (pipetted, if accuracy is important) into a graduated container (a volumetric flask, if high precision is important) and the solvent is added so that the total volume is the volume of solution that is desired. Notice that the identity of the solute does not enter into the calculation or process (as with weight percent).

## Example 10.13

How would you prepare 500.0 mL of a 15% (v/v) ethanol solution in water?

## Solution 10.13

$$\text{volume to measure} = (15\%/100) \times 500.0 \text{ mL} = 75\text{mL}$$

Thus, 75 mL of ethanol are measured into the vessel to contain the solution. Water is then added to the 500.0 mL level and the solution is stirred, shaken, or swirled.

## 10.7.4  Solid Solute and Weight-to-Volume Percent

The fact that the weights of solid solutes are more conveniently measured than their volumes, and the fact that volumes of liquid solutions are more conveniently measured than their weights, likely gave rise to the

---

*Chemistry Professionals at Work*          CPW Box 10.7

# WHY DOESN'T THIS WORK?

A new procedure that a chemistry technician was using called for 50 mL of a 50% solution of NaOH, but did not specify weight %, volume %, or weight/volume %. She proceeded to prepare the solution, assuming weight/volume %, which was the easiest to prepare—25 grams of NaOH dissolved and diluted to 50 mL of solution. She soon discovered that the seemingly large pile of NaOH weighed on the balance caused the volume to reach 50 mL with just a small amount of water added and would not dissolve completely, even with ultrasonic agitation.

The purpose of the solution was to neutralize a relatively high concentration of sulfuric acid and there were no apparent consequences of lack of accuracy in the solution concentration. In fact, after giving the situation some thought, she decided that it was the 25 grams of solid NaOH that was actually important for this neutralization, and not the solution concentration. She then proceeded to add more water to the solution until the NaOH dissolved. The sulfuric acid was subsequently neutralized and the experiment completed successfully.

At the technician's suggestion, the portion of the procedure indicating the concentration to be 50% was rewritten to read 25 grams of NaOH with sufficient water added to dissolve the solid.

*For Homework: There are some federal laws governing the alteration of standard operating procedures (SOPs) in laboratories that are under the jurisdiction of the EPA or the FDA. These laws have come to be known as GLP. In the reference below, find and report the specific requirements for such changes.*

*Kenkel, J.V., A Primer on Quality in the Analytical Laboratory, CRC Press/Lewis Publishers, Inc. (1999).*

---

method of weight/volume percent that is sometimes used for expressing solution concentration. The amount of solute to be weighed is a percentage of the total solution volume, rather than total solution weight.

$$\text{grams of solute} = (\% \text{ desired}/100) \times \text{solution volume desired} \qquad (10.29)$$

Thus, the solute can be weighed, placed in the container, and the solvent added up to the total solution volume desired.

## Example 10.14
How would you prepare 500.0 mL of a 5.0% (w/v) solution of NaCl?

## Solution 10.14

$$\text{weight to measure} = (5.0\%/100) \times 500.0 \text{ mL} = 25 \text{ grams}$$

Thus, 25 grams of NaCl are weighed, placed in the container, and dissolved in some water. More water is then added such that the total solution volume is 500.0 mL and the solution is shaken.

The compositions of water solutions prepared with weight/volume percent concentrations are nearly identical to water solutions prepared with weight/weight percent concentrations if the weight/volume percent and weight/weight percent are numerically the same. This is because the density of water and dilute water solutions is very nearly equal to 1. For example, 300.0 mL of a dilute water solution is approximately equal to 300.0 grams of that solution. The amounts of water used during the acts of preparing the two solutions are nearly the same, and so the compositions are nearly the same.

Frequently it is unclear whether a certain percent concentration called for in a given method is w/w or w/v. Since for dilute water solutions, solutions prepared by the two methods would be nearly identical anyway, it is best to simply choose the most convenient method of the two. In the event that the water solution is not a dilute solution, then the density of the solution may not be equal to 1. In this case, 300.0 mL of the solution would not equal 300.0 grams of the solution, and the two compositions would differ. In order to be accurate, specific concentration units (w/v or w/w) would need to be provided, or other information must be given, such as the number of grams of the solute that are needed in the application in which the solution is used. (See CPW Box 10.7.)

## 10.7.5 Solid Solute and Molarity

To prepare a given volume of a solution of a given molarity of solute when the weight of the pure, solid solute is to be measured, it is necessary to calculate the grams of solute that are required. This number of grams can be calculated from the desired molarity and the volume of solution that is desired. Grams can be calculated by multiplying moles by the formula weight, as we often saw in Chapter 8 [see also Eq. (8.2)],

$$\text{moles} \times \frac{\text{grams}}{\text{mole}} = \text{grams} \tag{10.30}$$

and the moles that are required can be calculated from the volume and molarity [compare Eq. (10.31) with Eq. (8.11)].

$$\text{moles} = \text{liters} \times \frac{\text{moles}}{\text{liter}} \tag{10.31}$$

Combining Eqs. (10.30) and (10.31) gives us the necessary formula.

$$\text{grams} = \text{liters} \times \frac{\text{moles}}{\text{liter}} \times \frac{\text{grams}}{\text{mole}} \tag{10.32}$$

The liters in Eq. (10.32) are the liters of solution that are desired; the moles per liter is the molarity desired; and the grams per mole is the formula weight of the solute. Thus, we have

$$\text{grams to weigh} = L_D \times M_D \times FW_{SOL} \tag{10.33}$$

in which $L_D$ refers to the liters that are desired, $M_D$ to desired molarity, and $FW_{SOL}$ to the formula weight of solute. The grams of solute thus calculated are weighed out, placed in the container, and water is added to dissolve and dilute the solution to volume.

## Example 10.15

How would you prepare 500.0 mL of a 0.20 M solution of NaOH from pure, solid NaOH?

## Solution 10.15

Using Eq. (10.33), we have the following.

$$\text{grams to weigh} = L_D \times M_D \times FW_{SOL}$$
$$\text{grams to weigh} = 0.5000 \times 0.20 \times 40.00 = 4.0 \text{ grams}$$

The chemist would weigh 4.0 grams of NaOH, place it in a container with a 500 mL calibration line, add water to dissolve the solid, and then dilute to volume and shake, stir, or swirl.

If the solute is a liquid (somewhat rare), the weight calculated from Eq. (10.33) can be measured on a balance, but rather inconveniently. However, this weight can also be converted to mL by using the density of the liquid (Section 7.6.5). In this way the volume of the liquid can be measured, rather than its weight, and it can be pipetted into the container.

Sometimes it may be necessary to prepare a solution of a certain molarity by diluting another solution, the concentration of which is known in units such as w/v percent. As noted earlier in Section 10.7.1, the dilution formula can only be applied when the two concentrations have the same units. In this case, the w/v percent can be converted to molarity and the volume to be diluted can be calculated in the usual way.

## Example 10.16

How would you prepare 500.0 mL of a 0.20 M solution of NaOH from a NaOH solution that is 10% (w/v)?

## Solution 10.16

$$10\% \text{ (w/v)} = 10 \text{ grams per } 100 \text{ mL} = 10 \text{ grams per } 0.10 \text{ L}$$
$$= 100 \text{ grams per L}$$

As in Example 10.5, the moles are calculated by dividing the grams by the FW and molarity is calculated by dividing these moles by the liters.

$$\text{Molarity} = \frac{\text{grams/FW}}{1L} = \frac{100/(39.997) \text{ moles}}{1L} = 2.50019 \text{ M}$$

Applying the dilution equation gives the following:

$$C_B \times V_B = C_A \times V_A$$

$$2.50019 \text{ M} \times V_B = 0.20 \text{ M} \times 500.0 \text{ mL}$$

$$V_B = 40 \text{ mL (2 sig figs)}$$

Thus, 40 mL (2 sig fig) of the 10% NaOH would be measured into a 500 mL container, diluted to volume, and made homogeneous.

## 10.7.6   Solid Solute and Molality

Molality was defined in Section 10.6.3 as the number of moles of dissolved solute per kilogram of solvent. Solutions of a particular molality are prepared by dissolving the required number of grams of solute in the required number of kilograms of solvent. The fundamental difference between molality and all other units of concentration is that the *amount of solvent* is measured, and not the amount of solution. Thus, the solutions are not prepared by diluting with the solvent to the specified volume of solution. Rather, a specified amount (weight) of solvent is added to the weight of solute.

A version of Eq. (10.11) is useful here.

$$\text{moles of solute} = \text{kg of solvent} \times \text{molality} \qquad (10.34)$$

Utilizing the fact that molality is moles of solute per kg of solvent plus the fact that formula weight is the grams per mole, we have

$$\text{grams of solute} = \cancel{\text{kg of solvent}} \times \frac{\text{moles of solute}}{\cancel{\text{kg of solvent}}} \times \frac{\text{grams of solute}}{\cancel{\text{mole of solute}}} \qquad (10.35)$$

or,

$$\text{grams of solute} = \text{kg}_S \times m_D \times \text{FW}_{SOL} \qquad (10.36)$$

in which $\text{kg}_S$ is the kilograms of solvent, $m_D$ is the molality desired, and $\text{FW}_{SOL}$ is the formula weight of the solute.

## Example 10.17
How would you prepare a solution in which NaCl must be dissolved in 250.0 grams of water to produce a solution that is 0.20 m in NaCl?

## Solution 10.17
Using Eq. (10.36), we have:

$$\text{grams of solute} = \text{kg}_D \times m_D \times \text{FW}_{SOL}$$
$$\text{grams of NaCl} = 0.2500 \text{ kg} \times 0.20 \text{ m} \times 58.443 \text{ g/mol}$$
$$= 2.9 \text{ grams}$$

Thus, 0.15 grams of NaCl would be weighed and dissolved in 250.0 kg of water. The solution would be rendered homogeneous by shaking, swirling, or stirring.

The advantage of preparing molal solutions instead of molar or percent-based solutions is that the concentrations of molal solutions are free of the effects of temperature. While volumes of solutions and liquid solutes change with temperature, their weights do not. For molar and percent-based units, the temperature of a solution must be 20°C during preparation (the volume calibration on most glassware is for 20°C). With the molal unit, the solvent and solute can be at any temperature, since their weights would not change as the temperature changes. Also, it is the molality of solutions on which colligative properties depend (see CPW Box 10.6).

## 10.7.7 Solid Solute and ppm

In order to prepare a given volume of a solution of a given ppm of solid solute, or a given ppm of an element in the solute formula, the weight (grams) of the solute must be determined. If the expressed ppm is of the solid solute itself (e.g., ppm Fe prepared from iron metal or ppm $CaCO_3$ prepared from pure solid $CaCO_3$, etc.), then the following formula is useful:

$$\text{grams to weigh} = \frac{\text{liters of solution desired} \times \text{ppm}}{1000} \qquad (10.37)$$

In this formula, multiplying ppm (mg per L) by liters gives milligrams and dividing by 1000 converts millgrams to grams.

If the expressed ppm is of an element within the formula of the solute, such as ppm F prepared from the compound NaF, or ppm Cu prepared from the compound $CuSO_4 \cdot 5H_2O$, then the weight fraction discussed in Section 10.6.4 [Eq. (10.13)] is useful. We must divide the desired ppm by the weight fraction in order to convert the weight of the element to the weight of the solute so that the required weight of solute can then be determined:

$$\text{grams to weigh} = \frac{\text{liters of solution desired} \times \text{ppm/weight fraction}}{1000} \qquad (10.38)$$

### Example 10.18

How many grams of iron are required to prepare 500.0 mL of a 15 ppm Fe solution?

### Solution 10.18

Iron metal itself is the solute, so Eq. (10.37) applies.

$$\text{grams to weigh} = \frac{\text{liters of solution desired} \times \text{ppm}}{1000}$$

Thus we have the following.

$$\text{grams to weigh} = \frac{0.5000 \text{ L} \times 15 \text{ ppm}}{1000} = 0.0075 \text{ grams Fe}$$

### Example 10.19

How would you prepare 500.0 mL of a 25 ppm Cu solution using $CuSO_4 \cdot 5H_2O$ as the solute?

### Solution 10.19

Since $CuSO_4 \cdot 5H_2O$ is the solute, and not Cu metal, the weight fraction must be determined so that Eq. (10.38) can be used to determine the weight of $CuSO_4 \cdot 5H_2O$ needed. The weight fraction is calculated as in Eq. (10.13).

$$\text{weight factor} = \frac{\text{atomic weight of Cu} \times \text{number of Cu atoms in } CuSO_4 \cdot 5H_2O \text{ formula}}{\text{formula weight of } CuSO_4 \cdot 5H_2O}$$

$$= \frac{63.546 \text{ grams per mole} \times 1 \text{ atom per formula}}{249.675 \text{ grams per mole}} = 0.2545149$$

Then, substituting into Eq. (10.38),

$$\text{grams to weigh} = \frac{\text{liters of solution desired} \times \text{ppm/weight fraction}}{1000}$$

$$= \frac{0.5000 \text{ L} \times 25 \text{ ppm}/0.2545149}{1000} = 0.051 \text{ grams } CuSo_4 \cdot 5H_2O$$

Thus, 0.051 grams of $CuSO_4 \cdot 5H_2O$ are weighed and placed into a flask with a 500.0 mL calibration line. Water is added to dissolve this salt, then more water is added to reach the calibration line and the solution shaken.

## 10.7.8   Solid Solute and Normality

To determine the weight of a solid that is needed to prepare a solution of a given normality, we can derive an equation in a manner similar to that in Section 10.7.5 for molarity.

$$\text{grams to weigh} = \text{liters} \times \frac{\text{equivalents}}{\text{liter}} \times \frac{\text{grams}}{\text{equivalent}} \tag{10.39}$$

$$\text{grams to weigh} = L_D \times N_D \times EW_{SOL} \tag{10.40}$$

in which $L_D$ is the liters of solution desired, $N_D$ is the normality desired, and $EW_{SOL}$ is the equivalent weight of the solute.

As in Section 10.6.5, acid/base neutralization reactions will be illustrated here. In order to calculate the equivalent weight of an acid, the balanced equation representing the reaction in which the solution is to be used is needed so that the number of hydrogens lost per molecule in the reaction can be determined. The equivalent weight of an acid is the formula weight of the acid divided by the number of hydrogens lost per molecule (see Section 10.6.5).

### Example 10.20

How many grams of $KH_2PO_4$ are needed to prepare 500.0 mL of a 0.200 N solution if it is to be used as in the following reaction?

$$KH_2PO_4 + 2\ KOH \rightarrow K_3PO_4 + 2\ H_2O$$

### Solution 10.20

There are two hydrogens lost per molecule of $KH_2PO_4$. Therefore the equivalent weight of $KH_2PO_4$ is the formula weight divided by 2. Utilizing Eq. (10.40), we have the following.

$$\text{grams to weigh} = 0.5000\ L \times 0.200\ \text{equiv per L} \times \frac{136.09}{2}\ \text{grams per equiv} = 6.80\ \text{grams}$$

## 10.7.9   Summary of Calculations

The various equations derived in this section for use in solving solution preparation problems are summarized in Table 10.3.

**TABLE 10.3**   The Various Equations Derived in Section 10.7 for Use in Calculating the Amount of Solute Needed for Solution Preparation

| | |
|---|---|
| Dilution | Equation 10.26 |
| Solid solute and weight percent | Equation 10.27 |
| Liquid solute and volume percent | Equation 10.28 |
| Solid solute and weight to volume percent | Equation 10.29 |
| Solid solute and molarity | Equation 10.33 |
| Solid solute and molality | Equation 10.36 |
| Solid solute and ppm | Equation 10.37 or 10.38 |
| Solid solute (acid) and normality | Equation 10.40 |

# TECHNICIAN SPECIALIST IN A POLYMER DIVISION

My name is V. Michael Mautino. I am a chemistry professional employed by Bayer Corporation at their North American headquarters in Pittsburgh, Pennsylvania. I am a Technician Specialist working in the applications and development area for the Technical Insulation Group in Bayer's Polymer Division.

I have been involved in the development of polyurethane rigid foam insulation for various market applications including refrigerators, entry doors, overseas cargo containers, and water heaters. My responsibilities have included formulating and evaluating polymer chemistry (primarily in the bench laboratory area), overseeing machine laboratory evaluations of rigid foam systems, and conducting production scale trials at customer facilities.

*Homework: Check out the Internet site for the Bayer Corporation and write a report on this company's activities.*

http://www.bayer.com/ (as of 6/4/00)

## 10.8   Homework Exercises

1. Describe and compare each of the following terms.
   (a) Solute
   (b) Solvent
   (c) Solution
2. Define each of the following terms that relate to solutes.
   (a) Soluble
   (b) Insoluble
   (c) Solubility
3. Define each of the following terms that relate to solutions.
   (a) Saturated
   (b) Unsaturated
   (c) Supersaturated
4. Point out the differences between the following pairs of terms.
   (a) Miscible and immiscible
   (b) Dilute and concentrated
5. State whether each statement is true or false.
   (a) A solution is a homogeneous mixture.
   (b) The solute is the solution component with the highest concentration.
   (c) A solution can contain more than one solute and more than one solvent.
   (d) The physical state of solutions can only be liquid.
   (e) The solute in a solution can be a solid, liquid, or gas.
   (f) If two liquids are miscible, a mixture of the two will be homogeneous.
   (g) Nonpolar substances tend to be soluble in water.
   (h) Two liquids that are not capable of mixing and forming a solution are said to be miscible.
   (i) An increase in temperature increases the solubility of most solids.
   (j) A dilute solution contains more solute than a concentrated solution.
   (k) Evidence of saturation is some undissolved solute in the container with the solution.
   (l) Since supersaturated solutions have more than the maximum amount of dissolved solute than is possible to be dissolved, one cannot make them without raising the temperature.
   (m) Carbon tetrachloride, a nonpolar liquid, is immiscible with water.
   (n) The solubility of NaCl in water is 36.0 grams per 100 grams of water. Therefore, a solution of NaCl that contains 35.2 grams in 100 grams of water is an unsaturated solution.
6. Answer each of the following with just one word.
   (a) What word describes a solution that has some undissolved solid solute present, even after shaking for a period of time?
   (b) What word describes two liquids that form two distinct liquid layers after being placed in the same container and shaken?
   (c) What is the solution component that is present in the lesser amount in a solution?
   (d) The maximum amount of $Li_2CO_3$ that can be dissolved in 100 grams of water at 20°C is 1.33 grams. This is the known as the _____ of $Li_2CO_3$ .
   (e) What is a word that describes a solution that has more solute dissolved than the maximum amount possible at a given temperature?
   (f) If you were to take a crystal of sodium nitrate and place it in a beaker of water and stir, it would dissolve. We say that the sodium nitrate is _____ in water.
   (g) _____ sulfuric acid, as purchased from a chemical supply company is 18 M.
7. A new chemistry technician who has come to work in the same lab that you work in says that when she tested a compound for its solubility in water (a pea-sized crystal in 5 mL), she found it to be insoluble. She is confused, however, because a chemistry handbook says that its solubility is

0.0135 grams per 100 mL. How would you explain to her that while her conclusion is correct, so is the handbook?

8. Describe three different methods of agitation that would help speed up the rate of dissolving a solid solute in a solvent.

9. Why is particle size a factor in dissolution rate?

10. In terms of sodium ions, chloride ions, and water molecules, describe exactly what happens when a crystal of NaCl dissolves in water.

11. Suppose a scientist is close to a breakthrough on a new microscope through which he says he will be able to see atoms, ions, and molecules. Suppose he wants to try it out by focusing on the process of an ionic solid dissolving in water, just to see what happens to the ions and water molecules as the solid dissolves. What would you tell him he should expect to see.

12. Based on the general rules listed in Section 10.4.1 for ionic solids, which of the following would be soluble in water and which would not: $NaBr$, $K_2SO_4$, $Fe(OH)_3$, $NH_4Cl$, $AgNO_3$, $Ag_2CO_3$, $Ba(C_2H_3O_2)_2$, $Na_3PO_4$.

13. What general class of compounds dissolve and ionize in water, even though they are not ionic? Give three examples.

14. Why does carbon tetrachloride, a nonpolar organic liquid, not dissolve in water?

15. What are "London forces?" How are London forces considered when discussing the solubility of a nonpolar solute in a nonpolar solvent?

16. What is meant by the phrase "like dissolves like?" Explain the chemical basis for this statement.

17. Why does a solution of NaCl conduct electricity while pure distilled water does not?

18. What is wrong with this statement: "Solutions conduct electricity because electrons can flow through them?"

19. Distinguish between strong electrolytes, weak electrolytes, and nonelectrolytes. Give examples of each.

20. Explain a way you may be able to determine whether an unknown compound is ionic or covalent.

21. What general process occurs at the cathode in a conductivity apparatus? Name some specific examples and give a description of the chemicals that are formed in each case.

22. Define each of the following terms.
    (a) Molarity
    (b) Molality
    (c) Normality
    (d) ppm

23. If 33.2 mL of ethyl alcohol are dissolved in water so that the total volume of solution is 350.0 mL, what is the concentration of this solution when expressed as a volume percent?

24. Consider a solution in which 30.0 mL of acetone are dissolved in 70.0 mL of water. Is this the same as a solution in which 30.0 mL of acetone are dissolved in 100.0 mL of a solution with water as the solvent? Explain.

25 What is the weight percent of each of the following solutes in the solutions given? Explain the similarities and differences in your answers.
    (a) 0.883 grams of NaCl dissolved in 90.0 grams of water.
    (b) 0.883 grams of NaCl dissolved in 90.0 grams of solution.
    (c) 0.883 grams of $Na_2SO_4$ dissolved in 90.0 grams of water.
    (d) 0.883 grams of $Na_2SO_4$ dissolved in 90.0 grams of solution.

26. What is the weight/volume percent concentration of each of the following solutes in the solutions given? Explain the similarities of your answers.
    (a) 2.337 grams of $K_2SO_4$ dissolved in 200.0 mL of solution.
    (b) 2.337 grams of KCl dissolved in 200.0 mL of solution.

27. If you prepared a solution in which 2.337 grams of KCl were dissolved in 200.0 mL of water, would this solution be exactly the same as the one described in #26(b)? Explain.

28. Calculate the molarity of each of the following solutes in the solutions given.
    (a) 0.339 moles of a solute are dissolved in 250.0 mL of solution.
    (b) 0.449 grams of $KNO_3$ are dissolved in 350.0 mL of solution.
    (c) 0.281 moles of $NH_4OH$ are dissolved in 500.0 mL of solution.
    (d) 1.227 grams of $BaCl_2 \cdot 2H_2O$ are dissolved in 200.0 mL of solution.

29. A chemist finds 250.0 mL of a solution in which the container label says 7.33 grams of $BaCl_2$ are dissolved. What is the molarity of this solution?

30. (a) If you had a solution of KOH that is 0.450 M, what is the concentration expressed in weight/ volume %?
    (b) If you had a solution of KOH that is 2.93% (weight to volume), what is the concentration in molarity?

31. What is the molality of a solution of NaI if 0.2445 moles are dissolved in 1.500 kg of water?

32. What is the molality of a solution of NaI if 8.384 grams are dissolved in 750.0 grams of solvent?

33. What is the ppm of iron in each of the following solutions?
    (a) 5.4 mg of iron metal are dissolved in 0.200 L of solution.
    (b) 0.0228 grams of iron metal are dissolved in 750.0 mL of solution.
    (c) 88.2 mg of $FeCl_3 \cdot 6H_2O$ are dissolved in 0.500 L of solution.
    (d) 0.0321 grams of $Fe(NO_3)_3 \cdot 9H_2O$ dissolved in 1500.0 mL of solution.

34. What is the ppm Cl in each of the following solutions?
    (a) 44.5 mg NaCl dissolved in 800.0 mL of solution.
    (b) 0.227 grams of KCl dissolved in 0.250 L of solution.
    (c) 0.145 grams of $BaCl_2 \cdot 2H_2O$ dissolved in 500.0 mL of solution.
    (d) 78.11 mg of $AlCl_3 \cdot 6H_2O$ dissolved in 0.200 L of solution.

35. The concentration of Hg in drinking water should not exceed 0.0020 ppm. If a 5.0-mL sample of water has 0.000050 mg of Hg dissolved in it, what is the concentration in ppm? Does it exceed the limit?

36. What is the normality of each of the following solutions?
    (a) 0.268 equivalents are dissolved in 0.150 L of solution.
    (b) 0.337 equivalents are dissolved in 250.0 mL of solution.

37. Answer the following questions pertaining to the reaction below between $H_3PO_4$ and NaOH.

$$H_3PO_4 + 3 \ NaOH \rightarrow Na_3PO_4 + 3 \ H_2O$$

    (a) What is the equivalent weight of $H_3PO_4$?
    (b) What is the normality of a solution of $H_3PO_4$ if there are 0.302 moles dissolved in 250.0 mL of solution?

38. What is the normality of HBr if it is used as in the following equation and 0.75 moles are dissolved in 1000 mL?

$$HBr + NaOH \rightarrow NaBr + H_2O$$

39. A 500.0 mL solution of HCl contains 2.0 moles. What is the normality of the acid?

40. If 800.0 mL solution of $H_2SO_4$ contains 0.15 moles, what is the normality of the acid? See Eq. (10.16).

41. What is the dilution equation? How is it used in the preparation of solutions?

42. How many mL of concentrated HCl (12.0 M) are needed to make each of the following?
    (a) 250.0 mL of 0.35 M HCl    (b) 1.50 L of 0.60 M HCl

43. How many mL of concentrated (18.0 M) sulfuric acid are needed to prepare each of the following:
    (a) 500.0 mL of a 0.35 M $H_2SO_4$? (b) 2.50 L of 0.050 M $H_2SO_4$?

44. How would you prepare the following?
    (a) 500.0 mL of a 0.20 M solution of nitric acid starting with concentrated nitric acid that is 16.0 M?
    (b) 1.20 L of a 0.50 M solution of acetic acid starting with concentrated acetic acid that is 17 M?

45. A chemist gives a laboratory technician a flask with a 2.00 L calibration line and a bottle of 12 M hydrochloric acid and asks her to prepare 2.00 L of a 1.5 M solution. If you were the technician, tell how you would do it.

46. In each of the following, tell how many mL of the more concentrated solution are needed to prepare 500.0 mL of the less concentrated solution.
    (a) The more concentrated solution: 1000.0 ppm Fe
       The less concentrated solution: 30.0 ppm Fe
    (b) The more concentrated solution: 36.0 N $H_2SO_4$
       The less concentrated solution: 5.0 N $H_2SO_4$
    (c) The more concentrated solution: 5.0 w/v %
       The less concentrated solution: 1.0 w/v %

47. When would you use a volumetric pipet and a volumetric flask to prepare a solution by dilution, as opposed to a graduated cylinder and a beaker?

48. How many grams of solute and how many grams of solvent are needed to prepare the following.
    (a) 50.0 grams of a 25.0 weight percent solution?
    (b) 1.20 kg of a 15.0 weight percent solution?

49. How many mL of ethyl alcohol are needed to prepare the following?
    (a) 300.0 mL of a 15.0 volume percent solution.
    (b) 24.0 L of a 1.50 volume percent solution.

50. If a person's blood alcohol level is determined to be 0.15% by volume and this person has 5.0 liters of blood in his/her body, how many liters of alcohol are in his/her blood?

51. A 15% (weight-to-volume) solution of $Ba(OH)_2$ is needed. How many grams of solid $Ba(OH)_2$ would be used if you prepared 250.0 mL of the solution?

52. How many grams of NaOH are needed to prepare 500.0 mL of a 10.0% weight-to-volume solution?

53. A 3.5% (weight-to-volume) NaCl solution is a solution that is corrosive to many metals and chemists often use it to simulate sea water. If a chemist who is studying the corrosion rate of a metal alloy exposed to this solution asks you to make a 1.5 liters of it using pure solid NaCl, how many grams would you need?

54. How many grams of solute are required for each of the following? Since the volume of solution and concentrations are the same, why are your answers different?
    (a) 500.0 mL of a 0.25 M solution of $KH_2PO_4$.
    (b) 500.0 mL of a 0.25 M solution of $K_2HPO_4$.

55. Imagine yourself as a laboratory technician. Your supervisor asks you to prepare 250.0 mL of a 0.60 M solution of KOH and hands you a container of pure solid KOH and a piece of glassware that has a 250.0 calibration line. Tell how you would prepare this solution using these materials.

56. A chemist needs a 0.75 M solution of sodium carbonate and asks you to make it. You find that you have a bottle of pure solid $Na_2CO_3$ and a solution that is 5.5 M in the lab. What are two different ways to prepare 500.0 mL of the solution with the materials at hand and a flask with a 500.0 mL calibration line?

57. A chemist needs a 500.0 mL of a 0.35 M solution of potassium dihydrogen phosphate ($KH_2PO_4$). Suppose you are assigned to prepare this solution. You have the pure solid available, but you also have a solution that is 4.0 M.
    (a) If you decide to use the pure solid, how would you prepare the solution?
    (b) If you decide to use the 4.0 M solution instead, how would you prepare the solution?

58. How would you prepare 750.0 mL of a 0.35 M solution of $Na_2SO_4$ using pure, solid $Na_2SO_4$ and a container with a 750.0 calibration line?

59. How would you prepare a solution of solid $NH_4Cl$ if 400.0 grams of solvent are to be used for a solution that is 0.40 m?

60. How many grams of copper metal are needed to prepare 250.0 mL of a 5.0 ppm solution?

61. How many grams of NaCl are needed to prepare 500.0 mL of a 25 ppm solution of sodium?

62. How would you prepare 500.0 mL of a 50.0 ppm phosphorus solution if you started with pure, solid $KH_2PO_4$?

63. How many grams of $H_3PO_4$ are need to prepare 200.0 mL of a 0.25 N solution if it is to be used as in the following reaction:

$$H_3PO_4 + 3KOH \rightarrow K_3PO_4 + 3H_2O$$

64. How many grams of $NaH_2PO_4$ are needed to prepare 1.00 liter of a 1.5 N solution if two hydrogens are lost per formula unit in the reaction?

## For Class Discussion and Reports

65. In Example 10.10, we indicated that the molarity of concentrated hydrochloric acid is 12 M. Concentrated solutions of other acids and ammonium hydroxide are of similar molarity. From the list below, choose two and look them up in a chemical supply catalog (ask your instructor for a copy) or on a chemical company's Internet site, such as that of the Fisher Chemical Company (www.fishersci.com) (as of 6/4/00), or any other source you may want. See if you can locate the molarity, the weight percent, and/or the density (g/mL) of the concentrated solutions. If you discover only the molarity, calculate the density and weight percent. If you find only the weight percent, calculate the density and the molarity, etc. Write a report on how easy or how difficult it was to find the information and/or to calculate these quantities.

List:   HCl, $H_2SO_4$, $HNO_3$, $HC_2H_3O_2$, $H_3PO_4$, $NH_4OH$

66. Solutions of metals are available from chemical supply companies as 1000 ppm solutions. They are called "Atomic Absorption Standards." Try to find these solutions in a chemical supply catalog. Also locate the inventory of such solutions in your laboratory and check out the label. Is there any indication from either source on how we can be sure that the solution is really 1000 ppm? For example, is the solution "certified?" If so, by whom? Do you find the word "traceable" or "NIST" in the catalog or on the label? Look up these words and see if you can tell what they have to do with being sure that the solutions are really 1000 ppm. Write a report.

67. Look in *Standard Methods for the Examination of Water and Wastewater* for a method for analyzing an environmental water sample for the metal of your choice. Identify all solutions that must be prepared in order to carry out the method. Write a report detailing exactly how to prepare these solutions. Do the required calculations, if necessary, as part of your report.

# 11

# Chemical Equilibrium

## 11.1 Introduction

Consider an experiment in which all the students in your class line up against the walls of the classroom, half on one side and half on the other. Then consider what would happen when one student on each side walks to the opposite side at exactly the same speed, passing each other in the middle. As soon as each of these students arrives at the other side, another student walks back to the opposite side, continuing at the same speed. Now imagine continuing this process indefinitely. It would be a very dynamic process, two students always moving from one side to the other in opposite directions, but there would never be any net change in the count of students on each side. If we were unable to see the students walking and were only able to count the number of students in the lines, we would never know that anything was happening. This is an example of what is known as equilibrium. *Equilibrium* is a process in which **two** opposing processes occur at the same time and at the same rate such that there is no net change.

The phenomenon of equilibrium occurs in chemical systems. Such systems are said to be *reversible*, which means that a process occurs in one direction, but the reverse process can also occur at the same time and at the same rate. We encountered one example of this in Chapter 9 (Section 9.14) when we discussed vapor pressure. The two opposing processes occurring in that discussion were the evaporation of water and the condensation of water in a container partially filled with liquid water and sealed from its surroundings. The evaporation of liquid water is reversible and the water vapor can condense and become liquid water. The evaporation and condensation occur at the same time and at the same rate so that there is no net change in the quantity of liquid water and no net change in the quantity of water vapor in the air space above the liquid water. The constant quantity of water vapor in the air space led us to define the vapor pressure of a liquid as the partial pressure exerted by the vaporized liquid in the air space. This quantity is a constant at a given temperature because it represents an equilibrium situation.

---

# WHY STUDY THIS TOPIC?

U ntil now in this text, we have spoken of chemical reactions as if the reactants are totally converted to products. In reality, if the products are not physically removed from the container, the reactants are *not* totally converted to products. Rather, the reaction vessel actually contains a mixture of both reactants and products. This phenomenon occurs because reactions are "reversible" and chemical equilibrium occurs. In many of these situations, the quantities of products present do far exceed the quantities of reactants, such that the conversion to products is essentially complete. This is certainly a requirement for most analytical techniques that rely on the completeness of reactions. However, it is important to realize that the reactions in which the reactants predominate over the products in the reaction vessel, or in which the product and reactant quantities are nearly equal, can also be very important.

Virtually all reactions performed in the analysis laboratory and in laboratories in which new substances are synthesized, as well as some reactions that are part of a manufacturing process, do not have their products physically removed from the container immediately and, thus, are reversible. The concept of chemical equilibrium is therefore very important. Chemistry professionals must be able to utilize and understand techniques that affect chemical equilibrium in order to optimize the results in an analytical laboratory, in a synthesis laboratory, and in manufacturing.

*For Class Discussion: Turn back to Chapter 2 and examine the chemical equations found in the Homework Exercises. Discuss as a class what type of container would be required at minimum for each to keep all reactants and products together so that chemical equilibrium would be encouraged.*

---

Chemists utilize a double arrow symbol ( $\rightleftharpoons$ ) to designate systems at equilibrium. For example, the equilibrium occurring in the beaker of water can be expressed as follows.

$$H_2O(\ell) \rightleftharpoons H_2O(g) \tag{11.1}$$

It is important to remember, as was pointed out in Chapter 9, that this equilibrium only occurs when the container is sealed. If it is not sealed, the water vapor would escape into the air and would not be available for the reverse reaction to occur.

The melting of ice is also reversible. An equilibrium occurs in a beaker containing liquid water and crushed ice in a uniform slushy mixture in surroundings that are held at a constant temperature of 0°C. Under these conditions, liquid water is freezing at the same time and at the same rate as the ice is melting. The water never completely freezes and the ice never completely melts. This equilibrium can be expressed as follows:

$$H_2O(s) \rightleftharpoons H_2O(\ell) \tag{11.2}$$

A third example is the behavior of a solute when given a choice of two solvents in which to dissolve. The dissolving of a solute in a solvent is reversible. Imagine a water solution of iodine, $I_2$, and imagine

---

*Chemistry Professionals at Work* *CPW Box 11.2*

# DECAFFEINATION OF COFFEE

The use of methylene chloride, $CH_2Cl_2$, as described here is called **extraction**. The equilibrium that occurs allows solutes dissolved in water to be "extracted" into methylene chloride either for the purpose of removing an unwanted solute in the water, or to get the solute into a solvent that will allow more convenient analysis. A number of organic liquids besides methylene chloride, are also useful for this. Extraction can be a small-scale laboratory process or a large-scale industrial process.

Extraction may also mean removing a component of a solid material. An example of such a solid material is coffee. One large scale industrial use of methylene chloride is the extraction of caffeine from coffee to create decaffeinated coffee. The green coffee beans are first steamed to increase the moisture content. This process brings the caffeine to the surface of the beans where it can be extracted upon contact with the methylene chloride. The beans are then dried, roasted, and sold as decaffeinated coffee beans.

*For Homework: A laboratory analysis in which a component of a solid material needs to be removed from the solid and analyzed may involve an apparatus known as a Soxhlet extractor. Research the design, operation, and application of a Soxhlet extractor, and find the cost of the equipment in a chemical supply catalog. Write a report.*

---

introducing a water-immiscible organic solvent, such as methylene chloride, $CH_2Cl_2$, into the container. The methylene chloride, because it is nonpolar, does not mix with the water (see Chapter 10, Sections 10.2 and 10.4) and forms a separate layer of liquid in the container. After vigorous shaking, the layers separate again and the methylene chloride layer appears pink due to iodine being dissolved in it. It is actually an equilibrium system because the iodine is moving into the methylene chloride at the same time and at the same rate as it is moving back into the water layer, as expressed in the following.

$$I_2(w) \rightleftharpoons I_2(o) \tag{11.3}$$

in which "w" refers to water and "o" refers to organic, meaning the methylene chloride layer.

## 11.2 Chemical Equilibrium

The two opposing processes in the definition of equilibrium may be chemical reactions. Imagine placing two chemicals that react with each other, "A" and "B," into a reaction vessel. Let's say that the products of the reaction are "C" and "D" and that the balancing coefficients are "a," "b," "c," and "d", as indicated in Eq. (11.4).

$$aA + bB \rightarrow cC + dD \tag{11.4}$$

As the reaction proceeds, "C" and "D" begin to form and their concentrations increase while the concentrations of "A" and "B" decrease. Now let's say that "C" and "D" also react with each other and that "A" and "B" are the products.

$$cC + dD \rightarrow aA + bB \tag{11.5}$$

This means that the reverse of reaction Eq. (11.4) occurs at the same time. After a period of time, imagine that these two opposing chemical reactions occur not only at the same time but also at same rate. When this happens, chemical equilibrium occurs. **Chemical equilibrium** refers to two opposing chemical reactions occurring at the same time and at the same rate with no net change. Such a reaction is written with a double arrow ($\rightleftharpoons$).

$$aA + bB \rightleftharpoons cD + dD \tag{11.6}$$

As with the vapor pressure example, if any of the chemicals, reactants, or products escape or are removed from the container, then equilibrium will not occur.

## 11.3    Important Examples of Chemical Equilibrium

Important examples of chemical equilibrium systems include (1) the Haber process for the manufacture of ammonia from hydrogen gas and nitrogen gas, (2) the ionization of weak electrolytes in water, and (3) the ionization and dissolution of ionic solids in saturated solutions.

In the Haber process, hydrogen gas and nitrogen gas are brought together at relatively high temperature and pressure and in the presence of a catalyst in a closed container. The chemical equilibrium that occurs is expressed as follows.

$$N_2(g) + 3H_2(g) \rightleftharpoons 2NH_3(g) \tag{11.7}$$

Here, nitrogen and hydrogen in the same container will react to give ammonia, but the reverse reaction also occurs and the reaction will reach a point where the forward and reverse reactions occur at the same time and at the same rate. We will be using this as an example throughout the discussion in this chapter.

Examples of weak electrolytes are the carboxylic acids, such as acetic acid, $HC_2H_3O_2$, and some bases, such as ammonium hydroxide, $NH_4OH$. When these chemicals are introduced into water, they "partially ionize," meaning that some fraction of the molecules or formula units split apart into ions. This occurs at the same time and at the same rate as the ions are coming together to form the un-ionized molecules or formula units. They are therefore examples of chemical equilibria. The reactions for these are written as follows.

$$HC_2H_3O_2(aq) \rightleftharpoons H^+(aq) + C_2H_3O_2^-(aq) \tag{11.8}$$

$$NH_4OH(aq) \rightleftharpoons NH_4^+(aq) + OH^-(aq) \tag{11.9}$$

Both of these are examples of equilibria that are very important in real-world analytical laboratories. Solutions of weak electrolytes are used to control the acidity of solutions in laboratory analyses. Such solutions are known as **buffer solutions**.

An equilibrium also exists at the surface of the undissolved ionic solid in a saturated solution of the solid. An example is a saturated solution of salt, NaCl, in water (see Chapter 10, Section 10.4.1). The sodium ions and chloride ions are breaking away from the ionic array at the same time and at the same rate as these same ions in solution are coming together to become part of the ionic array. This equilibrium can be expressed as follows.

$$NaCl(s) \rightleftharpoons Na^+(aq) + Cl^-(aq) \tag{11.10}$$

# 11.4   The Equilibrium Constant

In Section 11.1, we introduced the concept of equilibrium by suggesting a classroom experiment in which students in a class split into two groups and line up on two sides of the classroom, with one group on each side. We stated that as students pass from one side of the room to the other at the same time and at the same rate, there is no net change in the count of students on either side. Because there is no change in these counts, we can divide the number of students on one side by the number of students on the other side and always obtain the same number. This same concept applied to a chemical equilibrium system results in a number called the **equilibrium constant.**

In a chemical equilibrium system, the "count" is the molar concentration of the chemical species involved. For the purpose of symbolizing this molar concentration (and the equilibrium constant), we now introduce the symbolism of the brackets, "[]." The molar concentration (molarity) of a chemical species is symbolized by enclosing the formula for this species within brackets . For example, $[H^+]$ refers to the molar concentration of the hydrogen ion, $[NH_3]$ refers to the molar concentration of ammonia, and $[Cl^-]$ refers to the molar concentration of the chloride ion.

The **equilibrium constant** for a chemical equilibrium is defined as the molar concentrations of the products of the reaction, raised to the power of their respective balancing coefficients and multiplied together, divided by the molar concentrations of the reactants of the reaction, raised to the power of their balancing coefficients and multiplied together. Thus, for the general reaction

$$aA + bB \rightleftharpoons cD + dD \tag{11.11}$$

the equilibrium constant, symbolized as $K_{eq}$, is defined as in Equation (11.12).

$$K_{eq} = \frac{[C]^c \, [D]^d}{[A]^a \, [B]^b} \tag{11.12}$$

Knowing the molar concentrations of all the chemicals involved in the equilibrium, we can calculate the number that is given by the equilibrium constant. For Eq. (11.7), $K_{eq}$ is as shown in Equation (11.13).

$$K_{eq} = \frac{[NH_3]^2}{[H_2]^3 \, [N_2]} \tag{11.13}$$

and for Eq. (11.8), $K_{eq}$ is shown below,

$$K_{eq} = \frac{[H^+][C_2H_3O_2^-]}{[HC_2H_3O_2]} \tag{11.14}$$

and for Eq. (11.9) it is as follows.

$$K_{eq} = \frac{[NH_4^+][OH^-]}{[NH_4OH]} \tag{11.15}$$

Equations (11.8) and (11.9) represent "ionization" equilibria in which an uncharged chemical species (molecule or formula unit) splits apart into ions. Equation (11.8) represents the ionization of a weak acid, acetic acid, and Eq. (11.9) represents the ionization of a weak base, $NH_4OH$. We often refer to the equilibrium constants of ionization equilibria as **ionization constants** and symbolize them as $K_i$. Additionally, if the ionizing species is a weak acid, such as $HC_2H_3O_2$, we often refer to the ionization constant

# CARBONATE AND BICARBONATE EQUILIBRIA IN WATER

**C**arbon dioxide gas present in air dissolves in falling rain and other water exposed to air. This initiates a series of equilibrium reactions in the water, beginning with the following.

$$H_2O(\ell) + CO_2(g) \rightleftharpoons H_2CO_3(aq) \tag{1}$$

Since carbonic acid is a weak acid, the following equilibria also occur.

$$H_2CO_3(aq) \rightleftharpoons H^+(aq) + HCO_3^-(aq) \tag{2}$$

$$HCO_3^-(aq) \rightleftharpoons H^+(aq) + CO_3^{2-}(aq) \tag{3}$$

This means that water exposed to air is slightly acidic and contains small concentrations of carbonate and bicarbonate ions. This includes the distilled water used by chemistry professionals in their laboratories. If necessary, the situation can be remedied by boiling the water and then storing it in sealed containers after cooling. The presence of these equilibria in environmental water, such as in rain, lakes, streams, and wells, often means that such water must be monitored for **alkalinity**, or the acid neutralizing capacity of the water. More information on the alkalinity of environmental water is presented in CPW Box 12.5 in Chapter 12.

*For Homework: Eqs. (1), (2), and (3) also represent the equilibria that occur in sealed cans and bottles of carbonated beverages. Research the soda pop industry and write a report on the manufacturing/bottling processes, especially noting how the level of carbonation is achieved.*

as the **acid dissociation constant**, $K_a$ ("a" referring to "acid"). If the ionizing species is a weak base, such as $NH_4OH$, we refer to the ionization constant as $K_b$ ("b" referring to "base").

In these two ionization examples, all species in the chemical equilibrium are dissolved in water. This means that the molar concentrations needed in the equilibrium constant expression are all real numbers that can be used in the calculations. This includes the un-ionized species ($HC_2H_3O_2$, and $NH_4OH$) as well as the ions. The ionization of a salt, such as NaCl [see Eq. (11.10)], poses a special problem. The un-ionized chemical species in this equilibrium is *not* in solution. It is the undissolved solid that is present in all saturated solutions. Since it is undissolved, it does not make sense to refer to its molar concentration. An interesting and important fact about this undissolved solid is that regardless of how much is present, the concentrations of the dissolved ions are constants (at a given temperature). Thus, while the molar concentrations of the ions are real numbers, the molar concentration of the un-ionized species is a nonsensical term. In this case a special "equilibrium constant" must be defined that utilizes only the molar concentrations of the dissolved ions in its definition. This special equilibrium constant is called the ***solubility product constant***, which is symbolized "$K_{sp}$" and is defined as the mathematical product of the molar concentrations of the ions raised to the power of their balancing coefficients. For NaCl, $K_{sp}$ would be as shown below.

$$K_{sp} = [Na^+][Cl^-] \qquad (11.16)$$

An example of an ionic solid that has subscripts in its formula (and thus will have balancing coefficients other than "1") is $Mg_3(PO_4)_2$. The equation for the ionization of $Mg_3(PO_4)_2$ is

$$Mg_3(PO_4)_2(s) \rightleftharpoons 3Mg^{2+}(aq) + 2PO_4^{3-}(aq) \qquad (11.17)$$

and the $K_{sp}$ for $Mg_3(PO_4)_2$ is as follows.

$$K_{sp} = [Mg^{2+}]^3 [PO_4^{3-}]^2 \qquad (11.18)$$

## 11.5 The Magnitude of the Equilibrium Constant

At this point, we want to remind you that regardless of what constant we are speaking of (the equilibrium constant, the ionization constant, or the solubility product constant), each is a number. As indicated previously, if we know the molar concentrations of the reactants and products, this number may be calculated. We will not be concerned with such calculations at this point. However, it is useful to gain a feel for the magnitude that this number can have and of what it means when this number is large or small. Depending on the individual equilibrium process involved, this number can range from an extremely small number to an extremely large number.

An extremely small number results when the molar concentrations of the products (the numerator of the equilibrium constant expression) are very small values and the molar concentrations of the reactants (in the denominator) are very large. Dividing a small number by a large number, like when dividing the number "1" by the number "1000," results in a very small number. So an equilibrium constant is small when the concentration of the product is small relative to the concentration of the reactant. This occurs when the reverse reaction rate becomes equal to the forward reaction rate before the forward reaction has a chance to proceed to any appreciable extent, or when the reaction mixture consists mostly of the reactants. Conversely, the equilibrium constant is large when the concentration of the product is large compared to the concentration of the reactant. This occurs when the reverse reaction rate becomes equal to the forward reaction rate after the reaction has consumed nearly all the reactants and is nearly "complete," or when the reaction mixture consists mostly of the products.

Weak acid and weak base ionization equilibria have extremely small ionization constants. For example, the ionization constant of acetic acid is $1.8 \times 10^{-5}$. The ionization constant of ammonium hydroxide is $1.7 \times 10^{-5}$. This means that there are very few ions present at equilibrium—most of the acid or base is present as the un-ionized species. This is the reason they are described as "weak" acids or bases.

Similarly, the solubility product constants of precipitates (substances chemists usually describe as insoluble) are also extremely small. An apparently insoluble substance is never totally insoluble. In actuality, an equilibrium occurs in which there are an extremely small number of ions present. An example of this is $Mg_3(PO_4)_2$. If you were to drop a few grains of this solid into water and shake, it would appear not to dissolve. However, small concentrations of magnesium and phosphate ions would be found in the solution. The solubility product constant of this compound is $1 \times 10^{-25}$, an extremely small number.

Finally, consider the nitrogen, hydrogen, ammonia equilibrium—the Haber Process that is shown in Eq. (11.7). This is an industrial process for the manufacture of ammonia. The equilibrium constant depends on both temperature and pressure. It can be large or small depending on the conditions. However, to optimize the production of ammonia, chemical engineers control the conditions of temperature and pressure to give a large equilibrium constant because a large equilibrium constant means much more product (ammonia) than reactants (hydrogen and nitrogen) will be present in the reaction vessel.

## 11.6   Le Châtelier's Principle

As mentioned for the Haber process, temperature and pressure can help control the position of an equilibrium, the amount of product and reactant present at equilibrium, and thus the equilibrium constant. Temperature and pressure are parameters that, when varied, create a "stress" on the equilibrium. Stresses on equilibrium systems cause changes in the system. The nature of these changes are described by the principle of Le Châtelier.

**Le Châtelier's Principle** can be expressed as follows: when a stress is placed on a system at equilibrium, the equilibrium shifts to partially relieve the stress.

An *equilibrium shift* is defined as a spontaneous and momentary increase in the rate of an equilibrium reaction in one direction or the other, causing a temporary situation in which the reaction is not at equilibrium since the rates of the two opposing reactions are no longer equal. This phenomenon is temporary, since equilibrium is re-established at some later moment. When the equilibrium is reestablished, the concentrations of the reactants and products are not what they were previously because the temporary rate imbalance caused a surge toward one side and away from the other. The equilibrium constant may or may not change during this process depending on what caused the shift. This will be discussed below in conjunction with the various stresses that can be placed on a chemical equilibrium and cause it to shift.

## 11.7   Effect of Concentration Change

The change in concentration of a chemical participating in a chemical equilibrium is a stress on the equilibrium system. Thus, such a change would cause the equilibrium to shift, as stated by Le Châtelier's principle. Relieving the stress in this case means that if the concentration increases, the equilibrium shifts so as to partially decrease it. If the concentration decreases, the equilibrium shifts so as to partially increase it.  An increase in concentration may be caused simply by opening the reaction vessel and adding more of the chemical. A decrease in concentration may be caused by opening the reaction vessel and adding a chemical that would react with one of the chemicals involved in the reaction, thus removing this chemical from the system.

Let us use the Haber process for the manufacture of ammonia as an example. This reaction is shown again below.

$$N_2(g) + 3H_2(g) \rightleftharpoons 2NH_3(g) \tag{11.19}$$

Let us open the reaction chamber and pump in some nitrogen. This act places a concentration stress on the system consisting of an increase in the concentration of the $N_2$. According to Le Châtelier's principle, the equilibrium must shift to relieve this stress. This would mean consuming some of the $N_2$ in order to return to equilibrium. Thus, the equilibrium **shifts to the right**, as indicated below.

$$N_2(g) + 3H_2(g) \rightarrow\rightarrow 2NH_3(g) \tag{11.20}$$

before returning to the equilibrium state indicated in Eq. (11.19).

Similarly, if we were able to somehow open the container and selectively pump out some of the hydrogen, this too would be a concentration stress. In this case, however, the equilibrium would shift to replace the hydrogen that was removed. This would mean a shift to the left, or toward the hydrogen in the equation so that more would be produced.

$$N_2(g) + 3H_2(g) \leftarrow\leftarrow 2NH_3(g) \tag{11.21}$$

before once again returning to the equilibrium state indicated in Eq. (11.19).

In both of these examples, the return to the original equilibrium state does *not* mean that everything is the same as before the stress was imposed. The concentrations of all the chemicals involved have changed. If there is a shift to the right, some hydrogen gets consumed, and its concentration becomes lower. Also, some additional ammonia is produced, and its concentration is increased. And, some nitrogen has been added and only *partially* consumed, so its concentration has increased. The concentrations of all three chemicals are different than before the stress was imposed. Two have increased while one has decreased. You may think that because these concentrations have changed, the value of the equilibrium constant for this reaction must also have changed since these concentrations are used in the calculation of $K_{eq}$. In reality, the value of $K_{eq}$ does *not* change under these conditions. This indicates the true constancy of $K_{eq}$. In fact, we can say that in the event of a concentration stress, *the equilibrium shifts so as to maintain the constancy of the equilibrium constant.* Thus we don't look for the equilibrium constant to change; rather, we look for the shift to occur to the extent that would maintain its constancy.

## 11.8   Effect of Temperature Change

Another stress that we can impose on a chemical equilibrium system is an increase or decrease in temperature. All chemical equilibrium systems are affected by temperature changes. An increase in temperature causes a shift in one direction or the other, either to the left or to the right. Also, a decrease in temperature causes a shift in one direction or the other, either to the left or to the right. Reactions that shift to the right when the temperature is increased are described as **endothermic** reactions. Reactions which shift to the left when the temperature is increased are described as **exothermic** reactions. Endothermic reactions require heat to make them go, while exothermic reactions release heat as the reaction proceeds.

An increase in temperature means the addition of heat to the system. A decrease in temperature means removal of heat from the system. The temperature stress may be thought of as a heat stress. One of the equilibrium reactions, either the forward reaction (the one from left to right) or the reverse reaction (the one from right to left), is favored by the addition of heat. The situation is simplified if we simply think of heat as a reactant or product in the equation. If the forward reaction is favored by the addition of heat, then we can place the word "heat" on the left side to indicate that the addition of heat will make the reaction proceed from left to right. If the reverse reaction is favored by the addition of heat, we can place the word "heat" on the right side to indicate that the addition of heat will make the reaction proceed from right to left. A temperature stress then becomes like a concentration stress in which the "concentration" of heat either increases or decreases.

Let us look again at the Haber process. The reverse reaction of the Haber process is favored by adding heat. Thus, we place the word "heat" on the right side of Eq. (11.22).

$$N_2(g) + 3H_2(g) \rightleftharpoons 2NH_3(g) + heat \qquad (11.22)$$

Heat is released as the reaction proceeds from left to right. The reaction is exothermic. Increasing the temperature of the reaction vessel (the addition of heat) causes a shift from right to left to partially consume the heat that was added, as indicated below.

$$N_2(g) + 3H_2(g) \longleftarrow\!\longleftarrow 2NH_3(g) + heat \qquad (11.23)$$

Decreasing the temperature of the reaction vessel (the removal of heat) causes a shift from left to right to replace the heat that was removed.

$$N_2(g) + 3H_2(g) \longrightarrow\!\longrightarrow 2NH_3(g) + heat \qquad (11.24)$$

In each case, the system returns to the equilibrium state of Eq. (11.22).

*Chemistry Professionals at Work*                         *CPW Box 11.4*

# BUFFER SOLUTIONS

T he pH of a solution is a measure of its acid content or hydrogen ion concentration. Solutions in which hydrogen ions predominate over hydroxide ions have a pH between 0 and 7 and are said to be **acidic**. Solutions in which hydroxide ions predominate over hydrogen ions have a pH between 7 and 14 and are said to be **basic**. This topic is dealt with in more detail in Chapter 12.

When a weak acid or weak base ionization equilibrium is occurring in a solution, the pH is said to be **buffered**, which means the pH is resistant to change even when a strong acid or base is added. Buffer solutions have many applications in real-world laboratories because there are many instances in which a given experiment only works because the pH is constant.

The pH of such solutions depends on the $K_a$ or $K_b$ that is involved. If a weak acid has an extremely small $K_a$ (less than $10^{-3}$, for example), the pH of the solution will be closer to 7 than to 0. If the $K_a$ is small, but not extremely small (greater than $10^{-3}$), the pH will be closer to 0. Weak bases maintain pH levels between 7 and 14 . If a weak base has an extremely small $K_b$, the pH will be near 7. If the $K_b$ is small but not extremely small, the pH will be near 14. Acetic acid ($K_a = 1.8 \times 10^{-5}$) solutions maintain pH values around 5. Ammonium hydroxide ($K_b = 1.7 \times 10^{-5}$) solutions maintain pH values around 9.

Chemistry professionals often prepare and use buffer solutions in their labs and need to be aware of the effect of the $K_a$ or $K_b$ values that are involved.

*For Homework: Obtain specifications sheets for some commercially available buffer solutions, such as can be found at the following Internet site (as of 6/4/00):*

http://www.sigma.sial.com/sigma/proddata/b5020.htm

*Write a report summarizing the available information including company name, the pH of the solution, product number, composition, the pH at various temperatures, color code, etc.*

---

While the word "heat" can be added to one side or the other, depending on which reaction is favored by heat, it is important to recognize that this is only a gimmick that helps us determine what happens during the temperature stress. Of course, the word "heat" does not appear in the equilibrium constant expression—the "molar concentration" of heat does not have any meaning. This means that shifts caused by temperature stresses do cause the equilibrium constant to change. The concentrations change during a temperature stress in a way in which the constancy of $K_{eq}$ cannot be maintained. Thus all equilibrium constants are temperature dependent—they do change with temperature.

Perhaps a better way to show that heat is either consumed or generated is to indicate the enthalpy change that is involved rather than to just use the word "heat" on one side or the other of the equation. **Enthalpy change**, symbolized as "$\Delta H$," is defined as the heat of reaction at constant pressure. It is a number expressed in energy units, such as joules (J) or kilojoules (kJ). If the value of $\Delta H$ is negative, the reaction is exothermic (analogous to "heat" being placed on the right of the equation). If $\Delta H$ is positive, the reaction is endothermic (analogous to "heat" being placed on the left of the equation). The Haber Process, for example, is exothermic and can be written as follows with the $\Delta H$ symbolism.

$$3H_2(g) + N_2(g) \rightleftharpoons 2NH_3(g) \quad \Delta H = -92.22 \text{ kJ} \tag{11.25}$$

This indicates that 92.22 kJ of heat are released when 3 moles of hydrogen react with 1 mole of nitrogen to form 2 moles of ammonia. See Chapter 13 for more discussion of enthalpy and enthalpy change.

## 11.9   Effect of Pressure Change

The final stress we will consider is that caused by either an increase or decrease in the pressure within the reaction chamber. This pressure stress is akin to the pressure changes discussed for systems of gases in Chapter 9. In fact, within reasonable limits, the only time a pressure change is usually a stress on an equilibrium system is when one or more gases are involved in the equilibrium. If there are no gases involved, then there will be no shifting of the equilibrium in either direction.

Pressure stresses can be thought of as either pushing down or pulling up on the piston in the piston/cylinder arrangement in Chapter 9 when the equilibrium reaction is occurring within a volume enclosed by the piston and cylinder. Pushing down on the piston means a pressure increase in the system, while pulling up on the piston means a pressure decrease in the system.

If the stress is an increase in the pressure, then the system responds so as to decrease it again. If the stress is a decrease in the pressure, the system responds so as to increase it again. For the system to respond by decreasing the pressure, the equilibrium must shift in the direction that would produce less gas—less gas means lower pressure. For the system to respond by increasing the pressure, the shift must be in the direction that would produce more gas—more gas means higher pressure. In order to predict which shift would produce more or less gas, we need to determine which side contains more or less gas. This can easily be done by looking at the equation. First of all, which chemicals are gases would need to be expressed, such as with the "(g)" designation. Second, the equation must be balanced because the balancing coefficients for those designated with the "(g)" would tell us how many moles of each gas are involved. We can then determine which side of the equation consists of more or less gas.

Let us look again at the Haber process as an example.

$$N_2(g) + 3H_2(g) \rightleftharpoons 2NH_3(g) \tag{11.26}$$

All of the reactants and products in this reaction are gases. There are four moles of gas on the left and two moles of gas on the right. If the pressure is increased, the equilibrium would shift toward to the side where there is less gas. This relieves the stress by decreasing the pressure due to less gas being present. This means that the equilibrium will be a shift to the right in this case.

$$N_2(g) + 3H_2(g) \rightarrow\rightarrow 2NH_3(g) \tag{11.27}$$

If the pressure is decreased, the equilibrium would shift toward the side where there is more gas. This relieves the stress by increasing the pressure due to more gas being present. This means that the equilibrium would shift to the left in this case.

$$N_2(g) + 3H_2(g) \leftarrow\leftarrow 2NH_3(g) \tag{11.28}$$

If an equilibrium reaction has an equal numbers of moles of gas on both sides of the equation, then there would be no shift in either direction and the pressure change would have no effect.

---

*Chemistry Professionals at Work*                                          *CPW Box 11.5*

# OPTIMIZING THE MANUFACTURE OF AMMONIA

$$3H_2(g) + N_2(g) \rightleftharpoons 2NH_3(g) \quad \Delta H = -92.22 \text{ kJ}$$

**W**e have mentioned the Haber Process numerous times in this chapter. It is a good example of the application of LeChâtelier's Principle because it is a real-world process affected by concentration, heat, and pressure. It is also affected by the use of a catalyst.

If we wanted to optimize the quantity of ammonia produced, how exactly should this process be handled? Since there are fewer gas molecules on the right in this equation, a high pressure would be appropriate. Since it is exothermic, a heat-removal system would be appropriate. Also, it seems that we would want to remove ammonia periodically from the reaction chamber, since our purpose is to generate pure ammonia. This would seem to be a positive thing for the equilibrium too, since removing ammonia would shift it to the right and produce more ammonia. The addition of a catalyst should also help.

Indeed, for this process, a very high pressure is used (200 atm). However, a high temperature is also used (400 °C). The reason for the high temperature is that the rate at which the equilibrium is reached is slow and a higher temperature assists with that. The ammonia formed is removed by liquifying it in a condenser while the unreacted nitrogen and hydrogen are recycled. The catalyst is magnetite, $Fe_3O_4$, mixed with other metal oxides.

*For Homework: Fritz Haber won the Nobel Prize in chemistry in 1918 for the Haber Process. Write a report on his life. You will find the following Internet site useful.*

http://nobelprizes.com (as of 6/4/00)

---

## 11.10  Effect of Catalysts

A **catalyst** is a material that is present in the reaction vessel in order to increase or decrease the rate of the reaction, or the speed by which the reaction proceeds. It does not play any other role. Thus, it does not appear in the equation of the reaction (except sometimes over the arrow, as was shown in Chapter 2, Section 2.11). As far as equilibria are concerned, the presence of a catalyst does not affect the position of the equilibrium at all, meaning that adding a catalyst does not impose a stress on the system and the equilibrium does not shift in either direction. It only affects the speed with which the equilibrium is reached.

## 11.11  Equilibrium Calculations

The next step in the understanding of equilibrium theory is to apply the concepts that we have discussed to actual equilibrium calculations. In this section we will discuss equilibrium chemistry relating to medical diagnosis, the recovery of metals, and weak acids and bases. The important concepts to understand relating to these calculations are (1) how we can determine the amount of a substance that is in solution, and (2) how we can modify the amount of substance that is in solution.

## 11.11.1 How Much Exists in Solution?

A sludgy, white mixture of barium sulfate, $BaSO_4$, and water is used in medical testing because it is "radiodense." This means that it doesn't permit X-rays to pass through. So after the patient swallows the mixture, radiology technicians can follow the scattered x-rays in "realtime" with a fluoroscope, a device that displays the shadows of optically opaque substances as the solution travels through the digestive system. Using this "barium swallow," doctors can check for a variety of conditions such as cancerous growths in the esophagus, hiatal hernia, and acid reflux from the stomach. It is therefore important that the barium sulfate remain as a solid and not dissolve in the solution as barium and sulfate ions. Exactly how much of this material does dissolve per liter (the **quantitative solubility** defined in Chapter 10, Section 10.2) can be determined via a calculation using the solubility product constant.

The key question that we must ask when assessing equilibrium chemistry in solution is "What are the important equilibria?" The dissociation of barium sulfate into its ions in water is given by the following equation,

$$BaSO_4(s) \rightleftharpoons Ba^{2+}(aq) + SO_4^{2-}(aq) \quad K_{sp} = 1.5 \times 10^{-9} \quad (11.29)$$

The magnitude of the solubility product constant is quite small. As discussed in Section 11.5, this means that the reaction will not produce many ions, and most of the barium sulfate will remain as undissolved barium sulfate. But how much will dissolve? It is the amount represented by the quantitative solubility we defined in Section 10.2. We can calculate it using the solubility product constant expression.

Structured problem solving involving solubility includes examining the concentration changes that occur when an ionic solid dissolves. Initially, there are no $Ba^{2+}$ or $SO_4^{2-}$ ions in solution, so their initial concentrations are 0 M. At equilibrium, the barium sulfate will have ionized (dissolved) to a small extent. The moles of $BaSO_4$ that have ionized per liter is the same as the moles of $Ba^{2+}$ and the moles of $SO_4^{2-}$ that have formed per liter, and this will be the solubility. Let us call this solubility "s." It is useful to summarize this information in a table.

|  | $BaSO_4$ | $\rightleftharpoons$ | $Ba^{2+}(aq)$ | + | $SO_4^{2-}(aq)$ |
|---|---|---|---|---|---|
| Initial | – |  | 0 M |  | 0 M |
| Change | – |  | +s |  | +s |
| Equilibrium | – |  | +s |  | +s |

We can now substitute the equilibrium values into the solubility product expression to solve for "s" (and by using a solubility product constant of $1.5 \times 10^{-9}$),

$$K_{sp} = [Ba^{2+}][SO_4^{2-}] \quad (11.30)$$

$$1.5 \times 10^{-9} = (s)(s) = s^2 \quad (11.31)$$

$$s = [Ba^{2+}] = [SO_4^{2-}] = 3.9 \times 10^{-5} \text{ M} \quad (11.32)$$

Thus, the solubility of $BaSO_4$ in water is approximately $3.9 \times 10^{-5}$ M.

We can check our results by substituting the calculated concentrations back into the solubility product expression to determine $K_{sp}$,

$$K_{sp} = (3.9 \times 10^{-5})(3.9 \times 10^{-5}) = 1.5 \times 10^{-9} \quad (11.33)$$

The answers check. Checking your answers in this way is an important problem-solving step because it is an easy way to uncover math errors.

## Example 11.1

What is the solubility of silver iodide, AgI, in water if $K_{sp} = 1.5 \times 10^{-16}$ for AgI?

## Solution 11.1

We can proceed as described previously by writing our solubility product expression for the dissolution of the solid and noting that $[Ag^+] = [I^-] =$ "s." Note that this solubility equals the amount of AgI that dissolves per liter and is equal to the $[Ag^+]$ and $[I^-]$ concentrations.

|  | AgI(s) | $\rightleftharpoons$ | $Ag^+(aq)$ | + | $I(aq)$ |
|---|---|---|---|---|---|
| Initial | – |  | 0 M |  | 0 M |
| Change | – |  | + x |  | + x |
| At equilibrium | – |  | + x |  | + x |

$$K_{sp} = [Ag^+][I^-]$$

$$1.5 \times 10^{-16} = (s)(s) = s^2$$

$$s = [Ag^+] = [I^-] = 1.2 \times 10^{-8} \, M$$

Thus, the solubility, s, is equal to approximately $1.2 \times 10^{-8}$ M. You can check this by substituting back into the solubility product expression, as described previously. You will get an answer of $1.4 \times 10^{-16}$, which is acceptably close. The difference is due to our rounding off to two significant figures.

## Example 11.2

What is the solubility of a saturated solution of strontium fluoride, $SrF_2$, if $K_{sp} = 7.9 \times 10^{-10}$?

## Solution 11.2

The problem-solving strategy is the same here as in the previous problem. However, note that two fluoride ions form for every one strontium ion and for every $SrF_2$ formula unit dissolved.

$$SrF_2(s) \rightleftharpoons Sr^{2+}(aq) + 2F^+(aq)$$

The solubility product constant expression is

$$K_{sp} = [Sr^{2+}][F^-]^2$$

The table we will create to solve this problem will look a bit different from those used in the previous examples in that the amount of fluoride that will be formed will not be "s" but rather "2s".

|  | $SrF_2(s)$ | $\rightleftharpoons$ | $Sr^{2+}(aq)$ | + | $2F^-(aq)$ |
|---|---|---|---|---|---|
| Initial | – |  | 0 M |  | 0 M |
| Change | – |  | + s |  | + 2s |
| Equilibrium | – |  | + s |  | + 2s |

$$K_{sp} = [Sr^{2+}][F^-]^2; \quad 7.9 \times 10^{-10} = (s)(2s)^2 = 4s^3$$

$$s = [Sr^{2+}] = 5.8 \times 10^{-4} \text{ M}$$

Thus, $[Sr^{2+}] = s = 5.8 \times 10^{-4}$ M and the $[F^-] = 2s = 1.16 \times 10^{-3}$ M $= 1.2 \times 10^{-3}$ M Checking, we find that $K_{sp} = (5.8 \times 10^{-4})(1.16 \times 10^{-3})^2 = 7.8 \times 10^{-10}$ as expected.

## 11.11.2 Modifying the Amount that Dissolves

Silver nitrate, $AgNO_3$, is soluble in water and has uses varying from photography and silver plating to etching ivory and coloring porcelain. It is an expensive reagent because it contains silver. In the analytical laboratory, silver nitrate can be converted to silver chloride, AgCl, via reaction with sodium chloride. This is one way that silver can be recovered from solution.

$$AgNO_3(aq) + NaCl(aq) \rightleftharpoons AgCl(s) + NaNO_3(aq) \tag{11.34}$$

This reaction can also be written in its "net ionic" form, showing only the ions coming together to form the AgCl.

$$Ag^+(aq) + Cl^-(aq) \rightleftharpoons AgCl(s) \tag{11.35}$$

We can calculate the solubility of the AgCl as we have in previous examples, noting that this solubility, s, is equal to the $[Ag^+]$. Solving as usual gives the following result.

$$K_{sp} = [Ag^+][Cl^-] = (s)(s) = 1.6 \times 10^{-10}$$
$$s = [Ag^+] = [Cl^-] = 1.3 \times 10^{-5} \text{M} \tag{11.36}$$

In this discussion, we will use our understanding of Le Châtelier's Principle to decrease the solubility of AgCl even further, thus getting even more of the silver ion out of solution. We can begin by adding enough *sodium* chloride to the solution to raise the concentration of $Cl^-$ to 0.10 M. What does this do to the equilibrium? As we have seen, a concentration stress does not change the solubility product constant. That is, the $K_{sp}$ ($[Ag^+][Cl^-]$) will still be equal to $1.6 \times 10^{-10}$ at 25 °C. Therefore, if we *sharply increase* the $[Cl^-]$ by adding sodium chloride so that $[Cl^-]$ is about 0.10 M, the equilibrium reaction shifts to the right and the $[Ag^+]$ lowers. This is due to the fact that the product of the two is constant, $K_{sp}$, $1.6 \times 10^{-10}$. Thus, when we stress the system by adding an ion that is the same as one of the dissolution products (chloride or silver ion, in this case), the system responds by shifting the position of equilibrium *toward the reactants* as indicated by Le Châtelier's Principle.

We can calculate how the solubility of an essentially insoluble salt is affected by the addition of an ion that is common to one of the reaction products (chloride ion is therefore called a **common-ion**). This can be done by using the following type of table to set up the problem.

|  | AgCl (s) | $\rightleftharpoons$ | $Ag^+(aq)$ | + | $Cl^-(aq)$ |
|---|---|---|---|---|---|
| Initial | – | | 0 M | | 0.10 M |
| Change | – | | +s | | +s |
| Equilibrium | – | | +s | | 0.10 + s |
| With assumption | – | | +s | | 0.10 |

We have added an item to our previous table that is called "with assumption." This item shows what would happen if we make the assumption that the solubility, "s", was so low ($1.3 \times 10^{-5}$ M) that it is negligible compared to the concentration of the common-ion, $[Cl^-]$, which is present here at a concentration of 0.10 M. That is,

$$0.10 + s \approx 0.10. \tag{11.37}$$

However, *we must test any assumptions that we make*, and we will do so below. We may now solve the problem using our solubility product expression,

$$K_{sp} = [Ag^+][Cl^-] = (s)(0.10) + 1.6 \times 10^{-10} \tag{11.38}$$

$$s = [Ag^+] = 1.6 \times 10^{-9} \text{ M} \tag{11.39}$$

By the addition of the common ion, the solubility has been reduced by a factor of $10^4$! The **common-ion effect** can therefore be used in the laboratory setting to make precipitations more complete. Was our assumption that the solubility of the salt is negligible compared to 0.10 M valid? The solubility of $1.6 \times 10^{-9}$ M is far less than 0.10, so our assumption is fine. We can generally assume that "s" is negligible compared to the initial concentration whenever the solubility product constant is small (in many cases, $10^{-3}$ to $10^{-4}$ or less). You may extend this assumption to any equilibrium system.

## Example 11.3

Compare the solubility of the strontium fluoride solution in pure water (see Example 11.2) with its solubility in a 0.30 M NaF solution.

## Solution 11.3

We saw in Example 11.2 that the dissociation of the solid can be represented by

$$SrF_2(s) \rightleftharpoons Sr^{2+}(aq) + 2F^-(aq) \tag{11.40}$$

We also calculated the solubility in pure water to be $5.8 \times 10^{-4}$ M. When we change the fluoride ion concentration by adding NaF, the following reaction also occurs.

$$NaF(s) \rightarrow Na^+(aq) + F^-(aq) \tag{11.41}$$

The common-ion here is the fluoride ion which has a concentration of 0.30 M. We can set up a table to examine the impact of the common-ion,

|                   | $SrF_2(s)$ | $\rightleftharpoons$ | $Sr^{2+}(aq)$ | + | $F^-(aq)$ |
|-------------------|------------|----------------------|---------------|---|-----------|
| Initial           | –          |                      | 0 M           |   | 0.30 M    |
| Change            | –          |                      | +s            |   | +2s       |
| Equilibrium       | –          |                      | +s            |   | 0.30 + 2s |
| With assumption   | –          |                      | +s            |   | 0.30      |

$$K_{sp} = [Sr^{2+}][F^-]^2$$

$$7.9 \times 10^{-10} = (s)(0.30)^2$$

$$s = [Sr^{+2}] = 8.8 \times 10^{-9} \text{ M}$$

$$[F^-] = 0.30 \text{ M}$$

The solubility of the salt was decreased by a factor of about $10^5$ by adding the common-ion. Also, our assumption was valid because the value of $1.8 \times 10^{-8}$ M that we obtained for 2s is much smaller than 0.30 M.

## 11.11.3  Calculations Involving $K_a$ and $K_b$

We can apply our understanding of equilibrium, including the meaning of the solubility product expression, the equilibrium constant, and Le Châtelier's Principle, to solve a variety of problems. One of the most important of these is determining the acidity level of an aqueous solution. The basis of the work is the concept of equilibrium that we have discussed throughout this chapter.

We begin by calculating the value for $[H^+]$ in a solution that is 0.50 M in formic acid, HCOOH. Formic acid is used in tanning, electroplating, as a solvent in perfume, and in several kinds of chemical analysis. Remember the key question that we posed at the beginning of the section, *"What are the important equilibria?"* In this case, the significant equilibrium is the dissociation of the acid, which we can express in the following "shorthand" form.

$$HCOOH(aq) \rightleftharpoons H^+(aq) + HCOO^-(aq) \quad K_a = 1.8 \times 10^{-4} \tag{11.42}$$

The equilibrium expression for the reaction is given below.

$$K_a = \frac{[H^+][HCOO^-]}{[HCOOH]} \tag{11.43}$$

The value for $K_a$ is fairly small, as is characteristic of weak acids. This means that relatively little of the acid will dissociate, an amount we can call "x." We can set up our table,

|                 | HCOOH(aq)  | $\rightleftharpoons$ | H⁺(aq) | + | HCOO⁻(aq) |
| --------------- | ---------- | --- | ------ | --- | --------- |
| Initial         | 0.50       |     | 0 M    |     | 0 M       |
| Change          | −x         |     | +x     |     | +x        |
| Equilibrium     | 0.50 − x   |     | +x     |     | +x        |
| With assumption | 0.50       |     | +x     |     | +x        |

We assumed that "x" is negligible compared to 0.50 because formic acid is a weak acid.

$$1.8 \times 10^{-4} = \frac{(x)^2}{0.50} \tag{11.44}$$

$$x = [H^+] = 9.5 \times 10^{-3} \text{ M} \tag{11.45}$$

We can prove that substituting the concentration values back into the equilibrium expression does, in fact, result in $K_a$. Therefore, our answer mathematically checks. Was our assumption that "x" is negligible compared to the original formic acid concentration valid?

$$\frac{(9.5 \times 10^{-3})}{0.50} \times 100\% = 1.9\%$$

Very little ionization of the acid has occurred (1.9%) and thus our assumption is satisfactory.

Equation (11.44) can be written in a generalized form applicable to all weak acids as follows.

$$K_a = \frac{x^2}{C_{HA}} \tag{11.46}$$

In this equation, "x" is the hydrogen ion concentration and $C_{HA}$ is the total acid concentration, both ionized and un-ionized. The hydrogen ion concentration for any weak acid system can thus be calculated knowing the $K_a$ and the total acid concentration (see Example 12.5 in Chapter 12).

As we can use our understanding of equilibrium to determine $[H^+]$, we can also calculate $[OH^-]$, the production of which is characteristic of a base. In water, ammonia, $NH_3$, is a base. It reacts, though not extensively (note the small value for its equilibrium constant, $K_b$).

$$NH_3(aq) + H_2O(\ell) \rightleftharpoons NH_4^+(aq) + OH^-(aq) \quad K_b = 1.8 \times 10^{-5} \tag{11.47}$$

## *Example 11.4*

Calculate the concentration of $[OH^-]$ at equilibrium in a solution that has an initial $NH_3$ concentration of 0.20 M.

---

*Chemistry Professionals at Work*  *CPW Box 11.7*

# METABOLISM AND SAFETY EVALUATION AT MONSANTO

**M**y name is Sue Dudek. I am a chemistry technician employed by Monsanto at its Metabolism and Safety Evaluation (MSE) site in St. Louis, Missouri. My main job functions are sample preparation and analytical chemistry in support of toxicology studies performed at MSE. I receive samples from a variety of EPA and FDA studies and verify the homogeneity and concentration of the test material in the samples. High performance liquid chromatography and gas chromatography are the main analytical tools I use to analyze samples.

I began working at MSE in 1986, transferring there from the Chemical Division of Monsanto where I started working in 1976. I received an A.A. degree in 1979. I have been active in my local American Chemical Society section, having served on the Board of Directors and on various committees.

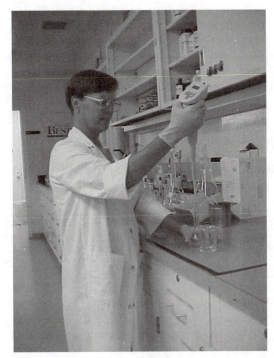

*For Homework: Check out the Internet site for Monsanto and write a report on the activities of this company.*

http://monsanto.com/ (as of 6/4/00)

---

*Solution 11.4*

As with our other equilibrium problems, we need to set up both the equilibrium expression and our table. For ammonia, a weak base, these are as follows.

$$K_b = \frac{[NH_4^+][OH^-]}{[NH_3]} \tag{11.48}$$

| | NH$_3$(aq) | + | H$_2$O($\ell$) | $\rightleftharpoons$ | NH$_4^+$ (aq) | + | OH$^-$ (aq) |
|---|---|---|---|---|---|---|---|
| Initial | 0.20 | | | | 0 M | | 0 M |
| Change | $-$x | | | | +x | | +x |
| Equilibrium | 0.20 $-$ x | | | | +x | | +x |
| With assumption | 0.20 | | | | +x | | +x |

We assume that "x" is negligible compared to 0.20 because K$_b$ is small.

$$1.8 \times 10^{-5} = \frac{(x)^2}{0.20} \tag{11.49}$$

$$x = [OH^-] = 1.9 \times 10^{-3} \text{ M} \tag{11.50}$$

We can prove that substituting the concentration values back into the equilibrium expression does, in fact, result in K$_b$. Therefore, our answer mathematically checks. Was our assumption that "x" is negligible compared to the original ammonia concentration valid?

$$\frac{1.9 \times 10^{-3}}{0.20} \times 100\% = 0.95\%$$

We will explore the special relationship between [H$^+$] and [OH$^-$] in the next chapter, which focuses on acids and bases.

## 11.12   Homework Exercises

1. (a)  Why do we speak of equilibrium when we define vapor pressure?
   (b)  Why do we speak of equilibrium when we describe what happens to a solute in a container with two immiscible liquid solvents?
2. What does it mean when we say that a chemical reaction is "reversible?"
3. What does a double arrow separating the reactants and products in a chemical equation mean? (To simply say "equilibrium" is not sufficient.)
4. What is chemical equilibrium?
5. When sulfur dioxide gas (SO$_2$) is mixed with oxygen gas (O$_2$), a reaction occurs and sulfur trioxide gas (SO$_3$) forms. After a certain length of time, this reaction reverses, which means that the sulfur trioxide decomposes to form oxygen gas and sulfur dioxide, and these two reactions then occur at the same time and rate. Write a single chemical equation that would describe this phenomenon.
6. Describe what is meant by an "equilibrium constant." Provide one specific example.
7. Write the equilibrium constant expression for the following equilibria.
   (a)  $2NO(g) + Cl_2(g) \rightleftharpoons 2NOCl(g)$
   (b)  $N_2(g) + O_2(g) \rightleftharpoons 2NO(g)$
   (c)  $2Cl_2(g) + 2H_2O(g) \rightleftharpoons 4HCl(g) + O_2(g)$
   (d)  $H_2SO_3(aq) \rightleftharpoons 2H^+(aq) + SO_3^{2-}(aq)$
   (e)  $H_3PO_4 \rightleftharpoons (3H^+ + PO_4^{3-})$
   (f)  $2SO_2(g) + O_2(g) \rightleftharpoons 2SO_3(g)$
8. What type of general reaction is described by an ionization constant, K$_i$? Give an example.
9. What type of general reaction is described by a solubility product constant, K$_{sp}$? Give an example.
10. Write equilibrium reactions and solubility product constant expressions for the following ionic compounds dissolving in water.   (a) Ag$_2$S  (b) PbI$_2$  (c) Bi$_2$S$_3$

11. The acid $H_3PO_3$ ionizes only very slightly when dissolved in water. Is the equilibrium constant a very small number, a number about equal to one, or a very large number? Explain.

12. The numerical value of the equilibrium constant for the reaction in Problem #7(d) is 0.0000000392. What does this say about the concentration of the $PO_4^{3-}$ in this equilibrium compared to the concentration of the $H_3PO_4$?

13. Consider the following equilibrium:

$$H_3BO_3 \rightleftharpoons 3H^+ + BO_3^{3-}$$

(a) Write the equilibrium constant expression for this reaction.

(b) Would you call this equilibrium constant $K_i$, $K_a$, $K_b$, or $K_{sp}$? Explain.

(c) If the value of the equilibrium constant is $1.6 \times 10^{-13}$, would you say that $H_3BO_3$ is highly ionized or only slightly ionized? Explain.

14. The equilibrium constant for the equilibrium in Problem #7(c) is $1.02 \times 10^{-7}$. In this reaction, is the concentration of $H^+$ at equilibrium large or small? Explain.

15. HCHO is a weak acid and a weak electrolyte.

(a) Write an equation representing the ionization that occurs when HCHO, dissolves in water.

(b) Write the equilibrium constant expression for this ionization.

(c) Would the $K_a$ for this weak acid be a large number, a small number, or a number close to 1.0? Explain.

16. What is Le Châtelier's Principle?

17. Why is Le Châtelier's Principle important in the description of chemical reactions?

18. What is meant by an "endothermic reaction?"

19. How is an endothermic reaction affected by the addition of heat?

20. What is meant by an "exothermic reaction?"

21. How is an exothermic reaction affected by the addition of heat?

22. Given the following equilibrium, fill in the blanks with "increase", decrease," or "no change."

$$Heat + CaCO_3(s) \rightleftharpoons CaO(s) + CO_2(g)$$

(a) If the temperature were decreased, the amount of CaO present would _____.

(b) If the pressure were increased, the amount of CaO present would _____.

(c) If the amount of $CaCO_3$ present increased, the amount of CaO would _____.

23. Consider the following reaction.

$$N_2(g) + 3H_2(g) \rightleftharpoons 2NH_3(g) + heat$$

Tell in what direction (left or right) the equilibrium will shift and what will happen to the concentration of $N_2$ (increase or decrease) when (a) some $H_2$ is added, (b) the pressure is decreased, and (c) the temperature is increased.

24. Which of the examples in Problem #7 would be affected by a change in pressure? Which would an increase in pressure not affect?

25. Consider the following equilibrium.

$$N_2(g) + 3H_2(g) \rightleftharpoons 2NH_3(g) + heat$$

In which direction will the equilibrium shift (right, left, or neither) and what will happen to the concentration of $H_2$ (increase, decrease, or no change) when each of the following occurs? (a) The reaction vessel is heated. (b) The pressure is increased. (c) More $N_2$ is added to the reaction vessel.

26. Consider the following equilibrium.

$$\text{Heat} + H_2(g) + F_2(g) \rightleftharpoons 2HF(g)$$

In which direction will the equilibrium shift (right, left, or neither) and what will happen to the concentration of HF (increase, decrease, or no change) when each of the following occurs? (a) The concentration of $F_2$ is decreased. (b) The pressure is increased. (c) The reaction vessel is cooled.

27. Describe what is meant by the term "enthalpy change?"

28. How is the enthalpy change related to the effect that temperature has on a reaction?

29. Consider the following equilibrium.

$$2H_2(g) + O_2(g) \rightleftharpoons 2H_2O(g) \quad \Delta H = -484 \text{ kJ}$$

In which direction will the equilibrium shift (right, left, or neither) and what will happen to the concentration of $H_2$ (increase, decrease, or no change) when each of the following occurs? (a) The reaction vessel is heated. (b) The pressure is decreased. (c) More $O_2$ is added to the reaction vessel.

30. Consider the following equilibrium:

$$C_6H_{12}(g) + \text{heat} \rightleftharpoons C_6H_6(g) + 3H_2(g)$$

In which direction will the equilibrium shift (right, left, or neither) and what will happen to the concentration of $H_2$ (increase, decrease, or no change) when each of the following occurs? (a) The concentration of $C_6H_6$ is decreased. (b) The pressure is increased. (c) The reaction vessel is cooled.

31. Consider the following equilibrium.

$$N_2(g) + 3H_2(g) \rightleftharpoons 2NH_3(g) + \text{heat}$$

In which direction will the equilibrium shift (right, left, or neither) and what will happen to the concentration of $NH_3$ (increase, decrease, or no change) when each of the following occurs? (a) The reaction vessel is cooled. (b) The pressure is decreased. (c) More $H_2$ is added to the reaction vessel.

32. If you were managing an industrial process in which $PH_3$ is made by the following reaction and you wanted to increase the output of this product (shift the equilibrium to the right) so that you could make more money, what would you do in terms of a high or low concentration of $H_2$, high or low pressure, and high or low temperature? Explain.

$$2P(s) + 3H_2(g) \rightleftharpoons 2PH_3(g) + \text{Heat}$$

33. Calculate the solubility of AgBr ($K_{sp} = 5.0 \times 10^{-13}$) in (a) pure water, and (b) 0.10 M NaBr.

34. Calculate the solubility of $Ag_2CrO_4$ ($K_{sp} = 1.9 \times 10^{-12}$) in (a) pure water, and (b) 0.20 M $AgNO_3$.

35. What is the solubility product ($K_{sp}$) of NiS if its molar solubility is $5.5 \times 10^{-11}$ M?

36. What is the solubility product constant for $CaF_2$ if its molar solubility is $2.1 \times 10^{-4}$ M?

37. Find the solubility of $Fe(OH)_3$ ($K_{sp} = 4 \times 10^{-38}$) in each of the following solutions.
    (a)  Pure water at 25°C, $[OH^-] = 1.0 \times 10^{-7}$ M
    (b)  $[OH^-] = 1.0 \times 10^{-12}$ M
    (c)  $[OH^-] = 1.0 \times 10^{-3}$ M

38. Which of the following precipitates is less soluble in water?
    (a)  $CaF_2(s)$, $K_{sp} = 1.8 \times 10^{-18}$
    (b)  $BaF_2(s)$, $K_{sp} = 2.4 \times 10^{-5}$

39. Find the solubility of $PbI_2$ ($K_{sp} = 1.4 \times 10^{-8}$) when it is dissolved in each of the following solutions.
    (a)  Water
    (b)  0.50 M $PbNO_3$
    (c)  2.0 M KI solution

40. How many grams of $Ca_3(PO_4)_2(s)$ ($K_{sp}$ = 1.3 × 10$^{-32}$) will dissolve in 1 L of 0.30 M $Na_3PO_4$ solution?

41. $PbBr_2$ is dissolved in water. If the concentration of $Pb^{2+}$ in a saturated solution is 1.05 × 10$^{-2}$ M. What is the $K_{sp}$ of $PbBr_2$?

42. What is the $[H^+]$ in a solution that is 0.20 M in acetic acid, $CH_3COOH$ ($K_a$ = 1.8 × 10$^{-5}$)?

43. What is the $[H^+]$ in a solution that is 0.15 M in a weak acid with a $K_a$ of 5.2 × 10$^{-6}$?

44. What is the $[OH^-]$ in a 0.25 M solution of a weak base with a $K_b$ of 3.77 × 10$^{-5}$?

## For Class Discussion and Reports

45. It can be quite disastrous in an industrial chemical plant when an equilibrium reaction present in a pressurized vessel is disturbed by faulty equipment, such as temperature controllers. A sudden increase or decrease in temperature can shift an equilibrium reaction such that more gases form and the vessel becomes overpressurized. In that event, the vessel may explode (see CPW Box 9.2) or a relief valve may be activated resulting in toxic or flammable gases being released into the air. Explore incidents of accidents and explosions at chemical plants and write a report giving details of specific causes and how they may have been prevented.

46. In this chapter, we discussed the manufacture of ammonia as an industrial process that involves a chemical equilibrium. We discussed how concepts of Le Châtelier's Principle can be applied to this process to optimize production (see CPW Box 11.5). Chemical equilibria are also involved in the manufacture of nitric acid and urea. Research the manufacture of these chemicals and report on what chemical equilibrium reactions are involved and what conditions of temperature and pressure optimize their production and why. Also report on the the amounts produced annually and what major uses these chemcals have.

# 12

# Acids and Bases

## 12.1    Introduction

We have noted several times in previous chapters that over thirteen million chemical compounds exist. Because of this, the study of the chemistry (structure, composition, and properties) of these compounds can indeed be a formidable task. To simplify this task, the usual procedure is to divide and subdivide the compounds into various classifications and then to study individual classifications separately. This was our procedure with organic chemicals in Chapter 6.

We have seen many classifications of compounds. We have seen gases, liquids, and solids, ionic and covalent compounds, inorganic and organic substances, polymeric and monomeric compounds, soluble and insoluble chemicals, and electrolytes and nonelectrolytes. In this chapter, we will study acids and bases.

## 12.2    Formulas and Strengths of Acids and Bases

We recognize that a compound is an **acid** when it has one or more hydrogens listed first in its formula. For instance, we have the well-known laboratory acids that are listed Table 12.1. In addition, there is a host of others that are less well known. Some of these are listed in Table 12.2.

*Chemistry Professionals at Work*                                    *CPW Box 12.1*

# WHY STUDY THIS TOPIC?

A cids and bases may very well be the most important chemicals used in chemical process industry laboratories. They are widely used as dissolving agents for difficult analysis samples. Examples are grain and grain products, metals and metal ores, vegetables and food products, and organic matter in environmental waters and wastes. They must also be present in many solutions that a laboratory worker prepares in order for such solutions to exhibit the properties that are required. Also, the acidity level of unknown solutions is often the object of experiments that laboratory workers perform.

Various safety issues regarding acids and bases often affect the approach a chemistry professional takes to the tasks he/she routinely performs. A fundamental knowledge of acids, bases, and their properties is vitally important.

*For Homework and Class Discussion: Interview a chemistry professional in your area and determine what acids and bases are used in his/her work area and for what purpose. Report back to the class.*

**TABLE 12.1**   Formulas and Names of Some Well-Known Laboratory Acids

| Formula | Name |
|---------|------|
| HCl | Hydrochloric acid |
| $H_2SO_4$ | Sulfuric acid |
| $HNO_3$ | Nitric acid |
| $H_3PO_4$ | Phosphoric acid |
| $HC_2H_3O_2$ | Acetic acid |

**TABLE 12.2**   Some Less Well-Known Acids

| Formula | Name |
|---------|------|
| HF | Hydrofluoric acid |
| HBr | Hydrobromic acid |
| HI | Hydroiodic acid |
| $H_2CO_3$ | Carbonic acid |
| $HNO_2$ | Nitrous acid |
| $H_2SO_3$ | Sulfurous acid |
| $H_2S$ | Hydrosulfuric acid |
| HCN | Hydrocyanic acid |

With the exception of acetic acid, all the acids listed the Tables 12.1 and 12.2 are **inorganic acids**, sometimes also known as **mineral acids**. Acetic acid is an example of an **organic acid**. It is a carboxylic acid (Chapter 6).

$$\text{HC}_2\text{H}_3\text{O}_2 \qquad \begin{array}{c} \text{O} \\ \| \\ \text{CH}_3\text{C--OH} \end{array} \qquad \text{acetic acid}$$

The hydrogen bonded to the oxygen in this structure is the hydrogen listed first in the formula. It is the hydrogen that makes acetic acid, or any other carboxylic acid, an acid. This is the "acidic" hydrogen in the formula. The other hydrogens are not acidic.

To be acidic, a hydrogen must be relatively loosely held, so that it is released under certain conditions. We have seen one such condition in Chapter 10 when we studied electrical conductivities of acid solutions. When acids dissolve in water, the acidic hydrogen can break away from the rest of the molecule creating a solution that has ions in it, which makes the solution electrically conductive. This ionization reaction for hydrochloric acid is as follows.

$$\text{HCl} \rightarrow \text{H}^+ + \text{Cl}^- \tag{12.1}$$

These hydrogens may also be released when the acid enters into some chemical reactions, such as when it reacts with a base or with an active metal. Hydrogens that are not acidic are tightly held and are not released in these reactions. Examples of these are the methyl-group hydrogens in acetic acid, the three hydrogens that are part of the acetate ion formed when acetic acid ionizes.

$$\text{HC}_2\text{H}_3\text{O}_2 \rightleftharpoons \text{H}^+ + \text{C}_2\text{H}_3\text{O}_2^- \tag{12.2}$$

Acids are often classified as "weak" or "strong." **Strong acids** are those that ionize completely when dissolved in water (and thus are strong electrolytes—Chapter 10). Strong acids are particularly corrosive and damaging to metals and organic materials such as plant and animal tissue. Hydrochloric acid, sulfuric acid, and nitric acid are three strong acids. Special precautions are required when handling them, especially when they are at high concentration levels. **Weak acids** are those that do not ionize completely when dissolved in water. Ionization constants (see Chapter 11) that are extremely small are often associated with weak acids and solutions of weak acids are also weak electrolytes (Chapter 10). Weak acids, although able to attack some metals and some organic materials, and thus may require some special handling precautions, are, with some exceptions, not considered to be as dangerous as strong acids. Acetic acid is a weak acid, as are most of the other acids listed in Tables 12.1 and 12.2. Most common strong and weak acids are sold as concentrated solutions.

There are a few compounds that do release hydrogen ions under certain conditions, although they do not have hydrogen first in their formulas. These are the salts of acids that have more than one hydrogen in their formulas and only some of these hydrogens have been replaced by metals. Recall the discussion of nomenclature in Chapter 4. Examples of acids with more than one hydrogen in their formulas include sulfuric acid, $\text{H}_2\text{SO}_4$, phosphoric acid, $\text{H}_3\text{PO}_4$, and carbonic acid, $\text{H}_2\text{CO}_3$. Examples of salts in which only some of the hydrogens have been replaced (and therefore are capable of releasing hydrogens) are $\text{NaHSO}_4$, $\text{NaH}_2\text{PO}_4$, $\text{Na}_2\text{HPO}_4$, and $\text{NaHCO}_3$.

A comment about water, $\text{H}_2\text{O}$, is appropriate here. While water has hydrogen first in the formula, we don't usually think of it as an acid. A small fraction of the water molecules in a sample of pure distilled water does release hydrogen ions, however, and we will return to this phenomenon as our discussion proceeds.

We recognize that a compound is a **base** when there is one or more hydroxide ions in the formula. Some of the common bases we have encountered are listed in Table 12.3. The behavior of bases with their hydroxide ions is similar to the behavior of acids with their hydrogen ions. Hydroxide ions are easily released under certain conditions. With sodium hydroxide, for example, we can write

$$\text{NaOH} \rightarrow \text{Na}^+ + \text{OH}^- \tag{12.3}$$

# FACTS ABOUT SULFURIC ACID

S ulfuric acid is by far the highest volume chemical produced in the United States. In 1995, 95.36 billion pounds were produced, compared to 68.04 billion pounds of the second-ranked chemical, nitrogen gas. Most of the sulfuric acid is produced in chemical plants utilizing what is called the "contact process." In this process, sulfur dioxide gas ($SO_2$), which is produced from either pure sulfur or from the iron ore $FeS_2$, is oxidized to sulfur trioxide gas ($SO_3$) by a catalyst (typically iron, vanadium oxide, or platinum) at a high temperature. Sulfuric acid is then produced by the reaction of sulfur trioxide with water.

Most of the sulfuric acid that is produced is used to make fertilizers, like ammonium sulfate and the superphosphates. It is used to make organic products, paints and pigments, and rayon. It is also used in the refining of petroleum and in processing metals.

Chemistry professionals are needed for all aspects of production and are used both as chemical plant operators and as laboratory technicians for quality control and waste monitoring.

*For Homework: Look up the manufacturing process for ammonium sulfate and superphosphates in the* Concise Encyclopedia of Chemical Technology *by Kirk-Othmer and write a report providing as many details as you can, including the use of sulfuric acid.*

**TABLE 12.3**   Some Examples of Bases that are Hydroxides

| Formula | Name |
|---------|------|
| NaOH | Sodium hydroxide |
| $NH_4OH$ | Ammonium hydroxide |
| KOH | Potassium hydroxide |
| $Ca(OH)_2$ | Calcium hydroxide |
| $Ba(OH)_2$ | Barium hydroxide |
| $Mg(OH)_2$ | Magnesium hydroxide |

to symbolize the ionization behavior when the base dissolves in water. Similarly, hydroxide ions are released when the base reacts with an acid.

Bases are also described as "strong" and "weak." Sodium hydroxide and potassium hydroxide fit into the category of **strong bases** and, at high concentration levels, are very corrosive and damaging to metals and plant and animal tissue. Again, special handling procedures are in order. Contrary to strong acids, these bases are normally purchased as solid chemicals.

Ammonium hydroxide is a good example of a **weak base**, meaning that it only ionizes slightly when dissolved in water.

$$NH_4OH \rightleftharpoons NH_4^+ + OH^-$$                    (12.4)

It is usually purchased as a concentrated solution like the acids. This concentrated solution smells very strongly of ammonia, $NH_3$, because there is a second equilibrium reaction, in addition to ionization, that occurs in ammonium hydroxide/water solutions.

$$NH_4OH \rightleftharpoons NH_3 + H_2O$$

Thus, these solutions are saturated solutions of ammonia. Because of this, concentrated solutions of ammonia are also concentrated solutions of ammonium hydroxide and vice versa.

## 12.3    Properties of Acids and Bases

### 12.3.1    General Properties

The most recognizable physical property associated with acids is their sour taste. The common household product vinegar is approximately 5% acetic acid and has a sour taste. A common property of citrus fruits (oranges, grapefruit, lemons, pineapple, etc.) is their sour taste due to the presence of citric acid, which is an organic tricarboxylic acid.

$$\text{Citric Acid} \quad \begin{array}{ccc} COOH & COOH & COOH \\ | & | & | \\ CH_2 \!\!-\!\!\!-\!\!\!-\!\! C\text{-}OH \!\!-\!\!\!-\!\!\!-\!\! CH_2 \end{array}$$

Bases have a bitter taste and a slippery, soapy feel. Often the terms alkali, alkaline, lye, and caustic are associated with bases. Sodium hydroxide and potassium hydroxide have, for example, been referred to as "caustic soda" and "caustic potash," respectively. The reason is, of course, that these compounds are caustic, corrosive compounds that can chemically attack a variety of materials, including plant and animal tissue. The slippery, soapy feel is not a coincidence. These strong bases are used in the production of soaps.

Even to a person not educated in the properties of acids, the word "acid" conjures up thoughts of the possibility of serious chemical burns on skin. Acids in general are quite harmful to skin and must be handled with extreme caution. If an acid does contact skin, we often feel a burning sensation quickly, indicating that rinsing with water and possibly some neutralizing treatment is in order. Bases can be equally harmful to the skin and the problem is more acute in this case because we often don't feel the burning sensation as quickly. It should be emphasized that while one should never handle any chemical, liquid or solid, directly with fingers and/or hands, it is especially true of solid acids and bases because of the potential for chemical burns.

The chemical properties that acids and bases exhibit are mostly due to the easy release of hydrogens (in the case of acids) and hydroxides (in the case of bases). The reason we study them together is that hydrogen ions and hydroxide ions react with each other to form water.

$$H^+ + OH^- \rightarrow H_2O \tag{12.5}$$

Thus, whenever hydrogen ions are released, hydroxide ions get consumed, creating **acidic conditions** in the solution, meaning that hydrogen ions predominate over the hydroxide ions. Whenever hydroxide ions are released, hydrogen ions get consumed, creating **basic conditions** in the solution, meaning that hydroxide ions predominate over hydrogen ions.

The presence of excess $H^+$ or excess $OH^-$ ions in a solution gives the solution certain colorful chemical properties. First, an excess of hydrogen ions in a solution causes litmus dye to turn pink. An excess of hydroxide ions causes litmus dye to turn blue. Paper saturated with litmus dye known as **litmus paper**, is often used as an indicator of acidic and basic conditions. There are a large number of other **indicators** which change color under various acidic and basic conditions. Phenolphthalein solutions are pink under basic conditions and colorless under acidic conditions. Methyl orange solutions are yellow under basic and slightly acidic conditions, and red under very acidic conditions. It is possible to mix various indicators and thus prepare an indicator that will change colors at a number of levels of acidity and basicity. Again we can saturate paper with this indicator mixture and have what is called ***pH paper***, which is paper that is saturated with a mixture of various indicators and can thus detect specific acidity and basicity levels in a solution simply by displaying different colors. Litmus paper, pH paper, and the other indicators are commonly used to determine the acidity level of a solution, i.e., whether hydrogen ions or hydroxide ions are predominating and to what extent. The acidity level is commonly expressed as the ***pH*** of the solution. The concept of pH is discussed in Section 12.7 and the use of pH meters to measure it is examined in Section 12.8.

In addition to the properties of acids and bases that affect the color of other chemicals, there are other important chemical properties of these compounds. One is ***neutralization***, which is the negation of the chemical properties of acids by their reaction with bases, and vice versa. In other words, it is the reaction of acids with bases. A base can be neutralized by an acid and an acid can be neutralized by a base. The reason that neutralization occurs is that hydrogen ions react with hydroxide ions to form water, as we discussed previously. Water is one product of such a reaction while a salt is the other. This will be discussed in more detail in Section 12.6.

## 12.3.2    Properties of Some Specific Acids and Bases

Some specific acids and bases are so commonly used in laboratories that it behooves us to briefly discuss some noteworthy properties of each individually. The acids we will focus on here are hydrochloric acid, HCl, sulfuric acid, $H_2SO_4$, nitric acid, $HNO_3$, and acetic acid, $HC_2H_3O_2$. We will also mention two bases, sodium hydroxide, NaOH, and ammonium hydroxide, $NH_4OH$. Containers for these chemicals are color-coded (both cap and label) as indicated in the following paragraphs. ***Safety is a primary concern*** with each of these, as will be obvious as we proceed through the following discussion. The discussion will center on concentrated solutions of these acids and bases as they are purchased from the chemical vendor. For these, at minimum, safety glasses and/or shields, latex gloves, and aprons should be used. Neutralizing solutions should be readily available in the event of a spill. Safety showers should be available in the event of a spill on one's person.

**Hydrochloric Acid.** Hydrochloric acid is a strong acid. Contact with skin and clothing should be avoided and appropriate safety gear used when handling it. Hydrochloric acid solutions are actually solutions of the gas hydrogen chloride, HCl. Concentrated HCl, which is how it is most often purchased, is a saturated solution of this gas, and, as such, exhibits significant fuming. The fumes are quite pungent and stifling, and thus concentrated HCl must be used in a good fume hood. Hydrochloric acid has a blue color code.

**Sulfuric Acid.** Sulfuric acid is also a strong acid. Concentrated sulfuric acid is a viscous, dense (about 1.8 times the density of water) liquid that is very damaging, almost on contact, to skin and clothing. The color code for sulfuric acid is yellow. When concentrated sulfuric acid is mixed with water, much heat is produced, sometimes causing the water to boil. In the preparation of 50-50 mixtures of sulfuric acid and water, the container can become so hot that heat burns may result. A means for cooling the container is often necessary.

**Nitric Acid.** Nitric acid is also a strong acid. Concentrated nitric acid is also a very dangerous acid in terms of possible damage to skin and clothing. Shortly after contact, skin turns yellow. Nitric acid is an oxidizing acid, which means it will produce stifling gases that are oxides of nitrogen upon contact with metals and other chemicals. Red is the color code for this acid.

**Acetic Acid.** As noted earlier, acetic acid is a weak acid ($K_a = 1.76 \times 10^{-5}$). Concentrated acetic acid, which is actually 100% acetic acid (not a water solution) also exhibits fuming and should be used in a good fume hood. While it is correctly described as a weak acid, in the pure form it is no less dangerous than the others and can produce serious chemical burns to the skin. Concentrated acetic

# ACID-BASE TITRATIONS—NEUTRALIZATION REACTIONS USING INDICATORS

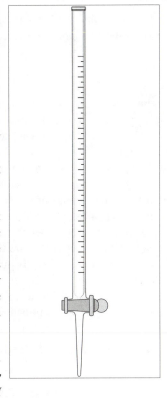

A *titration* is an experiment in which a chemical reaction occurs, but in a rather controlled manner. One reactant is added to another with the use of a *buret*, a special graduated cylinder like that shown at the right (see also CPW Box 7.5) . The solution in the buret is added to another solution in a container below the buret and the reaction between the components of the solutions occurs. Such an experiment is usually performed when a chemist wishes to analyze the solution in the receiving container by measuring how much solution in the buret it takes to react completely with it. It is very common for the reaction involved to be a neutralization reaction—an acid in the buret and a base in the receiving container, or vice versa. Such a titration is often called an *acid-base titration*.

One very important aspect of any titration experiment is the fact that the chemist must know precisely when the component in the receiving container is completely consumed by the reaction (the *end point*). In acid-base titrations, an *indicator* often serves this purpose. This is because the acidity level of the solution sharply changes at the end point, causing the color of the solution to change at precisely the correct time. The chemist then stops the titration and the results are calculated.

*For Homework: Find the table of acid-base indicators in the* Handbook of Chemistry and Physics *published by CRC Press, Inc. In view of the discussion above, explain the meaning of the "Approximate pH Range" column and the "Color Change" column in this table.*

acid is often referred to as *glacial acetic acid*, probably because it has a relatively low melting point (about 17°C) and can freeze in a cooler laboratory environment. Containers for acetic acid are color-coded brown.

**Hydrofluoric Acid**. While hydrofluoric acid is a weak acid ($K_a = 6.5 \times 10^{-4}$), it is a very dangerous acid because it readily penetrates the skin. It also dissolves glass. Thus, extreme caution is important both in handling and in storage of this acid. Gloves and face shields are the minimum requirements for safety in handling and hydrofluoric acid must be stored in plastic bottles.

**Sodium Hydroxide.** Sodium hydroxide is purchased as a pure white solid. It is neither powdered nor granular, but in the form of pellets. It is extremely hygroscopic (adsorbs water readily) and a single pellet can dissolve in its own adsorbed water if left exposed on a humid day. The pellets often appear shiny, due to the rapidly absorbed water. Concentrated solutions are just as dangerous as concentrated acid solutions in terms of their effect on skin and clothing. Solutions of NaOH are stored in plastic containers because it can dissolve glass to a small extent. The color code for sodium hydroxide is orange.

**Ammonium Hydroxide.** Ammonium hydroxide is a weak base ($K_b = 1.79 \times 10^{-5}$) and it, too, is dangerous. Concentrated $NH_4OH$ is a saturated solution of $NH_3$ gas, and thus exhibits an extremely strong ammonia odor. It must be handled in a good fume hood. It is often used to decrease the acidity level of solutions and in the creation of buffer solutions that must have a very low acidity level. Ammonium hydroxide has a green color code.

### 12.3.3 "Add Acid to Water" Rule

As we mentioned in the previous section, much heat is produced when sulfuric acid is mixed with water. This is because the ionization reaction

$$H_2SO_4 \rightarrow 2H^+ + SO_4^{2-} \qquad (12.6)$$

is highly exothermic. Actually, ionization reactions of all strong acids, and also of many weak acids, are exothermic. Extreme care must be exercised when any acid is mixed with water. A nontrivial question is this: "When mixing an acid with water, should you add the acid to the water, or the water to the acid?" An important consideration here is that when the exothermic ionization reaction occurs, some splashing may take place due to the high heat evolution and possible boiling of the liquid at the point of contact. A serious safety hazard is therefore present because the hot acid may be splashed on one's face and clothing. To greatly minimize the problem, one should *always add the acid to the water*. The reason for this is that the splashed liquid in this scenario consists mostly of the liquid in the receiving container and if it is mostly water, then it is, of course, much less harmful to the face and skin.

One final point about this has to do with the viscosities of the concentrated acids. We stated earlier that sulfuric acid is a rather viscous acid. This property actually works to a lab worker's advantage to a small extent because a highly viscous liquid is less likely to splash. Thus, while the problem of splashing is certainly present with sulfuric acid due to the extreme heat evolution, we need to be equally cautious with other acids as well, especially nitric acid because its viscosity is very low.

## 12.4    Theories of Acids and Bases

### 12.4.1    The Arrhenius Theory

The view of acids and bases as we have been discussing them thus far is known as the **Arrhenius Theory**, named after **Svante Arrhenius**, a Swedish scientist who introduced this theory in 1884. In this theory, an acid is defined as a substance that releases hydrogen ions when dissolved in water. A base is defined as a substance that releases hydroxide ions when dissolved in water. All Arrhenius acids have the symbol for hydrogen first in the formula. See Tables 12.1 and 12.2 again for examples. All Arrhenius bases have hydroxide ions in the formula. All bases listed in Table 12.3 are Arrhenius bases.

### 12.4.2    The Bronsted-Lowry Theory

There are substances other than Arrhenius bases that turn litmus paper blue, feel slippery in water solution, and/or neutralize acids, etc. Thus, **Johannes Bronsted**, a Danish chemist, and **Thomas Lowry**, an English chemist, presented a broader definition of acids and bases in 1923. The **Bronsted-Lowry**

**TABLE 12.4** Important Bronsted-Lowry Bases Other than Hydroxides

| Name | Formula |
|------|---------|
| Ammonia | $NH_3$ |
| Sodium carbonate | $Na_2CO_3$ |
| Potassium carbonate | $K_2CO_3$ |
| Calcium carbonate | $CaCO_3$ |
| Sodium bicarbonate | $NaHCO_3$ |
| Potassium bicarbonate | $KHCO_3$ |

**Theory** states that an acid is a proton donor and a base is a proton acceptor. This definition of an acid is similar to that in the Arrhenius Theory (a hydrogen ion is a proton) and, thus, all Arrhenius acids are also Bronsted-Lowry acids and vice versa. However, the Bronsted-Lowry definition of a base is broader. There are substances other than Arrhenius bases that accept hydrogen ions. In other words, the hydroxide group is a hydrogen ion acceptor, but there are other groups that also do this.

What exactly is meant by a hydrogen ion acceptor? A hydrogen ion acceptor is a chemical species to which a hydrogen ion can attach and form a covalent bond. The most important examples, in addition to the Arrhenius bases (hydroxides), are ammonia, $NH_3$, and the carbonates (see Table 12.4). Examples of reactions in which hydrogen ions are accepted are the following.

$$NaOH + H^+ \rightarrow H_2O + Na^+ \tag{12.7}$$

$$NH_3 + H^+ \rightarrow NH_4^+ \tag{12.8}$$

$$Na_2CO_3 + H^+ \rightarrow NaHCO_3 + Na^+ \tag{12.9}$$

$$Na_2CO_3 + 2H^+ \rightarrow H_2CO_3 + 2Na^+ \tag{12.10}$$

$$KHCO_3 + H^+ \rightarrow H_2CO_3 + K^+ \tag{12.11}$$

Notice that when hydroxides accept hydrogen ions, water is always the product. Water is not a product in the cases of ammonia and the carbonates and bicarbonates. However, in the case of carbonates and bicarbonates, water often does ultimately form. The carbonic acid, $H_2CO_3$, in the above reactions is unstable and will spontaneously decompose into water and carbon dioxide. This reaction is often accompanied by vigorous gas ($CO_2$) evolution and will be discussed in more detail in Section 12.6.

The list of Bronsted-Lowry bases is actually much longer than the list that is given here. There are many anions like carbonate and bicarbonate that will form a covalent bond with hydrogen ions and form a weak acid like the carbonic acid above. Thus, the anion of any weak acid qualifies as a Bronsted-Lowry base. This list would include sulfide and cyanide, as well as acetate and anions of other carboxylic acids.

In Section 12.2 we mentioned that water can be thought of as an acid because in a sample of pure distilled water, water molecules release hydrogen ions to a small extent. In the Bronsted-Lowry Theory, water can also be thought of as a base because water molecules accept hydrogen ions to form the *hydronium ion*, $H_3O^+$:

$$H^+ + H_2O \rightarrow H_3O^+ \tag{12.12}$$

Many chemists, in fact, consider the product of acid ionization in water solution to be the hydronium ion and not the hydrogen ion.

**TABLE 12.5**    Examples of Acid/Base Reactions in which the Conjugate Acid and Conjugate Base Form

| Acid | + | Base | → | Conjungate Acid | + | Conjungate Base |
|------|---|------|---|-----------------|---|-----------------|
| $HCl$ | + | $OH^-$ | → | $H_2O$ | + | $Cl^-$ |
| $2HC_2H_3O_2$ | + | $CO_3^{2-}$ | → | $H_2CO_3$ | + | $2C_2H_3O_2^{1-}$ |
| $H_2SO_4$ | + | $2CN^-$ | → | $2HCN$ | + | $SO_4^{2-}$ |
| $HF$ | + | $H_2PO_4^-$ | → | $H_3PO_4$ | + | $F^{1-}$ |
| $HPO_4^{2-}$ | + | $C_2H_3O_2^-$ | → | $HC_2H_3O_2$ | + | $PO_4^{3-}$ |

### 12.4.3   The Lewis Theory

There is a third theory by which acids and bases are defined. It is called the **Lewis Theory**. In this theory, an acid is defined as a chemical species that accepts a pair of electrons in a chemical reaction, while a base is a chemical species that donates a pair of electrons in a chemical reaction. This considerably broadens the concepts of acids and bases to include many substances that do not donate or accept hydrogens. Further discussion of this theory is beyond the scope of this text.

## 12.5   Conjugate Acids and Bases

It may seem strange that the reaction of a base with hydrogen ions, the source of which is an acid, may also produce an acid such as carbonic acid and the other weak acids mentioned in the previous section. Even though the anion involved may easily accept hydrogen ions, the acid thus formed may also, under appropriate conditions, lose the hydrogen ion it gained and therefore display acid properties. The acids formed in these reactions are known as *conjugate acids*. In addition, the anion that accompanied the hydrogen ion in the original reaction may, under appropriate conditions, accept the hydrogen ion back again and thus display base properties. Such a base is known as the *conjugate base*. There is an acid, a base, a conjugate acid, and a conjugate base in all hydrogen ion donation/acceptance reactions. Some examples are listed in Table 12.5.

## 12.6   Reactions Involving Acids and Bases

There are some specific reactions of acids and bases that are of some importance. In this section, we summarize them and give examples.

### 12.6.1   Neutralization

After acids lose or donate hydrogens, we say that they have been *neutralized*. Similarly, after bases accept hydrogens, we say that they have been neutralized. The reaction of a hydrogen ion donor with a hydrogen ion acceptor readily occurs and neutralizes both the acid and base. The reaction of an acid with a base has thus come to be known as a "neutralization" reaction. The reaction is an example of a *double replacement* reaction in which the acid and base simply "trade partners" as in the following general scheme.

$$AB + CD \rightarrow AD + BC \tag{12.13}$$

Trading partners in the case of an acid and a base, in which the base is a hydroxide, results in a salt and water being formed.

$$Acid + Base \ (hydroxide) \rightarrow Salt + Water \tag{12.14}$$

---

*Chemistry Professionals at Work*  CPW Box 12.4

# SCALE BUILDUP IN WATER DISTILLERS

Chemists and chemistry laboratory technicians who perform critical analyses of samples of bulk materials avoid serious error by using distilled water, rather than tap water, for solutions that they prepare. The reason is that the minerals that are present in tap water, commonly known as water **hardness**, can interfere with the analysis and give incorrect results. All laboratories must be equipped with water distillers, or deionizers, so that pure water can be used for these applications.

The most abundant dissolved mineral in hard water is calcium carbonate, $CaCO_3$. In water distillers, the concentration of $CaCO_3$ increases with time until its solubility is finally exceeded and it begins to precipitate on the walls of the boiler chamber as a "scale." Over time, this build up can become quite heavy and impair the efficiency of the distiller. At that point, the boiler chamber must be cleaned, meaning that the $CaCO_3$ scale must be removed.

The method by which the $CaCO_3$ is removed is by its reaction with an acid such as HCl. This reaction is exactly the kind of reaction described here—the reaction of an acid with a carbonate. Due to the formation of carbon dioxide during this reaction, there is considerable foaming during the cleaning procedure. When the reaction stops, however, the boiler chamber is again clean and able to produce pure, distilled water with high efficiency.

*For Homework: Ask your instructor to allow you to watch the next time he/she cleans the hardness scale from the water distiller in the laboratory. Also, ask him/her to allow you to see the instructions for this cleaning. Write a report on the process.*

---

Trading partners in the case of an acid and a base, in which the base is a carbonate, results in a salt and carbonic acid being formed.

$$\text{Acid + Base (carbonate)} \rightarrow \text{Salt + Carbonic Acid} \qquad (12.15)$$

As mentioned in Section 12.4, carbonic acid ($H_2CO_3$) is very unstable and spontaneously decomposes to form carbon dioxide and water.

$$H_2CO_3 \rightarrow CO_2 + H_2O \qquad (12.16)$$

Thus we can also write the following.

$$\text{Acid + Base (carbonate)} \rightarrow \text{Salt + Carbon Dioxide + Water} \qquad (12.17)$$

The neutralized acid may be either "partially neutralized" or "fully neutralized." In the latter, all acidic hydrogens shown in the formula are donated, while in the former, only one or two, etc., acidic hydrogens are donated. Thus, the salt product may still show some undonated hydrogens remaining in the formula. Some examples are shown in Table 12.6.

**TABLE 12.6**    Examples of Neutralization Reactions

(a)   $HCl + NaOH \rightarrow NaCl + H_2O$
(b)   $H_2SO_4 + 2KOH \rightarrow K_2SO_4 + 2H_2O$
(c)   $2H_3PO_4 + Mg(OH)_2 \rightarrow Mg(H_2PO_4)_2 + 2H_2O$
(d)   $H_3PO_4 + 2NaOH \rightarrow Na_2HPO_4 + 2H_2O$
(e)   $H_3PO_4 + 3KOH \rightarrow K_3PO_4 + 3H_2O$
(f)   $2HCl + CaCO_3 \rightarrow CaCl_2 + H_2CO_3$
(g)   $H_2SO_4 + Na_2CO_3 \rightarrow Na_2SO_4 + CO_2 + H_2O$
(h)   $HCl + Na_2CO_3 \rightarrow NaCl + NaHCO_3$

## 12.6.2   Reaction of Acids with Metals

Acids react with some metals to form hydrogen gas in what is termed a ***single-replacement*** reaction. In a single-replacement reaction, one element is released from a formula and the remaining element combines with the reacting element.

$$AB + C \rightarrow AC + B \tag{12.18}$$

Water is an acid of sufficient strength to react with the very active alkali metals, such as sodium and potassium. In fact, the reaction of these metals with water is quite violent, releasing hydrogen gas and leaving the metal hydroxide in solution. The reaction of sodium with water is as follows.

$$2H_2O + 2Na \rightarrow 2NaOH + H_2 \tag{12.19}$$

This reaction also occurs with the alkaline earth metals calcium and magnesium, although it is considerably less violent.

The less active metals, such as zinc and iron, require a stronger acid. In these cases, the products are the salt of the acid and hydrogen gas. For example, zinc reacts with hydrochloric acid as follows.

$$2HCl + Zn \rightarrow ZnCl_2 + H_2 \tag{12.20}$$

Very inactive metals such as copper and silver, do not undergo this reaction. These metals do react with nitric acid and other oxidizing acids, however, giving products different from those shown above.

## 12.6.3   Reaction of Metal and Nonmetal Oxides with Water

Metal oxides react with water to form the metal hydroxides, or bases.

$$\text{metal oxide} + \text{water} \rightarrow \text{metal hydroxide} \tag{12.21}$$

Examples include reactions of the oxides of sodium and magnesium with water.

$$Na_2O + H_2O \rightarrow 2NaOH \tag{12.22}$$

$$MgO + H_2O \rightarrow Mg(OH)_2 \tag{12.23}$$

Nonmetal oxides, such as oxides of carbon, nitrogen, sulfur, and phosphorus react with water to give acids.

$$\text{nonmetal oxide} + \text{water} \rightarrow \text{acid} \tag{12.24}$$

Examples include $CO_2$ and $SO_3$.

$$CO_2 + H_2O \rightarrow H_2CO_3 \qquad (12.25)$$

$$SO_3 + H_2O \rightarrow H_2SO_4 \qquad (12.26)$$

The formula of the acid can be predicted from the fact that the number of oxygen atoms in the acid is one more than the number of oxygen atoms in the oxide.

This type of reaction occurs when rain water reacts with nonmetal oxide pollutants in the air to form what has been commonly referred to as **acid rain**. Acid rain has been a troublesome phenomenon in regions surrounding heavily industrialized areas. Vegetation can be killed, metal and granite statues corroded, and the health of people can be adversely affected by acid rain.

### 12.6.4   Reaction of Metal Oxides with Acids

Since metal oxides react with water to form metal hydroxides, it is predictable that the products of the reaction of metal oxides with acids are the same as the reaction of metal hydroxides with acids. The products are a salt and water.

$$\text{acid } + \text{ metal oxide } \rightarrow \text{ salt } + \text{ water} \qquad (12.27)$$

Let us again use $Na_2O$ and MgO as our examples.

$$2HCl + Na_2O \rightarrow 2NaCl + H_2O \qquad (12.28)$$

$$2HCl + MgO \rightarrow MgCl_2 + H_2O \qquad (12.29)$$

## 12.7   Acidity Level: pH and pOH

For a variety of reasons, it is very important to be able to express the acidity level of solutions. We now proceed to define the common method for doing this—the concept of pH. This method, as we will see, has application in a chemistry lab as well as in our homes and environment.

### 12.7.1   Water Ionization and the Ion Product Constant

The fact that a small number of water molecules in pure distilled water ionize was mentioned in Section 12.2. This ionization is an equilibrium and can be written as follows:

$$H_2O \rightleftharpoons H^+ + OH^- \qquad (12.30)$$

This equilibrium lies far to the left (indicating that barely any ionization at all has taken place). This means that the concentrations of hydrogen ions and hydroxide ions are very small. At 25°C, $[H^+] = [OH^-] = 1.0 = 10^{-7}$ M. A special equilibrium constant, $K_w$, known as the *ion product constant* for water, is based on this ionization and is defined as follows at 25°C.

$$K_w = [H^+][OH^-] = 1.0 \times 10^{-14} \qquad (12.31)$$

Because it is an equilibrium, the addition of acids or bases to water shifts this equilibrium to the left. The removal of acids or bases through neutralization reactions shifts the equilibrium to the right. Addition of acids increases the $[H^+]$ and the resulting shift causes $[OH^-]$ to decrease. Addition of Arrhenius bases (hydroxides) increases $[OH^-]$ and the resulting shift causes $[H^+]$ to decrease. Removal of acids decreases $[H^+]$ and the resulting shift causes $[OH^-]$ to increase. Removal of Arrhenius bases decreases $[OH^-]$ and the resulting shift causes $[H^+]$ to increase. Thus, whenever $[H^+]$ increases, $[OH^-]$ decreases, and vice versa. However, the multiplication product of $[H^+]$ and $[OH^-]$, or $K_w$, like the equilibrium constants in Chapter 11, does not change. So knowing either $[H^+]$ or $[OH^-]$, one can calculate the other from Eq. (12.31). We will assume 25°C for all calculations involving $K_w$.

### Example 12.1

What is the $[OH^-]$ in a solution in which the $[H^+]$ is $5.2 \times 10^{-3}$ M?

### Solution 12.1

Rearranging Eq. (12.31), we have the following.

$$[OH^-] = \frac{K_w}{[H^+]} = \frac{1.0 \times 10^{-14}}{5.2 \times 10^{-3}} = 1.9 \times 10^{-12} \text{ M}$$

### 12.7.2  pH

Much of the chemistry that occurs in water solution depends on what the acid content, or acidity level of the solution is. For example, a procedure for the determination of caffeine in soda pop calls for a solution that has a $[H^+]$ of $1.00 \times 10^{-3}$ M. A procedure for the extraction of potassium from soil requires a solution that has a $[H^+]$ of $1.00 \times 10^{-7}$ M. Indeed, many activities in our everyday lives depend on the acidity of solutions. Gardeners who grow rhododendrons, for example, should use soil that has a $[H^+]$ between about $1 \times 10^{-4}$ M and $3 \times 10^{-6}$ M. The fluid in our stomachs should have $[H^+]$ of between $1 \times 10^{-1}$ M and $1 \times 10^{-3}$ M. Tropical fish in an aquarium survive only if the $[H^+]$ is between $1 \times 10^{-6}$ M and $1 \times 10^{-8}$ M. Fish in our lakes and rivers have been threatened by acid rain. The $[H^+]$ in lakes and rivers should not drop below $1 \times 10^{-5}$ M if fish and other aquatic life are to survive. If the $[H^+]$ in our blood is not within a very narrow range ($1.00 \times 10^{-7}$ M to $1.58 \times 10^{-8}$ M), we die.

If the $[H^+]$ in a solution is too high, a base may be added to decrease it. If the $[H^+]$ is too low, an acid may be added to increase it. The adjustment of the acidity level in real-world solutions such as those mentioned above has often been performed by persons with no formal training in acid/base chemistry. For example, gardeners often adjust the acidity level of soil by applying lime (CaO) to the soil. Individuals ill with indigestion, heartburn, etc., can adjust the $[H^+]$ of their stomach fluid by taking an antacid medication. The nonchemist is not likely to have an understanding of scientific notation of the often extremely small number by which $[H^+]$ is expressed. They also may not be comfortable with the numbers in a way that would allow them to solve problems related to acidity level.

For these reasons, and for their own convenience, scientists devised a much simpler method of expressing $[H^+]$. It is called the pH scale. The *pH scale* expresses the acidity level of a solution in easily understood and manageable numbers. The *pH* of a solution is represented by the negative logarithm of the $[H^+]$.

$$pH = -\log[H^+] \tag{12.32}$$

### 12.7.3  Logarithms

The *logarithm of a number* is the power to which 10 must be raised to give the number. For example, the power to which 10 must be raised to give the number 100 is 2. Thus, the logarithm of 100 is 2. Similarly, the power to which 10 must be raised to give the number 1000 is 3. Thus, the logarithm of 1000 is 3. See Table 12.7 for other examples.

# ALKALINITY

Chemistry professionals employed to assure that public drinking water is of the highest quality, or that a wastewater treatment facility is operating properly, are often concerned with a parameter called *alkalinity*. Alkalinity is the capacity of a water to neutralize acids. The word "alkalinity" is derived from the word "alkali," which is another word for "base."

Drinking water as well as wastewater have dissolved carbonates and bicarbonates due to the equilibria initiated by the dissolving of carbon dioxide from the air (see CPW Box 11.3). Carbonates and bicarbonates are also dissolved from soil and rock as the water percolates through it prior to getting into the ground water supply. Carbonate and bicarbonate react with acids to form carbonic acid and so their presence is the chief reason that the water can help neutralize acids.

Processes used by chemistry professionals to rid water of turbidity require a minimum amount of alkalinity in order to be effective. In addition, sufficient alkalinity in wastewater is needed to provide the buffering capacity for the anaerobic digestion processes used to treat the wastewater. Hence the need to measure it.

*For Homework: Find a reference on water and wastewater chemistry and look up the details of the procedure used to determine alkalinity. Write a report.*

All positive numbers have a logarithm, not just those that are multiples of 10. The power to which 10 must be raised in most cases is not a whole number like it is for the multiples of 10 in Table 12.7. For example, the logarithm of 2 is 0.30103. The logarithm of 3 is 0.4771213. The logarithm of 5.11932 is 0.7092123 (see Table 12.8).

Equation (12.32) indicates that pH is the *negative* logarithm of $[H^+]$. The *negative logarithm* of a number is the logarithm multiplied by $-1$. The $[H^+]$ is typically an extremely small positive number. All positive numbers have a logarithm, no matter how small the number is. Thus, even the extremely small numbers that are the hydrogen ion concentrations have logarithms (and negative

**TABLE 12.7**  Examples of Some Easily Determinable Logarithms

| | |
|---|---|
| $10^2 = 100$ | $\log 100 = 2$ |
| $10^3 = 1000$ | $\log 1000 = 3$ |
| $10^{-2} = \dfrac{1}{100} = \dfrac{1}{100} = 0.01$ | $\log 0.01 = -2$ |
| $10^{-3} = \dfrac{1}{10^3} = \dfrac{1}{1000} = 0.001$ | $\log 0.001 = -3$ |

**TABLE 12.8** Examples of Logarithms of Numbers that are not Multiples of 10

| | |
|---|---|
| $10^{0.30103} = 2$ | Log 2 = 0.30103 |
| $10^{0.4771213} = 3$ | Log 3 = 0.4771213 |
| $10^{0.7092123} = 5.11932$ | Log 5.11932 = 0.7092123 |

logarithms), and the negative logarithms are easily understood and manageable numbers. For example, the $[H^{1+}]$ in blood is typically $4.0 \times 10^{-8}$ M, or 0.000000040 M. The negative logarithm of 0.000000040 is 7.40.

The complete scale of pH values in water ranges from 0 to 14, with pH = 7 being the neutral point represented by pure distilled water or by any strong acid solution that has been exactly neutralized by a strong base. In pure, distilled water, or in a neutralized water solution, the $[H^+]$ and the $[OH^-]$ are equal to $1.0 \times 10^{-7}$ M. The pH of a neutral solution such as pure distilled water, is thus 7.00. Any pH values between 0 and 7 would indicate an "acidic" solution, one in which hydrogen ions predominate over hydroxide ions. Any pH values between 7 and 14 would indicate a "basic" solution, one in which hydroxide ions predominate over hydrogen ions. We see from this that the relationship between $[H^+]$ and pH is an inverse one—as $[H^+]$ increases, pH decreases, and vice versa. A low pH indicates a high $[H^+]$ (compared to $[OH^{1-}]$), while a high pH indicates a low $[H^+]$.

How is the logarithm of a number determined? Since every positive number has a logarithm specific to that number, tables of logarithms have been created, and certainly logarithms can be found by looking in these tables. However, logarithms can also be determined by using the LOG key on a scientific calculator. That will be the method used here. The pH of a solution that has a $[H^+]$ of $4.59 \times 10^{-5}$ M is 4.338. See Fig. 12.1 and 12.2 for the steps involved in determining this pH using a scientific calculator. (*Note:* If your scientific calculator has keys different from those indicated in Figure 12.1, you should study your calculator manual to determine how to solve this problem.)

It is also important to be able to calculate the $[H^+]$ given the pH. For example, if the pH of a solution is 3.51, what is the $[H^+]$? See Fig. 12.3 and 12.4 for the steps involved in determining this pH using a scientific calculator. Additional examples for practice are given below.

---

**To calculate pH given [H⁺]:**

1. **Type in [H⁺] using the following steps.**
   - **Type in the first part of the number**
   - **Press the "EE" or "EXP" key.**
   - **Press the "+/-" key.**
   - **Type in the power of 10.**

2. **Press the "LOG" key.**

3. **Press the "+/-" key.**

---

**FIGURE 12.1** The use of a scientific calculator for calculating the pH, given $[H^+]$.

**Example: What is the pH if [H$^+$] = 4.59 X 10$^{-5}$ M?**

    1. **Type in 4.59 X 10$^{-5}$ as follows.**
        – **Type in 4.59**
        – **Press the "EE" or "EXP" key.**
        – **Press the "+/-" key.**
        – **Type "5"**

    2. **Press the "LOG" key.**

    3. **Press the "+/-" key.**         **Answer 4.338**

**FIGURE 12.2** An example of the process shown in Fig. 12.1.

**To calculate [H$^+$] given a pH:**

    1. **Type in the pH.**

    2. **Press the "+/-" key.**

    3. **Press the "INV" or "2ND" key.**

    4. **Press the "LOG" key.**

**FIGURE 12.3** The use of a scientific calculator to calculate [H$^+$] given pH.

**Example: What is the [H$^+$] if the pH = 3.51?**

    1. **Type in 3.51.**

    2. **Press the "+/-" key.**

    3. **Press the "INV" or "2ND" key.**

    4. **Press the "LOG" key.**

        **Answer: 3.1 X 10$^{-4}$ M**

**FIGURE 12.4** An example of the process shown in Fig. 12.3.

*Example 12.2*

What is the pH of a solution that has a $[H^+]$ of $2.91 \times 10^{-4}$ M?

*Solution 12.2*

$$pH = -\log[H^+]$$
$$= -\log(2.91 \times 10^{-4})$$
$$= 3.536$$

*Example 12.3*

What is the $[H^+]$ in a solution if the pH of the solution is 9.551?

*Solution 12.3*

$$pH = -\log[H^+]$$

$$9.551 = -\log[H^+]$$

$$[H^+] = 2.81 \times 10^{-10} \text{ M}$$

Sometimes it may be necessary to calculate the pH of a solution when the concentration of an acid is given (and not the $[H^+]$). If the acid is a strong acid, it completely ionizes in water solution and, thus, the acid concentration and the $[H^+]$ are the same because the number of hydrogen ions present equals the number of acid molecules that are dissolved.

*Example 12.4*

What is the pH of a 0.0283 M solution of HCl?

*Solution 12.4*

HCl is a strong acid that completely ionizes when it dissolves in water.

$$HCl \rightarrow H^+ + Cl^-$$

Thus, the concentration of HCl (0.0283) is also the $[H^+]$.

$$pH = -\log[H^+] = -\log(0.0283) = 1.548$$

If the acid in the problem is a weak acid, the problem is more involved since the $[H^+]$ cannot be determined directly. The determination of $[H^+]$ in that case can be calculated by using the equilibrium constant expression for weak acid ionization (see Chapter 11, Section 11.11.3). The pH is then calculated from the $[H^+]$ using the method shown in Figure 12.1.

*Example 12.5*

What is the pH of a 0.150 M acetic acid solution? The $K_a$ for acetic acid is $1.76 \times 10^{-5}$.

*Solution 12.5*

Utilizing Eq. (11.44), in which "x" is the hydrogen ion concentration,

$$K_a = \frac{x^2}{C_{HA}}$$

we get the following.

$$1.76 \times 10^{-5} = \frac{x^2}{0.150}$$

$$x = [H^+] = 1.62 \times 10^{-3} \text{ M}$$

Then, calculating pH, we have the following.

$$pH = -\log [H^+] = -\log (1.62 \times 10^{-3}) = 2.789$$

## 12.7.4 Significant Figures Rule

The rule of significant figures when determining the logarithm of a number is as follows: the number of significant figures after the decimal point in the logarithm of a number is the same as the total number of significant figures in the number. You should notice that this rule was used in the examples in Figs. 12.2 and 12.4 and in Examples 12.2, 12.3, 12.4, and 12.5.

### 12.7.5  pOH

The hydroxide ion concentration may also be expressed in easily manageable numbers by defining the pOH in a way similar to pH.

$$pOH = -\log[OH^-] \tag{12.33}$$

Thus, pOH and $[OH^-]$ may be calculated in the same way as pH and $[H^+]$ in the previous examples.

*Example 12.6*
What is the pOH if the $[OH^-]$ is $7.19 \times 10^{-12}$ M?

*Solution 12.6*

$$
\begin{aligned}
pOH &= -\log[OH^-] \\
&= -\log(7.19 \times 10^{-12}) \\
&= 11.143
\end{aligned}
$$

*Example 12.7*
What is the $[OH^-]$ if pOH is 6.44?

*Solution 12.7*

$$
\begin{aligned}
pOH &= -\log[OH^-] \\
6.44 &= -\log[OH^-] \\
[OH^-] &= 3.6 \times 10^{-7} \text{ M}
\end{aligned}
$$

### 12.7.6  Relationship between pH and pOH

Beginning with Eq. (12.31), we can now derive a new equation that simplifies some of the calculations involving the water ionization process. This derivation recognizes the logarithmic identity that when

$$A = B \times C \tag{12.34}$$

the logarithm of A is the sum of the logarithms of B and C.

$$\log A = \log B + \log C \tag{12.35}$$

Also, the negative logarithm of A is the sum of the negative logarithms of B and C,

$$-\log A = -\log B - \log C \tag{12.36}$$

so that Eq. (12.31) can be rewritten as follows.

$$-\log K_w = -\log[H^+] - \log[OH^-] \tag{12.37}$$

or

$$14.00 = pH + pOH \tag{12.38}$$

**TABLE 12.9** Equations That Can Be Used to Interconvert $[H^+]$, $[OH^-]$, pH, and pOH

| | |
|---|---|
| Equation (1) | $K_w = [H^+][OH^-]$ |
| Equation (2) | $pH = -\log[H^+]$ |
| Equation (3) | $pOH = -\log[OH^-]$ |
| Equation (4) | $pH + pOH = 14.00$ |

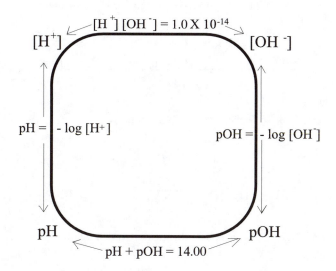

**FIGURE 12.5** Different pathways for interconverting $[H^+]$, $[OH^-]$, pH, and pOH.

*Example 12.8*

What is the pH of a solution that has a pOH of 5.39?

*Solution 12.8*

$$pH + pOH = 14.00$$
$$pH = 14.00 - pOH$$
$$= 14.00 - 5.39 = 8.61$$

## 12.7.7 Calculations Summary

We have learned four formulas in this section that can be utilized for the interconversion of $[H^+]$, $[OH^-]$, pH, and pOH. They are listed in Table 12.9. Given one of these four parameters, any of the other three may be calculated in two different ways. For example, given the $[H^+]$, the pOH may be calculated by either using Eqs. (1) and (3) in Table 12.9, or by using Eqs. (2) and (4). Similarly, given $[OH^-]$, $[H^+]$ may be calculated by either utilizing Eq. (1), or by using Eqs. (2), (3), and (4). Figure 12.5 shows the different pathways possible. An example appears below.

*Example 12.9*

What is the pH of a solution if the $[OH^-]$ is $6.3 \times 10^{-4}$ M?

## Solution 12.9(a)

Choosing Eq. (3) from Table 12.9, we have the following.

$$pOH = -\log[H^-]$$
$$= -\log(6.3 \times 10^{-4}) = 3.20$$

Then, choosing Eq. (4) from Table 12.9, we have the answer given below.

$$pH + pOH = 14.00$$
$$pH = 14.00 - pOH = 10.80$$

## Solution 12.9(b)

Choosing Eq. (1) from Table 12.9, we have the following.

$$[H^+][OH^-] = 1.0 \times 10^{-14}$$
$$[H^+] = \frac{1.0 \times 10^{-14}}{6.3 \times 10^{-4}} = 1.6 \times 10^{-11} M$$

Choosing Eq. (2) from Table 12.9, we obtain the answer shown below.

$$pH = -\log[H^+]$$
$$= (-\log(1.6 \times 10^{-11})) = 10.80$$

# 12.8  Measurement of pH

In Section 12.3, we discussed the use of litmus paper, pH paper, and indicators as means for determining the acidity level, or pH of a solution. While these have been widely used and are convenient and inexpensive for many applications, they do not provide the laboratory worker with the accuracy that is often needed. The use of pH paper is the most accurate of the three, but it can only determine pH to within 1 pH unit. Often the pH of a solution needs to be measured to the first or second decimal place.

A much more precise method of measuring pH has been developed. It utilizes what is called a **pH electrode** and a **pH meter**. The pH electrode is a small glass or plastic enclosure the size of a small test tube, that is electrically sensitive to the hydrogen ions in a solution. When this electrode is immersed into the solution to be tested, an "electrical potential" or voltage is produced across the glass membrane at the tip that separates the interior of the electrode from the solution. This voltage can be measured with a simple voltage measuring device, a voltmeter. The concentration of the hydrogen ions in the solution determines the voltage level that develops and, thus, this voltage is proportional to $[H^+]$. Since pH is also proportional to $[H^+]$, it is possible to relate this voltage directly to the pH of the solution. The pH meter is a voltmeter that utilizes an electrical circuit that automatically converts the voltage level to pH and displays the pH on its readout. Thus, simply stated, the pH of a solution can be measured or monitored by immersing this pH probe, connected to the pH meter, into the solution of interest and reading the pH on the readout of the meter.

Since the meter is a sort of modified voltmeter, the measurement is a "relative" measurement. This means that the voltage developed at the tip of the probe is measured "relative" to some other voltage. This other voltage is the voltage of a **reference electrode**. Modern pH probes have a reference electrode

---

*Chemistry Professionals at Work*                          *CPW Box 12.7*

# MOVING THROUGH THE RANKS

**M**y name is D. Richard Cobb. I am a Senior Research Technician employed by Eastman Kodak Company at their Kodak Park Site in Rochester, New York.

I have spent most of my career at Kodak working on the formation of new film products for the marketplace. Within the last 15 years, these have been mainly reversal film products. As a Senior Technician, I have experienced a shift in responsibility over the years from a job that consisted mostly of laboratory work to one involving design, organization, data analysis, and team leadership.

Aside from the regular work load, I have also been the Chemical Hygiene Officer for my division for about 8 years. My observation is that additional duties of this type are not unusual for chemistry technicians as they move through the ranks.

*For Homework: Check out the Internet site for Eastman Kodak and report on this company's activities.*

http://www.kodak.com/ (as of 6/4/00)

---

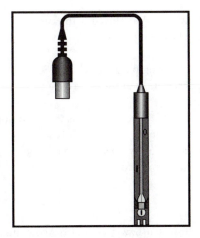

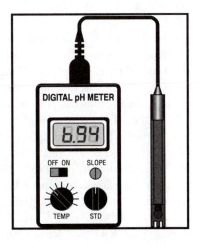

**FIGURE 12.6** A drawing of a pH probe with electrical connection.

**FIGURE 12.7** A pH meter with probe connected and pH displayed.

built in and are called ***combination pH electrodes***. The interior design of these is beyond the scope of this text. A drawing of a combination pH probe, including the electrode connection, is shown in Fig. 12.6. Figure 12.7 shows a pH meter with the probe connected.

## 12.9  Homework Exercises

1. Write down the names of at least four common laboratory acids and give their chemical formulas.
2. What is a carboxylic acid and why is acetic acid classified as a carboxylic acid? Why is hydrochloric acid not a carboxylic acid?
3. Sodium acetate, $NaC_2H_3O_2$, has several hydrogens in its formula, but it is not an acid. Explain. (It is not sufficient to say that the hydrogens are not listed first in the formula.)
4. What is the difference between strong acids and weak acids? Give examples of each.
5. Is the ionization constant of a weak acid typically a very small number, a very large number, or something in between? Explain.
6. Give two examples of compounds that are acids in spite of the fact that they do not have a hydrogen first in the formula. Explain why they are acids.
7. Name three compounds that are common laboratory bases and give their formulas.
8. Why are concentrated solutions of the weak base $NH_4OH$ also concentrated solutions of ammonia, $NH_3$? Why is $NH_4OH$ considered a weak base?
9. Why are weak acids and weak bases also weak electrolytes?
10. Why do citrus fruits have a sour taste?
11. What are some general properties of acids? Include both chemical and physical properties in your answer.
12. What are some general properties of bases? Include both chemical and physical properties in your answer.
13. What is an "indicator?" How are these used with solutions that contain acids or bases?
14. What is pH paper? How is it used with acid or base solutions?
15. What does it mean to say that $H^+$ ions predominate over $OH^-$ ions? What properties does this give to a solution?
16. What is meant by a neutralization reaction? Give an example.
17. Give one notable property of concentrated solutions of each of the following compounds that causes laboratory workers to take special precautions when they use them.
    (a) HCl  (b) $H_2SO_4$  (c) $HNO_3$  (d) $HC_2H_3O_2$  (e) NaOH  (f) $NH_4OH$
18. Write the formula or name of a concentrated acid or base that fits the given description.
    (a) Diffuses through skin
    (b) Has a blue color code
    (c) Can turn skin yellow
    (d) Gets extremely hot when mixed with water
    (e) Smells strongly of ammonia
    (f) An acid that has stifling, pungent fumes
    (g) Is almost twice as dense as water
    (h) Is often described as "glacial" because it has a low freezing point
19. What is the "add acid to water" rule? Give a detailed description of the procedure you would use to prepare a 50-50 mixture of sulfuric acid and water.
20. What is the definition of an acid according to the Arrhenius Theory? What is the definition of a base according to this theory. Give one example of an Arrhenius-type acid or base.
21. What is the definition of an acid according to the Bronsted-Lowry Theory? What is the definition of a base according to this theory? Give one example of a Bronsted-Lowry acid or base.
22. Consider the following compounds.
    (a) $HNO_3$  (b) $Na_2CO_3$  (c) $HC_2H_3O_2$  (d) $H_2SO_4$
    (e) $NH_3$  (f) KOH  (g) $CaCO_3$  (h) NaCl
    From this list, tell which are (1) Arrhenius acids, (2) Bronsted/Lowry acids, (3) Arrhenius bases, or (4) Bronsted/Lowry bases.
23. What is the hydronium ion? What does its formation have to do with water being a Bronsted/Lowry base?
24. What is the Lewis Theory of acids and bases? How is this different from the Arrhenius Theory and Bronsted-Lowry Theory?
25. What is the conjugate acid of the following bases: $CN^-$, $OH^-$, $C_2H_3O_2^-$, and $PO_4^{3-}$?

26. What is meant by a "double-replacement reaction?" What is a "single-replacement reaction?" Give one example of each.
27. Write the general reactions that occur when the following types of chemicals are combined together.
    (a)  Acid + Metal
    (b)  Metal Oxide + Acid
    (c)  Nonmetal Oxide + Water
    (d)  Metal Oxide + Water
28. Give the products and balance the following equations.
    (a)  $K_2O + H2O \rightarrow$
    (b)  $HCl + CaCO_3 \rightarrow$
    (c)  $NH_4OH + H_2SO_4 \rightarrow$
    (d)  $NO_2 + H_2O \rightarrow$
    (e)  $HCl + Mg \rightarrow$
    (f)  $CaO + HBr \rightarrow$
29. Fill in the product side and balance the following reactions.
    (a)  $H_2SO_4 + K_2CO_3 \rightarrow$
    (b)  $NO_2 + H_2O \rightarrow$
    (c)  $HNO_3 + Mg(OH)_2 \rightarrow$
    (d)  $K + H_2O \rightarrow$
    (e)  $Ca + HCl \rightarrow$
30. What would you expect to happen if you placed a piece of potassium metal in a beaker of water?
31. $Mg(OH)_2$ is a base contained in Phillips' Milk of Magnesia. It reacts with the excess HCl in our stomach walls to form what products?
32. What is acid rain? How does it form? Give an example of an acid that may be dissolved in acid rain.
33. What is the definition of $K_w$? Why is this important when dealing with water solutions?
34. Define pH. Why is pH important when dealing with water solutions?
35. What is the pH when $[H^+]$ has the following values.
    (a)  $5.02 \times 10^{-3}$ M
    (b)  $9.29 \times 10^{-5}$ M
    (c)  $3.81 \times 10^{-8}$ M
    (d)  $4.88 \times 10^{-11}$ M
    (e)  $6.06 \times 10^{-13}$ M
    (f)  0.00123 M
36. What is the $[H^+]$ when the pH is as follows.
    (a)  4.119    (b) 2.911    (c) 10.449    (d) 12.383    (e) 6.229
37. Calculate the pH of the following.
    (a)  A solution in which $[H^+]$ is $9.42 \times 10^{-12}$ M
    (b)  A solution in which $[OH^-]$ is $2.04 \times 10^{-3}$ M
    (c)  A solution in which the pOH is 7.4
    (d)  A solution in which the HCl concentration is 0.077 M
    (e)  A 0.150 M solution of chloroacetic acid, which has a $K_a$ of $1.40 \times 10^{-3}$
    (f)  A 0.300 M solution of iodoacetic acid, which has a $K_a$ of $7.5 \times 10^{-4}$
38. What is the $[H^+]$ in each of the following situations?
    (a)  pOH = 4.51
    (b)  $[OH^-] = 1.88 \times 10^{-2}$ M
    (c)  pH = 7.00
    (d)  pH = 3.29
39. Tell which of the solutions in Problems 35, 36, 37, and 38 are acidic, basic, and neutral.
40. What is meant by a (a) pH meter, and a (b) combination pH probe?

## For Class Discussion and Reports

41. Manufacturers and distributors of pH electrodes usually include a set of instructions for storage, care, and maintenance of the electrodes. Ask your instructor to give you a copy of an example of these instructions. Using this and any other sources you can find, write a report on how to care for pH electrodes. Include discussions of long-term storage, short-term storage, and care and maintenance while in use.

42. Find a copy of the September 1999, issue of the journal *Today's Chemist at Work* in a library and flip to the page where the column "Lighter Elements" is found. Read the first segment (titled "Oh, Those F-Words"), Item #3. Explain why there was an explosion. Write the chemical equations to support your explanation, both for what the student was intending and for what actually happened.

43. Explore the roles of acids and bases in the art of papermaking. Suggested references include an article entitled "Sizing Up Paper" in the April 1998, issue of the journal *ChemMatters*, and another entitled "The Chemistry of Paper Preservation, Part 4, Alkaline Paper" in the May 1997, issue of the *Journal of Chemical Education*. Write a report.

# 13

# Introduction to Thermodynamics

## 13.1   Introduction

Nitric acid ($HNO_3$) ranks 14th among all the chemicals produced by the chemical industry, with nearly 8.3 billion kilograms (9.1 million tons) being manufactured in 1997. Its primary uses are in the production of ammonium nitrate-based fertilizers and explosives. It is used to a lesser extent in photoengraving and metals processing. Laboratory work is done with a 15.9 M solution of this acid, although more concentrated solutions can be prepared by dehydration with sulfuric acid.

The manufacture of nitric acid is a multistep method, called the **Ostwald Process**, in which ammonia, $NH_3$, is first converted to nitrogen dioxide, $NO_2$, which is then reacted with water to form nitric acid. These steps are shown below.

$$4NH_3(g) + 5O_2(g) \rightarrow 4NO(g) + 6H_2O(g) \tag{13.1}$$

$$2NO(g) + O_2(g) \rightarrow 2NO_2(g) \tag{13.2}$$

$$3NO_2(g) + H_2O(\ell) \rightarrow 2HNO_3(aq) + NO(g) \tag{13.3}$$

$$\text{(overall)} \quad NH_3(g) + 2O_2(g) \rightarrow HNO_3(aq) + H_2O(\ell) \tag{13.4}$$

# WHY STUDY THIS TOPIC?

Simply stated, every process that occurs in the universe is accompanied by energy changes. This was true at the time of the "Big Bang," after which energy changes led to the formation of the atoms that are the elements. This is also true now, as our bodies undergo the reactions that allow us to live. We require food that provides the energy for us to continue to live. Plants require energy from the sun to produce the fruits, vegetables, and grain that is our food. We are at a stage now in our social and scientific development in which we can use our understanding of chemistry and energy exchange in the industrial production of goods. We can also use chemicals to generate energy, such as in an automobile, train, or plane. Understanding the nature of energy exchanges, therefore, gives us insight into how and why chemical processes occur. We will use this knowledge, along with our study of kinetics in the next chapter, to learn how we can affect the success of these chemical processes.

*For Homework: Use the Internet search engine "metacrawler" to find the Web addresses of several major chemical process industries, such as Eastman Chemical Company and Procter & Gamble. Find out about some of the products they manufacture. How is energy exchange important in the manufacture of each one?*

Note that Eqs. (13.1)–(13.3) do *not* add up to the overall reaction described in Eq. (13.4)! This is because there are a number of other side reactions that occur that form nitrogen gas and several nitrogen oxides.

We open this chapter discussing the manufacture of nitric acid for three reasons. First, it is a vital industrial chemical. Second, its synthesis can be made chemically and economically viable *only* if there are fairly strict temperature controls in several parts of the process. Finally, a substance called a ***catalyst*** (see Chapter 11), is added to speed up the rate of the process. The catalyst in the industrial manufacture of nitric acid is made up of platinum and rhodium.

The impact of temperature and the catalyst on chemical reactions is part of the broader study of thermodynamics and kinetics. *Thermodynamics* is concerned with the energy exchanges that take place in chemical processes. *Kinetics* is the study of how fast, and by what mechanism, these changes occur. We will revisit the Ostwald Process on occasion in our study of thermodynamics (this chapter) and kinetics (Chapter 14).

## 13.2   What is Energy?

We have already noted that temperature control is vital to the success of the Ostwald Process. This is because so much heat energy is exchanged in the chemical reactions. But we are getting ahead of ourselves. Before we can talk about energy exchange, we need to define "energy." We discussed this previously in Section 3.3.1, but will now take a slightly different approach in its application.

When you pick up a very heavy book, do you use energy? When you run fast, are you using energy? When you pull two strong magnets apart, do you require energy? What about when you jump in the air? Is energy required to melt a cake of ice and is more required to boil the water that results? Your experience tells you that each of these processes requires energy. And in each case, we are using energy to oppose a force that comes from natural attraction. When we pick up the book, run fast, or jump, we are opposing the force of gravity. When we pull two magnets apart, we are opposing the natural force of opposite magnetic poles attracting. Melting and boiling water requires the breaking of hydrogen bonds that result from the attraction of oppositely charged dipoles among water molecules.

We can therefore define energy as "that which is used to oppose natural attractions." This often requires us to move an object (a magnet, ourselves, or a molecule) a certain distance. We call that movement **work (w)**, which is defined as **force (F) $\times$ distance (D)**.

$$w = F \times D \tag{13.5}$$

So **energy** is often defined as *the capacity to do work*.

**Force** is the mass (m) of an object times the acceleration (a) of the object.

$$F = m \times a \tag{13.6}$$

This means that if you have a mass of 87 kg and are being pulled toward the Earth with a gravitational attraction of 9.8 meters per second (9.8 m s$^{-2}$), the force is as follows.

$$F = 87 \text{ kg } (9.8 \text{ m s}^{-2}) = 8.5 \times 10^2 \text{ kg m s}^{-2} = 8.5 \times 10^2 \text{ Newtons(N)}$$

We would call a downward force of 850 N the **weight** of an object that is equivalent to about 190 pounds. If someone pushes you and your 850 N force a distance of 0.50 meters, the work done (neglecting air resistance, friction of your shoes, and such) would be as shown below (to two significant figures).

$$w = F \times D = 850 \text{ kg m s}^{-2}(0.50\text{m}) \approx 430 \text{ kg m}^2\text{s}^{-2} \approx 430 \text{ joules}$$

The **joule** is the SI unit of energy. This shows that work is expressed in units of energy.

## Example 13.1

In Chapter 9, one unit of pressure that was mentioned was psi, or pounds per square inch, lb/in$^2$. Thus, pressure can be thought of as the **force per unit area** and can also be expressed in metric units.

$$P = (F/A) = (\text{kg m s}^{-2}/\text{m}^2) = \text{kg m}^{-1}\text{s}^{-2}$$

Volume can be expressed in meters$^3$ (m$^3$). Using dimensional analysis, prove that pressure $\times$ volume can be expressed in joules, and is therefore a unit of energy.

## Solution 13.1

We can multiply P $\times$ V and look at the units.

$$\text{kg m}^{-1}\text{s}^{-2} \times \text{m}^3 = \text{kg m}^2\text{s}^{-2} = \text{joules}$$

Therefore, P $\times$ V is also a unit of energy. We will make use of this a bit later to understand the changes in energy that occur in gases.

## 13.2.1 Kinetic and Potential Energy

All of the different forms in which energy exists can be classified into two groups—kinetic energy and potential energy. **Potential energy** is stored energy. It has the potential to be released and used. For example,

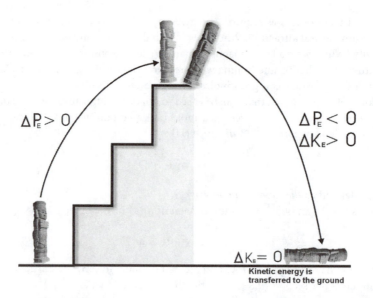

**FIGURE 13.1** A statue being raised and dropped. This is an example of the conversion from kinetic energy to potential energy and back again to kinetic energy.

as shown in Fig. 13.1, if you have a stone statue that is ready to be dropped from a mountaintop, it contains potential energy because when the statue is dropped, it converts that energy to kinetic energy, which it *transfers* to the ground when the statue hits it. **Kinetic energy** is the energy of motion. The statue, which has mass, moves with a certain velocity. Ultimately, energy is being released because instead of *opposing* a natural attraction (gravity), which requires energy, the statue is moving *with* the natural attraction.

*Example 13.2*

Recalling that kinetic energy (K.E.) = $\frac{1}{2} mv^2$ and that "v" = velocity in m/s, show that kinetic energy has units of energy.

*Solution 13.2*

We know that energy (joules) has units of kg m$^2$ s$^{-2}$. Our goal is to prove that the units that we get from kinetic energy are, in fact, in these units. While this seems to be the same type of exercise as 13.1, we think it is quite important to be able to manipulate units to derive these sorts of relationships. Kinetic energy = $\frac{1}{2} mv^2$ = kg m$^2$ s$^{-2}$ = joules. This is why we can talk about heat as a form of kinetic energy.

Using the same concept, chemical bonds store potential energy. When the sugar glucose, $C_6H_{12}O_6$, is burned to form carbon dioxide and water,

$$C_6H_{12}O_6(s) + 6O_2(g) \rightarrow 6CO_2(g) + 6H_2O(g) \tag{13.7}$$

some of the potential energy is converted to kinetic energy and is released to the surroundings, as shown in Fig. 13.2. If the glucose is polymerized as cellulose in wood, then the chemical reaction in which the potential energy is converted to kinetic energy (wood burning) results in a fire. Using our industrial example, some of the potential energy stored in the chemical bonds of ammonia and oxygen is released as kinetic energy when nitric acid is formed in the Ostwald Process,

$$NH_3(g) + 2O_2(g) \rightarrow HNO_3(aq) + H_2O(l) \tag{13.8}$$

---

*Chemistry Professionals at Work*      *CPW Box 13.2*

# THERMAL ANALYSIS

The identity and purity of drugs can be indicated by precisely controlling thermodynamic processes, such as melting or boiling. Thus the determination of melting points and boiling points can be involved in the standard methods for analyzing for drugs. Broadly speaking, thermal analysis is the observation of properties while the temperature is changed. Techniques include ***differential scanning calorimetry*** (DSC), ***differential thermal analysis*** (DTA), and ***thermogravimetric analysis*** (TGA).

*For Homework: Research the three thermal methods of analysis mentioned above. Write a report giving brief but thorough descriptions of each, detailing exactly what properties are measured and what observations are made by the chemistry professional.*

---

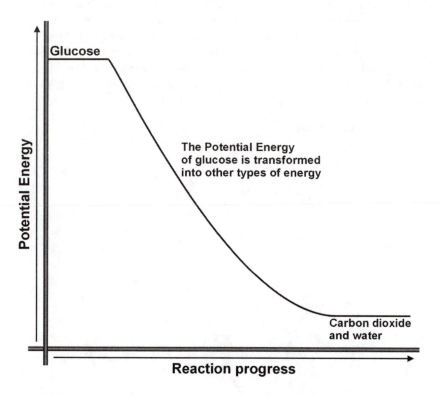

**FIGURE 13.2** When glucose is burned, some of the potential energy stored in the bonds is converted to kinetic energy and released as kinetic energy.

## 13.3   Types of Energy and Energy Exchange

### 13.3.1   Heat and Its Relationship to Temperature

Kinetic energy can be manifested in a variety of ways. When particles move, they possess **heat energy.** A measure of the average kinetic energy of the particles of a substance is called *temperature.* Temperature is *not* a measure of the total amount of heat in a system. Rather, it is a measure of the *average* kinetic energy of the system. Thus, a hot penny at a temperature of 1000°C has less heat energy than a lake at 20°C, because the penny is very small compared to the lake, even though it has a higher temperature. Please review the temperature scale conversions in Section 7.7—we will use temperature often, but not as a measure of heat content. Instead, we will use it as a measure of *energy exchange* between systems.

### 13.3.2   Light—Frequency, Wavelength, and Energy

Equation (13.7) describes the burning of glucose. We discussed the burning of a polymer of glucose, cellulose, which results not only in heat, but also in light from the fire. Light, as discussed in Chapter 3, is therefore another form of energy. Let us review the basic concepts of light from Chapter 3.

Light is **electromagnetic radiation** because electric and magnetic fields exist as part of the radiation. Energy can be absorbed by chemicals. When this happens, electrons can move from their ground states to higher, excited states. One of the ways that chemicals can get rid of this excess energy is by emitting electromagnetic radiation in the form of light.

Electromagnetic radiation travels in waves that have a particular **wavelength,** as shown in Fig. 13.3. Wavelength is expressed in meters or related units such as nanometers or cm. The waves pass by a point at a certain rate which we call the **frequency.** Frequency can be expressed in waves/second, or often simply 1/sec = $s^{-1}$. Visible light is but a small portion of the entire continuous set of electromagnetic radiation, and just as light can have all kinds of different colors, there also exists a huge range of wavelengths (and, therefore frequencies) of electromagnetic radiation. This range, known as the **electromagnetic spectrum**, is given in Fig. 13.4. Electromagnetic radiation has many practical uses. X-rays are used in medical diagnosis. Microwaves are used in cooking. Radio waves are used to transmit music to your stereo.

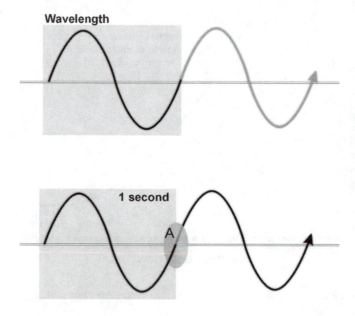

**FIGURE 13.3**   Electromagnetic radiation travels in waves that have the property of wavelength and frequency. In this illustration, one wave passes a point in one second.

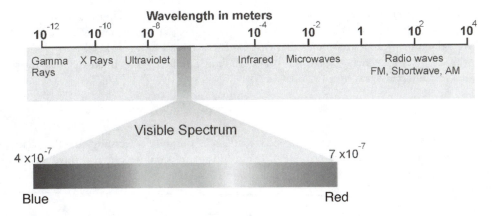

**Figure 13.4** The wavelength range and regions that comprise the electromagnetic spectrum. The visible wavelengths are only a small part of the entire spectrum.

We speak of electromagnetic radiation in terms of wavelength, frequency, or energy and these three are interrelated. Chemists will speak of frequency, or wavelength, depending on the instruments used, so it is important to be conversant in both units.

*Frequency ($\nu$) and wavelength ($\lambda$)*

$$\text{Frequency} = \frac{\text{speed of light (c)}}{\text{wavelength}} \tag{13.9}$$

$$\nu = \frac{c}{\lambda} \tag{13.10}$$

The speed of light is $3.00 \times 10^8$ m/s. This means that red light with a wavelength of 620 nanometers (nano $= 10^{-9}$) will have a frequency of $4.84 \times 10^{14}$ s$^{-1}$, as follows.

$$\nu = \frac{3.00 \times 10^8 \text{m s}^{-1}}{6.20 \times 10^{-7} \text{m}} = 4.84 \times 10^{14} \text{s}^{-1}$$

## Example 13.3

Determine the wavelength of radiation in meters and nanometers, that has a frequency of $2.63 \times 10^{15}$ s$^{-1}$. What region of the electromagnetic spectrum is represented by radiation of this wavelength (see Fig. 13.4).

## Solution 13.3

We can modify Eq. (13.10) to solve for the wavelength.

$$\lambda = \frac{c}{\nu} \tag{13.11}$$

Substituting our values for the frequency and the speed of light, we obtain the following result.

$$\lambda = \frac{3.00 \times 10^8 \text{ms}^{-1}}{2.63 \times 10^{15} \text{s}^{-1}} = 1.10 \times 10^{-7} \text{m}$$

This wavelength is in the *ultraviolet* (UV) region.

*The relationship of wavelength and frequency to energy (E)* We can appreciate the energy that electromagnetic radiation has when we think of its effects. We get sunburned by UV radiation. Our rooms are lighted by incandescent bulbs that emit both visible and infrared (heat) energy. Our food is heated with relatively low-energy microwaves. The relationship of energy to frequency was described in Section 3.3 and is stated again here in Eq. (13.12).

$$E = h\nu \tag{13.12}$$

in which "h" = $6.63 \times 10^{-34}$ J s, a value known as *Planck's constant*. If we substitute the relationship between wavelength and frequency [Eq. (13.10)] into this expression, we can also solve for energy in terms of wavelength.

$$E = \frac{hc}{\lambda} \tag{13.13}$$

When we solve for energy, notice that the units cancel to give joules.

## Example 13.4

An atom emits UV radiation with a wavelength of 110 nm, as described in Example 13.3. What is the energy of this radiation?

## Solution 13.4

We can solve the problem using the wavelength [Eq. (13.13)] or the frequency of the UV radiation, [Eq. (13.12)]. We'll use the frequency here, as determined in Example 13.3.

$$E = h\nu$$

$$E = 6.63 \times 10^{-34} \text{ J s} \times 2.63 \times 10^{15} \text{ s}^{-1} = 1.74 \times 10^{-18} \text{ J}$$

This is a very small amount of energy when compared to the roughly 400 kJ we could get from, for example, a bowl of cereal. However, a bowl of cereal contains much more matter than a single atom. Keep the relative amounts of energy in mind as we examine chemical reactions in the rest of the chapter.

## Example 13.5

Calculate the amount of energy released when *one mole* of photons ($6.022 \times 10^{23}$) is emitted as UV radiation with a wavelength of 110 nm.

## Solution 13.5

We calculated the energy released as one photon in Example 13.4. To determine the energy released per mole of photons, we note that there are Avogadro's number ($6.02 \times 10^{23}$) of photons in a mole.

$$\text{Energy released per mole} = \frac{1.74 \times 10^{-18} \text{ J}}{\text{photon}} \times \frac{6.02 \times 10^{23} \text{ photons}}{\text{mole}}$$

$$= \frac{1.05 \times 10^{6} \text{ J}}{\text{mole}} = \frac{1.05 \times 10^{3} \text{ kJ}}{\text{mole}}$$

This is a substantial amount of energy. When solving problems involving energy conversions, it is important to keep in mind how much starting material we have. This will help us to determine if our answer makes sense.

The energy in Example 13.4 was given off when an electron moved closer to the nucleus and emitted a bundle of energy called a *photon.* Example 13.5 compares the energy released from a single photon with that emitted from a mole of photons.

### 13.3.3 Nuclear Processes Give Off Energy

Many atomic nuclei are energetically unstable. They will give off energy in the form of particles as they work their way toward stability. For example, plutonium-239 loses a particle that is equal in mass to a helium nucleus (more commonly known as an "alpha" particle) to form uranium-235. The equation for this "nuclear decay" is shown below.

$$^{239}\text{Pu} \quad \rightarrow \quad ^{235}\text{U} \quad + \quad ^{4}\text{He} \tag{13.14}$$
$$\text{mass (g/mol)} \, 239.0522 \quad 235.0439 \quad 4.00260$$

The sum of the product masses is 239.0465 g. This is 0.0057 g *less* than the mass of the starting material. What happened to this mass? It was released *as energy.* This notion that energy and mass are interconvertible is what is meant by Einstein's famous equation,

$$E = mc^2 \tag{13.15}$$

and, equivalently,

$$\Delta E = \Delta mc^2 \tag{13.16}$$

in which $\Delta E$ = the change in energy in joules, $\Delta m$ = the change in mass in kg and c = the speed of light in (meters per second)$^2$. We spent considerable time at the beginning of the chapter working with the units of energy. Let's visit units one more time,

$$\Delta mc^2 = \text{kg} \times \text{m}^2\text{s}^{-2} = \text{joules} \tag{13.17}$$

This presents a consistent picture. The formulas that we have introduced each deal with different quantities, but they all have units that work out to give energy in joules.

### Example 13.6
What is the energy change for the decay of the plutonium-239 nucleus?

### Solution 13.6

$$\Delta E = \Delta mc^2$$

The change in mass, $\Delta m$, is 0.0057 grams per mole of plutonium. In order to have our answer in joules, we must change the mass to kilograms, so $\Delta m = 5.7 \times 10^{-6}$ kg. The energy change associated with this decay is given below.

$$\Delta E = \Delta mc^2 = 5.7 \times 10^{-6} \text{ kg} \, (3.00 \times 10^8 \text{ms}^{-1})^2$$
$$= 5.1 \times 10^{11} \text{ kg m}^2\text{s}^{-2} = 5.1 \times 10^{11} \text{ J} = 5.1 \times 10^8 \text{ kJ (per mole)}$$

This is *considerably* larger than the energy given off by a mole of photons. As we shall discuss in the next section, it is also many orders of magnitude more than the energy per mole given off by the Ostwald Process or any other chemical reaction. Nuclear processes involve exceptionally large amounts of energy exchange. This is one reason why they are used as a municipal energy source.

## 13.3.4   Chemical Reactions Involve Energy Exchange

A chemical reaction occurs when bonds are broken and different bonds are formed. When this occurs, more energetically stable systems are often formed. One reaction of tremendous industrial importance is that of hydrogen with oxygen to form water.

$$2H_2(g) + O_2(g) \rightarrow 2H_2O(g) \tag{13.18}$$

This reaction is used on the space shuttle as an energy source that supplies useable water. It is also used in prototype automobiles as a "clean fuel." Recent work indicates that the reaction might be useful on a large scale to supply power for cities. Why does this reaction supply energy? We can look at the reaction as being composed of two parts—**bond breaking** and **bond forming**.

*Bond breaking:* The reactants are composed of hydrogen and oxygen molecules that have the following structures

$$H–H \qquad\qquad H–H \qquad\qquad O=O$$

Each of these bonds consists of shared electrons. An input of energy is required to break these bonds. The amount of energy (in kJ/mol) that is required to break the bond(s) between atoms is called the **bond energy.** Table 13.1 lists bond energies for several bonds. Note that for two of the same kind of atoms (carbon, for example), the bond energy increases from single to double to triple bonds (347, 614, and 839 kJ/mole, respectively).

The total amount of energy needed to break these bonds is the sum of the bond energies.

$$
\begin{array}{ll}
H–H & 432 \text{ kJ} \\
H–H & 432 \\
O=O & 495 \\
\hline
& 1359 \text{ kJ}
\end{array}
$$

Because energy has to be *put into the system* to break the bonds, we call bond breaking an **endothermic** process. We give the energy needed for this process a "+" sign, indicating that energy is *added* to the system (see Section 11.8 for a review of these terms). *Bond breakage* is an **endothermic process.** In this case, the change in energy, $\Delta E$, is $+1359$ kJ. Measuring the *total* energy actually present in a molecule is not practical. There are too many variables and interactions to account for! So we measure *changes* in energy (including heat energy, as we discuss below).

*Bond forming:* Bond formation brings energetic stability to a system. Therefore, *bond formation releases energy. It is an* **exothermic process.** We give its value a "−" sign, indicating that energy is *released* or *subtracted* from the system. We can find out how much energy is released by determining the bond energies of the products. Since the bond energy represents the amount of energy required to break bonds, this also represents the energy released when the bonds themselves are formed.

**TABLE 13.1** Average Bond Energies (kJ/mol) for Several Bonds

| Bond | Energy | Bond | Energy |
|------|--------|------|--------|
| H–H | 432 | N–H | 391 |
| C–H | 413 | N–N | 160 |
| C–C | 347 | N–O | 201 |
| C=C | 614 | N=O | 607 |
| C≡C | 839 | N=N | 418 |
| C–O | 358 | N≡N | 941 |
| C=O | 745 | O–H | 467 |
| C=O (CO$_2$) | 799 | O–O | 146 |
| C–Cl | 339 | O=O | 495 |

*Chemistry Professionals at Work*       *CPW Box 13.3*

# ENERGY MANAGEMENT IN THE CHEMICAL INDUSTRY

T he manufacture of chemicals through industrial chemical processes involves the management of energy going *into* the process as well as the management of energy coming *from* the process. Industrial companies consider energy an essential element of their economics—as much as, or more than other obvious factors such as land, labor, capital, and raw materials. Indeed, the proper management of energy dictates how the other factors are handled.

The goal of the chemical process industry, like any other enterprise in the private sector, is to make money. Thus the cost of energy and quantity of energy required enter into the equation of a superior operation. Energy is often recycled, meaning that excess energy created in one part of the operation is often used in another part. For example, heat exchangers are often used to transfer the heat from flue gases to preheat the water going into a boiler. Also, the heat exchangers used to cool a reactor can be used to preheat the raw material entering the reactor.

*For Homework: Look up "Heat-Exchange Technology" in the* Concise Encyclopedia of Chemical Technology *and other sources and report on the specific methodologies used for the recycling of energy in the chemical industry.*

---

The total amount of energy released from the formation of the covalent bonds of two moles of water (noting that each water contains two O–H bonds) is shown below.

$$4 \text{ moles of O–H bonds} \times 467 \text{ kJ/mole of O–H bonds} = 1868 \text{ kJ} \qquad (13.19)$$

Bond formation is an exothermic process, so we can say that the energy change, $\Delta E$, is $-1868$ kJ.

The energy change for the entire reaction then, is the sum of the energy changes for breaking reactant bonds and forming product bonds. Fig. 13.5 illustrates this.

This means that the formation of two moles of water from the reaction of two moles of hydrogen and one mole of oxygen will result in the release of 500 kJ of energy. Just about all of the energy will be released as heat, though a little bit will also be done as work (see Section 13.2 for a review of "work"). For the remainder of this discussion, we will assume that our energy is heat energy.

The term for heat energy at constant pressure (such as atmospheric pressure) is called **enthalpy, H**. We symbolize **changes in enthalpy**, which can be measured, by $\Delta H$. We will therefore say that the enthalpy change, or heat exchange, $\Delta H$, is $-500$ kJ for our reaction. We now see from the standpoint of energy exchange why this process can be used as an energy source. If we compare the energy exchange in the nuclear process to that in the chemical process, we see that changes in the nucleus provide vastly more energy per mole than chemical processes. This is why nuclear processes are so valued as power sources and in warfare, in spite of the attendant risks.

Two final notes on determining energy exchange using bond energy values:

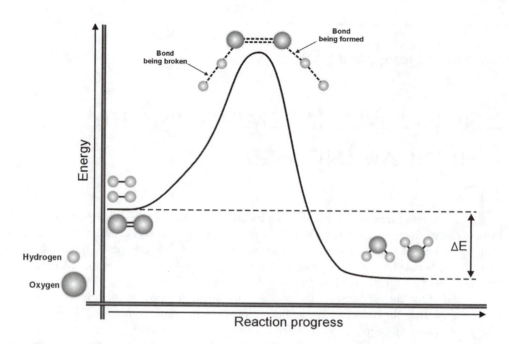

**FIGURE 13.5** Breaking bonds requires energy. Forming bonds releases energy. In this figure, the overall process is exothermic, meaning that the heat energy (roughly equal to the total energy, $\Delta E$) is released. $\Delta E = +1359$ kJ + $-1868$ kJ $= -509$ kJ $\approx -500$ kJ.

1. The values are *averages* of the energies of the bonds in many compounds. The energy for a particular bond will depend on the compound in which it is a part. So the answers we get will be reasonable but not especially accurate. That is why we rounded the value for the formation of water to $-500$ kJ for 2 moles, or $-250$ kJ per mole.

2. Bond energy calculations are best done for gases because they do not take into account the energy required to break intermolecular bonds, such as hydrogen bonds in liquid water. While these bonds are much weaker than the *intra*molecular bonds used in the calculations, they are, nonetheless, contributions that we are neglecting.

## Example 13.7

Methane, $CH_4$, is the gas used to heat many U.S. homes in winter. It is also used as the fuel in gas stoves. Many people in rural areas use propane, $C_3H_8$, for their cooking needs. Calculate the energy exchange for the combustion of one mole of methane with oxygen and one mole of propane with oxygen.

$$CH_4(g) + 2O_2(g) \rightarrow CO_2(g) + 2H_2O(g)$$
$$C_3H_8(g) + 5O_2(g) \rightarrow 3CO_2(g) + 4H_2O(g)$$

## Solution 13.7

Our first goal is to determine the bonds that are being broken and formed. We then look up the bond energy values and determine the energy required to break bonds and the energy released in bond formation. The sum of the two values gives us the net energy exchange for the reaction.

a. **Methane combustion**

| Bonds Broken (moles) | $\Delta E$ (kJ) | Bonds Formed (moles) | $\Delta E$ (kJ) |
|---|---|---|---|
| 4 C–H | $4 \times 413 = 1652$ | 2C=O | $2 \times 805$ |
| 2 O=O | $2 \times 495 = \underline{990}$ | 4O–H | $4 \times \underline{467}$ |
| TOTAL | $+ 2642$ kJ | | $-3478$ |

---

*Chemistry Professionals at Work*

*CPW Box 13.4*

# Making Predictions Using Enthalpy

Wen fuels are selected to power vehicles and other machines, it is important to be able to predict their individual energy outputs. This knowledge will help predict how far a car will drive, how large the fuel tank must be to obtain a particular amount of energy, and how the cost and power of different fuels compare. These predictions can be made using enthalpy. Since the energy from a fuel is obtained when it is burned (reacted with oxygen), standard enthalpies of formation can be used to determine the number of kilojoules emitted from each mole of fuel. In some cases, it is useful to convert the enthalpy value to kJ/g, particularly when making standard comparisons between fuels that concern mass or volume. The quantity is known as the fuel value, and requires only a simple conversion to obtain. Of course, other factors must be considered when designing a fuel system, but energy output is a key characteristic that may determine which fuels could be a viable energy source for the future.

*For Homework: Use bond energy data to compare the energy per mole and per gram released by octane, $C_8H_{18}$, and methane, $CH_4$, when each is reacted with oxygen in the air. Based on your answers, which is the better fuel for use in cars and homes? What other considerations besides energy exchange enter into this decision?*

---

$$\Delta E \approx \Delta H \approx +2642 \text{ kJ} + -3478 \text{ kJ} \approx -836 \text{ kJ} \approx -840 \text{ kJ}$$

Thus, the combustion of methane is an exothermic process, which agrees with our experience.

### b. *Propane combustion*

| Bonds Broken (moles) | | | $\Delta E$ (kJ) | Bonds Formed (moles) | | $\Delta E$ (kJ) |
|---|---|---|---|---|---|---|
| 2 | C–C | $2 \times 347 =$ | 694 | 6C=O | $6 \times 805 =$ | 4830 |
| 8 | C–H | $8 \times 413 =$ | 3304 | 8O–H | $8 \times 467 =$ | 3736 |
| 5 | O=O | $5 \times 495 =$ | 2475 | | | |
| TOTAL | | | + 6473 kJ | | | −8566 kJ |

$$\Delta E \approx \Delta H \approx +6473 \text{ kJ} - 8566 \text{ kJ} \approx -2100 \text{ kJ}$$

The combustion of propane gives off considerably more energy per mole than that of methane. Why, then, is methane used as the fuel of choice in so many cities and towns? Part of the reason can be found by determining the energy released on a *per gram* basis. Methane combustion releases 840 kJ/16 grams per mole of methane, or 52 kJ/g $CH_4$. Propane releases 2093 kJ/44 grams per mole of propane = 48 kJ/g $C_3H_8$. Not only is methane quite readily available from natural gas deposits in the Earth, but it also *releases more energy per gram* than propane.

# THE EVER-THINNING ALUMINUM CAN

When we drink soda from a can, we don't often think of the can as being made from a chemical process that has been around for over 100 years, and that is being updated all the time. The Hall-Heroult process for manufacturing aluminum involves the reaction of the mineral *bauxite*, which is largely $Al_2O_3$, with large carbon rods.

$$2Al_2O_3 + 3C \rightarrow 4Al + 3CO_2$$

This process requires substantial amounts of energy to generate the millions of tons of aluminum needed in aluminum cans, cookware, and other industrial applications. One goal of chemists is to design an aluminum can that weighs as little as possible so less aluminum will be needed, saving lots of energy. According to the American Can Manufacturers (http://www.cancentral.com [as of 6/4/00]), the first beverage can, made in 1935, weighed three ounces, or about 90 grams. Today's aluminum cans weigh about 16 grams and contain a variety of lightweight metals including small amounts of magnesium, manganese, iron, and copper. Even more energy savings can be achieved by recycling, which uses less than 10% of the energy required to win aluminum from bauxite. In 1996, 62.8 billion aluminum cans were recycled. So the combination of new, lightweight alloys for the can and recycling by the public is saving energy—the equivalent of 20.6 million barrels of oil in 1996 alone!

*For Homework: Go to the American Can Manufacturers site on the Internet. What other information regarding the relationship of energy to can manufacturing can you find? What decisions have manufacturers made based on energy that have affected our eating or drinking habits?*

## 13.3.5   Where Does the Energy Go?

What happens to the energy that is released in the combustion of hydrocarbons or any other exothermic reaction? It goes from the reaction itself to whatever is surrounding that reaction. It could be a large reaction vessel or perhaps the air. If you touch the reaction vessel, then the energy (in the form of heat) has gone to your hand. We call the reaction itself the **system**. We call everything that is surrounding the reaction the **surroundings**. The system and the surroundings comprise the entire **universe**. In an exothermic chemical reaction, energy is transferred from the system to the surroundings. In an endothermic chemical reaction, energy is transferred from the surroundings to the system. In both cases, energy is being transferred, not created nor destroyed. We can state this more formally as **The First Law of Thermodynamics:** The energy of the universe is constant. This is also known as **The Law of Conservation of Energy.** Energy is neither created nor destroyed, although it can change form (heat into light, for example).

In this section, we have been introduced to several types of energy and we determined the amount of energy released or absorbed in some important processes. In the next section, we will discuss a useful method of determining the heats of reaction, or enthalpies of formation.

# 13.4 Enthalpies of Formation

As a student, you can study the energy changes in many reactions by using the *standard enthalpy of formation* of each of the compounds in the reaction. This value is defined as the change in enthalpy when forming one mole of a compound from its elements with all of the substances in their standard states. The *standard state* is defined:

- For a gas: a pressure of 1 bar, or 100,000 Pascals (some scientists still use the old measure of 1 atmosphere, which equals 101,325 Pascals).
- For a solute, it is an activity (which can be close to a concentration) of 1 M.
- For an element, it is the form in which it exists at 25°C and 100,000 Pascals.

The standard state of oxygen is the form in which it exists at standard conditions, 25°C and one atmosphere, which is $O_2$ in the gaseous state. The standard state of copper is $Cu(s)$. Since the definition of the enthalpy of formation relates to the energy *change* when a substance is formed from its elements in the standard states, the enthalpy of formation of an element in its standard state is zero, because it already exists in that state–there is no change needed. Thus, for $O_2(g)$ and $H_2(g)$, we have the following.

$$\Delta H_f^\circ(O_2(g)) = 0 \text{ kJ/mol} \tag{13.20}$$

$$\Delta H_f^\circ(H_2(g)) = 0 \text{ kJ/mol} \tag{13.21}$$

$$\Delta H_f^\circ(Cu(s)) = 0 \text{ kJ/mol} \tag{13.22}$$

The symbol "$\Delta$" means "change in," the "$H^\circ$" means "enthalpy in the standard state," the "f" means "formation"; therefore, the entire symbol is $\Delta H_f^\circ$. Table 13.2 lists the heats of formation for several substances. Compare the heats of formation of the elements in their standard states with other values. Many compounds form because they are more energetically stable than the individual elements that make them up. And an exothermic reaction is one of two important factors that favor that stability. For example, the production of water from $H_2(g)$ and $O_2(g)$ is an exothermic reaction.

$$\Delta H_f^\circ(H_2O(g)) = -242 \text{ kJ/mol} \tag{13.23}$$

Since water is formed from the reaction of hydrogen and oxygen,

$$2H_2(g) + O_2(g) \rightarrow 2H_2O(g) \tag{13.24}$$

**TABLE 13.2**  Heats of Formation for Several Substances

| Substance | $\Delta H_f^\circ$ (kJ/mol) |
|---|---|
| $C_2H_2(g)$ | 227 |
| $C_2H_4(g)$ | 52 |
| $C_2H_6(g)$ | −84.7 |
| $CO_2(g)$ | −393.5 |
| $HCl(g)$ | −92 |
| $H_2O(g)$ | −242 |
| $H_2O(l)$ | −286 |
| $H_2O_2(l)$ | −187.8 |
| $NO(g)$ | 90 |
| $NO_2(g)$ | 34 |
| $O_3(g)$ | 143 |
| $SO_2(g)$ | −297 |
| $SO_3(g)$ | −396 |

the change in energy for the reaction is equal to the heat of the formation of the products minus the reactants. In general,

$$\Delta H^\circ_{reaction} = \Sigma\Delta H^\circ_{f(products)} - \Sigma\Delta H^\circ_{f(reactants)} \tag{13.25}$$

where the symbol "$\Sigma$" means "sum of" or "total of."

The sum of the heat of formation of the products is just that of the water, $-286$ kJ/mole $\times$ 2 moles $=$ $-572$ kJ. The sum of the heat of formation of the reactants $= 0$ kJ, because both reactants are in their standard states, in which $\Delta H^\circ_f = 0$. So the heat of reaction (that is, the heat energy change upon the formation of water from its elements) is $-572$ kJ for 2 moles of water, or $-286$ kJ per mole of water. Heat of formation data are measured accurately for each compound, so when they are available, they are better to use than bond energy data. Thus, the value for the heat of reaction for the formation of water calculated this way, $-286$ kJ, is more appropriate to use than the value for the formation from the elements calculated via bond energies, which is about $-250$ kJ. Also, bond energy calculations do not account for the formation of liquid water, where extra energy is required to break hydrogen bonds between water molecules.

## Example 13.8

Use enthalpy of formation data to determine the energy change occurring during the Ostwald Process.

## Solution 13.8

$$NH_3(g) + 2O_2(g) \rightarrow HNO_3(aq) + H_2O(l)$$

The heat of reaction can be found from the heats (enthalpies) of formation, given in Table 13.2.

$$\Delta H^\circ_{reaction} = \Sigma\Delta H^\circ_{f(products)} - \Sigma\Delta H^\circ_{f(reactants)}$$

$$\Sigma\Delta H^\circ_{f(products)} = \Delta H^\circ_{f(HNO_3(aq))} + \Delta H^\circ_{f(H_2O(l))} = -207 \text{ kJ} + -286 \text{ kJ} = -493 \text{ kJ}$$

$$\Sigma\Delta H^\circ_{f(reactants)} = \Delta H^\circ_{f(NH_3(g))} + \Delta H^\circ_{f(O_2(g))} = -46 \text{ kJ} + 0 \text{ kJ} = -46 \text{ kJ}$$

$$\Delta H^\circ_{reaction} = -493 \text{ kJ} - (-46 \text{ kJ}) = (-447 \text{ kJ})$$

## Example 13.9

Use enthalpy of formation data to determine the heat of reaction for the combustion of propane, $C_3H_8$. Compare this with the bond energy calculation for propane in Example 13.7.

## Solution 13.9

The reaction for the combustion of propane was given in Example 13.7.

$$C_3H_8(g) + 5O_2(g) \rightarrow 3CO_2(g) + 4H_2O(g)$$

Using heats of formation, we obtain the following.

$$\Sigma\Delta H^\circ_{f(products)} = 3\Delta H^\circ_{f(CO_2(g))} + 4\Delta H^\circ_{f(H_2O(g))} = 3 \times -394 + 4 \times -242 = -2150 \text{ kJ}$$

$$\Sigma \Delta H^{\circ}_{f(reactants)} = \Delta H^{\circ}_{f(C_3H_8(g))} + 5\Delta H^{\circ}_{f(O_2(g))} = -108 + 5 \times 0 = -108 \text{ kJ}$$

$$\Delta H^{\circ}_{reaction} = -2150 \text{ kJ} - (-108 \text{ kJ}) = -2042 \text{ kJ}$$

This is quite close to the value of $-2100$ kJ we determined from bond energy calculations.

## 13.5  Why Do Chemical Reactions Happen?

We now have a sense of what energy is and we have compared several types of energy. We also have some tools that we can use to calculate the amount of energy that is given off or absorbed. From the standpoint of chemical reactions, why do some reactions occur while others do not? We already know that the potential to release energy is one thing that favors a spontaneous reaction. But it is not the only thing. For example, the chemical cold packs that sports enthusiasts use on injuries work by combining an ammonium salt such as ammonium nitrate, $NH_4NO_3$, with water. The dissolution occurs readily, with a **heat of solution** (that is the heat exchange when the salt dissolves) of about $+26$ kJ per mole of ammonium nitrate.

$$NH_4NO_3(s) \xrightarrow{\text{H}_2\text{O}} NH_4^+ (aq) + NO_3^-(aq) \qquad \Delta H^{\circ} = +26 \text{ kJ} \qquad (13.26)$$

This dissolution is endothermic, yet it still occurs on its own once the ammonium nitrate and water mix. A process such as this that occurs without continuous outside intervention is called a **spontaneous process**. How can an endothermic reaction be spontaneous? There must be something else that is going on that makes it possible for the process to proceed spontaneously in spite of the energy gain.

### 13.5.1  The Importance of Probability

To understand the other factor that favors spontaneous reactions, let's think about blowing the seeds off of a dandelion. How do they distribute themselves? That is, do they all fall in a space the size of a dandelion or, do they spread out? Spreading out is more likely than concentrating together. We can see this when we sneeze or even when we sit in a movie theater, where people who don't know each other sit as far apart as possible.

Why do processes tend to result in spreading apart? Mathematically, there are more ways to be apart than there are to be together. We can look at the movie theater analogy to see this. Let's say that you and a friend are alone in a movie theater with 40 seats. How many ways can you sit together? You can sit in seats #1 and 2, or 2 and 3, or 3 and 4, and so forth, but you are still together. There are 39 different places to sit and still be together. You can switch seats and do the same thing, so there are really 39 × 2, or 78 ways to be together. This is illustrated in Fig. 13.6a.

How many ways can you sit apart? You can sit in seat #1 and your friend can sit in seat #3, or #4, or #5, or #26, or #40. He or she cannot sit in seat #2, because she would be next to you. So when you are seat #1, your friend can pick any one of 38 seats so that you are apart. This is illustrated in Fig. 13.6b. There are only one or two seats that she can pick in which you will be together. The bottom line is that whether considering people in a movie theater, seeds of a dandelion, or molecules in a gas, *the mathematical likelihood favors spreading apart rather than consolidating*. The mathematical measure of disorder is called **entropy, S**.

Increasing the disorder of a system favors a spontaneous reaction. For example, we know from Examples 13.7 and 13.9 that the combustion of propane is a spontaneous process. It is exothermic, releasing 2042 kJ per mole of $C_3H_8$ burned.

$$C_3H_8(g) + 5O_2(g) \rightarrow 3CO_2(g) + 4H_2O(g) \qquad (13.27)$$

We also note that more moles of gas are formed (7) than moles reacted (6). This represents a slight gain in the entropy of the system and favors a reaction which proceeds vigorously.

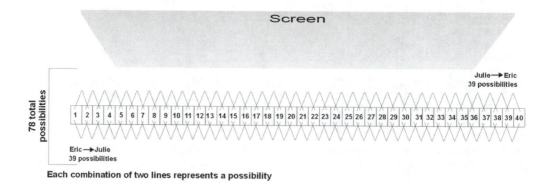

**Each combination of two lines represents a possibility**

**FIGURE 13.6(a)** There are a total of 78 ways that two friends, Eric and Julie, can sit together, side-by-side, in a 40-seat movie theater. If that seems like quite a few, compare that to the ways that they can sit apart!!

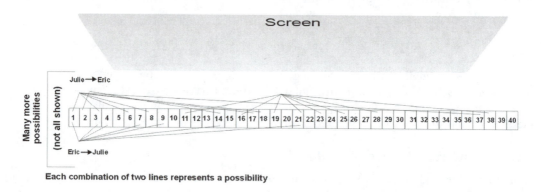

**Each combination of two lines represents a possibility**

**FIGURE 13.6(b)** There are 38 seats for Julie to sit in that are apart from Eric, for any seat in which Julie sits. Since there are 40 seats, that represents 1524 possibilities for separation.

So we have two measures that favor spontaneous reaction—release of energy to the surroundings and an increase in entropy of the system. Not all spontaneous reactions are both exothermic and show an increase in their entropy. For example, the chemical cold pack dissolution we discussed previously does proceed even though the reaction is endothermic. Why is it spontaneous?

$$NH_4NH_3(s) \xrightarrow{H_2O} NH_4^+(aq) + NO_3^-(aq) \qquad \Delta H = +26 \text{ kJ/mol} \qquad (13.28)$$

The solid ammonium nitrate has split apart into ions that are dispersed in the solution. In a descriptive sense, this has increased the disorder of the system. What was once a solid crystal has mixed with water to become dispersed ions. This increase in the entropy of the system has compensated for the endothermic nature of the process, so the dissolution proceeds. Forming ions in solution increases the entropy of a system. Forming gases from the reaction of solids or liquids really increases the entropy of a system.

## 13.5.2   The Entropy of the Universe

Let's look a bit further at why energy release from a system favors its spontaneous reaction. When energy is released in the form of heat, this increases the kinetic energy of the particles in the surroundings that gain the energy. Since kinetic energy $= \frac{1}{2} mv^2$, the velocity of the particles is also increased, resulting in more movement, *leading to higher disorder* of the surroundings.

An exothermic reaction increases the entropy of the surroundings. We also discussed previously that an increase in entropy of the system favors a spontaneous reaction. Since the system plus the surroundings equals

**TABLE 13.3**   Entropy Values for Common Substances

| Substance | $S°$ (J/K mol) |
|---|---|
| $C_2H_2(g)$ | 201 |
| $C_2H_4(g)$ | 219 |
| HBrg) | 199 |
| HCl(g) | 187 |
| $H_2O(g)$ | 189 |
| $H_2O(l)$ | 70 |
| $H_2O_2(l)$ | 109.6 |
| $NH_3(g)$ | 193 |
| $NH_4Cl(s)$ | 96 |
| NO(g) | 211 |
| $NO_2(g)$ | 240 |
| $SO_2(g)$ | 248 |
| $SO_3(g)$ | 257 |

the universe, we can say more generally that a spontaneous process results in an increase in the entropy of the universe. This is known as **The Second Law of Thermodynamics** and this is why reactions happen.

At this point, then, in order to assess whether or not a reaction is spontaneous, we need to measure both its energy exchange and its entropy change. These are the factors that contribute to the change in the entropy of the universe. We already know two ways of measuring the energy exchange in a system, via bond energies and heats of formation, with the heat of formation being the preferred method. How do we measure the entropy change in a system?

### 13.5.3   Measuring Entropy Change of the System in a Reaction

The **Third Law of Thermodynamics** may be stated as follows: the entropy of a perfect crystalline substance at 0K is zero. Using this as a reference point, the entropy of substances can be measured. Since all substances at any temperature above 0 Kelvin are more disordered than a perfect crystal at 0K, their entropy values in J/K mol will be greater than zero. Entropy values for some common substances are given in Table 13.3. Note that these are *not changes* in entropy, but are actual measures of entropy. **We measure S, not $\Delta S$ of substances.** However, when determining what has happened to the system disorder in a reaction, we measure the changes in entropy. As with enthalpy changes, we are looking at products minus reactants.[1]

$$\Delta S^°_{reaction} = \Sigma S^°_{(products)} - \Sigma S^°_{(reactants)} \tag{13.29}$$

This is illustrated in Example 13.10, which involves the determination of entropy changes at standard conditions.

*Example 13.10*

We said above that the dissolution of ammonium nitrate is an endothermic process that is nonetheless spontaneous. It is necessary, then, that the reaction (the system) have an increase in entropy to overcome the decrease in the entropy of the surroundings because the reaction is endothermic. Calculate $\Delta S$ for this dissolution.

$$NH_4NO_3(s) \rightarrow NH_4^+(aq) + NO_3^-(aq) \tag{13.30}$$

---

[1]The disorder of a compound will increase with temperature, so the values of entropy generally change quite a bit with temperature. Enthalpy values also change with temperature. We will work assuming standard temperature conditions, even though corrections are possible for nonstandard conditions.

| Substance | $S°$(J/K mol) |
|---|---|
| $NH_4NO_2$ | 151 |
| $NH_4^+$ | 113 |
| $NO_3^-$ | 147 |

## Solution 13.10

The entropy change for the reaction is the sum of the entropy values of the products minus the sum of the entropy values of the reactants.

$$\Delta S°_{reaction} = \Sigma S°_{(products)} - \Sigma S°_{(reactants)}$$

$$\Sigma S°_{(products)} = S°(NH_4^+) + S°(NO_3^-) = 1 \text{ mole} \times 113 \text{ J/K mol} + 1 \text{ mol} \times 147 \text{ J/K mol}$$
$$= +260 \text{ J/K}$$

$$\Sigma S°_{(reactants)} = S°(NH_4NO_3) = 1 \text{ mol} \times 151 \text{ J/K mol} = +151 \text{ J/K}$$

$$\Delta S°_{reaction} = +260 - (+151) = +109 \text{ J/K}$$

The entropy of the system has increased as a result of this reaction. The reaction does proceed spontaneously, so we know that the increase is large enough to overcome the decrease in the entropy of the surroundings via this endothermic reaction. But the more general question is, "How do we know how much is enough?" More generally, is there a single measure that takes into account the heat exchange of the reaction *and* the entropy change of the reaction?

## 13.5.4　Introducing Free Energy

The single measure that combines both the enthalpy and entropy of the system is called the **Gibbs Free Energy**, or more often just *free energy*. It is named after **Josiah Willard Gibbs** (1839–1903), an American physicist who, according to writer H. Arthur Klein,[2] "once referred to entropy as a measure of 'mixed-upness.'" The free energy of a system is symbolized by $\Delta G$ (after Gibbs) and is related to the entropy, enthalpy, and temperature *of the system*.

$$\Delta G = \Delta H - T\Delta S \tag{13.31}$$

Since the combination of an exothermic reaction($\Delta H$ is negative) and greater system disorder ($\Delta S$ is positive, so $T\Delta S$ is negative) favor a spontaneous reaction, we can conclude the following.

**$\Delta G$ is negative for a spontaneous reaction.**

**$\Delta G$ is positive for a nonspontaneous reaction.**

**$\Delta G$ = zero for a reaction at equilibrium.**

So it is possible for a reaction to be spontaneous and endothermic if the $T\Delta S$ term is large enough to compensate. It is also possible for a reaction to cause a decrease in system entropy and still be spontaneous if it is sufficiently exothermic. Let's look at a couple of examples at standard conditions.

---

[2] *The World of Measurements*, Simon and Schuster, New York, 1974, p. 396.

## Example 13.11

Calculate the free energy change for the dissolution of ammonium nitrate in water at standard conditions. Does your answer make sense?

## Solution 13.11

The formula for the calculation of the free energy of the reaction requires that we know the enthalpy and entropy changes for the reaction. From Eq. (13.28) and Example 13.10, we have the following values. Note the units on each value. Especially important is that $\Delta H°$ has units of **kJ** while the $\Delta S°$ value is in **J/K.**

$$\Delta H° = +26 \text{ kJ}$$

$$\Delta S° = +109 \text{ J/K}$$

We can use our equation for $\Delta G$, keeping in mind that temperature is in K. To keep our units consistent, we can convert $\Delta H°$ to J, so +26 kJ becomes 26,000 J (with 2 significant figures).

$$\Delta G° = \Delta H° - T\Delta S° = +26,000 \text{ J} - 298 \text{ K} (109 \text{ J/K})$$

$$= 26,000 - 32482 = -6482 \text{ J}$$

$$\Delta G° = -6.5 \text{ kJ}$$

The free energy for the reaction is less than zero, as is expected for a spontaneous reaction. The answer, therefore, makes sense.

## Example 13.12

We know that the Ostwald Process is spontaneous with $\Delta H° = -447$ kJ (see Example 13.8). Using the data in Table 13.3, prove that the free energy of the reaction is less than zero at standard conditions.

$$NH_3(g) + 2O_2(g) \rightarrow HNO_3(aq) + H_2O(l)$$

## Solution 13.12

We have the enthalpy change ($-447$ kJ $= -447,000$ J) and the temperature. We can calculate the entropy change. Keep in mind the importance of including the number of moles of each compound when you calculate its contribution to the entropy change of the reaction.

$$\Delta S°_{reaction} = \Sigma \Delta S°_{(products)} - \Sigma \Delta S°_{(reactants)} \qquad (13.32)$$

$$\Sigma S°_{(products)} = S°(HNO_3) + S°(H_2O) = 1 \text{ mole} \times 146 \text{ J/K mol} + 1 \text{ mol} \times 70 \text{ J/K mol}$$
$$= +216 \text{ J/K}$$

$$\Sigma S°_{(reactants)} = S°(NH_3) + S°(O_2) = 1 \text{ mol} \times 193 \text{ J/K mol} + 2 \text{ mol} \times 205 \text{ J/K mol}$$
$$= +603 \text{ J/K}$$

$$\Delta S°_{reaction} = +216 - (+603) = -387 \text{ J/K}$$

Why does the entropy of the system decrease? The reactants, gases, form products that are a liquid and a solid. We expect there to be a significant decrease in disorder when forming a solid and liquid from a gas.

The decrease in the entropy value of the system indicates this. We can now calculate the free energy.

$$\Delta G^\circ = \Delta H^\circ - T\Delta S^\circ = +447{,}000 \text{ J} - 298 \text{ K} (-387 \text{ J/K})$$
$$= 447{,}000 - (+115326) = -331674 \text{ J}$$
$$\Delta G^\circ = -332 \text{ kJ}$$

The reaction is spontaneous under standard conditions.

It is possible to determine the spontaneity of chemical reactions at nonstandard conditions. This would require us to use a new temperature as well as values of enthalpy and entropy at nonstandard conditions for each compound in the reaction. We will not do these calculations in this text.

## 13.5.6 Free Energy of Formation Values

Tables of standard free energy values are available. Table 13.4 gives these values for some common substances. Use the data in the same way that you use enthalpy and entropy data.

$$\Delta G^\circ_{\text{reaction}} = \Sigma \Delta G^\circ_{f(\text{products})} - \Sigma \Delta G^\circ_{f(\text{reactants})} \tag{13.33}$$

Free energy changes in reactions can be determined either by using enthalpy and entropy values [Eq. (13.31)] or directly via free energy values [Eq. (13.33)]. Let's prove this point in Example 13.13. As we do this example, note that as is true with standard enthalpy values, the standard free energy of formation for elements, pure liquids, and solids at their standard reference states, is defined as being equal to zero, $\Delta G^\circ = 0$.

## *Example 13.13*

Calculate the free energy change for the Ostwald process using standard free energy of formation data. Compare your answers with those you calculated in Example 13.12.

## *Solution 13.13*

$$NH_3(g) + 2O_2(g) \rightarrow HNO_3(aq) + H_2O(\ell)$$

$$\Delta G^\circ_{\text{reaction}} = \Sigma \Delta G^\circ_{f(\text{products})} - \Sigma \Delta G^\circ_{f(\text{reactants})}$$

We can use the values in the standard free energy of formation tables.

$$\Sigma \Delta G^\circ_{(\text{products})} = \Delta G^\circ(HNO_3) + \Delta G^\circ(H_2O) = 1 \text{ mole} \times -111 \text{ kJ/mol} +$$
$$1 \text{ mol} \times -237 \text{ kJ/mol} = -348 \text{ kJ}$$

**TABLE 13.4**   Standard Free Energy
Values for Some Common Substances

| Substance | $\Delta G^\circ_f$ (kJ/mol) |
|---|---|
| $C_2H_2(g)$ | 209 |
| $C_2H_4(g)$ | 68 |
| $C_2H_6$ (g) | −32.9 |
| $CO_2(g)$ | −394 |
| $H_2O(g)$ | −229 |
| $H_2O(l)$ | −237 |
| $H_2O_2(l)$ | −120.4 |
| $NO(g)$ | 87 |
| $NO_2(g)$ | 52 |
| $SO_2(g)$ | −300 |
| $SO_3(g)$ | −371 |

$$\Sigma \Delta G^{\circ}_{(reactants)} = \Delta G^{\circ}(NH_3) + \Delta G^{\circ}(O_2) = 1\ mol \times -17\ kJ/mol + 2\ mol \times 0$$
$$= -17\ kJ$$

$$\Delta G^{\circ}_{reaction} = -348\ kJ - (-17\ kJ) = -331\ kJ$$

This value agrees well with the $\Delta G^{\circ}$ value determined by using the enthalpy and entropy data.

## 13.6 Where Have We Been and Where Are We Going?

Thermodynamics is concerned with energy exchanges that occur in chemical reactions. We have seen that every process is, at its root, driven by energy exchange, and we can predict whether or not processes will occur based on their free energy change. What we cannot predict is how fast spontaneous processes are likely to occur. There are other factors that we need to know about before we can assess the rate and the mechanism of a reaction, which are known as the *kinetics* of a reaction. That will be the focus of our next chapter.

In this chapter, we learned that:

1. Energy is the capacity to do work, and (in a less opaque sense) that which is required to oppose natural attractions.
2. We can express energy using several measures, including force × distance, pressure × volume, and mass ×velocity$^2$.
3. Energy can be emitted or absorbed in reactions as light and heat, in several forms.
4. Energy and mass are equivalent.
5. Bond breakage is an endothermic process. Bond formation is an exothermic process.
6. We can determine the net energy exchange using several measures, including heats of formation and bond energy calculations.
7. Reactions occur because the entropy of the universe increases as a result.
8. We can determine the entropy change of the universe from a reaction via the free energy of a reaction.

## 13.7  Homework Exercises

1. Regarding the electromagnetic spectrum:
   (a) List the regions.
   (b) List the wavelengths associated with each region.
   (c) Search journals such as *Analytical Chemistry* or *Clinical Chemistry* and come up with laboratory analysis applications that use the different regions of the electromagnetic spectrum, and the role played by the electromagnetic radiation (for example, infrared spectroscopy, in which the I.R. radiation induces the molecule to bend or stretch in ways that allow us to identify it).
2. The concentration of copper can be determined in a solution by exciting the electrons in an atom with a lamp that emits only wavelengths specific to copper. If a lamp emits 325.4 nm, what is the frequency of the light?
3. When we look at blue light, we are seeing wavelengths in the range of about 400 to 475 nm. What frequency range does this include?
4. What is the wavelength associated with a photon of energy = $3.61 \times 10^{-19}$ J?
5. What is the energy of a photon of radiation with a wavelength of 812 nm? In what region of the electromagnetic spectrum is this wavelength found?
6. What is the energy of a mole of photons emitted at a wavelength of 812 nm? How does this compare to the heat energy emitted by the combustion of 1 mole of butane in the presence of an excess of oxygen?

$$C_4H_{10}(l) + 13/2 O_2(g) \rightarrow 4CO_2(g) + 5H_2O(g)$$

7. Calculate the energy change when one mole of $^{14}C$ nuclei decay to form one mole of $^{14}N$ nuclei. The atomic mass of $^{14}C$ = 14.003241 g/mol. The atomic mass of $^{14}N$ = 14.003074 g/mol. How does this compare to the energy given off by the combustion of one mole of methane?
8. Calculate the enthalpy change under standard conditions for the decomposition of hydrogen peroxide.

$$2H_2O_2(l) \rightarrow 2H_2O(l) + O_2(g)$$

9. Calculate the enthalpy change under standard conditions for the hydrogenation of acetylene.

$$C_2H_2(g) + H_2(g) \rightarrow C_2H_4(g)$$

10. Determine the entropy and free energy changes for the reactions in the previous two problems. Are the reactions spontaneous?

11. Explain why the enthalpy of formation of $O_2$ is zero, while the enthalpy of formation of $O_3(g)$ is not.
12. Use bond energies to find $\Delta H°$ for the following reaction.

$$2C_2H_6(g) + 7O_2(g) \rightarrow 4CO_2(g) + 6H_2O(g)$$

13. Use the $\Delta H_f^o$ values from your text and compare the result from the answer in Question 12.
14. Calculate $\Delta H°$ if 0.500 grams of $H_2$ react with excess $Cl_2$ under standard conditions via the following reaction.

$$H_2(S) + Cl_2(g) \rightleftharpoons 2HCl(g)$$

15. Would the following reactions increase or decrease the entropy of the system?
    (a) $NH_4Cl(s) \rightarrow NH_3(g) + HCl(g)$
    (b) $H_2(g) + Br_2(g) \rightarrow 2HBr(g)$
16. Use your text to find the standard $\Delta S$ values for the reactions in Question 15 to determine the actual changes in entropy that are taking place.
17. When ice melts, does entropy increase or decrease? Justify your answer.
18. A chemical reaction decreases the entropy of the universe by 25 J/K. Is the process spontaneous? Why or why not?
19. If a reaction is exothermic, this means that
    (a) The process will be spontaneous under all conditions.
    (b) It will always increase the entropy of the universe.
    (c) The process tends to be spontaneous, but can be nonspontaneous under certain conditions.
    (d) The process will never be spontaneous.
20. Which of the following conditions must be met for a process to be spontaneous?
    (a) $\Delta G > 0$
    (b) $\Delta H < 0$
    (c) $\Delta S_{universe} > 0$
    (d) $\Delta S_{universe} < 0$
21. Which of the following system conditions indicate that a reaction must be spontaneous?
    (a) $\Delta S > 0$, $\Delta H < 0$
    (b) $\Delta S > 0$, $\Delta H > 0$
    (c) $\Delta S < 0$, $\Delta H > 0$
    (d) $\Delta S < 0$, $\Delta H < 0$
22. If a reaction is nonspontaneous, which of the following system conditions must be true?
    (a) $\Delta H > 0$
    (b) $\Delta S < 0$
    (c) $\Delta S > 0$
    (d) $\Delta G > 0$
23. Determine whether the reactions represented by the following sets of data are spontaneous or nonspontaneous.

|    | $\Delta H$(kJ) | $\Delta S$(J/K) | T(K) |
|----|------|------|------|
| a. | $-250$ | $+50$ | 298 |
| b. | 22 | $-25$ | 300 |
| c. | $-250$ | 100 | 200 |
| d. | $+15$ | $-16$ | 200 |
| e. | $-53$ | $-30$ | 500 |

24. If $\Delta H = -246$ kJ, $\Delta S = +52$ J/K, and $\Delta G = -256$ kJ, at what temperature did this reaction occur?

25. Using the standard enthalpy and entropy data from Table 13.3, determine whether the following reaction is spontaneous at a temperature of 298 K.

$$SO_2(g) + NO_2(g) \rightarrow SO_3(g) + NO(g)$$

26. Use the the $\Delta G°$ values in Table 13.4 for the reactants and products in Question 25. How does this value compare to those you found using the formula $\Delta G° = \Delta H - T\Delta S$?

27. Using the $\Delta G°$ values in Table 13.4, find $\Delta G°$ for the following reaction under standard conditions.

$$2C_2H_6(g) + 7O_2(g) \rightarrow 4CO_2(g) + 6H_2O(l)$$

Is the reaction spontaneous at this temperature? If not, should the temperature be increased or decreased to make the reaction occur spontaneously?

## For Class Discussion and Reports

28. Automobile gasoline is a complex mixture of hydrocarbons. One of the important compounds in gasoline is octane, $C_8H_{18}$. Look up the enthalpy of formation for this compound and for methane, $CH_4$, in the *CRC Handbook of Chemistry and Physics*. Which compound releases more heat energy per mole when burned in oxygen? Which releases more per gram? Compare these values to that achieved when combining hydrogen and oxygen. Based on energy considerations, which of these is the "fuel of the future" in automobiles? What other factors must be taken into account when deciding on a fuel for the future? Based on these other factors, which combustion reaction is best for 21st-century cars? Discuss the answers to these questions as a class.

29. One of the important spectrophotometric analysis procedures is called *Inductively Coupled Plasma Spectrophotometry (ICP)*, in which many metals can be detected simultaneously (see Section 3.10). Search textbooks on analytical chemistry and past issues of the journal *Analytical Chemistry* in order to find out more about ICP. Determine how it is possible to manipulate light energy in such a way as to quantitate the amounts of elements in a sample. Write a report.

# 14

# Introduction to Kinetics

## 14.1  Introduction

In the last chapter we discussed a number of reactions that are ***spontaneous***—that is, reactions that can happen without continued outside intervention. We know that it is possible to calculate a ***free energy*** change for a reaction, and that if it is negative, the reaction will be spontaneous. The Ostwald Process for the production of ammonia was one example of a spontaneous reaction.

$$NH_3(g) + 2O_2(g) \rightarrow HNO_3(aq) + H_2O(\ell) \quad \Delta G^{\circ}_{reaction} = -331 \text{ kJ} \qquad (14.1)$$

We discussed some other spontaneous reactions, including the combustion of propane and the formation of water from hydrogen and oxygen gas. We did not calculate the $\Delta G^{\circ}_{reaction}$ values for these in Chapter 13, but they are given below.

$$C_3H_8(g) + 5O_2(g) \rightarrow 3CO_2(g) + 4H_2O(g) \quad \Delta G^{\circ}_{reaction} = -1303 \text{ kJ} \qquad (14.2)$$

$$2H_2(g) + O_2(g) \rightarrow 2H_2O(g) \quad \Delta G^{\circ}_{reaction} = -458 \text{ kJ} \qquad (14.3)$$

Although each of these reactions is thermodynamically spontaneous at room temperature, *none* of them will happen without some outside intervention to get them started. For example, the combustion of propane requires the temperature of the molecules to be raised high enough so that they will react on their own. One way that this can be accomplished is with a heat source (such as a flame). The hydrogen and oxygen reaction can also be started this way, or it can be initiated at room temperature by a metal catalyst. If there is no heat source or no catalyst, the reactants in these processes remain unreacted.

---

*Chemistry Professionals at Work* CPW Box 14.1

# WHY STUDY THIS TOPIC?

Your experience with medicines can show why the topic of kinetics is so useful to the chemical industry. The chemicals in our body—sugars, proteins, lipids, nucleic acids, vitamins, and minerals—undergo thousands of reactions that, when properly synchronized, keep us alive. Many of these reactions happen quickly, requiring **enzymes** (biological catalysts) to speed them up. Every reaction in the body happens at a particular rate. When you are not feeling well, you take a medicine that contains a chemical that interacts with the body at a certain rate. The pharmaceutical company develops the product so that a certain dose is delivered in a given time, such as by "timed-release" capsules. Studies done in the laboratory determine the rate at which the capsule itself dissolves. Further studies, often done on animals, examine drug **metabolites** (the chemical products resulting from the breakdown of a substance in a living system) as a function of time, along with the physiologic responses of the animal to the drug. Human clinical trials then take place. If the product works as hoped, with few important side effects, the product can be given Food and Drug Administration approval and be marketed. The dosage and frequency with which you take the medicine are all related to the *kinetics* of the interaction of the medicine with your body.

*For Class Discussion: This one is a field trip. The next time you go food shopping, visit the medicine aisle. Compare the timed-release capsules with other types in terms of their potency (the amount of the chemical that fights the symptoms), the number of doses in a day, and the cost. Are there any conclusions that can be drawn? Which type of medicine do you use and why?*

---

From this we can see that **thermodynamics** is concerned with whether or not a process **can** occur. It is not explicitly concerned with **how fast** the process occurs. To address issues involving the speed of a reaction, we must discuss the topic of kinetics. **Kinetics** is the field of study concerned with how fast, and by what mechanism, processes occur.

## 14.2 Rates of Reaction

### 14.2.1 Calculating the Rate

In the Space Shuttle main engines, the reaction of hydrogen with oxygen occurs at several thousand degrees Celsius. The reaction is fast and violent, releasing enough energy to help launch the craft. In fact, the mole ratio of hydrogen to oxygen that reacts is *not* 2:1 but is closer to 4:1, due to the need for great thrust and other nonchemical considerations. The reaction of hydrogen and oxygen is used on the shuttle to produce water and occurs slowly, in a controlled way. If we were to study this reaction in a laboratory setting, the amounts of each reactant and product could be measured at any time.

Let's say that we begin our water formation reaction with 0.500 moles each of hydrogen and oxygen. After 5.00 minutes of reaction, 0.400 moles of oxygen remain. How much water was formed under these conditions (see Section 8.8 for a review of mole–mole conversions)?

The reaction here is as follows.

$$2H_2(g) + O_2(g) \rightarrow 2H_2O(g) \tag{14.4}$$

A total of $0.500 - 0.400$ moles $= 0.100$ moles of $O_2$ was consumed. According to Eq. (14.4), there are two moles of $H_2O$ formed for every mole of oxygen reacted. We can use this to determine the amount of water formed.

$$0.100 \text{ mol } O_2 \times \frac{2 \text{ mol } H_2O}{1 \text{ mol } O_2} = 0.200 \text{ moles of } H_2O \tag{14.5}$$

*But how many moles of hydrogen were consumed in this same reaction?* The mole ratio of hydrogen to water is 1:1, which means that the same amount of hydrogen is consumed as the amount of water that was formed. Therefore, 0.200 moles of $H_2$ were used. A total of 0.300 moles of hydrogen remain. In kinetics, we are concerned with how fast reactions occur. We call this speed the *rate*, and it has units of amount, or concentration per unit time.

$$\text{Rate} = \frac{\text{amount of subtance reacted (produced)}}{\text{time change}} \tag{14.6}$$

Let us now see how this pertains to our particular reaction. For example, what is the rate of reaction of hydrogen? A total of 0.200 moles were reacted in 5.00 minutes, so the rate of consumption of hydrogen is given by the following equation.

$$\text{Rate } (H_2) = \frac{-0.200 \text{ mol}}{5.00 \text{ min}} = -0.0400 \text{ mol min}^{-1}$$

The "$-$" sign here indicates that hydrogen was consumed rather than produced. We can similarly determine the rate of reaction of oxygen and the rate of production of water.

$$\text{Rate } (O_2) = \frac{-0.100 \text{ mol}}{5.00 \text{ min}} = -0.0200 \text{ mol min}^{-1}$$

$$\text{Rate } (H_2O) = \frac{+0.200 \text{ mol}}{5.00 \text{ min}} = +0.0400 \text{ mol min}^{-1}$$

Note that these rates are related to the coefficients of the chemical equation.

## Example 14.1

If 0.060 moles of propane, $C_3H_8$, are burned in the first 20.0 seconds of the reaction shown below, what is the rate of production of water in that time?

$$C_3H_8(g) + 5O_2(g) \rightarrow 3CO_2(g) + 4H_2O(g)$$

## Solution 14.1

According to the above equation, there are four moles of water produced for each mole of propane burned, so the *rate* of production of water is four times the *rate* of consumption of propane. Mathematically, we say this by using the equation shown below.

$$\frac{-4\Delta(\text{moles } C_3H_8)}{\Delta \text{ time}} = \frac{\Delta(\text{moles } H_2O)}{\Delta \text{ time}}$$

This means the above statement says that if there are two moles of propane consumed (hence the "−" sign) per second, then there will be four times that, or eight moles of water produced. The "$\Delta$" means "change" (as we discussed in Chapter 13, in which "$\Delta H$" meant "change in enthalpy").

   The rate of consumption of the propane is

$$\text{Rate } (C_3H_8) = \frac{-0.060 \text{ mol}}{20.0 \text{ s}} = -0.0030 \text{ mol s}^{-1}$$

and the rate of production of $H_2O$ is

$$\text{Rate } (H_2O) = \frac{+0.24 \text{ mol}}{20.0 \text{ s}} = +0.012 \text{ mol s}^{-1}$$

which is a rate four times that of the consumption of propane.

   We can also consider the rates of reaction in solution, where it is convenient to express changes in *concentration* by using *molarity*, rather than amount.

## Example 14.2

When ethanol, $C_2H_5OH$, is added to a solution of potassium dichromate, $K_2Cr_2O_7$, no reaction occurs. However, when hydrochloric acid is added to the mixture, the chromium is readily changed to $Cr^{3+}$ ion. The net ionic reaction is as follows.

$$16H^+(aq) + 2Cr_2O_7^{2-}(aq) + C_2H_5OH(aq) \rightarrow 4Cr^{3+}(aq) + 2CO_2(g) + 11H_2O(\ell)$$

If the initial concentration of $H^+$ is 0.500 M and, after 3.00 minutes, $[H^+] = 0.230$ M, what is the rate of $Cr^{3+}$ production?

## Solution 14.2

There are two things that will help us to determine the rate of $Cr^{3+}$ production. The first is the rate of reaction of the hydrogen ion. The second is the relationship of the $Cr^{3+}$ concentration to the $H^+$ concentration.

$$\text{Rate } (H^+) = \frac{-(0.500 - 0.230)M}{3.00 \text{ min}} = \frac{-0.270 \text{ M}}{3.00 \text{ min}} = -0.0900 \text{ M s}^{-1}$$

According to this chemical equation, there are four moles of $Cr^{3+}$ produced for each sixteen moles of $H^+$ consumed. So, $H^+$ is consumed four times faster than $Cr^{3+}$ is produced. We can express this mathematically as follows.

$$\frac{\Delta([H^+])}{\Delta \text{ time}} = \frac{4\Delta([Cr^{3+}])}{\Delta \text{ time}}$$

Therefore, if the rate of consumption of $H^+$ is 0.0900 M s$^{-1}$, the rate of formation of $Cr^{3+}$ is 1/4 of this.

$$\text{Rate } (Cr^{3+}) = \frac{-0.0900 \text{ M s}^{-1}}{4} = 0.0225 \text{ M s}^{-1}$$

## 14.2.2 Factors That Influence the Rate of a Reaction

We discussed why reactions happen from a *thermodynamic* viewpoint in Chapter 13. We said there that a spontaneous reaction may occur if the entropy of the universe is increased as a result. From a *kinetic* standpoint, we can look at the process that enables reactions to occur. This is best done at the molecular level. If we view two molecules, each of which has a specific shape, there are several requirements in order for them to react.

*The molecules must come in contact with each other* (Fig. 14.1a). The more reactant molecules there are, the more often contact will occur. Therefore, all other things being equal, **reaction rate will increase with higher reactant concentrations**. The exception to this is when reactions happen on a catalyst surface, such as in a catalytic converter in an automobile, where the restricted catalyst surface area limits the extent of reaction.

*The molecules must collide with sufficient energy to react* (Fig. 14.1b). The kinetic energy of a system is proportional to its temperature. Therefore, the higher the temperature, the higher the velocity of the particles (K.E. = $\frac{1}{2}$ mv$^2$). At higher velocities, particles collide more often and a higher fraction of them have enough energy for successful reaction. The amount of energy needed to initiate a reaction is called the **activation energy, $E_a$**. We will discuss activation energy later. A rule of thumb is that **the reaction rate will double for every 10°C increase in system temperature**. Ammonium salts often react endo-thermically, so that the temperature of the reaction vessel actually *decreases* with the reaction. Yet even this reaction has an activation energy barrier that must be overcome in order for the reaction to proceed. This energy input goes to breaking bonds.

*The molecules must collide with the proper orientation for reaction* (Fig. 14.1c). The molecules must meet in such a way that they can react. An analogy can be made between molecules and long-lost friends going to give each other a hug. If they are facing away from each other, the hug will not happen. Nor will it happen if they are beside each other. Only if they have the proper orientation, facing one another, will they be able to hug. So it is with particles colliding.

*A catalyst provides an alternative pathway for a reaction to occur* (Fig. 14.1d). The chemical reactions that we discuss are really summaries of what are often very complex processes. For example, the process of photosynthesis is often summarized with this reaction.

$$C_6H_{12}O_6(s) + 6O_2(g) \rightarrow 6CO_2(g) + 6H_2O(l) \tag{14.7}$$

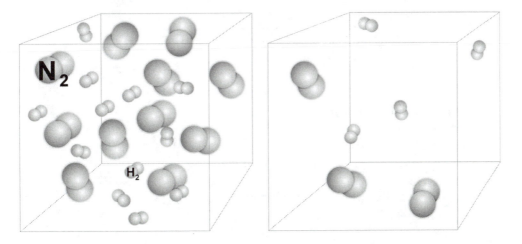

**FIGURE 14.1(a)**   In order for molecules, such as N$_2$ and H$_2$ to react, they must come in contact with each other. More contact can come with higher concentration, as on the left. Less contact, and therefore a slower reaction, are a result of a lower concentration on the right.

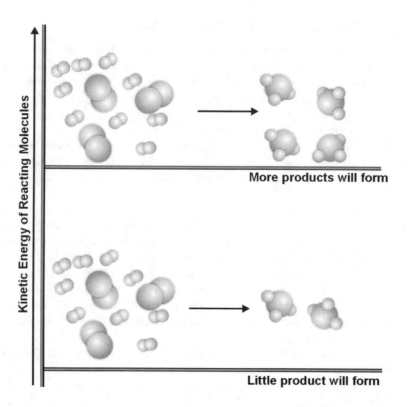

**FIGURE 14.1(b)**    If $NH_3$ is to form from $N_2$ and $H_2$, then the collisions that occur must have sufficient energy to initiate the reaction.

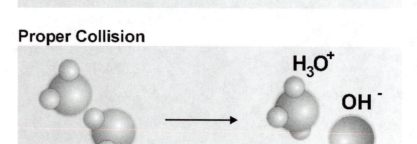

**FIGURE 14.1(c)**    Molecules must have the proper orientation for a reaction to occur.

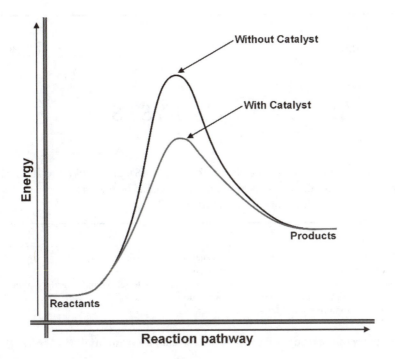

**FIGURE 14.1(d)** A catalyst lowers the activation energy of a reaction by providing an alternate reaction pathway.

In fact, photosynthesis is a process that includes over a hundred different reactions, some of which need light (they are part of the "light cycle" of photosynthesis), while others occur in darkness (the "dark cycle" of photosynthesis). Reactions happen via specific "routes," called **mechanisms**, that are unique for each reaction. A catalyst works to speed up reactions by getting involved in the chemistry. In doing so, it alters the mechanism—the route—by which a reaction proceeds. The catalyst ends up in the same chemical form in which it began. Although it may change *during* the reaction, it will change back.

An example of this is the decomposition of hydrogen peroxide to form water and oxygen,

$$2H_2O_2(l) \rightarrow 2H_2O(l) + O_2(g) \tag{14.8}$$

This reaction is quite slow, with half of a 30% solution of hydrogen peroxide decomposing after only many months at room temperature. However, if some potassium bromide is added to the solution, decomposition occurs within seconds. The bromide changes form and reacts as follows.

$$H_2O_2(aq) + Br^-(aq) + 2H^+(aq) \rightarrow Br_2(aq) + 2H_2O(\ell) \tag{14.9}$$

$$Br_2(aq) + H_2O_2(aq) \rightarrow 2Br^-(aq) + 2H^+(aq) + O_2(g) \tag{14.10}$$

The bromine formed in the reaction in Eq. (14.9) later returns to the original form, $Br^-$, and is therefore a catalyst. Because the reaction speeds up so much in the presence of a catalyst, the *rate of the reaction* is far, far greater than when no catalyst is present.

We see, then, that there are four important factors that affect the rate of a reaction, or how quickly a reaction proceeds. We can readily control three of these—the concentration, the temperature, and the presence of a catalyst.

---

*Chemistry Professionals at Work*                    *CPW Box 14.2*

# MORE ABOUT CATALYSTS

Catalysts are a vital part of the chemical industry. (See also CPW Box 2.6.) Metal-based catalysts are especially useful in chemical synthesis. The table below lists a few catalysts and some of the reactions they make possible on an industrial scale.

| Catalyst | Reaction |
|---|---|
| Copper | Hydrogenation and dehydrogenation |
| Nickel or cobalt | Hydrogenation |
| Nickel | Conversion of primary to secondary or tertiary amines |
| $(C_6H_5)_3P$ (triphenylphosphine) | Polymerization reactions |
| $NaOCH_3$ (sodium methylate) | Isomerizations |

(Source:  http:/www.basf.com/businesses/chemicals/catalysts/menu.html) (as of 6/7/00)

*For Homework: We have listed catalysts and reactions provided by the Websites listed. Go to these sites and find 5 more catalysts and reactions, focusing especially on ones that help produce products with which you are familiar.*

---

## 14.2.3   How Does the Rate Change with Time?

The rate of a reaction often changes as the reaction proceeds. Here are some data for the iodide-catalyzed decomposition of hydrogen peroxide.[1] The data shown assume an excess of iodide ion, so that its concentration is essentially constant as $[H_2O_2]$ changes.

| Time (seconds) | $[H_2O_2]$ |
|---|---|
| 0.00 | 0.272 |
| 1.00 | 0.259 |
| 2.00 | 0.248 |
| 5.00 | 0.215 |
| 10.0 | 0.170 |
| 15.0 | 0.134 |
| 20.0 | 0.106 |
| 25.0 | 0.084 |
| 30.0 | 0.066 |

These data are shown graphically in Fig. 14.2. Note that the concentration of $H_2O_2$ vs. time does not decrease linearly. Rather, the concentration decrease slows down with time. This is typical of all chemical

---

[1]Hansen, J.C., *J. Chem. Educ.*, 73, 728–731, 1996.

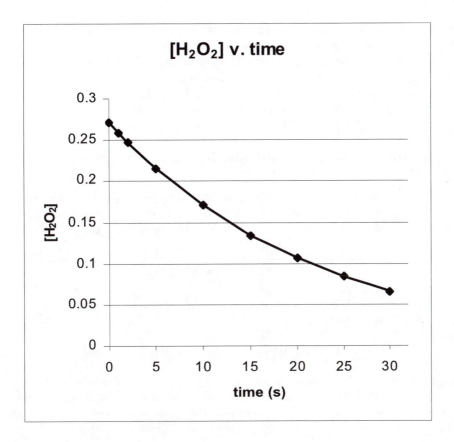

**FIGURE 14.2**    Note the decrease in concentration of hydrogen peroxide with time. These data are used to determine the rate law of the reaction.

reactions in that *the rate of reaction decreases with time*. Why is this so? Recall the first point in our discussion of the factors that change the rate of a reaction (Section 14.2.2). The first one was the ability of the reactants to collide. As the reaction proceeds and products are formed, the concentration of reactants decreases, leaving fewer to collide. The rate of reaction will therefore decrease.

Just as we can plot the decrease in hydrogen peroxide concentration with time, we can also look at the increase in oxygen concentration with time. As the rate of reaction of hydrogen peroxide decreases with time, the rate of oxygen production also decreases.

## 14.2.4   The Differential Rate Equation

The data from the decomposition of hydrogen peroxide indicate that the hydrogen peroxide concentration seems to change in a systematic way, as if it might fit a mathematical function. Let's try to come up with such a function. We know that the rate of the reaction $\left(\frac{-\Delta[H_2O_2]}{\Delta t}\right)$ at any time is related to the concentration of hydrogen peroxide at that time. This is summarized by the following equation.

$$\frac{-\Delta[H_2O_2]}{\Delta t} = k[H_2O_2]^n \qquad (14.11)$$

Equation (14.11) is called the **differential rate equation** and it relates the rate of the reaction to the concentration of the reactants. If the superscript "n" is 1, this means that *the rate is directly related to* $[H_2O_2]$. The reaction is said to be *first-order in hydrogen peroxide*. If $n = 2$, then the rate is related to $[H_2O_2]^2$. This would be a *second-order reaction with respect to hydrogen peroxide*. In practice, some reactions can have fractional orders, though we will not deal with such cases here. The term "k" is called the *rate constant* for the reaction, and takes into account the other factors that control the rate of reaction that we discussed in Section 14.2.2, including the temperature, the orientation of molecules, and the presence of a catalyst. For a first-order process, we often label the rate constant "$k_1$", where the "1" designates "first-order."

A vital step in understanding the way a reaction works is to determine the values of "n" and "k." This helps to understand how fast and by what route the reaction proceeds.

## 14.2.5   Determining k for a First-Order Reaction: The Integrated First-Order Rate Equation

A reasonable assumption that we can make concerning the decomposition of a substance is that the rate of the decomposition reaction is directly related to the concentration of the substance. This is the first-order case we discussed in Section 14.2.3. In this case, the differential rate equation is as follows.

$$\frac{-\Delta[H_2O_2]}{\Delta t} = k_1[H_2O_2]$$

This equation can be modified using a technique of calculus called *integration*. The result is called an *integrated first-order rate equation*.

$$\ln[H_2O_2]_t = -k_1t + \ln[H_2O_2]_o \qquad (14.12)$$

The mathematical operator "ln" means "the natural logarithm (log) of." The term "$[H_2O_2]_o$," means the concentration of $H_2O_2$ at time t = 0, in other words, the initial concentration of $H_2O_2$. The term "$[H_2O_2]_t$" is the concentration of $H_2O_2$ at time t. This equation is in the form y = mx + b, the equation for a straight line, where y, x, and b are represented by the following terms.

$$y = \ln[H_2O_2]_t \qquad (14.13)$$

$$x = time(t) \qquad (14.14)$$

$$m = \text{the slope of the line} = -k_1 \qquad (14.15)$$

$$b = \text{the intercept of the line} = \ln[H_2O_2]_o \qquad (14.16)$$

Here is how to use the natural log of a number [the natural log is the (base 10) log × 2.3026]. Let's use the first data point (t = 0) for the decomposition of $H_2O_2$, 0.272 M. On a *nongraphical scientific calculator*, type in the value "0.272." Then press the button labeled "LN" or "ln." The display should read (rounded to 4 figures) "−1.302." On a *graphical scientific calculator*, press "LN" or "ln," then type in the value "0.272" and press the "ENTER" button. The display should read (rounded to 4 figures) "−1.302."

We can now process our data in the proper way to construct a plot. Here is a list of the time, concentration, and $\ln[H_2O_2]$ data for our reaction.

| Time (seconds) | $[H_2O_2]$ | $\ln[H_2O_2]$ |
|---|---|---|
| 0.00 | 0.272 | −1.302 |
| 1.00 | 0.259 | −1.351 |
| 2.00 | 0.248 | −1.394 |
| 5.00 | 0.215 | −1.537 |
| 0.0 | 0.170 | −1.772 |
| 5.0 | 0.134 | −2.010 |
| 20.0 | 0.106 | −2.244 |
| 25.0 | 0.084 | −2.48 |
| 30.0 | 0.066 | −2.72 |

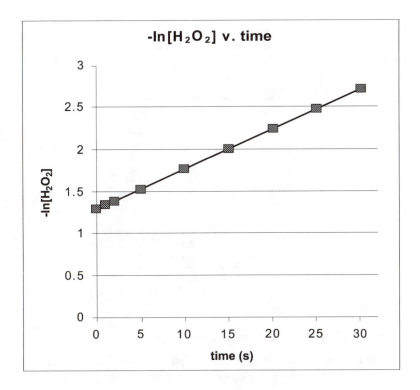

**FIGURE 14.3**  A plot of the natural log of the hydrogen peroxide concentration vs. time shows that this is a first-order process.

If the decomposition of hydrogen peroxide is first-order for $H_2O_2$, a plot of $\ln[H_2O_2]_t$ vs. time should look like the one shown in Fig. 14.3. From this type of plot, we can determine the first-order rate constant, $k_1$ from the slope of the line that goes through our data.

The line in Fig. 14.3 is straight. This is a characteristic tool for a first-order process, so we can confidently assert that "n" = 1. If the plot of $\ln[A]$ vs. time for a reactant A is *not* straight, then *the reaction is not first-order for A*. However, since we did get a straight line in this case, the slope ("rise/run") of the line, which has a value of $-0.047 \text{ s}^{-1}$, gives "$-k_1$", the negative value of the first-order rate constant. A first-order constant always has units of seconds$^{-1}$ or time$^{-1}$.

## Example 14.3

The following concentration vs. time data were obtained for the reaction below

$$2I(g) \rightarrow I_2(g)$$

| Time (nanoseconds) | [I] |
|---|---|
| 0.000 | 0.500 |
| 0.100 | 0.385 |
| 0.200 | 0.312 |
| 0.300 | 0.263 |
| 0.40 | 0.227 |
| 0.500 | 0.200 |
| 0.600 | 0.179 |
| 0.700 | 0.161 |
| 0.800 | 0.147 |
| 0.900 | 0.135 |
| 1.00 | 0.125 |

Is this reaction a first-order process?

## Solution 14.3

Our task is to prepare a plot of ln [I] vs. time. If the resulting data form a straight line, then our reaction can be considered to be first-order for iodine. If the plot of the data does not form a straight line, then the reaction is of some other order.

| Time (nanoseconds) | [I] | ln [I] |
|---|---|---|
| 0.000 | 0.500 | −0.693 |
| 0.100 | 0.385 | −0.955 |
| 0.200 | 0.312 | −1.165 |
| 0.300 | 0.263 | −1.336 |
| 0.400 | 0.227 | −1.483 |
| 0.500 | 0.200 | −1.609 |
| 0.600 | 0.179 | −1.720 |
| 0.700 | 0.161 | −1.826 |
| 0.800 | 0.147 | −1.917 |
| 0.900 | 0.135 | −2.002 |
| 1.00 | 0.125 | −2.079 |

When these data are plotted, the graph appears as in Fig. 14.4. The line is decidedly curved, so the reaction is not first-order.

## 14.2.6   The Integrated Second-Order Equation

The determination of the proper order of even fairly simple systems can be quite challenging. Scientists combine sophisticated experimental methods with complex data-handling programs to come up with

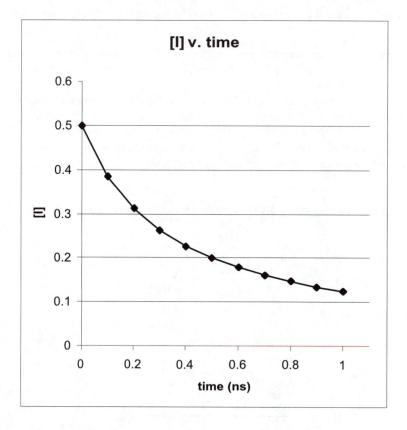

**FIGURE 14.4**   The curved line indicates that the reaction is not first-order.

rate expressions that are often complicated. What we are showing here is that there is a rational approach to the analysis, even if we can't deal with some of the fancy chemical systems. Yet even simple systems can demonstrate the thinking process.

We saw in Example 14.3 that the reaction for the formation of molecular iodine from iodine atoms is not first-order. What order could it be? The chemical equation indicates that 2 iodine atoms combine to form an iodine molecule. Perhaps the reaction is *second-order* (which we will symbolize by the rate constant "$k_2$"). This would give the following rate equation.

$$\frac{-\Delta[I]}{\Delta t} = k_2[I]^2 \tag{14.17}$$

The rate law gives us some information about the way the reaction proceeds, as we will discuss in Section 14.4. As with the first-order rate equation, the second-order rate expression can be integrated to get the *integrated second-order rate equation*.

$$\frac{1}{I_t} = \frac{1}{I_o} + k_2t \tag{14.18}$$

This is in the straight line form of $y = mx + b$, where

$$y = \frac{1}{I_t} \tag{14.19}$$

$$x = \text{time (t)} \tag{14.20}$$

$$m = \text{the slope of the line} = k \tag{14.21}$$

$$b = \text{the intercept of the line} = \frac{1}{I_t} \tag{14.22}$$

Therefore, in order to determine if a reaction is second-order in iodine, we can plot $1/[I]_t$ vs. time. If the resulting line is straight, our reaction is likely second-order. The slope of the line represents the second-order rate constant, $k_2$, and the intercept is $1/[I]_o$. Using our data from Example 14.3, we get the following results.

| Time (nanoseconds) | $[I]_t$ | $1/[I]_t$ |
|---|---|---|
| 0.00 | 0.500 | 2.00 |
| 0.10 | 0.385 | 2.60 |
| 0.20 | 0.312 | 3.20 |
| 0.30 | 0.263 | 3.80 |
| 0.40 | 0.227 | 4.40 |
| 0.50 | 0.200 | 5.00 |
| 0.60 | 0.179 | 5.60 |
| 0.70 | 0.161 | 6.20 |
| 0.80 | 0.147 | 6.60 |
| 0.90 | 0.135 | 7.40 |
| 1.00 | 0.125 | 8.00 |

Figure 14.5 shows that our data are represented by a straight line. This means that Eq. (14.17) is probably the correct rate expression for the iodine reaction. The rate constant, $k_2 = 6.0 \times 10^9 \text{ M}^{-1}\text{s}^{-1}$, is the slope of the line.

*Chemistry Professionals at Work*                    *CPW Box 14.3*

# ORGANIC SYNTHESIS

**I**t is apparent from the discussion here and in the rest of the chapter that the balanced equation for a chemical reaction is often a mere summary of the chemistry that occurs. Oftentimes, many steps are necessary to create the desired product that is now shown in the balanced equation. In industrial Research and Development laboratories in which new organic compounds are being produced and researched (called *organic synthesis*), the chemist and his/her team uses the knowledge of different organic reactions to decide the series of reactions that will be used to create the desired product. The factors that must be considered are cost, time, number of steps needed, and yield. An optimum synthesis will produce a high yield of product in a small number of simple, quick steps using the cheapest materials possible. But when an optimum synthesis is not possible, it is up to the individual industry or researcher to determine which reaction pathway to select. (See also CPW Box 8.7.)

*For Homework: Alternative fuels for automobiles is a pressing issue, given fluctuation in gasoline prices and environmental concerns. Look up the articles on fuel cells in the July 1999 issue of* Scientific American. *Write an essay describing which of the automobile fuel options make the most sense to you and why.*

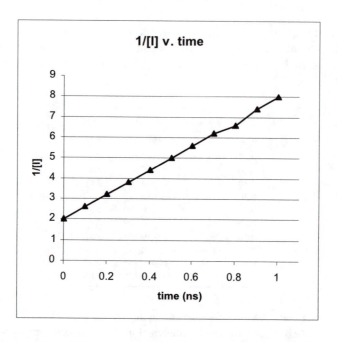

**FIGURE 14.5**   Second-order data will form a straight line when 1/concentration is plotted vs. time.

# 14.3   Applications of Rate Expressions

We now have the tools to decide whether a reaction is first- or second-order. That is, for a process with just one reactant, we can determine whether a reaction rate is related to the concentration of reactant species in a linear way (first-order) or, rather, to the square of the concentration (second-order). Once we establish the rate expression for a reaction, we can use that information to figure out some important quantities. The two most useful are the concentration of the reactant at any time and the half-life of the reaction.

## 14.3.1   Concentration at Any Time

We can determine the concentration at any time for a first- or second-order system in which there is only one reactant. We will discuss only first-order systems here. Recall the first-order rate expression for the decomposition of hydrogen peroxide.

$$\frac{-\Delta[H_2O_2]}{\Delta t} = k_1[H_2O_2] \tag{14.23}$$

The integrated rate law for this was given by Equation 14.12.

$$\ln[H_2O_2]_t = -k_1t + \ln[H_2O_2]_0 \tag{14.24}$$

If we *exponentiate* both sides (that is, take the *inverse* of the natural log), we come up with an expression for the concentration of $H_2O_2$ in terms of time and $k_1$, where "e" has the same mathematical relationship to "ln" as "10" has to "log."

$$[H_2O_2]_t = [H_2O_2]_0 \, e^{-k_1t} \tag{14.25}$$

For the general case in which we have a single reactant, "A," reacting by first-order kinetics, Eq. (14.25) can be re-written as shown below.

$$A_t = A_0 \, e^{-k_1t} \tag{14.26}$$

Let's illustrate the use of this equation with data from our hydrogen peroxide example, in which $[H_2O_2]_0 = 0.272$ M and $k_1 = 0.047$ $s^{-1}$. At t = 15.0 seconds, the $[H_2O_2]_t$ is calculated as follows.

$$[H_2O_2]_t = [H_2O_2]_0 \, e^{-k_1t} = 0.272 \, e^{(-0.047s^{-1} \times 15.0 \, s)} = 0.134 \text{ M}$$

You can do this by using the following process:
  If you are using a nongraphical scientific calculator:

  1. Enter "0.047," press the "+/−" key.
  2. Press the "×" key, enter "15.0" and press "=" The display will read "−0.705."
  3. Press the "2nd" button and then "LN" (or "ln"). The display will read 0.494 (to 3 figures).
  4. Multiply this value (which is $e^{-k_1t}$) by pressing "×" and then entering "0.272" (which is $[H_2O_2]_0$). Your display should read, "0.134" (to 3 figures).

  If you are using a graphical scientific calculator, the entire equation can be entered as one continuous line.

  1. Enter "0.272."
  2. Press the "2nd" and then the "LN" (or "ln") buttons.

3. Press the "(" button, then the "−" button.
4. Enter "0.047", "×", and "15". Press ")".

Only after the entire string of values has been put into the calculator should you hit the "ENTER" key. The result, "0.134" (to 3 figures) should be on the display, along with the equation you have entered.

### Example 14.4

What is the concentration of hydrogen peroxide after 7.00 seconds in a reaction with $[H_2O_2]_0$ = 0.272 M, $k_1$ = 0.047 s$^{-1}$? Does your answer make sense when compared with the data on page 379?

### Solution 14.4

The problem can be solved using Eq. (14.26).

$$[H_2O_2]_t = [H_2O_2]_0 \; e^{-k_1t} = 0.272 \; e^{(-0.047s^{-1} \times 7.00 \, s)} = 0.196 \; M$$

When comparing the data with those on page 379, note that the value for the concentration at 7.00 seconds is between the values at t = 5 and t = 10 seconds. Our answer therefore makes sense.

### Example 14.5

The radioactive nucleus gold-198 ($^{198}$Au) is used in diagnostic medicine to determine how well the kidneys are functioning. It decays via first-order kinetics with a rate constant $k_1$ = 0.258 d$^{-1}$ (d = days). If 3.00 mg of gold-198 are injected into the body, how much will remain after 2.69 days?

### Solution 14.5

We are given the initial amount of gold-198, the rate constant, and the time. We can solve by using Eq. (14.26).

$$A_t = A_0 \; e^{-k_1t} = 3.00 \; e^{(-0.258d^{-1} \times 2.69 \, d)} = 1.50 \; mg \; remaining$$

## 14.3.2   The Half-Life of a First-Order Reaction

The answer we determined for the amount of gold-198 remaining in Example 14.5 was interesting in that it was just half of the amount that we started with. The time that it takes for half of a radioactive isotope to undergo decay is called the *half-life* of the compound. The half-life of gold-198 is 2.69 days. Chemists sometimes talk about the decomposition of chemical compounds using half-lives. This is often used when considering the breakdown of chlorofluorocarbons (CFCs) in the stratosphere, where such compounds can have half-lives of more than 100 years.

We can relate the rate constant for a first-order reaction to the half-life ($t_{1/2}$) by using Eq. (14.26) and recognizing that at the half-life, the concentration (or amount) of the reactant remaining is, by definition, exactly ½ the original amount, or $A_t$ = ½ $A_0$. Therefore,

$$\tfrac{1}{2} A_0 = A_0 \; e^{-k_1t} \tag{14.27}$$

where $t_{1/2}$ is the half-life. Dividing through by $A_0$ gives the following result.

## SCIENTIFIC JOURNALS

There are many journals that publish articles important to chemistry professionals. Many are very specialized and publish the results of sophisticated chemical research performed by university professors and their research groups. Others focus on activities of interest to a broader constituency. In this latter group are three published by the American Chemical Society: *Chemical and Engineering News (C&E News)*, *Today's Chemist at Work*, and *Chemtech*. Articles relating to catalysts and kinetics appear regularly in these journals. For example, in the September 20, 1999, issue of *C&E News*, an article entitled "Catalyst Makers Look for Growth" appeared as the cover story. This article focused on modern problems facing catalyst producers and users.

*For Homework: Look through past issues of the journals mentioned above and find other articles relating to kinetics or catalysts. Select one such article and write a report on the major points made. You may also use other references if they relate to the chosen article.*

$$\tfrac{1}{2} = e^{-k_1 t_{1/2}} \tag{14.28}$$

Inverting the $\tfrac{1}{2}$ and $e^{-k_1 t_{1/2}}$ gives us "2" on the left side and changes the exponential sign on the right side.

$$2 = e^{-k_1 t_{1/2}} \tag{14.29}$$

Taking the natural log of both sides gives the following result.

$$\ln(2) = k_1 t_{1/2} \tag{14.30}$$

Since $\ln(2) = 0.6931$, the relationship between $t_{1/2}$ and $k_1$ can be written as follows.

$$t_{1/2} = 0.6931/k_1 \tag{14.31}$$

For the gold-198 isotope, $t_{1/2} = 0.6931/0.258 \text{ d}^{-1} = 2.69$ days.

### 14.3.2.1  How Much of Our Sample is Left after "n" Half-Lives?

We can use the analogy of an apple pie to look more at half-lives. Let's say that a bunch of people eat $\tfrac{1}{2}$ of the pie in 5 minutes. They are getting filled up, so they can only eat half of what remains in the next 5 minutes. So after the second 5 minutes, only $\tfrac{1}{2}$ of the $\tfrac{1}{2}$, or $\tfrac{1}{4}$, of the original pie is left. By now, they are pretty stuffed. So after a third 5 minute period, they can only eat $\tfrac{1}{2}$ of the remaining $\tfrac{1}{4}$, or $\tfrac{1}{8}$. Thus,

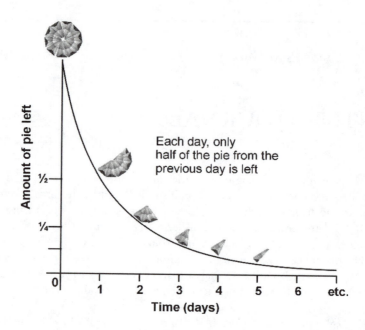

**FIGURE 14.6** The apple pie analogy to the half-life of a reactant. Each day you eat half of the *remaining* amount.

after three pie "half-lives" of 5 minutes (total time = 15 minutes), the pie that remains is

$$\frac{1}{2} \times \frac{1}{2} \times \frac{1}{2} = \left(\frac{1}{2}\right)^3 = \frac{1}{8}.$$

After 30 minutes, which is six half-lives, the amount of the pie remaining is

$$\frac{1}{2} \times \frac{1}{2} \times \frac{1}{2} \times \frac{1}{2} \times \frac{1}{2} \times \frac{1}{2} = \left(\frac{1}{2}\right)^6 = \frac{1}{64}.$$

In general, for "n" half-lives, the *fraction of an original sample remaining* is $(1/2)^n$, and the amount of a sample left is

$$A_n = A_o \left(\frac{1}{2}\right)^n \tag{14.32}$$

This is shown for the apple pie analogy in Fig. 14.6.

### Example 14.6
How much gold-198 is left from our 3.00 mg sample after 8.07 days?

### Solution 14.6
There are two ways to approach this problem and both will give us the correct answer.

1. *Via half-lives.* Each half-life is 2.69 days. A total of 8.07 days is three half-lives ($3 \times 2.69 = 8.07$). Each half-life represents a loss of half of what was left before. Three half-lives means that we

have $\frac{1}{2} \times \frac{1}{2} \times \frac{1}{2} = \frac{1}{8}$ of our sample left.

$$3.00 \text{ mg} \times \frac{1}{8} = 0.375 \text{ mg gold-198 remaining}$$

2. *Via the rate constant.* We know that $k_1 = 0.258 \text{ d}^{-1}$. We know that $A_o = 3.00$ mg and $t = 8.07$ days.

$$A_t = A_o e^{-k_1 t_{1/2}} = 3.0 \text{ e}^{-(0.258 \times 8.07)} = 3.00 \times 0.1247$$

$$= 0.374 \text{ mg gold-198 remaining}$$

The answers are slightly different because the half-life is actually 2.6866 days, which we rounded to 2.69. So 8.07 is actually a tiny bit longer than three half-lives.

## Example 14.7

Plutonium-239 is produced in breeder reactors. This nuclide is also used as a fuel source for interplanetary craft. $^{239}$Pu has a half-life of about $2.44 \times 10^4$ y (years). How many years must pass before 90% of a sample of plutonium has decayed?

## Solution 14.7

We need to find the time at which only 10% of the original sample remains. In this case, $^{239}$Pu$_t$ = 0.10 $^{239}$Pu$_o$. Using Eq. (14.26) we obtain the following result.

$$0.10 \ ^{239}\text{Pu}_o = \ ^{239}\text{Pu}_o \, e^{-k_1 t}$$

Dividing through by $^{239}$Pu$_o$ and inverting as in Eq. 14.12 then gives us

$$10 = e^{-k_1 t}$$

and if we take the natural log of both sides, the following equation is produced.

$$\ln (10) = 2.3026 = k_1 t \qquad (14.33)$$

We can now find $k_1$ by using Eq. (14.31).

$$t_{1/2} = 0.6931/k_1$$

$$k_1 = 0.6931/t_{1/2} = 0.6931 / 2.44 \times 10^4 \text{ y} = 2.84 \times 10^{-5} \text{y}^{-1}$$

Substituting into Eq. (14.7) gives us the following answer.

$$2.3026 = k_1 t$$

$$2.3026 = 2.84 \times 10^{-5} \text{ y}^{-1}(t)$$

$$t = 8.1 \times 10^4 \text{ y}$$

This is a little more than three half-lives and shows why plutonium is of concern as a long-term environmental hazard. This also shows why plutonium is used as a long-term energy source for interplanetary missions.

## 14.4 Reaction Mechanisms

Rate studies of reactions not only give us insight into how fast reactions occur, they also help us to understand **reaction mechanisms**—pathways by which reactions proceed. Such studies are vital in the pharmaceutical industry, where the specific mode of interaction of a compound with the body helps in the design of effective analogs (compounds with similar chemical structure). They also help us better understand the relationship between chemicals synthesized for consumer and industrial use and the environment. One example is the interaction of ozone in the stratosphere with chlorine atoms from CFCs, such as $CCl_2F_2$. These Cl atoms, separated from the CFCs as a result of the energy provided by the Sun's UV radiation, react with ozone and oxygen atoms as shown below.

$$Cl + O_3 \rightarrow ClO + O_2 \tag{14.34}$$

$$ClO + O \rightarrow Cl + O_2 \tag{14.35}$$

$$\overline{O_3 + O \rightarrow 2\,O_2} \tag{14.36}$$

The chlorine atoms, present when the reaction begins, greatly speed up the rate of reaction. Even though they change form during the reaction, they return to their original form as chlorine atoms. Chlorine is therefore a **catalyst** in this process. One chlorine atom can catalyze the decomposition of many ozone molecules. This is one explanation for the recent detection of sizable ozone holes in the stratosphere. The compound ClO is *not* present at the beginning or end of the reaction. It is *produced and then destroyed during* the reaction. It is known as a **reaction intermediate**.

Recent research[2] suggests that iodine, released naturally into the atmosphere from the world's oceans, may interact with chlorine to enhance the destruction of the ozone layer via the following mechanism.

$$Cl + O_3 \rightarrow ClO + O_2 \tag{14.37}$$

$$I + O_3 \rightarrow IO + O_2 \tag{14.38}$$

$$\overline{ClO + IO \rightarrow Cl + I + O_2} \tag{14.39}$$

$$2O_3 \rightarrow 3O_2 \tag{14.40}$$

Thus, kinetic studies help us to understand the details of a process that are not evident when looking just at the summary reaction.

### 14.4.1 The Rate-Determining Step

Remember the last time you were on a single-lane road traveling at 60 miles per hour, only to get behind a truck moving at 25 miles per hour? If it was not safe to pass, then you just had to sit back and enjoy the ride. Your original rate did not mean much. The rate of travel of the truck was the determining factor in the travel of both you and the truck. This analogy applies to chemical reactions in which there are several processes, one of which is often the slowest. Like the truck, the **rate-determining step** determines how fast the reaction will go. In that sense, the rate expression for the reaction can give us the rate-determining step. The other so-called "elementary steps" in the reaction can often, but not always, be surmised from this.

---

[2]Zurer, P., *Chem. Eng. News*, 8–9, November 14, 1994.

For example, the reaction of hypochlorite ion, $ClO^-$, to form chlorate ion, $ClO_3^-$, and chloride ion is summarized by the following equation.

$$3ClO^- \rightarrow ClO_3^- + 2Cl^- \qquad (14.41)$$

The rate law for the reaction has been determined to be

$$\text{rate} = \frac{-\Delta[ClO^-]}{\Delta t} = k[ClO^-]^2 = k[ClO^-][ClO^-] \qquad (14.42)$$

Therefore, the slow, or rate-determining step likely involves the interaction of two $ClO^-$ atoms. Based on this, the likely mechanism is as shown below.

$$ClO^- + ClO^- \rightarrow ClO_2^- + Cl^- \quad \text{(slow)} \qquad (14.43)$$

$$ClO^- + ClO_2^- \rightarrow ClO_3^- + Cl^- \quad \text{(fast)} \qquad (14.44)$$

$$\overline{3ClO^- \rightarrow ClO_3^- + 2Cl^-} \qquad (14.45)$$

## Example 14.8

Nitrogen dioxide, $NO_2$, can react with carbon monoxide, CO, as follows.

$$NO_2 + CO \rightarrow CO_2 + NO$$

The rate equation is

$$\text{rate} = k[NO_2]^2$$

Is the following mechanism consistent with this rate equation? Also, identify any catalyst or intermediate in this reaction.

$$NO_2 + NO_2 \rightarrow NO_3 + NO \quad \text{(slow)}$$

$$NO_3 + CO \rightarrow NO_2 + CO_2 \quad \text{(fast)}$$

## Solution 14.8

The mechanism is consistent with the rate law, because the rate-determining step (the first elementary reaction, in this case) is the slow step, and this involves the collision of two $NO_2$ molecules. Also, $NO_3$ is an intermediate in this reaction because it is neither present at the beginning nor at the end of the reaction.

## 14.4.2 The Activation Energy

We have talked often about the impact of a catalyst on a chemical reaction. A catalyst sharply increases the rate of the reaction because it lowers the energy barrier that the reaction must hurdle. Altering the reaction mechanism is the way the catalyst works. Therefore, to understand the role of a catalyst, we ought to examine the energy changes that take place during a chemical reaction.

Let's first look at a general reaction,

$$A + B \rightarrow C + D$$

As discussed in Section 14.2.2, there are a number of factors that are necessary for collisions between species to result in a chemical reaction. One of these is that the substances contain sufficient energy to react. In particular, when A+B combines with C+D, they form an ***activated complex***, also known as a ***transition state***, which is an energetically unstable state where the reactants and products are in transition. They have neither the stability of the reactants nor the products. Yet it is an energy barrier that must be overcome if the reaction is to occur. The barrier exists whether or not a reaction is spontaneous. The energy of the activated complex is called the ***activation energy, $E_a$***. Figures 14.7(a) and 14.7(b) show the energy profile for endothermic and exothermic reactions.

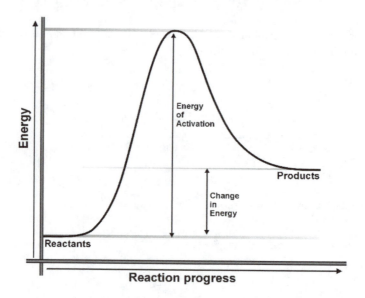

**FIGURE 14.7(a)**   The energy profile of an endothermic reaction. Note the key quantities: the energy of activation; the change in energy; and the relative energies of the reactants and products.

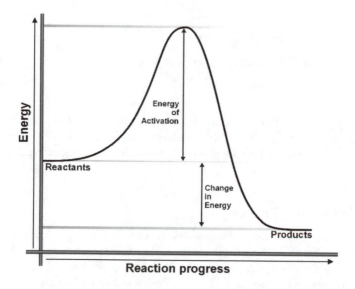

**FIGURE 14.7(b)**   The energy profile of an exothermic reaction. Note the relatively lower activation energy and the nature of the change in energy from reactants to products.

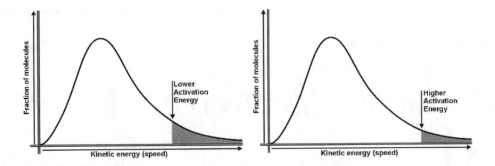

**FIGURE 14.8** Molecules have a wide range of energies that change with temperature.

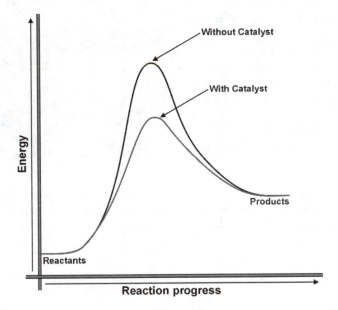

**FIGURE 14.9** A catalyst provides an alternate reaction mechanism that lowers the activation energy barrier for a chemical process.

What energy is sufficient to permit reaction? Every uncatalyzed reaction has its own activation energy. Reactant molecules have a wide range of speeds, and because the kinetic energy = $\frac{1}{2}$ mass $\times$ velocity, they will also have a wide range of energies, as shown in Fig. 14.8. Some of them have enough energy to overcome the activation energy barrier. So if the goal is to have a higher proportion of the reactants with enough energy to overcome $E_a$, we can do two things.

1. *Raise the temperature.* This will increase the speed of the reactants, increasing their average energy. More will contain sufficient energy to overcome the activation energy barrier.
2. *Use a catalyst.* The catalyst changes the reaction pathway, thus *lowering the activation energy barrier,* as shown in Fig. 14.9.

## 14.5 Where Have We Been?

We have learned that thermodynamics is concerned primarily with the energy exchanges that occur in chemical reactions and the conditions needed for spontaneity in the reactions. Our understanding of thermodynamics helps us to predict whether a reaction can occur. On the other hand,

*Chemistry Professionals at Work*                                 *CPW Box 14.5*

# COMMUNICATION SKILLS ARE VITAL

My name is Vena Adams. I am a chemistry technician employed by The Dow Chemical Company at their Texas Operations site in Freeport, Texas. I started as a pharmacy technician nineteen years ago. Communication skills are vital in my present work as a research technologist. I do product documentation and support for new product implementation. My career has included equipment automation, customer support, resin synthesis for TS&D, and pilot plant teamwork to develop innovative bleach feed technology.

*For Homework: Think of a new chemistry-based product that would tremendously improve some aspect of your life. Now, imagine that you, as an employee of a progressive company, were the one that developed this product for the company. Write a 2–3 page essay that would serve as the official company documentation for this product.*

kinetics is concerned with how fast and by what mechanism reactions occur. Thermodynamics and kinetics are two of the central topics in all of chemistry and, as we have seen, are vital to the chemical industry.

In this chapter, we have learned that:

1. The rate of chemical reactions can be used to characterize them.
2. This characterization takes the form of a rate equation (expression) that expresses how the rate of reaction changes with concentration.
3. This characterization also takes the form of the rate constant, which does not change at a given temperature unless a catalyst is added to the system.
4. We can use concentration vs. time data to test for the order of a reaction and to determine the rate constant.
5. We can use the integrated rate expression to determine the concentration of a reactant at any time.
6. The rate expression of a chemical reaction is a guide to the rate-determining step, by which we can often understand the elementary steps in a reaction.
7. Every reaction has an activation energy barrier that must be overcome in order for the reaction to proceed.
8. A catalyst changes the reaction mechanism so as to lower the activation energy barrier.

## 14.6  Homework Exercises

1. What can the reaction rate help us determine? Why are these things important?

2. At 20°C, $H_2O_2(aq)$ will decompose according to the following reaction.

$$2H_2O_2(aq) \rightarrow 2H_2O(l) + O_2(g)$$

The following data were collected for the concentration of $H_2O_2$ at various times.

| Time(s) | $[H_2O_2](M)$ |
|---------|---------------|
| 0 | 1.000 |
| $7.10 \times 10^4$ | 0.500 |
| $1.42 \times 10^5$ | 0.250 |

   (a) Calculate the average rate of decomposition for $H_2O_2$ between 0 and $7.10 \times 10_4$ seconds. Use this rate to calculate the rate of production of $O_2$.
   (b) What is the rate of decomposition of $H_2O_2$ from $7.10 \times 10^4$ s to $1.42 \times 10^5$ s?

3. Discuss what would happen to the rate of reaction if the hydrogen peroxide reaction in the previous problem was carried out at 60°C instead of 20°C.

4. The decomposition reaction in Question 2 is first-order. Calculate the half-life of hydrogen peroxide and the rate constant by using the given concentration data.

5. Using your answers from Questions 2 and 4, determine the concentration of hydrogen peroxide that would be present after 1 week.

6. Given the following hypothetical equation and data, find the coefficients q and r.:

$$qA \rightarrow rB + 3C$$

$$-\Delta[A]/t = 0.312 \text{ mol/L s}$$
$$+\Delta[B]/t = 0.052 \text{ mol/L s}$$
$$+\Delta[C]/t = 0.156 \text{ mol/L s}$$

7. Using the information in the following table, find the average rate of production of $O_2$ between times of 0 and 30.0 minutes.

| Time,( minutes) | $[O_2]$, M |
|-----------------|-----------|
| 0.0 | 0.50 |
| 10.0 | 0.66 |
| 20.0 | 0.78 |
| 30.0 | 0.87 |

8. Calculate the rate constant of a reaction where rate $= k[SO_3]^2$. The initial rate was $1.5 \times 10^{-5}$ M/min, and $[SO_3]_{initial} = 0.150$ M.

9. For the reaction,

$$2NH_3(g) \rightarrow N_2(g) + 3H_2(g)$$

the average rate of decomposition of $NH_3(\Delta[NH_3]/\Delta t)$ was $1.63 \times 10^{-4}$ M/min. Find the average rate of production of both nitrogen and hydrogen.

10. Consider the following reaction.

$$Ba^{2+} + CrO_4^{2-} \rightarrow BaCrO_4$$

(a) What would be the rate of consumption of the barium ion if $[Ba^{2+}]_o = 0.750$ M and its concentration after 5 seconds is 0.500 M?

(b) What would be the rate of production of $BaCrO_4$ if there were none present initially, and 0.250 moles were formed after 10 seconds?

11. The following data were obtained from the consumption of a compound, "A."

| Time (minutes) | [A], M |
|---|---|
| 0.0 | 0.500 M |
| 1.0 | 0.450 M |
| 2.0 | 0.405 M |
| 3.0 | 0.364 M |
| 4.0 | 0.328 M |

(a) What is the order of this reaction?

(b) What is the rate constant for this reaction?

12. Look at the following data to answer the given questions.

| Time (seconds) | [A], M |
|---|---|
| 0.0 | 0.150 M |
| 1.0 | 0.125 M |
| 2.0 | 0.107 M |
| 3.0 | 0.094 M |
| 4.0 | 0.084 M |

(a) What is the order of the reaction?

(b) What is the rate constant for this reaction?

13. Look at the data for the following reaction.

$$4PH_3(g) \rightarrow P_4(g) + 6H_2(g)$$

| $[PH_3]$ | Rate (M s$^{-1}$) |
|---|---|
| 0.06 | $1.2 \times 10^{-3}$ |
| 0.18 | $3.6 \times 10^{-3}$ |
| 0.12 | $2.4 \times 10^{-3}$ |

(a) Write the rate expression for this reaction.

(b) Calculate the rate constant and half-life for the decomposition of $PH_3$.

14. The following data were obtained from three different decomposition reactions of a compound AB.

| Reaction # | [AB] | Rate (M s$^{-1}$) |
|---|---|---|
| 1 | 0.2 | $1.0 \times 10^{-3}$ |
| 2 | 0.4 | $4.0 \times 10^{-3}$ |
| 3 | 0.6 | $9.0 \times 10^{-3}$ |

(a) What is the order of this reaction?

(b) Calculate the rate constant, k.

(c) If $[AB]_o = 0.500$ M, what would the concentration of AB be after 1 minute?

15. The gas IBr decomposes according to the following reaction:

$$IBr \ (g) \rightarrow I_2(g) + Br_2(g)$$

A plot of 1/[Br] vs. time is a straight line. What is the rate law for this reaction?

16. What would the rate-determining step be in the following process?

(a) Cream butter and sugar for 5 minutes.

(b) Add eggs and vanilla and chocolate chips and stir.

(c) Roll dough into balls.

(d) Bake at 375° for 20 minutes.

(e) Let cookies cool for 5 minutes before eating.

17. Using the following reaction,

$$2NO + Cl_2 \rightarrow 2NOCl$$

(a) Express the rate of reaction in terms of the formation of NOCl.

(b) Express the rate of disappearance of $Cl_2$ in terms of the formation of NOCl.

(c) If NOCl is formed at a rate of 4.5 M $min^{-1}$, how fast is the NO consumed?

18. The rate of decomposition of a substance is first order. Initially, its concentration is 0.500 M. If k = $3.5 \times 10^{-4}$ $s^{-1}$, what concentration will remain after 450 seconds of decomposition?

19. The rate at which a particular solid reacts is found to be first-order. If the initial concentration of the substance is 0.400 M, and 0.135 M remains after 2 days, what is the rate constant?

20. A particular pharmacy approves the sale of its medicines if less than 5.0% of the substances in the bottles have undergone chemical change. A particular antibiotic, which undergoes a reaction by first-order kinetics, has a rate constant of $1.4 \times 10^{-2}$ $days^{-1}$. What is the maximum number of days it can sit on the shelf before it is not approved for sale?

21. If an antibiotic that undergoes first-order decay with k = $1.4 \times 10^{-2}$ $days^{-1}$ is left on the shelf for 6 months, what percentage of the original drug will have undergone a chemical change? Assume 30 days in each month.

22. A chemical is left on the shelf of a chemistry classroom for five years and undergoes chemical change via first-order kinetics. If $[chemical]_o = 6.00$ M and its concentration at the end of the period is 4.50 M, what is the rate constant for this chemical change?

23. The formation of POCl (g) is second-order with respect to PO(g). If $[PO]_o = 0.500$ M and k = 0.0368 $M^{-1}$ $hr^{-1}$, how much time will it take for [PO] to reach 0.100 M?

24. For the same reaction and initial conditions as in Question 23, what would [PO] be after 2.00 days?

25. If a substance has a rate constant k = $5.0 \times 10^2$ $years^{-1}$ and decays by first-order kinetics, how long is its half-life?

26. The following reaction carried out in the liquid phase, follows second-order kinetics.

$$2ClO^- \rightarrow 2\ Cl^- + O_2$$

(a) If k = $2.66 \times 10^{-4}$ $M^{-1}$ $min^{-1}$ and the initial concentration of $ClO^-$ is 1.0 M, what is the half-life of $ClO^-$?

(b) How long will it take for 90.0% of the $ClO^-$ ion to decompose?

27. If k = $1.68 \times 10^{-6}$ $s^{-1}$ and $[NH_3]_o = 0.650$ M, and the $NH_3$ reacts via first-order kinetics, what will the concentration of ammonia be after 2.0 weeks?

28. A substance undergoes decomposition via first-order kinetics. Its rate constant is $5.98 \times 10^{-4}$ $s^{-1}$. What is its half-life?

29. The mechanism proposed for the decomposition of $N_2O_5$ (g) is as follows:

$N_2O_5(g) \rightleftharpoons NO_2(g) + NO_3(g)$

$NO_2(g) + NO_3(g) \rightarrow NO(g) + NO_2(g) + \frac{1}{2}O_2(g)$

$NO(g) + NO_3(g) \rightarrow 2NO_2(g)$

Identify the intermediates in this mechanism.

30. The initial concentration of a reactant is 0.350 M. It decomposes via first-order kinetics. After 5 hours, the concentration is 0.175 M. What will the concentration of the reactant be after 24 hours have passed?

31. Iodine combines to form molecular iodine by the reaction:

$$2I(g) \rightarrow I_2(g)$$

where Rate = $k[I]^2$ and k = $7.0 \times 10^9$.
(a) What should the units for k be?
(b) If $[I]_o$ = 0.25 M, how much molecular iodine will be produced after $4.5 \times 10^{-10}$ seconds?
(c) What is the half-life of the iodine in this reaction if $[I]_o$ = 0.25 M?
(d) How much time will it take for 95% of the molecular iodine to react?

32. Consider the following reaction.

$$2NOBr(g) \rightarrow 2NO(g) + Br_2(g)$$

$$Rate = k[NOBr]^2, \quad where \ k = 0.80 \ L/mol \ s$$

(a) Determine $t_{1/2}$ when $[NOBr]_o$ = 0.300 M.
(b) Calculate [NOBr] at t = 0.45 s if $[NOBr]_o$ = 0.300 M.
(c) If $[NOBr]_o$ = 0.300 M, how long will it take for 99.9% of the NOBr to react?

33. The following decomposition reaction occurs via second-order kinetics with k = $5.0 \times 10^5$ seconds and $[C_4H_6]_o$ = 1.0 M .

$$2C_4H_6 \rightarrow C_8H_{12}$$

(a) How long will it take for 80% of the original amount of $C_4H_6$ to decay?
(b) How long is the half-life of $C_4H_6$?
(c) How long will it take for the concentration to reach 0.15 M?

34. A radioactive isotope, carbon-14, has a half-life of 5730. If a cloth had 0.0040 moles of carbon-14 when it was first woven and contained 0.0030 moles in 2000, in what year was it woven?

35. Radon decays to polonium according to the following equation.

$$^{222}Rn \rightarrow {}^{218}Po + {}^4He$$

The first-order rate constant for this decay is 0.181 days$^{-1}$. If you begin with a 4.16 g sample of pure $^{222}Rn$, how much will be left after 1.96 days? After 3.82 days?

36. Initially there are 50.0 g of $^{111}In$, which has a half-life of 67 hours. How much is left after 6.0 days?

37. The radioactive isotope $^{15}O$ is used in medical diagnosis to help determine how well the lungs are functioning. Its half-life is 122 seconds. How long will it take for 90.0% of a sample of $^{15}O$ to decay in the lungs?

38. Discuss several reasons why $^{15}O$ is a better choice for use in lung function tests than $^{239}Pu$.

39. Why do catalysts lower the activation energy in chemical reactions? Why does that lead to reactions occurring more quickly than without catalysts? What are **enzymes**, and how do they relate to catalysts? Search the chemical literature to find examples of enzymes. Do they speed up reactions as much as industrially prepared catalysts?

40. What could be done to increase the rate of a reaction in terms of temperature and concentration? Explain why this would be effective on a molecular level.

## For Class Discussion and Reports

41. An industrial chemist monitors the course of a one-reactant reaction of "A" going to products, and plots the concentration vs. time data. He or she plots the data in an effort to determine whether the reaction is first- or second-order. Neither the ln[A] vs. time nor a plot of 1/[A] vs. time gives a straight line. What could be the possible explanations for these results? Discuss as a class.

42. In this chapter we said that the catalysts used in catalytic converters in automobiles are exceptions to the idea that concentration is related to rate. Find out what metals are used in catalytic converters and why they are chosen. The reactions that occur on a catalytic converter are considered to be "zeroth-order" in relation to the amount of catalyst. What does this mean? Write a report on the design and construction of a catalytic converter.

# 15

# Nuclear Chemistry

## 15.1   Introduction

In just over two generations, or about 60 years, the term "nuclear energy" has gone from the realm of science fiction to become an indelible part of our daily vocabulary. The term conjures up different images to each of us depending upon where we live, what our opinions are, and what we do for a living. An important use of nuclear energy is in the generation of electricity. There are about 340 operating nuclear power plants throughout the world, and 113 plants currently operating in the United States. Three countries, France, Belgium, and Sweden, generate the majority of their electricity from nuclear reactors, while the U.S. meets about 22% of its electricity needs in this way. The energy derived from nuclear processes also has important uses in the chemical and medical fields, especially in diagnostic medicine in assessing how well our organs are working.

In Chapter 13 we discussed the energy exchange that occurs with nuclear changes ($\Delta E = \Delta mc^2$). This energy is given off as extraordinarily high-energy electromagnetic waves known as **radiation** or **radioactivity**. In Chapter 14, we looked at how the amount of an unstable isotope changes with time (first-order kinetics). In this chapter, we will examine the specific changes that can occur in nuclei, why they occur, and the uses to which these changes can be put.

## 15.2   The Nature of the Nucleus

### 15.2.1   Review of the Nucleus

Recall from Chapter 1 that every nucleus except for hydrogen is made up of protons and neutrons (the hydrogen nucleus has no neutrons). A more detailed model[1] suggests that protons and neutrons are, themselves, made up of "subatomic" particles known as **quarks**. Three so-called "up" and "down" quarks

---

[1]Quinn, Helen R. and Witherell, Michael S., *Scientific American*, 76–81, October 1998.

Chemistry Professionals at Work CPW Box 15.1

# WHY STUDY THIS TOPIC?

Chemical reactions involving the atomic nucleus are of great concern to us as citizens as well as to chemistry professionals in the workplace. As citizens in modern society, we are concerned about nuclear warfare. We are concerned about nuclear energy. We are concerned about radioactivity from radon in our homes, or from nuclear waste. Some of us are concerned about medical treatments that involve radioactive isotopes. Chemists and chemistry technicians may encounter radioactive isotopes in their laboratory work. For example, chemistry professionals are employed by nuclear power plants where they take radioactive samples from the reactor coolant systems or control the nuclear processes that go on in the plant. They may also be employed by government laboratories at which new and better nuclear fuel is being researched, or in quality assurance laboratories in which they must analyze the quality of radioactive materials based on the intensity of dangerous radioactivity that they possess. Or they may be researching the latest advances of radioactive materials for medical applications. In some instances, radioactive materials may be a part of the equipment used for routine work in the laboratory (see CPW Box 15.2). The study of nuclear processes and radioactivity can thus be very important as we progress in our study of chemistry.

*For Class Discussion: As the semester progresses, be on the lookout for articles in newspapers and news magazines in which radioactivity or nuclear chemistry is an issue. Report to the class anything that you find giving details that you feel are of great interest to us as citizens or to the chemist.*

combine to make a proton or a neutron. This "Standard Model" is too complex to deal with in this text, so we will stick with atomic particles, which serve our needs nicely. The number of protons in an atom is known as the *atomic number* and the sum of the number of protons and neutrons is equal to the *atomic mass number*, or just *mass number*. The nucleus of an iron atom has 26 protons, so this is its atomic number. If the nucleus has 30 neutrons, then the atomic mass number of the iron atom is 26 + 30 = 56. We call this "iron-56" and it is designated by the chemist's shorthand shown below, with the atomic number being given in the lower left-hand corner and the atomic mass in the upper left-hand corner.

$$_{26}^{56}\text{Fe}$$

Another, less stable type of iron nucleus contains 33 neutrons and can be written "iron-59", which is symbolized as follows.

$$_{26}^{59}\text{Fe}$$

## Example 15.1

Fill in the blanks with the missing information.

| Symbol | Protons | Neutrons | Mass Number |
|---|---|---|---|
| $^{57}_{25}\text{Mn}$ | — | — | — |
| $^{182}\_\_$ | 79 | — | — |
| $^{\_}_{\_}\_\_$ | 84 | 126 | — |

## Solution 15.1

Remember that the number of protons plus the number of neutrons equals the mass number. This means that the mass number should always be larger than either the number of protons or neutrons (and equal to their sum).

| Symbol | Protons | Neutrons | Mass Number |
|---|---|---|---|
| $^{57}_{25}\text{Mn}$ | 25 | 32 | 57 |
| $^{182}_{79}\text{Au}$ | 79 | 103 | 182 |
| $^{210}_{84}\text{Po}$ | 84 | 126 | 210 |

Forms of an element that contain different numbers of neutrons are known as *isotopes* of the element. Hydrogen, for example, has three isotopes: $^1\text{H}$; $^2\text{H}$; and $^3\text{H}$, which are called hydrogen, deuterium, and tritium, respectively. Every element in the periodic table has several isotopes (we will often call each isotope of an element a *nuclide*). Many of the isotopes of the 92 naturally occurring elements[2] are energetically unstable because they have too many protons for the number of neutrons, or too many neutrons for the number of protons in their nuclei. The nuclei of all the elements with atomic numbers greater than 83 (bismuth) are unstable. So *unstable nuclei* will modify themselves in such a way as to acquire a more energetically stable ratio of neutrons to protons. There is no specific ratio of neutrons to protons that is best for all elements. Rather, there is a *zone (or "band") of stability* within which stable nuclides are present (see Fig. 15.1). Unstable nuclides will change themselves in such a way as to get within the band of stability.

## 15.2.2 Nuclear Decay Processes

When unstable nuclei change to become more energetically stable, they release energy. This is the basis for the use of certain nuclides in nuclear power plants. That is, these nuclides release particles and energy as they become more stable. This energy is what powers the plant to produce electricity. We will focus on this in Section 15.5. What particles are released (or sometimes gained) as isotopes change identity? That depends on the isotope. Here are seven different outcomes in the nuclear pathway to stability.

**1. Alpha (α) particle production** results in the release from the nucleus of a particle that has 2 protons and 2 neutrons, just like a helium nucleus. An alpha particle is symbolized as $^4_2\text{He}$. Energy is also released as the nucleus becomes more stable. Since the number of protons changes (as in the loss of 2 protons in plutonium), and the identity of the element is based strictly on the number of protons, losing an alpha particle has changed the element! For example, we can convert radon to polonium with a loss of two

---

[2]All of the first 93 elements, with the exception of #43, technetium (used in medical diagnosis) occur naturally. Only trace amounts of some, such as plutonium and francium, have ever been detected.

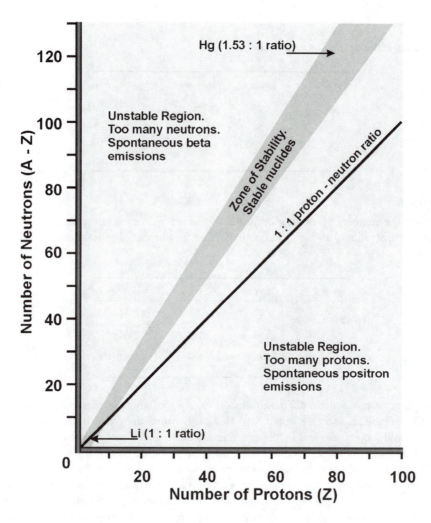

**FIGURE 15.1**   The zone, or band, of stability represents the ratio of neutrons to protons for stable nuclides. Nuclei will decay in such a way as to get within the zone. Note how the ratio of neutrons to protons increases with increasing atomic number.

protons. These types of conversions are a regular part of nature and have been occurring nearly since the beginning of the universe.

$$^{222}_{86}\text{Rn} \rightarrow {}^{4}_{2}\text{He} + {}^{218}_{84}\text{Po} \tag{15.1}$$

In many parts of the U.S., homes must be tested for radon concentration before they can be sold. This is because the radon-222 nuclide is often found in low concentrations in the basements of homes in certain geographic areas. Even though radon is chemically inert (i.e., it does not exchange electrons readily with other elements), the nucleus decays. So when radon-222 gas is inhaled in a house, it decays by loss of an alpha particle to form polonium-218 [see Eq. (15.1)] and then decays again to form bismuth-214, both of which are solids which can adhere to the lungs. There is some concern that even at low concentrations, exposure to these substances may lead to lung cancer. Alpha particles are relatively slow moving, and are not especially dangerous outside the body. But when they are present inside the body, such as when radon gas is inhaled and decays, they can be harmful.

**2. Beta (β⁻) particle production** results in the release of an electron from the nucleus. A beta particle is symbolized as ${}^{0}_{-1}\text{e}$, as shown in Table 15.1. Electrons do not reside in the nucleus. But in order to become

---

*Chemistry Professionals at Work*                    *CPW Box 15.2*

# A RADIATION DETECTOR

Chromatography is one of the most important instrumental techniques available in the chemical laboratory (see also CPW Boxes 6.8, 9.7, and 10.2) . The wide variety of chromatographic instruments have in common their ability to separate and electronically detect many compounds and ions in a mixture. And one of the standard and highly reliable electronic detectors for gas chromatography (CPW Box 9.7) is based on nuclear decay. The detector, called an "ECD" for "Electron Capture Detector" is based on the current that exists as a result of the emission of β-particles from a radioactive source, such as $^{63}Ni$. When a substance that can absorb (or "capture") some of the electrons passes through the ECD, the current flow is reduced by a certain amount depending upon the nature and amount of the substance. So ECDs are especially good at detecting compounds that capture electrons readily, such as chlorinated hydrocarbon pesticides. These compounds can be detected at the parts per billion level using ECDs. Other compounds of environmental concern are "polyaromatic hydrocarbons," or PAHs (see Section 6.6.4 for a discussion of aromatic compounds), which can also be readily monitored using this detector.

*For Homework: Find a book in your science library, or from your instructor, that discusses gas chromatography. Look in the section on detectors and find information on the advantages and disadvantages of the various types that exist, including the ECD. Write a report.*

---

**TABLE 15.1**    Some Nuclides Used in Diagnostic Medicine

| Nuclide | Decay Mode | Application |
|---|---|---|
| $^{131}Ba$ | Electron capture | Bone tissue location |
| $^{18}F$ | Beta decay | Bone scanning |
| $^{197}Hg$ | Electron capture | Spleen function |
| $^{15}O$ | Positron emission | Lung function test |
| $^{131}I$ | Beta decay | Kidney cyst location |

more energetically stable (or lie within the band of stability), some nuclides will undergo beta particle emission where a *neutron is converted to a proton*. This results in an increase of one in the atomic number, with no change in the mass number. For example, osmium-194 becomes iridium-194 via ejection of a β⁻particle.

$$^{194}_{76}Os \rightarrow \, ^{194}_{77}Ir + \, ^{0}_{-1}e \qquad (15.2)$$

Note that the total mass number and number of protons on the left of the equation equals that on the right side of the equation. As we discussed in Chapter 13, some of the mass is converted to huge amounts of energy in nuclear changes. But the mass changes caused by conversion to energy are not large enough to affect the mass number. Beta particles are ejected from the nucleus at high speeds and are highly energetic. They are therefore, hazardous, especially if taken into the body.

**3. Emission of gamma (γ) rays** occurs as the excess energy that accompanies α or β emission. Gamma rays are highly energetic and can penetrate seemingly impervious material such as concrete walls and lead. Most health concerns related to radioactive materials are due to gamma rays that penetrate the body from external radioactive sources.

**4. Positron ($_1^0e$) emission,** is the release of a positron (a nearly mass-less particle with the opposite charge of an electron). The outcome of positron emission is the *conversion of a proton to a neutron.* One example of positron emission is the conversion of carbon-11 to boron-11.

$$_6^{11}C \rightarrow {}_5^{11}B + {}_1^0e \tag{15.3}$$

Note how this puts the nuclides closer to the band of stability. One important medical application of this mode of nuclear decay is ***positron emission tomography*** in which carbon-11 or fluorine-18 atoms are substituted for atoms in, for example, glucose in the body. Emitted positrons strike nearby electrons, annihilate each other, and release gamma rays. The gamma rays can then be detected, allowing particular parts of the body to be scanned as the glucose travels within.

**5. Electron capture** occurs when the nucleus captures an inner orbital electron. The electron combines with a proton in the nucleus to form a neutron. So the effect is the same as with positron emission—*the conversion of a proton to a neutron.* It is merely a different mechanism. An example is the conversion of silver-100 to palladium-100.

$$_{47}^{100}Ag + {}_{-1}^0e \rightarrow {}_{46}^{100}Pd \tag{15.4}$$

**6. Spontaneous fission** is the decomposition of the nucleus into two smaller nuclei plus several neutrons. This occurs often with man-made elements such as nobelium-252, which are especially unstable.

**7. Induced fission,** fission in which a large nucleus blasts apart to form some smaller nuclei and neutrons, is vital to power generation in nuclear power plants (Section 15.5) and atomic bombs. But the fission is induced—that is, made to happen—by bombardment of the nucleus with a fast-moving particle such as a neutron. This makes the nucleus especially unstable, which leads to fission.

## Example 15.2

Write equations for each of the following processes.

a. $_{96}^{241}Cf$ undergoes electron capture.

b. $_{95}^{241}Am$ produces an α particle.

c. $_{54}^{121}Xe$ produces a positron.

d. $_{53}^{138}I$ produces a β particle.

## Solution 15.2

Remember that the ***sum*** of protons and neutrons, as well as the ***number*** of protons, must be ***consistent on both sides*** of the equation. The key is to calculate these values and then determine the elements that have the values associated with them.

a. $_{96}^{241}Cf + {}_{-1}^0e \rightarrow {}_{95}^{241}Am$

b. $_{95}^{241}Am \rightarrow {}_{93}^{237}Np + {}_2^4He$

c. $_{54}^{121}Xe \rightarrow {}_{53}^{121}I + {}_1^0e$

d. $_{53}^{138}I \rightarrow {}_{54}^{138}Xe + {}_{-1}^0e$

## Example 15.3
Fill in the missing particle in each of the following equations:

a. $^{164}_{67}\text{Ho} + ^{0}_{-1}\text{e} \rightarrow$ ?

b. $^{158}_{67}$? $\rightarrow ^{158}_{66}\text{Dy} + ^{0}_{1}\text{e}$

c. $^{242}_{94}\text{Pu} \rightarrow$ ? $+ ^{238}_{92}\text{U}$

## Solution 15.3

a. $^{164}_{67}\text{Ho} + ^{0}_{-1}\text{e} \rightarrow ^{164}_{66}\text{Dy}$

b. $^{158}_{67}\text{Ho} \rightarrow ^{158}_{66}\text{Dy} + ^{0}_{1}\text{e}$

c. $^{242}_{94}\text{Pu} \rightarrow ^{4}_{2}\text{He} + ^{238}_{92}\text{U}$

## 15.2.3 Decay Series

Although alpha or beta emission helps an unstable nuclide edge toward the band of stability in many cases, especially with heavier nuclei, one emission isn't enough. Transuranium elements (those with atomic numbers greater than uranium's 92) always have several decays before they find a stable nuclide. For example, $^{238}\text{U}$ decays via a series of $\alpha$ and $\beta$ decays, called a *decay series* (as shown in Fig. 15.2) until it reaches $^{206}\text{Pb}$, which is stable and undergoes no further decay.

## Example 15.4
What nuclide will $^{252}\text{Es}$ form if it has the following decay series?

$$^{252}\text{Es} \rightarrow \alpha, \beta^-, \alpha, \alpha, \alpha \rightarrow ?$$

## Solution 15.4

$$^{252}_{99}\text{Es} \rightarrow ^{4}_{2}\text{He} + ^{248}_{97}\text{Bk}$$

$$^{248}_{97}\text{Bk} \rightarrow ^{0}_{-1}\text{e} + ^{248}_{98}\text{Cf}$$

$$^{248}_{98}\text{Cf} \rightarrow ^{4}_{2}\text{He} + ^{244}_{96}\text{Cm}$$

$$^{244}_{96}\text{Cm} \rightarrow ^{4}_{2}\text{He} + ^{240}_{94}\text{Pu}$$

$$^{240}_{94}\text{Pu} \rightarrow ^{4}_{2}\text{He} + ^{236}_{92}\text{U}$$

The resulting $^{236}\text{U}$ nucleus will undergo the series of decays to form the stable $^{206}\text{Pb}$ nucleus, as shown in Fig. 15.2.

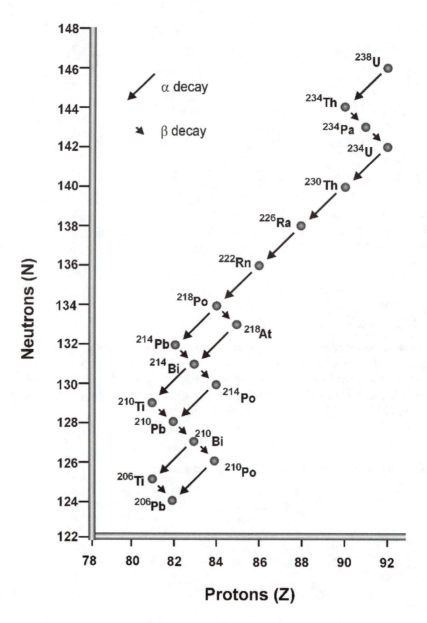

**FIGURE 15.2**  In this decay series, $^{238}U$ loses $\alpha$ and $\beta$ particles in its push for nuclear stability. The final, stable, nucleus is $^{206}Pb$.

## 15.3  Rates of Nuclear Decay

Radioactive nuclides are often used in diagnostic medicine because, as discussed with positron emission tomography, the gamma radiation given off by the nuclides can be monitored as the nuclides pass through the body. Additionally, such nuclides do not interfere with normal body processes and can leave the body efficiently, typically through sweating or urination. Table 15.1 lists some of these nuclides, their uses, their modes of decay, and their half-lives. Recall from Section 14.3.2 that the half-life ($t_{1/2}$) of an unstable nuclide is the amount of time it takes for exactly $1/2$ of the nuclide to decay. The half-lives of unstable nuclides vary from microseconds to $10^{15}$ years! In order for an unstable nuclide to be useful in medicine, the half-life must be long enough to follow the radiation being released, but short enough so that most of the sample does not release radioactivity for a long time. Note that most of the nuclides in Table 15.1 meet these criteria.

To review, the key point of Section 14.3.2 was that all energetically unstable nuclides decay via first-order kinetics. We then developed an equation for the amount of the starting nuclide, $A_t$, remaining after some time, t.

$$A_t = A_0 e^{-k_1 t} \tag{15.5}$$

Please review Section 14.3.2 and try the following example.

## *Example 15.5*

$^{125}I$ is used in diagnostic medicine to determine blood hormone levels. It is also used in radioactive "seeds" that are planted in cancerous prostate glands to kill the tumor and prevent recurrence of the cancer. Some physicians say that a patient who has had the $^{125}I$ seeds implanted must be careful to avoid close contact with children for 3 months after the contact. What fraction of the original $^{125}I$ sample is left after 93.0 days if the $t_{1/2}$ for the nuclide is 59.9 days?

## *Solution 15.5*

Before we explicitly solve the problem, let's determine what kind of answer makes sense. If the half-life of $^{125}I$ is about 60 days and the sample decays for 93 days, that is about 1.5 half-lives. Half of the sample will decay in 60 days, and half of the remainder $(\frac{1}{2} \times \frac{1}{2}) = \frac{1}{4}$ will be left over after two half-lives, or 120 days. So we would expect our final answer to indicate that between $\frac{1}{2}$ and $\frac{1}{4}$ of the original $^{125}I$ remains after 93.0 days.

*Via the half-life.* We discussed this type of calculation in Section 14.3.2, given here.

$$A_n = A_0(1/2)^n$$

The number of half-lives, $n = \dfrac{1 \text{ half-life}}{59.9 \text{ days}} \times 93.0 \text{ days} = 1.55$ half-lives

We are after the fraction left, which is the ratio of $A_0/A_t = (\frac{1}{2})^n$

$$A_0/A_t = (1/2)^{1.55} = 0.34$$

This answer is about what we expected, so it makes sense. We can also solve this problem by finding the first-order rate constant, $k_1$, as shown in Chapter 14. If you want to solve it this way, you should come up with the same answer as we have.

## 15.4   Review of the Energy Changes

In Section 13.3.3, we discussed the equality of mass and energy as defined by the equation

$$\Delta E = \Delta mc^2 \tag{15.6}$$

We solved a problem related to the $\alpha$ decay of $^{239}Pu$ to $^{235}U$ and we determined that the amount of energy given off in this *nuclear conversion* was many orders of magnitude larger than that given off by *chemical* reactions involving a comparable amount of any chemicals.

As a prelude to our discussion of nuclear power plants, let's calculate the energy changes associated with two examples of reactions that represent the present and future ways of meeting our personal and industrial energy requirements, fusion and fission. **Fusion** is the joining together of two nuclei into a larger, more energetically stable one. We discussed **fission** previously as the splitting apart of a nucleus

*Chemistry Professionals at Work*                                    *CPW Box 15.3*

# RESEARCHERS MAKE SUPERHEAVY ELEMENTS[1]

We said early in the chapter that there were 92 naturally occurring elements, yet the periodic table lists many more. How are these additional elements created and where are they now? The first man-made element was technetium, atomic number 43, one form of which is commonly used in medical imaging scans of the brain, lungs, and kidneys. This element was first made by Emilio Sergé in Berkeley, California, by bombarding molybdenum with deuterium nuclei. In the 1940s, and 1950s, researchers working at the Berkeley labs under the direction of Glenn T. Seaborg synthesized a host of elements with atomic numbers 93 through 100 by bombarding elements with neutrons, followed by beta decay (recall that this changes a neutron to a proton, thus increasing the atomic number). For example, $^{241}$Am is made by the following reaction.

$$^{239}_{94}\text{Pu} + 2\,^{1}_{0}\text{n} \rightarrow\, ^{241}_{94}\text{Pu} \rightarrow\, ^{241}_{95}\text{Am} + \,^{0}_{-1}\text{e} \tag{15.9}$$

For atomic numbers 100 and beyond, fusion of heavier nuclei proved successful. For instance, lead and argon nuclei fuse to make fermium (atomic number 100). For the elements with the highest atomic numbers, more massive nuclei have been fused, such as lead-208 and nickel-62 to give element 110.

$$^{208}_{82}\text{Pb} + \,^{62}_{28}\text{Ni} \rightarrow\, ^{270}_{110}\text{X} \tag{15.10}$$

These super-heavy elements are also super-unstable because the concentration of positive charge in the nucleus is so high, yielding a huge repulsive force that makes the super-heavy nuclei decay in micro- or milliseconds (with the exception of $^{269}$Hs, which has a $t_{1/2} = 9.3$ seconds). To date, elements up to atomic number 112 have been made, though they decay away almost immediately.

*For Homework: Research and write the reactions for the preparation of at least one isotope from each of the elements with atomic number 93 to 100. Write a report on anything else you discovered during your research, including any important uses of any of these elements.*

---

[1] Ambruster and Hessberger, *Scientific American*, 72–77, September 1998.

into smaller nuclei and neutrons. One example of an induced fission reaction is the splitting apart of the $^{235}$U nucleus by bombardment with neutrons.

$$^{235}_{92}\text{U} + \,^{1}_{0}\text{n} \rightarrow\, ^{87}_{35}\text{Br} + \,^{146}_{57}\text{La} + 3\,^{1}_{0}\text{n} \tag{15.7}$$

This reaction not only produces neutrons which can be used to split apart other $^{235}U$ atoms, but it also supplies energy.

## Example 15.6
How much energy is released by the induced fission of one mole of $^{235}U$ in Eq. (15.7)? Please review Section 13.3.3 for a review of this type of calculation.

$$^{235}_{92}U \quad + \quad ^{1}_{0}n \quad \rightarrow \quad ^{87}_{35}Br \quad + \quad ^{146}_{57}La \quad + \quad 3^{1}_{0}n$$

| mass(g/mol) | 235.0439 | 1.008665 | 86.920690 | 145.925530 | 3(1.008665) |
|---|---|---|---|---|---|

## Solution 15.6
In order to calculate the amount of energy released, we first need to calculate the change in mass when one mole of $^{235}U$ is split apart.

$$\Delta m = \text{mass of the products} - \text{mass of the reactants}$$
$$\Delta m = [86.920690 + 145.925530 + 3(1.008665)] - [235.0439 + 1.008665]$$
$$= 235.878425 - 236.052565 = -0.17414 \text{ grams} = -1.74 \times 10^{-4} \text{ kg}$$

We can now calculate the change in energy via Eq. (15.6),

$$\Delta E = \Delta mc^2$$
$$= -1.74 \times 10^{-4} \text{ kg} \ (3.00 \times 10^8 \text{m s}^{-1})^2$$
$$= 1.57 \text{ kg m}^2\text{s}^{-2} = 1.6 \times 10^{13} \text{ J} = -1.6 \times 10^{10} \text{ kJ (per mole)}$$

The "−" sign means that energy is being released, that is, the process is exothermic. And the amount of energy released is extraordinary. It meets our expectation that it is far greater than the amount of energy released in the chemical reaction between a mole of any chemical reactants. Thus, our answer makes sense.

Let's compare the amount of energy released in a fission process to the energy released in the process of fusion.

## Example 15.7
How much energy is released in the fusion of one mole of deuterium nuclei with a mole of tritium nuclei? How does this compare to the energy change in the fission reaction discussed in Example 15.6?

$$^{2}_{1}H \quad + \quad ^{3}_{1}H \quad \rightarrow \quad ^{4}_{2}He \quad + \quad ^{1}_{0}n$$

| mass(g/mol) | 2.0140 | 3.01605 | 4.00260 | 1.008665 |
|---|---|---|---|---|

## Solution 15.7
As in the previous example, let's first calculate the change in mass.

$$\Delta m = \text{mass of the products} \ - \text{mass of the reactants}$$
$$\Delta m = [4.00260 + 1.008665] - [2.0140 + 3.01605]$$
$$= 5.011265 - 5.03005 = 0.018785 \text{ grams} = -(1.88) \times 10^{-5}\text{kg}$$

We can now calculate the change in energy via Eq. (15.6),

$$\Delta E = \Delta mc^2$$
$$= -1.88 \times 10^{-5} \text{ kg } (3.00 \times 10^8 \text{ m s}^{-1})^2$$
$$= -1.69 \times 10^{12} \text{kg m}^2\text{s}^{-2} = -1.7 \times 10^{12}\text{J} = -1.7 \times 10^9 \text{ kJ (per mole)}$$

The fusion process yields less energy per mole of reactant than fission. However, on a "per gram" basis, fusion gives off much more energy than fission because the reactants are 100 times lighter. In addition, fusion is a "clean" process, yielding a stable helium nucleus, rather than radioactive products such as bromine-87 and lanthanum-146.

## 15.5    Nuclear Power Plants

We have shown that both fission and fusion provide tremendous amounts of energy. Unfortunately, in order to fuse nuclei, a huge deposit of energy is required (typically a temperature equivalent of about 100,000,000°C, as is found in the center of stars) in order to yield the dividend of even more energy after the fusion is complete. So for now, nuclear power plants are all based on fission processes. There are two kinds of nuclear power plants, conventional and breeder reactors.

### 15.5.1    Conventional Nuclear Reactors

Figure 15.3 shows a schematic of a conventional nuclear reactor. All such reactors have in their **core** (the central part that holds the fissionable material) about 100 to 200 **fuel rods** that contain fissionable $^{235}$U-based pellets. The fission process supplies a substantial amount of energy, as we saw in Example 15.6. Also in the core are cadmium or boron **control rods**, which absorb excess neutrons so the rate of fission of uranium is kept at a satisfactory level (called **"critical"**) and not allowed to reach explosive (**"super-critical"**) levels. Water in the core slows down neutrons so they can be captured by the uranium pellets.

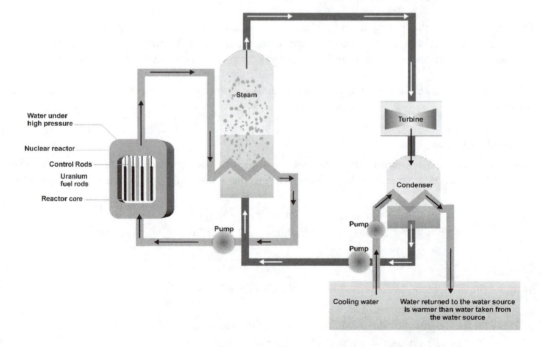

**FIGURE 15.3** A schematic drawing of a conventional nuclear fission. Among the key components are the core, which contains the fuel and control rods, circulating water, and the turbine.

The heat generated by fission is used to turn water surrounding the core into steam, which powers a **turbine**, a device that actually generates the electricity. Turbines are common to huge power plants, whether coal, hydroelectric (flowing water turns the turbine), or nuclear. So the key difference between nuclear and other types of power plants is the energy source.

### 15.5.2 Breeder Reactors

Only about 0.7% of all of the uranium on Earth is $^{235}U$, the kind of uranium that is used in conventional nuclear power plants. The other 99.3% is $^{238}U$, which does not undergo fission upon neutron bombardment. Breeder reactors make use of this relatively abundant supply of $^{238}U$ by creating, or **breeding,** fissionable $^{239}Pu$ by neutron bombardment. It is this plutonium that supplies the energy to heat the water that turns the turbines, producing electricity. The first great concern with breeder reactors is that plutonium-239 is a substance that is among the most dangerous known, and any accident might prove tragic if $^{239}Pu$ were dispersed into the atmosphere. The other concern is that because breeder reactors do not have control rods, more heat is generated than in conventional nuclear reactors. A more efficient coolant is therefore needed. Liquid sodium is often used, but the risk of explosion exists if the sodium contacts water in case of an accident. The high-risk design of breeder reactors makes them a less popular choice than conventional fission reactors for large-scale power generation.

The generation of electricity by nuclear power plants remains a controversial issue because of the large amounts of fissionable material required, concerns about how to dispose of the radioactive fission products, quality assurance issues relating to construction and management of the plants, and the short (typically 30 years or less) lifetimes of nuclear facilities. Chemists have a critical role in dealing with these problems. One example is the design of glass containers that are sufficiently rugged that they can seal in nuclear wastes when they are buried underground for countless years.

## 15.6   Where Have We Been?

We have learned that all elements have isotopes that are energetically unstable and will decay in such a way as to become more stable. We can calculate the amount of energy released for a given decay process and it is huge. The changes in nuclei make them especially useful in the medical field, as well as in commercial power generation.

In this chapter, we have learned that:

1. Some isotopes are unstable and will emit particles that will enable the resulting nuclei to be more stable.
2. Nuclei decay with a characteristic half-life.
3. Nuclei often achieve stability via a decay series.
4. We can determine the change in energy for a decay process.
5. Radioactive nuclei are used in many applications including medical and commercial power generation.
6. Nuclei can be made to fuse to create new elements.

## 15.1   Homework Exercises

1. Determine the correct number of protons and neutrons in the nucleus of an $^{125}_{53}I$ atom.
2. Fill in the blanks with the missing information.

| Symbol | Protons | Neutrons | Mass Number |
|---|---|---|---|
| $^{130}_{55}Cs$ | —— | —— | —— |
| $^{89}_{\_\_}$—— | 35 | —— | —— |
| $^{\_\_}_{\_\_}$—— | 42 | 49 | —— |

3. What element would result after a $^{224}_{88}Ra$ atom underwent electron capture? Write out the equation for this process.
4. Indicate the decay product that would result if plutonium-242 emitted an alpha particle. Write out the equation for this process.
5. Which of the following radioactive decay processes increases the proton to neutron ratio? Explain the general result of each process.
   (a) Gamma rays
   (b) Electron capture
   (c) Alpha decay
   (d) Positron emission
6 Which of the following radioactive decay processes decreases the proton to neutron ratio?
   (a) Gamma rays
   (b) Electron capture
   (c) Alpha decay
   (d) Positron emission

7. Which of the following radioactive decay processes does not alter the proton to neutron ratio?
   (a) Gamma rays
   (b) Electron capture
   (c) Alpha decay
   (d) Positron emission
8. What type of radioactive decay would convert $^{28}$Si to $^{32}$S?
9. Fill in the products of the following nuclear decays:
   (a) Rhenium-95 decays by positron emission.
   (b) Gallium-72 decays by beta-particle emission.
   (c) Osmium-186 decays by alpha emission.
10. What decay products would result from the following processes?
    (a) $^{13}_{6}$O undergoes positron emission.

    (b) $^{22}_{11}$Na undergoes electron capture.

    (c) $^{234}_{92}$U undergoes alpha particle production.
11. Write equations for each of the following processes:
    (a) $^{237}_{93}$Np undergoes alpha particle production.

    (b) $^{239}_{92}$U undergoes beta emission.

    (c) $^{22}_{13}$Al undergoes positron emission.
12. Fill in the blanks:
    (a) $^{251}$No $\rightarrow$ $^{4}$He + ――――

    (b) ―――― $\rightarrow$ $^{0}_{-1}$e + $^{183}$Ta

    (c) $^{212}$Po $\rightarrow$ ―――― + $^{208}$Pb
13. Complete the following equations with the appropriate particle:
    (a) $^{19}_{7}$N $\rightarrow$ _____ + $^{0}_{-1}$e

    (b) $^{52}_{25}$Mn $\rightarrow$ $^{52}_{26}$Fe + _____

    (c) $^{214}_{84}$Po $\rightarrow$ _____ + $^{4}_{2}$He
14. Complete the following equations with the appropriate particle:
    (a) _____ + $^{0}_{-1}$e $\rightarrow$ $^{71}_{32}$Ge

    (b) _____ $\rightarrow$ $^{196}_{85}$At + $^{4}_{2}$He

    (c) $^{212}_{86}$Rn $\rightarrow$ _____ + $^{4}_{2}$He

    (d) $^{191}_{80}$Hg $\rightarrow$ _____ + $^{0}_{1}$e
15. Which is more hazardous to human health, alpha, beta, or gamma emissions? Why?
16. Why do high atomic number nuclides generally decay via emission of an $\alpha$-particle while lower atomic number nuclides often decay via $\beta$-emission?
17. How much energy must be supplied to break a single silver-96 nucleus into separated protons and neutrons if the nucleus has a mass of 95.03083 amu? The mass of a proton is 1.0072765 amu and the mass of a neutron 1.0086655 amu.
18. $^{131}_{53}$I, has a half-life of 8.07 days. What is its rate constant?
19. The half-life for gold-110 is 24.6 seconds. Calculate the rate constant for this isotope.
20. Calculate the rate constants for the decay of the following radioactive nuclei:
    (a) 237 min$^{-1}$
    (b) 0.389 s$^{-1}$

(c)  0.354 min⁻¹

(d)  4.62 s⁻¹

21. The half-life of $^{84}_{35}$Br is 31.8 minutes. If the starting sample was 1.000 g, how long would it take to decay to a mass of 0.25 grams?

22. The half-life of $^{129}_{53}$I is $1.7 \times 10^7$ years. How long would it take for 75% of the original sample to decay?

23. The half-life of $^{188}_{79}$Au is 8.8 minutes. If the original sample has a mass of 0.50 grams, how much would remain after 1 day?

24. The half-life of $^{199}_{84}$Po is 5.2 minutes. How long would it take for 99.9999% of a sample to decompose?

25. If the current decay rate of carbon-14 in a cloth was 16 decays/g/minute, what would the decay rate be 5 years later? $t_{1/2} = 5730$ years

26. The half-life of $^{125}$I is about 60 days. How much $^{125}$I remains from a 5.00 gram sample after 180 days?

27. A sample is found to contain 0.350 grams of $^{224}$Ra. After 7.32 days, the sample is found to contain 0.0875 grams of $^{224}$Ra. What is the half-life of $^{224}$Ra?

28. How many years would it take for 90% of a sample of $^{238}$U to decay? $t_{1/2} = 4.46 \times 10^9$ years?

29. A jar of $^{131}$I ($t_{1/2} = 8.1$ days) was left open on the shelf of a stockroom for ten days before it was discovered by a lab technician. The iodine remaining in the jar has a mass of 15.2 grams. How much was present ten days before?

30. A vial containing a radioactive solid was left in a student's drawer for 20 years before it was discovered by a janitor. From the label, the half-life of the solid was determined to be 25.6 days. What percentage of the original material remains in the vial?

31. The half-life of $^{208}_{84}$Po is 2.83 years. How long will it take for 99% of a 10.0 g sample to decay?

32. The half-life of $^{38}_{19}$K is 7.71 minutes. If 0.050 g of this isotope remain after 2 hours of decay, how much was initially present?

33. $^{60}_{27}$Co has a half-life of 5.26 years. If a 25 gram sample is allowed to decay for 3.21 years, how much of the $^{60}_{27}$Co will remain?

34. Suppose a chemist is working with an unknown element X. If the chemist starts with 4.00 grams and 1.41 grams remains after 4 hours, what is the half-life of element X?

35. How much time is required for a sample of 3.52 grams of iodine-131 to decay to 0.01 grams? Its half-life is 8.06 days.

36. The half-life of indium-116 is 1.94 seconds. A total of 0.55 grams of indium-116 remain after 24 seconds. How large was the original sample?

37. The half-life of gold-188 is 8 minutes. Calculate how long it will take for 75% of a sample of gold-188 to decompose.

38. Cesium-137, a radioactive isotope implanted in medical radiation therapy, must be replaced when its radioactivity falls to 75 percent of the original sample. If the medical technicians purchased the sample in July of 1999, when will the sample have to be replaced? The half-life of cesium-137 is 30 years.

39. What is the product of a $^{238}$U nuclide that decays according to the following series?

$$\alpha, \beta, \beta, \alpha, \alpha, \alpha, \alpha, \alpha, \beta, \alpha, \beta, \beta, \alpha, \beta$$

40. What would the product be after $^{235}_{92}$U undergoes 7 alpha and 1 beta emissions?

41. The radioactive isotope $^{242}_{96}$Cm decays by the series:

$$\alpha, \alpha, \alpha, \alpha, \alpha, \alpha, \alpha, \beta, \beta, \alpha, \beta, \beta, \alpha$$

What nuclide will remain after the seventh step?

42. Continue the decay series from the previous problem. After which step will the particle $^{206}_{82}$Pb be emitted?

43. $^{107}$Sn undergoes electron capture.
    (a) Write out the equation for this process, including the symbols for all reactants and products.
    (b) The half-life of $^{107}$Sn is 2.92 minutes. What percentage of the original sample will remain after 5.0 minutes of decay?

## For Class Discussion and Reports

44. One reason that fusion is preferred over fission as a long-term energy solution is that the energy released per gram is greater with fusing light nuclei than with breaking apart heavy nuclei. Prove that this is so by finding one example of a fusion reaction and one of a fission reaction and calculating the energy released per gram of starting materials for each.

45. Check out the Website http://www.nrc.gov (as of 6/7/00) and other sources of information about the Nuclear Regulatory Commission. Write a report on its mission, its history, its recent initiatives, and how you think its activities affect chemists and chemistry technicians on the job.

46. Two of the most notorious accidents that have occurred in nuclear power plants were those that occurred at the Three Mile Island plant in Pennsylvania in 1979 and at Chernobyl in the former Soviet Union in 1986. Research these two accidents and write a report discussing exactly what happened in each case and how serious each was in terms of the effects on life in the surrounding areas.

# 16

# Reduction-Oxidation Processes

## 16.1   Introduction

There are many types of chemical reactions that can take place between two compounds. Some examples that we have already seen include those that involve the sharing of electrons to form new chemical bonds (Chapter 2), or the neutralization reaction that occurs when a proton is transferred from an acid to a base (Chapter 12). Another type of chemical reaction that can occur is one in which electrons are transferred from one chemical species to another, thus changing the oxidation numbers of atoms in both reactants. This type of process is known as an ***electrochemical reaction***, since it involves the transfer of electrons between chemical compounds. These types of reactions are essential to life on Earth, since they form part of the basis for photosynthesis in plants and the use of oxygen by our bodies to help us turn food into energy. Electrochemical reactions are also important in many industrial processes and in the generation of electricity by batteries.

In this chapter we will examine the different types of electrochemical reactions and will learn how chemists describe these reactions. We will also learn how to predict whether or not two particular chemicals will take part in this type of process and will see how these reactions can be used to generate electricity or make useful products.

## 16.2   What Is an Electrochemical Reaction?

Let's begin by taking a close look at one specific electrochemical process, the reaction that occurs when $Fe^{2+}$ is combined with $Ce^{4+}$ in an aqueous solution, which is a common titration reaction used in analytical laboratories for the measurement of $Fe^{2+}$ in water.

$$Fe^{2+} + Ce^{4+} \rightleftharpoons Fe^{3+} + Ce^{3+} \tag{16.1}$$

*Chemistry Professionals at Work*                                    *CPW Box 16.1*

# WHY STUDY THIS TOPIC?

Electrochemical reactions are encountered by chemists, chemical laboratory technicians, and process technicians in a wide variety of situations. For example, the production of galvanized steel utilizes an electrochemical reaction to coat steel with a protective layer of zinc. An electrochemical reaction is also at the heart of the process that produces aluminum from bauxite ore, $Al_2O_3 \cdot XH_2O$. The production and use of batteries (see CPW Box 4.4) is yet another example of where electrochemical reactions are important.

But electrochemical reactions also form the basis for many common methods used in analytical testing laboratories. For example, the electrochemical reaction between a pollutant and the dichromate ion $(Cr_2O_7^{2-})$ is a common way for determining the approximate level of the pollution in a water sample. Also, a popular analysis technique known as the Kjeldahl method is used for determining the amount of nitrogen-containing organic compounds in a sample by first using an electrochemical reaction to convert all of the nitrogen into ammonia, which is then measured. The examples are many and varied.

*For Homework: Look up information on how electrochemical reactions are used in industry to galvanize steel, to produce aluminum, or to test for water pollution. Prepare a report on one of these applications.*

In Section 2.4 of Chapter 2, we learned how to determine the oxidation number of elements that appear in ions or in chemical compounds. We can see in the reaction in Eq. (16.1) that the oxidation numbers for both iron and cerium change as we go from the left to the right-hand side of the reaction (see Chapter 2 for a review of how to determine the oxidation number for an element in a chemical compound). For iron, the oxidation number begins at $+2$ and ends up at $+3$, while the cerium goes from an oxidation number of $+4$ to $+3$. A change in oxidation number is characteristic of electrochemical reactions since electrons are actually being transferred from one reactant to the another. The process by which $Fe^{2+}$ *loses an electron* to form $Fe^{3+}$ is called **oxidation**, and the process by which $Ce^{4+}$ *gains an electron* to form $Ce^{3+}$ is called **reduction** (see Fig. 16.1). To help remember which electrochemical process is which, use the acronym **OIL RIG**. This stands for *oxidation involves loss* (of an electron); *reduction involves gain* (of an electron).

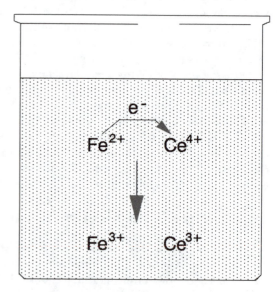

**FIGURE 16.1**    Transfer of an electron from $Fe^{2+}$ to $Ce^{4+}$ resulting in the oxidation of $Fe^{2+}$ and reduction of $Ce^{4+}$.

In this type of reaction, the chemical species that experiences an increase in oxidation number ($Fe^{2+}$, as it forms $Fe^{3+}$), or that has an oxidation number that becomes more positive, is said to be **oxidized**. And the reactant that has a decrease in oxidation number ($Ce^{4+}$, as it forms $Ce^{3+}$), or that has an oxidation number that becomes more negative, is said to be **reduced**. Notice that *oxidation and reduction always occur together in an electrochemical reaction*. It is for this reason that electrochemical reactions are also sometimes called **reduction-oxidation reactions**, or **redox reactions**.

## Example 16.1

Which of the following chemical reactions are also redox reactions?

(a) $2KClO_3 \rightleftharpoons 2KCl + 3O_2$

(b) $CaCO_3 \rightleftharpoons CaO + CO_2$

(c) $SnCl_2 + PbCl_4 \rightleftharpoons SnCl_4 + PbCl_2$

## Solution 16.1

Remember that redox reactions are chemical reactions that involve a change in the oxidation numbers of the reactants as they form products. The reaction in (a) is a redox reaction since the oxidation number for oxygen changes from $-2$ to $0$ as oxygen goes from a combined state on the left to its standard state on the right. In (b), calcium is in the $+2$ oxidation state on both sides of the reaction, oxygen is in the $-2$ oxidation state on both sides, and carbon is in the $+4$ oxidation state on both sides. This means that (b) is not a redox reaction. In (c), tin is in the $+2$ oxidation state on the left and in the $+4$ state on the right. The oxidation number for lead has also changed in this reaction, going from $+4$ on the left to $+2$ on the right. Thus, (c) is a redox reaction.

Besides involving the transfer of electrons, electrochemical reactions can result in (or require) the breaking and forming of chemical bonds in the compounds that are being oxidized or reduced. An example of this is shown below for the reaction of an aqueous solution of $Ce^{4+}$ with liquid mercury, Hg.

$$2Ce^{4+} + 2Hg + 2Cl^- \rightleftharpoons 2Ce^{3+} + Hg_2Cl_2 \tag{16.2}$$

As in the earlier example, $Ce^{4+}$ is being reduced to produce $Ce^{3+}$. But now the species that is being oxidized, Hg, not only experiences an increase in oxidation number (as it forms $Hg^+$), but it also undergoes a reaction, with chloride ions to form the new compound $Hg_2Cl_2$. This type of electrochemical reaction, in which both electron transfer and a change in chemical bonding occurs, is actually more common than the simpler type of electrochemical process that was shown earlier between $Fe^{2+}$ and $Ce^{4+}$.

In an electrochemical reaction, the reactant that is reduced is sometimes called the **oxidizing agent** (or **oxidant**), since its reduction results in the oxidation of some other chemical species. In the same manner, the reactant that is oxidized is known as the **reducing agent** (or **reductant**), because its oxidation leads to the reduction of some other chemical in the system. For the reaction between $Fe^{2+}$ and $Ce^{4+}$ in Eq. (16.1), $Fe^{2+}$ would be the reducing agent, since it causes the reduction of $Ce^{4+}$ to $Ce^{3+}$, and $Ce^{4+}$ would be the oxidizing agent, since it causes the oxidation of $Fe^{2+}$ to $Fe^{3+}$. In the reaction shown in Eq. (16.2) between $Ce^{4+}$ and Hg, $Ce^{4+}$ is the oxidizing agent and Hg is the reducing agent.

## Example 16.2

Identify the species that are being oxidized and reduced in each of the following redox reactions.

(a) $Cu^+ + Fe^{3+} \rightleftharpoons Cu^{2+} + Fe^{2+}$

(b) $I_2 + 2S_2O_3^{2-} \rightleftharpoons 2I^- + S_4O_6^{2-}$

(c) $Mg + Cl_2 \rightleftharpoons Mg^{2+} + 2Cl^-$

## Solution 16.2

The chemical species that are being oxidized, or losing electrons and increasing in their oxidation number, are: (a) $Cu^+$, which forms $Cu^{2+}$; (b) the sulfur in $S_2O_3^{2-}$, which goes on to form the sulfur in $S_4O_6^{2-}$; and (c) Mg, which is converted to $Mg^{2+}$. The species that are being reduced, or gaining electrons and decreasing in their oxidation number, are: (a) $Fe^{3+}$, which produces $Fe^{2+}$; (b) $I_2$, which forms $I^-$; and (c) $Cl_2$, which forms $Cl^-$.

## Example 16.3

List the reducing agents and oxidizing agents for each of the reactions in Example 16.2.

## Solution 16.3

The oxidizing agents are the chemicals that contain the species undergoing a reduction. These agents are (a) $Fe^{3+}$, (b) $I_2$, and (c) $Cl_2$. The reducing agents are the chemicals that contain the species undergoing an oxidation. These agents are (a) $Cu^+$, (b) $S_2O_3^{2-}$, and (c) Mg.

# 16.3   Half-Reactions

In order to make it more convenient to see what is happening during an electrochemical reaction, chemists will often take the reduction and oxidation parts of such a reaction and write these as two separate *half-reactions*. For example, the reaction between $Fe^{2+}$ and $Ce^{4+}$ that was given earlier in Eq. (16.1) can also be represented by the two half-reactions shown below.

$$\text{Reduction Half-Reaction:} \qquad Ce^{4+} + e^- \rightleftharpoons Ce^{3+} \qquad (16.3)$$

$$\text{Oxidation Half-Reaction:} \qquad Fe^{2+} \rightleftharpoons Fe^{3+} + e^- \qquad (16.4)$$

$$\text{Net Reaction:} \qquad Fe^{2+} + Ce^{4+} \rightleftharpoons Fe^{3+} + Ce^{3+}$$

Notice that the oxidation reaction for $Fe^{2+}$ going to $Fe^{3+}$ is written with an electron on the right hand side of the equation, showing that $Fe^{2+}$ is giving up an electron and is being oxidized. On the other hand, the reduction of $Ce^{4+}$ to $Ce^{3+}$ is written with an electron on the left, showing that $Ce^{4+}$ is gaining an electron and is being reduced.

## Example 16.4

Write the half-reactions and net reaction for each of the electrochemical reactions given in Example 16.2.

## Solution 16.4

$$\text{(a) Oxidation Half-Reaction:} \qquad Cu^+ \rightleftharpoons Cu^{2+} + e^-$$

$$\text{Reduction Half-Reaction:} \qquad Fe^{3+} + e^- \rightleftharpoons Fe^{2+}$$

$$\text{Net Reaction:} \qquad Cu^+ + Fe^{3+} \rightleftharpoons Cu^{2+} + Fe^{2+}$$

(b) Oxidation Half-Reaction: $\qquad 2S_2O_3^{2-} \rightleftharpoons S_4O_6^{2-} + 2e^-$

   Reduction Half-Reaction: $\qquad I_2 + 2e^- \rightleftharpoons 2I^-$

   Net Reaction: $\qquad I_2 + 2S_2O_3^{2-} \rightleftharpoons 2I^- + S_4O_6^{2-}$

(c) Oxidation Half-Reaction: $\qquad Mg \rightleftharpoons Mg^{2+} + 2e^-$

   Reduction Half-Reaction: $\qquad Cl_2 + 2e^- \rightleftharpoons 2Cl^-$

   Net Reaction: $\qquad Mg + Cl_2 \rightleftharpoons Mg^{2+} + 2Cl^-$

We can see in Example 16.4 that adding the oxidation and reduction half-reactions together should give the same net reaction with which we started. This occurs because the numbers of electrons involved in the oxidation and reduction processes are identical and cancel out as we add the oxidation and reduction half-reactions to get the final, net reaction. This process is yet another example of how the Law of the Conservation of Mass is applied to chemical reactions (see Chapter 2 for a review of this law). From Example 16.4, we observe that there can be more than one electron involved in an electrochemical process. For instance, in Examples 16.4(b) and 16.4(c) two electrons are actually transferred during oxidation and reduction.

As demonstrated in the above exercise, one important difference between a net electrochemical reaction and its half-reactions is that the half-reactions actually show the electrons that are being transferred from one chemical species to another. Although this approach is somewhat artificial since electrons generally do not exist by themselves in solution, this bookkeeping method is still valuable for examining what is actually being oxidized or reduced in an electrochemical process. In the next section we will see that another application of half-reactions is their use in balancing the net reactions that are predicted to occur as we combine together various types of oxidizing and reducing agents.

## 16.4   Balancing Electrochemical Reactions

One reason that chemists use half-reactions is that this gives them a convenient way of predicting how oxidizing and reducing agents will react with each other. To help do this, there are various lists of half-reactions that have been created for a large variety of electrochemically reactive chemicals. An example of such a list is provided in Table 16.1. By current convention, all half-reactions in such tables are written as *reduction processes* (that is, with the reaction arranged so that the electrons always appear on the left-hand side of the equation). For instance, notice in Table 16.1 that the half-reactions for $Fe^{2+}/Fe^{3+}$ and $Ce^{4+}/Ce^{3+}$ appear in the following forms.

$$Ce^{4+} + e^- \rightleftharpoons Ce^{3+} \qquad (16.5)$$

$$Fe^{3+} + e^- \rightleftharpoons Fe^{2+} \qquad (16.6)$$

There are some older texts that contain similar lists with all of the half-reactions written as oxidation processes (or with the electrons appearing on the right), but reduction half-reactions are now preferred for use in such tables.

Now that we are familiar with half-reactions, let's look at how these might be used to predict what will happen when a new combination of oxidizing and reducing agents are mixed together. To do this, we first need to know how to balance and combine half-reactions so that they make chemical sense (that is, both the number of atoms of each element and the overall charge are balanced on both sides of the chemical equation).

# An Ironic Procedure

In CPW Box 2.4 we noted that when ground water is pumped to the surface and exposed to air, any $Fe^{2+}$ present in the water is converted to $Fe^{3+}$. This is a reduction-oxidation reaction in which the oxygen in the air is the oxidizing agent. Ironically, when a chemist or chemistry technician analyzes the water for iron content via a popular colorimetric technique (see CPW Box 3.3), the iron must be in the +2 oxidation state. To convert the $Fe^{3+}$ back to $Fe^{2+}$, and to sustain the +2 oxidation state, a reducing agent must be added to the water as part of the sample preparation.

*For Homework: Look up the method described above and discover what reducing agents are useful for the sample preparation procedure described. Read the method in detail and write a report on what else the sample preparation and analysis involves.*

Although this process can be a bit confusing at first, it becomes straightforward if we use the procedure outlined below. To help illustrate this, we will use this approach to balance the following reduction-oxidation reaction between $H_2S$ and $Cl_2$ in an aqueous solution under acidic conditions.

$$H_2S + Cl_2 \rightleftharpoons S + Cl^-$$ (16.7)

**Step 1.** *Identify which chemical compounds are the reduced and oxidized species.*
   (a) If the oxidized and reduced products are known, then proceed to Step 2.
   (b) If the oxidized and reduced products are not known, then use a list of reduction half-reactions (like Table 16.1) to determine what some likely products might be for the oxidized and reduced compounds that we are studying.

In our example, chlorine has a decrease in its oxidation number from 0 to $-1$ as it goes from chlorine atom in $Cl_2$ to a chloride ion $(Cl^-)$, so $Cl_2$ is the chemical species that is being reduced. Similarly, the sulfur in $H_2S$ is the species that is being oxidized, with a change in oxidation number from $-2$ to 0 in going from $H_2S$ to S.

**Step 2.** *Write down the two initial half-reactions.* Remember to write the half-reaction for the reduced species so that the electrons appear on the left-hand side of the equation, and to place the electrons for the oxidation half-reaction on the right-hand side.
   Based on the oxidized and reduced species that we identified in Step 1 [and the known products that are given in Eq. (16.7)], we can write the following initial half-reactions for our example.

$$\text{Reduction:} \quad Cl_2 + e^- \rightleftharpoons Cl^-$$

$$\text{Oxidation:} \quad H_2S \rightleftharpoons S + e^-$$

**TABLE 16.1** Examples of Reduction Half-Reactions

| Half-Reaction | E° (in Volts) |
|---|---|
| $Co^{3+} + e^- \rightleftharpoons Co^{2+}$ | +1.81 |
| $2HOCl + 2H^+ + 2e^- \rightleftharpoons Cl_2 + 2H_2O$ | +1.63 |
| $Ce^{4+} + e^- \rightleftharpoons Ce^{3+}$ | +1.61 |
| $MnO_4^- + 8H^+ + 5e^- \rightleftharpoons Mn^{2+} + 4H_2O$ | +1.51 |
| $Cl_2$ (in aqueous solution) $+ 2e^- \rightleftharpoons 2Cl^-$ | +1.40 |
| $Cr_2O_7^{2-} + 14H^+ + 6e^- \rightleftharpoons 2Cr^{3+} + 7H_2O$ | +1.33 |
| $MnO_2 + 4H^+ + 2e^- \rightleftharpoons Mn^{2+} + 2H_2O$ | +1.23 |
| $O_2 + 4H^+ + 4e^- \rightleftharpoons 2H_2O$ | +1.23 |
| $Br_2$ (in aqueous solution) $+ 2e^- \rightleftharpoons 2Br^-$ | +1.09 |
| $Hg^{2+} + 2e^- \rightleftharpoons Hg$ | +0.85 |
| $Ag^+ + e^- \rightleftharpoons Ag$ | +0.80 |
| $Fe^{3+} + e^- \rightleftharpoons Fe^{2+}$ | +0.77 |
| $I_2$ (in aqueous solution) $+ 2e^- \rightleftharpoons 2I^-$ | +0.62 |
| $Cu^+ + e^- \rightleftharpoons Cu$ | +0.52 |
| $Ni(OH)_3 + e^- \rightleftharpoons Ni(OH)_2 + OH^-$ | +0.49 |
| $O_2 + 2H_2O + 4e^- \rightleftharpoons 4OH^-$ | +0.40 |
| $Cu^{2+} + 2e^- \rightleftharpoons Cu$ | +0.34 |
| $Ru^{3+} + e^- \rightleftharpoons Ru^{2+}$ | +0.25 |
| $Hg_2Cl_2 + 2e^- \rightleftharpoons 2Hg + 2Cl^-$ | +0.24 |
| $AgCl + e^- \rightleftharpoons Ag + Cl^-$ | +0.22 |
| $Cu^{2+} + e^- \rightleftharpoons Cu^+$ | +0.16 |
| $Sn^{4+} + 2e^- \rightleftharpoons Sn^{2+}$ | +0.15 |
| $S + 2H^+ + 2e^- \rightleftharpoons H_2S$ | +0.14 |
| $S_4O_6^{2-} + 2e^- \rightleftharpoons 2S_2O_3^{2-}$ | +0.09 |
| $NO_3^- + H_2O + 2e^- \rightleftharpoons NO_2^- + 2OH^-$ | +0.01 |
| $2H^+ + 2e^- \rightleftharpoons H_2$     (Reference Reaction) | 0.00 |
| $Fe^{3+} + 3e^- \rightleftharpoons Fe$ | −0.04 |
| $Pb^{2+} + 2e^- \rightleftharpoons Pb$ | −0.13 |
| $Sn^{2+} + 2e^- \rightleftharpoons Sn$ | −0.14 |
| $Ni^{2+} + 2e^- \rightleftharpoons Ni$ | −0.23 |
| $Co^{2+} + 2e^- \rightleftharpoons Co$ | −0.28 |
| $In^{3+} + 3e^- \rightleftharpoons In$ | −0.34 |
| $Tl + e^- \rightleftharpoons Tl$ | −0.34 |
| $PbSO_4 + 2e^- \rightleftharpoons Pb + SO_4^{2-}$ | −0.36 |
| $Cd^{2+} + 2e^- \rightleftharpoons Cd$ | −0.40 |
| $Cr^{3+} + e^- \rightleftharpoons Cr^{2+}$ | −0.41 |
| $Fe^{2+} + 2e^- \rightleftharpoons Fe$ | −0.44 |
| $Cr^{3+} + 3e^- \rightleftharpoons Cr$ | −0.74 |
| $Zn^{2+} + 2e^- \rightleftharpoons Zn$ | −0.76 |
| $Cd(OH)_2 + 2e^- \rightleftharpoons Cd + 2OH^-$ | −0.81 |
| $2H_2O + 2e^- \rightleftharpoons H_2 + 2OH^-$ | −0.83 |
| $Cr^{2+} + 2e^- \rightleftharpoons Cr$ | −0.91 |
| $Mn^{2+} + 2e^- \rightleftharpoons Mn$ | −1.18 |
| $Al^{3+} + 3e^- \rightleftharpoons Al$ | −1.66 |
| $Ce^{3+} + 3e^- \rightleftharpoons Ce$ | −2.34 |
| $Mg^{2+} + 2e^- \rightleftharpoons Mg$ | −2.38 |
| $Na^+ + e^- \rightleftharpoons Na$ | −2.71 |
| $Ca^{2+} + 2e^- \rightleftharpoons Ca$ | −2.87 |
| $K^+ + e^- \rightleftharpoons K$ | −2.93 |

To remind us of which process is reduction and which is oxidation, we've included one electron in each half-reaction on the appropriate side of the equation (we won't worry about balancing the number of electrons at this point, as this will be taken care of later).

**Step 3.** *Balance the chemical contents of each half-reaction* so that both sides of the half-reaction have the same number of atoms for all elements that appear in the reaction (except oxygen and hydrogen, which will be dealt with in the next step).

The reduction half-reaction needs to be balanced to give two chlorines on each side of the reaction. No change is needed in the oxidation half-reaction at this point, since there are the same number of sulfur atoms on each side and the hydrogen atoms will be balanced later.

$$\text{Reduction:} \quad Cl_2 + e^- \rightleftharpoons 2Cl^-$$

$$\text{Oxidation:} \quad H_2S \rightleftharpoons S + e^- \quad \text{(Unchanged)}$$

**Step 4.** *Determine whether the reaction is going to take place in an acidic or basic aqueous solution.*

(a) If the reaction is to be carried out in an *acidic* aqueous solution, then balance the number of oxygen atoms in the half-reactions by adding $H_2O$ to any side that needs additional atoms of this element. After this is done, balance the hydrogen atoms by adding $H^+$ to each side that needs more hydrogen. Notice here that we use $H^+$ and $H_2O$ to balance the hydrogen and oxygen atoms since we know that both hydrogen ions and water are present in our system. We know this since it was stated earlier that the reaction is being carried out in an aqueous solution under acidic conditions. We can't use other chemicals, like $O_2$ or $H_2$, to balance the oxygen and hydrogen atoms since we aren't told or don't know that these chemicals are present in our reaction container.

(b) If the reaction is to carried in a *basic* aqueous solution, first balance the number of oxygen atoms in each half-reaction by adding $H_2O$ to any side that needs additional atoms of this element. Next, balance the hydrogen atoms by adding one additional $H_2O$ for each hydrogen atom that is needed on a given side, along with the addition of one $OH^-$ per hydrogen atom to the other side of the half-reaction (Note: the simultaneous addition of one $H_2O$ and one $OH^-$ to opposite sides of the half-reaction results in the net addition of one hydrogen atom to the side that needs it). As in 4(a), water and hydroxide ions are used as our sources of extra hydrogen and oxygen atoms since we know that both of these compounds are present if it is stated that the reaction is taking place in an aqueous solution under basic conditions.

(c) If it is not specified or known whether the reaction is taking place under acidic or basic conditions, then the reaction may not be pH dependent. This is often true for aqueous reactions that do not have any hydrogen or oxygen atoms in their reactants or products. We can confirm this by using both the approaches given under Steps 4(a) or 4(b) and see if we get the same final answer.

We know that our example reaction is going to be carried out under acidic conditions, so we will use the procedure given in Step 4(a). The reduction half-reaction does not require any changes, but we do have to balance the number of hydrogens in the oxidation half-reaction. We can do this by adding two hydrogen ions to the right side of this equation. No further changes are required at this time.

$$\text{Reduction:} \quad Cl_2 + e^- \rightleftharpoons 2Cl^- \quad \text{(Unchanged)}$$

$$\text{Oxidation:} \quad H_2S \rightleftharpoons S + 2H^+ + e^-$$

**Step 5.** *Balance each half-reaction* so that the charge is the same on both sides of the equation. This is done by adding electrons to any side that has an excess of positive charge.

Two electrons are needed on the left of the reduction half-reaction to give both sides a net charge of $-2$. Two electrons are also needed on the right of the oxidation half-reaction to give both sides a net charge of 0.

$$\text{Reduction:} \quad Cl_2 + 2e^- \rightleftharpoons 2Cl^-$$

$$\text{Oxidation:} \quad H_2S \rightleftharpoons S + 2H^+ + 2e^-$$

**Step 6.** *Multiply one or both of the half-reactions by factors that cause them to have the same number of electrons* (but with one half-reaction having its electrons on the right and the other having them on the left).

Both half-reactions have the same number of electrons, so no change is required in this example.

$$\text{Reduction:} \quad Cl_2 + 2e^- \rightleftharpoons 2Cl^- \qquad \text{(Unchanged)}$$

$$\text{Oxidation:} \quad H_2S \rightleftharpoons S + 2H^+ + 2e^- \quad \text{(Unchanged)}$$

**Step 7.** *Add together the two half-reactions*, combine common terms, and cancel out any factors that appear on both sides of the reaction. Confirm that there is an equal number of each type of atom on both sides of the equation and that the net charge on both sides is also equal (the electrons should cancel out and no longer appear in the equation at this point). If all of this checks out, then we have successfully balanced the reduction-oxidation equation.

Once the two half-reactions have been combined, the only common term on both sides of the net reaction are the two electrons, which cancel each other out. Each side of the final equation has two chlorines, two hydrogens, and one sulfur, thus indicating that all atoms have been properly balanced. In addition, the final charge on both sides of the final reaction is the same, with a value of 0.

$$\text{Reduction:} \quad Cl_2 + 2e^- \rightleftharpoons 2Cl^-$$

$$\text{Oxidation:} \quad H_2S \rightleftharpoons S + 2H^+ + 2e^-$$

$$\overline{\phantom{XXXXXXXXXXXXXXXXXXXXXXXXXXXXXXXXXXX}}$$

$$Cl_2 + 2e^- + H_2S \rightleftharpoons 2Cl^- + S + 2H^+ + 2e^-$$

$$\text{Final Answer:} \quad Cl_2 + H_2S \rightleftharpoons 2Cl^- + S + 2H^+$$

## Example 16.5

Balance the following reduction-oxidation reaction carried out in an *acidic* aqueous solution. This reaction is used in industrial laboratories for the analysis of $Fe^{2+}$ in aqueous samples by titrating these samples with a standardized solution of dichromate $(Cr_2O_7^{2-})$.

$$Fe^{2+} + Cr_2O_7^{2-} \rightleftharpoons Fe^{3+} + Cr^{3+}$$

## Solution 16.5

**Step 1:** *Identify the oxidized and reduced species.* A quick look at the change in oxidation numbers shows us that the oxidized species is $Fe^{2+}$, with its corresponding product being $Fe^{3+}$. The reduced species is the chromium in $Cr_2O_7^{2-}$ and the resulting product is $Cr^{3+}$.

**Step 2:** *Write down the two initial half-reactions.*

$$\text{Oxidation:} \quad Fe^{2+} \rightleftharpoons Fe^{3+} + e^-$$

$$\text{Reduction:} \quad Cr_2O_7^{2-} + e^- \rightleftharpoons Cr^{3+}$$

**Step 3:** *Balance the chemical contents of each half-reaction* (except for O and H). No change is required for the oxidation half-reaction, but an additional chromium is needed on the product side of the reduction half-reaction.

$$\text{Oxidation:} \quad Fe^{2+} \rightleftharpoons Fe^{3+} + e^- \quad \text{(Unchanged)}$$

$$\text{Reduction:} \quad Cr_2O_7^{2-} + e^- \rightleftharpoons 2Cr^{3+}$$

**Step 4:** *Determine whether the reaction is taking place in acidic or basic solution, and balance the oxygen and hydrogen atoms.* We were told that the reaction will be carried out under acidic conditions. This means that $H_2O$ and $H^+$ will be used to balance the number of oxygen and hydrogen atoms, respectively. No change is needed in the oxidation half-reaction. However, seven oxygens must be added to the product side of the reduction half-reaction (added in the form of seven molecules of $H_2O$). Also, fourteen hydrogen ions must now be added to the reactant side of the same reaction in order to balance out the extra hydrogens that have been placed on the product side.

$$\text{Oxidation:} \quad Fe^{2+} \rightleftharpoons Fe^{3+} + e^- \quad \text{(Unchanged)}$$
$$\text{Reduction:} \quad Cr_2O_7^{2-} + 14H^+ + e^- \rightleftharpoons 2Cr^{3+} + 7H_2O$$

**Step 5:** *Balance the charge in each half-reaction.* No change is needed in the oxidation half-reaction, which has a net charge of $+2$ on both sides. But five more electrons must be added to the reactant side of the reduction half-reaction to give a net charge of $+6$ on each side.

$$\text{Oxidation:} \quad Fe^{2+} \rightleftharpoons Fe^{3+} + e^- \quad \text{(Unchanged)}$$
$$\text{Reduction:} \quad Cr_2O_7^{2-} + 14H^+ + 6e^- \rightleftharpoons 2Cr^{3+} + 7H_2O$$

**Step 6:** *Multiply through both half-reactions so that they have the same number of electrons.* The easiest way to do this is to find the least common product for the number of electrons in the half-reactions that we currently have written. There is one electron listed in the oxidation half-reaction and six in the reduction half-reaction, so the least common product for each is six. No change is needed in the reduction half-reaction to get this number, since it already has six electrons listed. However, the number of moles for all products and reactants in the oxidation half-reaction must be multiplied by six in order to also get six electrons in this reaction.

$$\text{Oxidation:} \quad 6Fe^{2+} \rightleftharpoons 6Fe^{3+} + 6e^-$$
$$\text{Reduction:} \quad Cr_2O_7^{2-} + 14H^+ + 6e^- \rightleftharpoons 2Cr^{3+} + 7H_2O \quad \text{(Unchanged)}$$

**Step 7.** *Add together the two half-reactions,* combine common terms, and cancel out any factors that appear on both sides of the reaction. Once the two half-reactions have been combined, the only common term on both sides of the net reaction is the six electrons, which cancel each other out. Each side of the final equation has six irons, two chromiums, seven oxygens, and fourteen hydrogens, thus indicating that all atoms have been properly balanced. In addition, the overall charge is $+24$ on both sides of the final reaction.

$$\text{Oxidation:} \quad 6Fe^{2+} \rightleftharpoons 6Fe^{3+} + 6e^-$$
$$\text{Reduction:} \quad Cr_2O_7^{2-} + 14H^+ + 6e^- \rightleftharpoons 2Cr^{3+} + 7H_2O$$

---

$$6Fe^{2+} + Cr_2O_7^{2-} + 14H^+ + 6e^- \rightleftharpoons 6Fe^{3+} + 2Cr^{3+} + 7H_2O + 6e^-$$

$$\text{Final Answer:} \quad 6Fe^{2+} + Cr_2O_7^{2-} + 14H^+ \rightleftharpoons 6Fe^{3+} + 2Cr^{3+} + 7H_2O$$

## Example 16.6

Write the balanced electrochemical reaction between $Sn^{4+}$ and $NO_2^-$ to give $Sn^{2+}$ and $NO_3^-$ in the presence of a *basic* aqueous solution.

$$Sn^{4+} + NO_2^- \rightleftharpoons Sn^{2+} + NO_3^-$$

## Solution 16.6

**Step 1:** *Identify the oxidized and reduced species.* The change in oxidation numbers indicates that the oxidized species is the nitrogen in $NO_2^-$, which forms the product $NO_3^-$. The reduced species is $Sn^{4+}$ and its product is $Sn^{2+}$.

**Step 2:** *Write down the two initial half-reactions.*

$$\text{Oxidation:} \quad NO_2^- \rightleftharpoons NO_3^- + e^-$$

$$\text{Reduction:} \quad Sn^{4+} + e^- \rightleftharpoons Sn^{2+}$$

**Step 3:** *Balance the chemical contents of each half-reaction* (except for O and H).

$$\text{Oxidation:} \quad NO_2^- \rightleftharpoons NO_3^- + e^- \quad \text{(Unchanged)}$$

$$\text{Reduction:} \quad Sn^{4+} + e^- \rightleftharpoons Sn^{2+} \quad \text{(Unchanged)}$$

**Step 4:** *Determine whether the reaction is taking place in acidic or basic solution, and balance the oxygen and hydrogen atoms.* It is stated that the reaction will be carried out under basic conditions, so $H_2O$ and $H_2O$ plus $OH^-$ will be used to balance the number of oxygen and hydrogen atoms, respectively.

$$\text{Oxidation:} \quad NO_2^- + H_2O + 2OH^- \rightleftharpoons NO_3^- + 2H_2O + e^-$$

$$\text{Reduction:} \quad Sn^{4+} + e^- \rightleftharpoons Sn^{2+} \quad \text{(Unchanged)}$$

**Step 5:** *Balance the charge in each half-reaction.* Two electrons are needed on the right of the oxidation half-reaction to give both sides a net charge of $-3$. Two electrons are needed on the left of the reduction half-reaction to give a net charge of $+2$ on each side.

$$\text{Oxidation:} \quad NO_2^- + H_2O + 2OH^- \rightleftharpoons NO_3^- + 2H_2O + 2e^-$$

$$\text{Reduction:} \quad Sn^{4+} + 2e^- \rightleftharpoons Sn^{2+}$$

**Step 6:** *Multiply through both half-reactions so that they have the same number of electrons.* Both half-reactions have the same number of electrons, so no change is required.

**Step 7.** *Add together the two half-reactions*, combine common terms, and cancel out any factors that appear on both sides of the reaction. Common terms that cancel out are the two electrons and one water molecule (one will be left over on the product side of the reaction). The final reaction should have one nitrogen, one tin, four oxygens, and two hydrogens on each side, with a net charge of $+1$ on both sides of the reaction.

$$\text{Oxidation:} \quad NO_2^- + H_2O + 2OH^- \rightleftharpoons NO_3^- + 2H_2O + 2e^-$$

$$\text{Reduction:} \quad Sn^{4+} + 2e^- \rightleftharpoons Sn^{2+}$$

$$\overline{Sn^{4+} + NO_2^- + H_2O + 2OH^- + 2e^- \rightleftharpoons Sn^{2+} + NO_3^- + 2H_2O + 2e^-}$$

$$\text{Final Answer:} \quad Sn^{4+} + NO_2^- + 2OH^- \rightleftharpoons Sn^{2+} + NO_3^- + H_2O$$

At this point, we have learned what a redox reaction is and how to balance redox reactions. We have seen several examples of how this balancing process is performed. In the following sections we will see how we can predict whether or not a particular redox reaction is likely to occur, and we will learn how to estimate the amount of energy that will be released by the reaction or that will be required to make the reaction occur. This will allow us to then consider some possible applications of reduction-oxidation reactions.

# 16.5    Standard Reduction Potentials

One question that we still need to address is how do we know in advance which chemicals will be reduced or oxidized in any given electrochemical reaction. This can be answered by using the **standard reduction potential** ($E°$) for a given chemical. A list of $E°$ values is included in Table 16.1 for the reduction half-reactions that are provided. The size of $E°$ for an electrochemical reaction is a measure of that reaction's tendency to proceed under *standard reaction conditions*, or a temperature of 25°C, a pressure of 1 atm for all gaseous reactants or products, a chemical activity (see below) of one for all solids, and 1 mol/L activity for all soluble reactants and products.

The term **chemical activity**, also known as simply "activity," that is used here is a type of concentration that takes into account the interactions that occur between neighboring molecules or ions in solution. Activity is expressed in the same units as ordinary concentrations (such as moles/L for dissolved compounds) and is closely related to these concentrations through a factor known as the **activity coefficient**. Chemical activities are often used in place of ordinary concentrations in electrochemistry since activities are a more accurate representation of the *effective concentration* of a compound and the ability of this compound to take part in chemical reactions. The chemical activity and the molar concentration of a compound are approximately the same when we are dealing with relatively dilute solutions. For the sake of convenience, in the remainder of this chapter we will assume that we are working at low enough concentrations so that molar concentrations can be used in place of chemical activity in our discussion.

In Table 16.1, the reduction half-reactions with large, positive values for $E°$ will be more likely to occur than reactions that have small or negative $E°$ values. Another way of looking at this is that *chemicals with large, positive $E°$ values will be easily reduced and will act as good oxidizing agents*. These types of compounds will tend to react as shown by the reduction half-reactions in Table 16.1, where the reactant on the left is reduced to form the product on the right. We can also see from this table that *chemicals with small or negative $E°$ values will be easily oxidized and will act as good reducing agents*. This second group of chemicals tends to react in the opposite direction of the reactions shown in Table 16.1, with the species on the right-hand side tending to be oxidized and giving up one or more electrons to form the chemicals shown on the left-hand side of this table.

Later we will discuss how the $E°$ values in Table 16.1 were actually determined, but it should be pointed out here that the calculation of these numbers always requires that a comparison be made to some reference reaction. In theory, this is done by using the reduction of hydrogen ions to hydrogen gas as the

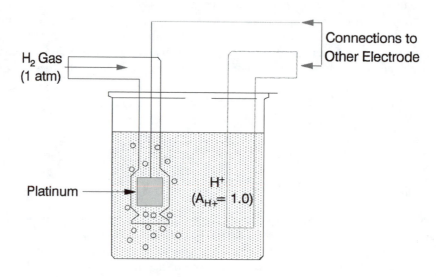

**FIGURE 16.2**    The standard hydrogen electrode.

---

*Chemistry Professionals at Work*                    *CPW Box 16.3*

# THE CHEMICAL TREATMENT
# OF DRINKING WATER

A general knowledge of whether a particular chemical is a strong oxidizing or reducing agent can be important in industrial applications in helping choose which chemicals might be best to use for a particular type of reduction-oxidation reaction. For example, chlorine ($Cl_2$) is often used as an agent for the treatment of drinking water. Table 16.1 shows us that one reason chlorine is a good choice for water treatment is that $Cl_2$ dissolved in water makes a fairly strong oxidizing agent. The standard reduction potential for dissolved $Cl_2$ in water is $+1.40$ volts, which makes chlorine a stronger oxidizing agent than even oxygen under the same conditions. This means that combining $Cl_2$ with many other compounds will tend to result in a reaction where $Cl_2$ is reduced and these other chemicals are oxidized, converting them into forms that are usually less hazardous to humans, plants, and animals.

*For Homework: Contact your local water treatment facility and ask what types of processes are used there for the treatment of drinking water. Prepare a report on your findings.*

---

reference, where $E°$ for this process is automatically assigned a value of 0.00 volts (see Fig. 16.2).

$$2H^+ + 2e^- \rightleftharpoons H_2 \qquad (16.8)$$

Because of this convention, any reduction half-reaction that has a positive value for $E°$ will occur more readily than the reaction of two hydrogen ions to form $H_2$ under standard reaction conditions. On the other hand, any half-reaction with a negative value for $E°$ will instead be more likely to occur as an oxidation process when compared to the reduction of $H^+$ to hydrogen gas.

## Example 16.7

Which of the chemicals in the following pairs is more easily *reduced* under standard reaction conditions? (Hint: see Table 16.1 for $E°$ values).

(a) $MnO_4^-$ (forming $Mn^{2+}$)   or   $Cr_2O_7^{2-}$ (forming $Cr^{3+}$)

(b) $Fe^{2+}$ (forming Fe)   or   $Fe^{3+}$ (forming Fe)

(c) $Cu^{2+}$ (forming Cu)   or   $Fe^{2+}$ (forming Fe)

## Solution 16.7

The chemical in each pair that will be most easily reduced is the one with the *largest* $E°$ value for its corresponding reduction half-reaction: (a) $MnO_4^-$, (b) $Fe^{3+}$, and (c) $Cu^{2+}$.

---

*Chemistry Professionals at Work*                                    *CPW Box 16.4*

# GALVANIC CELLS AND BATTERIES

The Daniell cell is just one of the many types of electrochemical cells that we use to supply us with energy everyday in batteries. Other examples are shown in Table 16.2. One of these is the lead storage battery that is used to start cars. To produce electricity, this battery uses the reduction of $PbO_2$ to form $PbSO_4$ in the presence of sulfuric acid, and the oxidation of Pb to also form $PbSO_4$. Smaller batteries, like those used in calculators, flashlights, radios, and electronic games, are based on a variety of other reactions. Some examples include alkaline cells, which use the oxidation of Zn and the reduction of $MnO_2$ in the presence of potassium hydroxide, and nickel-cadmium batteries, in which Cd is oxidized to $Cd(OH)_2$ and $NiO_2$ is reduced to $Ni(OH)_2$ in the presence of a hydroxide solution. The design and proper construction of all these batteries requires a good working knowledge of electrochemical cells and reduction-oxidation reactions. See also CPW Box 4.4.

*For Homework: Navigate around several Websites of some companies that market batteries. Examples are (as of 6/7/00) Duracell (www.duracell.com), Energizer (www.energizer.com), Rayovac (www.rayovac.com), and Sears Diehard™ (www.sears.com). Write a report on the information you discover, including the contents of some of the batteries and any new products that may be forthcoming.*

---

## Example 16.8

Which of the chemicals in the following pairs is more easily *oxidized* under standard reaction conditions?

(a) $Cl^-$ (forming $Cl_2$)   or   $Br^-$ (forming $Br_2$)

(b) $Fe^{2+}$ (forming $Fe^{3+}$)   or   Fe(forming $Fe^{3+}$)

(c) Pb (forming $Pb^{2+}$)   or   Cu(forming $(Cu^{2+}$)

## Solution 16.8

The chemical in each pair that will be most easily oxidized is the one with the *smallest* $E°$ value for its corresponding reduction half-reaction: (a) $Br^-$, (b) Fe, and (c) Pb.

## 16.6   Electrochemical Cells

In the previous examples we have seen various processes in which two half-reactions occur simultaneously in the same solution. However, electrochemical reactions can also be made to occur in systems in which the two half-reactions are physically separated from one another, but still allowed to have some contact so that *electrons can flow from one half-reaction to the other*. This type of arrangement is known as an **electrochemical cell**. The general group of electrochemical cells that are probably the most familiar to all of us are those used in batteries. This type of cell is called a **galvanic cell** and it is characterized by

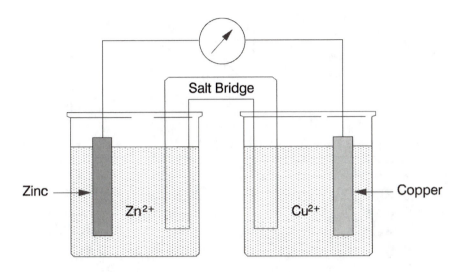

**FIGURE 16.3**   A zinc/copper galvanic cell.

the production of electricity (or electron flow) as a result of the spontaneous reaction between an oxidizing agent and a reducing agent (see Chapter 13 for a review of what is meant by a spontaneous chemical reaction). The flow of electrons through this system is useful as a means of generating electricity for driving motors or electronic devices.

We saw earlier in Section 10.5 of Chapter 10 that the basic design of an electrochemical cell consists of two metal conductors, known as *electrodes*, that are placed into a solution that contains chemicals that can be oxidized and reduced (see Fig. 16.3). At one of these electrodes (called the *anode*, or "–" electrode in a galvanic cell), oxidation will take place as electrons are released from a chemical and passed into the electrode. These electrons then flow through a wire or circuit to the other electrode (known as the *cathode*, or "+" electrode in a galvanic cell) where reduction takes place as the electrons are passed onto another chemical. To complete the electrical circuit, there must also be a flow of ions in solution between the cathode and anode in order to maintain a balance of charge in the overall system. These ions are supplied by a solution or medium known as the *electrolyte*. If the solutions that contain the anode and cathode are held in separate containers, as shown in Fig. 16.3, then an additional intermediate solution or medium known as the *salt bridge* is used to help ions flow from one electrode's solution to the other.

A simple example of a galvanic cell is shown in Figure 16.3. This particular cell is based on the reaction of zinc metal with copper +2 ions in solution to give zinc +2 ions and copper metal, as shown by Fig. 16.4 and the reactions given below.

$$\text{Oxidation Half-Reaction:} \quad \text{Zn} \rightleftharpoons \text{Zn}^{2+} + 2e^- \tag{16.9}$$

$$\text{Reduction Half-Reaction:} \quad \underline{\text{Cu}^{2+} + 2e^- \rightleftharpoons \text{Cu}} \tag{16.10}$$

$$\text{Net Reaction:} \quad \text{Zn} + \text{Cu}^{2+} \rightleftharpoons \text{Zn}^{2+} + \text{Cu} \tag{16.11}$$

This reaction forms the basis for a simple type of battery known as the **Daniell cell**, which was invented by the English chemist **John Frederick Daniell** in 1836. The two electrodes in this cell consist of the same metals that are taking part in the oxidation and reduction reactions, Zn and Cu. The zinc electrode (which acts as the anode) is placed into a solution containing $\text{Zn}^{2+}$ ions, while the copper electrode (which acts as the cathode) is in contact with a $\text{Cu}^{2+}$ solution. A wire that connects the zinc and copper electrodes is used to make electrical contact between these parts of the cell. In addition, the $\text{Zn}^{2+}$ and $\text{Cu}^{2+}$ solutions are kept separate but in contact with one another by using a gel-filled tube which acts as the salt bridge.

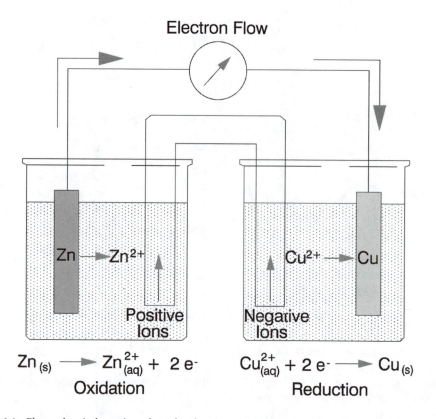

**FIGURE 16.4**   Electrochemical reactions that take place in a zinc/copper galvanic cell.

**TABLE 16.2**   Common Batteries

| Type of Battery | Reduction-Oxidation Reactions |
|---|---|
| Lead storage battery | Cathode: $PbO_2 + HSO_4^- + 3H^+ + 2e^- \rightleftharpoons PbSO_4 + 2H_2O$ |
| | Anode: $Pb + HSO_4^- \rightleftharpoons PbSO_4 + H^+ + 2e^-$ |
| Nickel-cadmium battery | Cathode: $NiO_2 + 2H_2O + 2e^- \rightleftharpoons Ni(OH)_2 + 2OH^-$ |
| | Anode: $Cd + 2OH^- \rightleftharpoons Cd(OH)_2 + 2e^-$ |
| Alkaline battery | Cathode: $2MnO_2 + H_2O + 2e^- \rightleftharpoons Mn_2O_3 + 2OH^-$ |
| | Anode: $Zn + 2OH^- \rightleftharpoons ZnO + H_2O + 2e^-$ |

Examples of familiar batteries and their anode and cathode reactions are given in Table 16.2.

Besides being useful in ranking the strength of various chemicals as reducing agents or oxidizing agents, the standard reduction potential ($E°$) can be employed to determine the overall change in potential that will result when oxidizing and reducing agents are combined to form an electrochemical cell under standard reaction conditions. This is done by simply taking the standard reduction potentials for the two half-reactions present in the cell and using these $E°$ values in the following equation to calculate the overall ***standard cell potential*** ($E°_{Cell}$).

$$E°_{Cell} = E°_{Cathode} - E°_{Anode} \qquad (16.12)$$

In this equation, $E°_{Cathode}$ is the standard reduction potential for the reduction half-reaction (which takes place at the cathode) and $E°_{Anode}$ is the standard reduction potential for the oxidation half-reaction (which takes place at the anode). The reason for subtracting rather than adding $E°_{Anode}$ in this equation is that the $E°$ values being used here are all based on reduction processes, and the $E°$ for an oxidation half-reaction would be opposite in sign to the $E°$ value obtained for the corresponding reduction half-reaction.

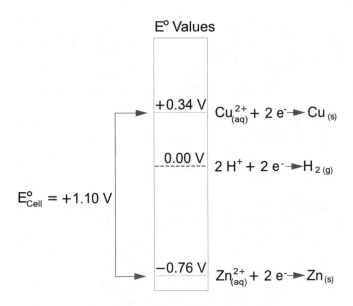

**FIGURE 16.5** Cell potential for a zinc/copper galvanic cell under standard reaction conditions.

## Example 16.9

What is the expected value for $E_{Cell}^{o}$ in the zinc/copper cell (see Fig. 16.3) under standard reaction conditions?

## Solution 16.9

From Table 16.1, the values of $E^{o}$ for the cathode and anode reactions are $+0.34$ volts and $-0.76$ volts, respectively. This gives the following result for $E_{Cell}^{o}$, which is further illustrated in Fig. 16.5.

$$
\begin{aligned}
E_{Cell}^{o} &= E_{Cathode}^{o} - E_{Anode}^{o} \\
&= (+0.34 \text{ volts}) - (-0.76 \text{ volts}) \\
&= +1.10 \text{ volts}
\end{aligned}
$$

We can see from Example 16.9 that a *positive* value for $E_{Cell}^{o}$ is present in the zinc/copper cell. This type of positive potential value is a characteristic of galvanic cells and is an indication that these cells can be used as a source of electrical energy. This is the general principle behind the operation of any battery, which is simply a combination of one or more electrochemical cells that are allowed to undergo a reduction-oxidation reaction to produce a flow of electrons and create electricity.

## 16.7  Measuring Standard Reduction Potentials

In addition to being used to supply energy, electrochemical cells form the basis for measuring the $E^{o}$ value for a half-reaction. This is done by placing the half-reaction of interest into one side of an electrochemical cell and placing into the other side a second, reference half-reaction that has a well-defined potential. Earlier we said that the reduction of $2H^{+}$ to $H_2$ is the conventional reaction against which, at least in theory, all other standard reduction potentials are compared [Eq. (16.8), restated below].

$$
2H^{+} + 2e^{-} \rightleftharpoons H_2 \qquad E^{o} = 0.00 \text{ volts}
$$

This type of comparison is performed experimentally by placing the components of the above half-reaction into the anode side of an electrochemical cell and the components for the reaction to be studied into the cathode side. The particular electrode that is used for the $H^+/H_2$ reference reaction consists of a platinum electrode in contact with bubbling hydrogen gas (at 1 atm pressure) and in a solution that contains hydrogen ions at 1 mol/L activity (see Fig. 16.2). This type of electrode is known as the **standard hydrogen electrode (SHE)**.

Since the standard hydrogen electrode is used as the anode of the cell, the net cell potential that is measured will be given by Eq. (16.13).

$$E^o_{Cell} = E^o_{Cathode} - E^o_{SHE} = E^o_{Cathode} - (0.00\ volts) = E^o_{Cathode} \qquad (16.13)$$

In other words, the value of $E^o_{Cell}$ that results for this system will automatically be equal to the standard reduction potential for the reaction taking place in the cathode side of the cell.

### Example 16.10

A chemist wishes to measure the standard reduction potential for the following reaction by using a standard hydrogen electrode as the reference.

$$Ru^{3+} + e^- \rightleftharpoons Ru^{2+}$$

A cell potential of $+0.25$ volts is measured for the resulting electrochemical cell under standard reaction conditions. What is the value of $E^o$ for the half-reaction that is being studied?

### Solution 16.10

$$E^o_{Cell} = E^o_{Cathode} - E^o_{SHE}$$

$$+0.25\ volts = E^o_{Cathode}$$

$$\text{Answer: } E^o = E^o_{Cathode} = +0.25\ volts$$

Although the standard hydrogen electrode is the reference listed in most tables of $E^o$ values, this type of electrode is not convenient to use on a routine basis. As a result, other reference reactions are used instead by industrial or research laboratories. Two common examples are the **saturated calomel electrode** (or **SCE**) and the **silver-silver chloride electrode** (or **Ag/AgCl electrode**). The reduction half-reactions for each of these two electrodes are shown in Eqs. (16.14) and (16.15), and diagrams of these electrodes are provided in Fig. 16.6 and 16.7.

Saturated Calomel Electrode:

$$Hg_2Cl_2 + 2e^- \rightleftharpoons 2Hg + 2Cl^- \quad E^o = +0.27\ volts \qquad (16.14)$$

Silver-Silver Chloride Electrode:

$$AgCl + e^- \rightleftharpoons Ag + Cl^- \quad E^o = +0.22\ volts \qquad (16.15)$$

Both of these alternative reference electrodes are much easier to make and use than the standard hydrogen electrode and are utilized daily by laboratories throughout the world for electrochemical

---

*Chemistry Professionals at Work*                    *CPW Box 16.5*

# POTENTIOMETRY

Y ou can see from the Nernst Equation that the measured potential depends on the concentration of dissolved species. Because of this, a number of techniques have been developed whereby chemistry laboratory workers can determine the concentration of a chemical species by measuring this potential. The resulting method of ***potentiometry*** has become a popular technique. An example is the measurement of pH with the use of a pH electrode (see Chapter 12, Section 12.8). Besides this measurement of hydrogen ions, other ions can also be determined by various ***ion-selective electrodes***.

*For Homework: Look up more of the theory of how a pH electrode or ion-selective electrode works and the procedure that is involved. Write a report.*

---

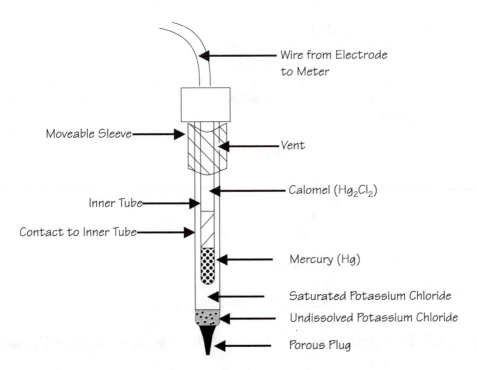

**FIGURE 16.6**   The saturated calomel electrode.

measurements. The only real change that is needed when using these other references is to remember that $E^\circ_{Anode}$ will no longer be equal to zero, which must be considered when determining the value of $E^\circ_{Cathode}$ from $E^\circ_{Cell}$ .

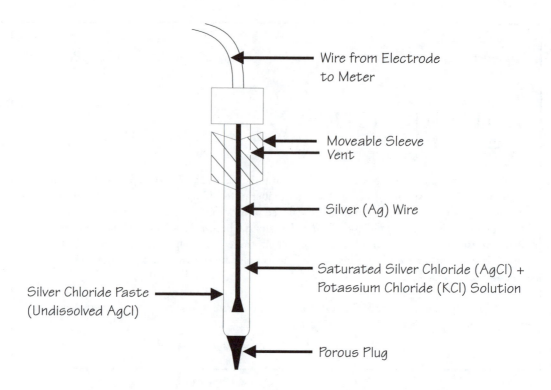

**FIGURE 16.7**   The silver-silver chloride electrode.

*Example 16.11*

Using a cell that has a total measured potential of $-0.63$ volts under standard reaction conditions, determine the standard reduction potential for the following reaction when a saturated calomel electrode (SCE) is used as the reference.

$$PbSO_4 + 2e^- \rightleftharpoons Pb + SO_4^{2-}$$

*Solution 16.11*

$$E^\circ_{Cell} = E^\circ_{Cathode} - E^\circ_{SHE}$$

$$-0.63 \text{ volts} = E^\circ_{Cathode} - (+0.27 \text{ volts})$$

$$\text{Answer: } E^\circ = E^\circ_{Cathode} = -0.36 \text{ volts}$$

## 16.8   Cell Potentials under Nonstandard Reaction Conditions

Up to this point we have only considered cells and half-reactions that have been operated under standard reaction conditions (1 atm pressure, 1 mol/L activity for each solution-phase species, and so on). But only rarely do we deal with reactions that actually occur under standard conditions, and even if we do start under such conditions, the concentrations of the oxidized and reduced species will begin to change as soon as the reactions starts. So what happens if the chemicals are present at nonstandard concentrations or pressures? As we might expect, this changes the potential that is measured for a cell or that is present for a half-reaction. In order to determine what this change will be, a relationship known as the **Nernst equation** is used.

Let's begin by looking at the following general half-reaction, in which "a" moles of a chemical "Ox" receives "n" moles of electrons to be converted into "b" moles of another chemical, "Red."

$$a \text{ Ox} + n e^- \rightleftharpoons b \text{ Red} \tag{16.16}$$

For this reaction, the measured reduction potential under nonstandard conditions (E) is related to the standard reduction potential (E°) through the following form of the Nernst equation.

$$E = E° + 2.303(R \cdot T)/(n \cdot F) \log([Ox]^a/[Red]^b) \tag{16.17}$$

In this relationship, R is the ideal gas law constant (8.314 volts · coulombs/mol · K, or 8.314 $m^3$ · pascals/mol · K), T is the absolute temperature of the system (in Kelvin), and F is a parameter known as the **Faraday constant** (a value equal to $9.649 \times 10^4$ coulombs/mol of $e^-$). The terms [Ox] and [Red] are the molar concentrations of Ox and Red in the system (which, as was said earlier, we will assume are roughly equal to the chemical activities of Ox and Red). Also, the terms "a", "b," and "n" are the same as the coefficients shown for Ox, Red, and $e^-$ in the above half-reaction. When working at 25°C (T = 298 K), the various constants in the Nernst equation are sometimes combined (where $(R \cdot T)/F = 0.0592$ volts at 298 K) to give the following alternative relationship.

$$E = E° + (0.0592 \text{ volts})/(n) \log([Ox]^a/[Red]^b) \tag{16.18}$$

## Example 16.12

Determine the value of E at 25°C for each of the following half-reactions. In doing so, use a "concentration" of 1.0 for any reactant or product that is a solid.

(a) $Ce^{4+} + e^- \rightleftharpoons Ce^{3+}$ ($[Ce^{4+}] = 0.10$ M; $[Ce^{3+}] = 0.01$ M)

(b) $Ce^{4+} + e^- \rightleftharpoons Ce^{3+}$ ($[Ce^{4+}] = 0.01$ M; $[Ce^{3+}] = 0.10$ M)

(c) $Cu^{2+} + 2e^- \rightleftharpoons Cu$ ($[Cu^{2+}] = 0.050$ M; "concentration" of Cu $= 1.00$,

   since Cu is a solid)

(d) $MnO_4^- + 8H^+ + 5e^- \rightleftharpoons Mn^{2+} + 4H_2O$ ($[MnO_4^-] = 0.15$ M; $[H^+] = 1.0 \times 10^{-5}$,

   or   pH $= 5.0$; $[Mn^{2+}] = 0.010$ M)

## Solution 16.12

(a) $E = E° + (0.0592 \text{ volts})/(n) \log([Ox]^a/[Ox]^b)$

   $= (+1.61 \text{ volts}) + (0.0592 \text{ volts})/(1) \log((0.10 \text{ M})^1/(0.10 \text{ M})^1)$

   $= +1.67$ volts

(b) $E = (+1.61 \text{ volts}) + ((0.0592 \text{ volts})/(1)) \log((0.01 \text{ M})^1/(0.01 \text{ M})^1)$

    $= +1.55 \text{ volts}$

(c) $E = (+0.34 \text{ volts}) + (0.0592 \text{ volts})/(2) \log((0.050 \text{ M})^1/(1.00)^1)$

    $= +0.30 \text{ volts}$

(d) $E = (+1.51 \text{ volts}) + (0.0592 \text{ volts})/(5) \log((0.15 \text{ M})^1(1.0 \times 10^{-5})^8/(0.010)^1)$

    $= +1.05 \text{ volts}$

For a reduction reaction that involves more than one chemical as a reactant, the top part of the chemical ratio in the Nernst equation (that is, the ratio that occurs within the log term) will be equal to the product of each reactant's concentration (or activity) raised to the power of that reactant's coefficient in the reduction half-reaction as in the equilibrium constant expressions we encountered in Chapter 11. For example, $[A]^{a} \bullet [B]^{b} \bullet [C]^{c}$ would be used to represent the reactants $\mathbf{a\,A} + \mathbf{b\,B} + \mathbf{c\,C} \rightleftharpoons ....$ Similarly, the bottom portion of the chemical ratio will be equal to the multiplication product of each chemical product's concentration (or activity) raised to the power of that product's coefficient in the reduction half-reaction. For instance, $[D]^{d} \bullet [E]^{e} \bullet [F]^{f}$ would be used to represent the products ... $\rightleftharpoons$ $\mathbf{d\,D} + \mathbf{e\,E} + \mathbf{f\,F}$.

Under standard reaction conditions, all of the concentrations and activities that are present in the log term of the Nernst equation will be equal to one and the log term itself will be equal to zero, since the logarithm of one is equal to zero. In this situation the Nernst equation predicts that the value of E for a half-reaction will be equal to $E^{\circ}$, as we would expect. However, this is not usually the case when concentrations or activities that are *not* equal to one are present in the system. For instance, if the concentrations of Ox and Red give a ratio for $[Ox]^{a}/[Red]^{b}$ that is greater than one, then E will be larger than $E^{\circ}$. If the concentrations of Ox and Red give a ratio for $[Ox]^{a}/[Red]^{b}$ that is less than one, then E will be smaller than $E^{\circ}$. Thus, we can see that the potential measured for a half-reaction will depend on the initial concentrations and activities of the oxidized and reduced species that are present for that half-reaction.

To use the Nernst equation to determine cell potentials under nonstandard reaction conditions, we first need to determine the corrected reduction half-cell potentials for each side of the cell. The resulting values of E for the anode and cathode are then used to calculate the overall cell potential through the same type of relationship that was given earlier for standard reaction conditions.

$$E_{Cell} = E_{Cathode} - E_{Anode} \qquad (16.19)$$

Notice that the only difference between Eqs. (16.19) and (16.12) is that Eq. (16.12) is based on *standard reduction potentials*, while Eq. (16.19) can be used with potentials determined under *either standard or nonstandard reaction conditions*.

### Example 16.13

Determine the cell potential at 25°C for a zinc/copper cell based on the following half-reactions.

$\text{Zn} \rightleftharpoons \text{Zn}^{2+} + 2e^{-}$ (at Anode; $[\text{Zn}^{2+}] = 0.020$ M; "concentration" of $\text{Zn} = 1.00$,

since Zn is a solid)

$\text{Cu}^{2+} + 2e^{-} \rightleftharpoons \text{Cu}$ (at Cathode; $[\text{Cu}^{2+}] = 0.010$ M; "concentration" of $\text{Cu} = 1.00$,

since Cu is a solid)

## Solution 16.13

For the $Zn^{2+}/Zn$ *reduction* half-reaction ($Zn^{2+}+2e^- \rightleftharpoons Zn$; $E = E_{Anode}$):

$$
\begin{aligned}
E_{Anode} &= E° + (0.0592 \text{ volts})/(n) \log([Ox]^a/[Red]^b) \\
&= (-0.76 \text{ volts}) + ((0.0592 \text{ volts})/(2)) \log((0.020 \text{ M})^1/(1.00)^1) \\
&= -0.81 \text{ volts}
\end{aligned}
$$

For the $Cu^{2+}/Cu$ *reduction* half-reaction ($Cu^{2+}+2e^- \rightleftharpoons Cu$; $E = E_{Cathode}$):

$$
\begin{aligned}
E_{Cathode} &= (+0.34 \text{ volts}) + (0.0592 \text{ volts})/(2) \log((0.010 \text{ M})^1/(1.00)^1) \\
&= +0.28 \text{ volts}
\end{aligned}
$$

Net Cell Potential:

$$
\begin{aligned}
E_{Cell} &= E_{Cathode} - E_{Anode} \\
&= (+0.28 \text{ volts}) - (-0.81 \text{ volts}) \\
&= +1.09 \text{ volts}
\end{aligned}
$$

# 16.9   Relationship between Cell Potential and Free Energy

Now that we've seen how to determine the overall potential for an electrochemical cell, let's consider how the potential of this cell is related to the amount of energy that can be obtained from the system. For instance, this type of information is important to consider when an electrochemical cell is being designed for use as a battery. This is done by using the relationship shown below between cell potential and the change in free energy for a reaction.

$$\Delta G = -(n\, FE_{Cell}) \tag{16.20}$$

As we learned in Chapter 13, $\Delta G$ is the total change in free energy that is possible for the system (given here in units of joules). The other terms in Eq. (16.20) are the same as used throughout this present chapter, where n represents the number of moles of electrons that are involved in the electrochemical reaction, F is the Faraday constant (in coulombs/mol of $e^-$), and $E_{Cell}$ is the cell potential (expressed in units of volts or joules/coulomb). For a galvanic cell (that is, one that can be used to generate electricity), we observed earlier that the cell potential will have a positive value. According to Eq. (16.20), this will result in a negative value for $\Delta G$. This represents a spontaneous chemical reaction, as discussed earlier in Chapter 13. In the next section of this chapter we will examine electrochemical cells in which energy must be put into the system in order for a reaction to occur. These types of cells, called ***electrolytic cells***, have negative values for $E_{Cell}$ and positive values for $\Delta G$.

As the value of $E_{Cell}$ becomes a larger positive value, Eq. (16.20) predicts that it will be possible to produce more energy from this system. Conversely, if $E_{Cell}$ becomes a more negative value, then more energy must be placed into the system in order for the desired electrochemical reaction to take place.

## Example 16.14

Determine the total change in free energy that is possible for the Zn/Cu cell in Example 16.13.

## Solution 16.14

It was calculated in Example 16.13 that the net cell potential ($E_{Cell}$) for this system is +1.09 volts (or +1.09 joules/coulomb). This value is then used along with Eq. (16.20) to obtain the total change in free

# APPLICATIONS OF ELECTROLYTIC CELLS

In Chapter 10 (Section 10.5), we learned about how electrolytic cells can be used in electroplating with copper or in the commercial preparation of chlorine and sodium hydroxide. Some other examples of commercial applications of electrolytic cells include the use of these cells to generate magnesium metal from molten magnesium chloride, refine copper from copper ore (chalcopyrite, $CuFeS_2$), and obtain aluminum metal from bauxite ($Al_2O_3 \cdot xH_2O$).

Chemists and chemistry technicians, making use of small-scale electrolytic cells in the laboratory, can analyze for nearly thirty different metals at concentration levels that are as low as 50 parts per billion using the techniques mentioned in the text. In addition, electrolytic cells can be used in instrumental liquid chromatography for detection and quantitation purposes.

*For Homework: Find an electroanalytical method that utilizes an electrolytic cell. Try looking in a book of methods, such as* Standard Methods for the Examination of Water and Wastewater, *published by the American Waterworks Association. Examples of techniques include those mentioned in the text as well as the method of "amperometric titration." Write a report on your selected method and its application.*

energy that is possible for this cell ($\Delta G$):

$$\Delta G = -(nFE_{Cell})$$

$$= -(2 \text{ mol e}^-)(9.649 \times 10^4 \text{coulombs/mol e}^-)(+1.09 \text{ joules/coulomb})$$

$$= -210 \times 10^3 \text{ joules}$$

$$= -210 \text{ kilojoules}$$

## 16.10   Electrolytic Cells

As stated earlier, electrolytic cells are simply electrochemical cells that have negative values for $E_{Cell}$ and therefore require that energy be placed into them in order for their electrochemical reactions to take place. Although this means that such cells cannot be used as energy sources, it still makes them useful in generating various types of chemicals through reduction-oxidation processes.

There are many examples of chemical products that are generated by electrolytic cells. One example is the chromium plating that is found on many metals, such as the exterior of cars. This is done by taking the metal to be plated and using it as one of the electrodes in a giant electrochemical cell. This metal is then placed in contact with a solution of $CrO_3$ in dilute sulfuric acid. As a current and energy are applied to this cell, the following reduction half-reaction occurs in the solution and at the metal's surface.

$$CrO_3 + 6H^+ + 6e^- \rightleftharpoons Cr + 3H_2O$$

---

*Chemistry Professionals at Work*                                 *CPW Box 16.7*

# ELECTROCHEMISTRY AT EASTMAN KODAK

**M**y name is Debra Butterfield. As a Chemical Technician employed by Eastman Kodak Company in Rochester, NY, I've worked in chromatography labs for the majority of my 20-year career at Kodak. Liquid chromatography with electrochemical detection (LC-EC) is my current area of expertise. By combining the selectivity of HPLC with the sensitivity of electrochemical detection, trace level (ng/g) electrochemically active species that cause unique photographic effects can be detected and correlated to sensito-metric results.

My daily job dynamics include a mixture of routine analysis, problem solving, troubleshooting, methods development, sup-port to routine production areas, and project management. I partner with a Ph.D. electrochemist and am continuously chal-lenged by day-to-day activities. Whether it's sample preparation, instrumentation, data interpretation, computer usage, customer interactions, training, meetings, presentations, Good Laboratory Practice, time management, or equipment purchasing, I'm constantly reevaluating priorities. I've always enjoyed my job as a chemical technician and I look forward to the ever-changing oppor-tunities it brings.

*For Homework: In a book on high-performance liquid chromatography (HPLC), or the HPLC chapter of an analytical chemistry textbook or instrumental analysis textbook, look up electrochemical detectors. Write a report giving the full details of the design of these detectors and how they work. Ms. Butterfield mentioned the sensitivity of these detectors. In your report, compare the sensitivities of these detectors with that of other conventional detectors.*

---

The result is a coating of chromium (Cr) that forms on the surface of the metal as a protective layer. This process is fairly expensive to perform because of the large amount of electricity required (6 moles of $e^-$ are needed per mol of deposited Cr), but it does result in a nice appearance and in good corrosion resistance for the coated surface.

Small-scale electrolytic cells are used in analytical testing laboratories for techniques that are called electroanalytical methods. Examples of such methods are polarography, voltammetry, amperometry, and electrochemical detection in liquid chromatography. These methods measure dissolved chemicals by applying a controlled voltage or current to an electrode and observing and measuring the electrolytic response, either a current or voltage. Such responses lead to the concentrations of the dissolved chemicals in the samples being analyzed.

# 16.11 Homework Exercises

1. Define the terms "oxidation" and "reduction." Explain why both of these must always be present in an electrochemical reaction.
2. Define each of the following terms.
   (a) Oxidizing agent          (b) Reducing agent
   (c) Oxidant                  (d) Reductant
3. What is a "half-reaction?" How are half-reactions used in describing reduction-oxidation reactions?
4. What is a "standard reduction potential?" How can this be measured?
5. How is the oxidizing or reducing ability of a chemical related to its standard reduction potential?
6. Dichromate ($Cr_2O_7^{2-}$) and permanganate ($MnO_4^-$) are two examples of common oxidizing agents. Explain the reasons for this based on the information that is given in Table 16.1.
7. Elemental sodium (Na) and calcium (Ca) are two examples of good reducing agents. Explain the reasons for this based on the information that is given in Table 16.1.
8. Rank the following chemicals in the order of their ability to be reduced: $Ag^+$ (forming Ag), $Ca^{2+}$ (forming Ca), $Ce^{4+}$ (forming $Ce^{3+}$), $Co^{3+}$ (forming $Co^{2+}$), $K^+$ (forming K), and $Na^+$ (forming Na).
9. Rank the following chemicals in the order of their ability to be oxidized: Ag (forming $Ag^+$), Ca (forming $Ca^{2+}$), Ce (forming $Ce^{3+}$), Co (forming $Co^{2+}$), K (forming $K^+$), and Na (forming $Na^+$)
10. Balance each of the following electrochemical reactions. If no pH range is specified, then balance the reaction by using either acidic or basic pH conditions.
    (a) $Fe^{3+} + Cu \rightleftharpoons Fe^{2+} + Cu^{2+}$
    (b) $Ni(OH)_3 + Cd \rightleftharpoons Ni(OH)_2 + Cd(OH)_2$ (*basic pH*)
    (c) $Cl_2 \rightleftharpoons HOCl + Cl^-$ (*acidic pH*) (Hint: consider $H_2O$ as a reactant)
    (d) $MnO_4^- + S^{2-} \rightleftharpoons MnO_2 + S$ (*basic pH*)
11. What is the difference between a galvanic cell and an electrolytic cell?
12. Define each of the following terms:
    (a) Cathode                  (b) Anode
    (c) Salt bridge              (d) Electrolyte
13. Explain the role played in an electrochemical cell by each of the items that are listed in Problem 12.
14. How are the concentrations of the chemicals in a reduction half-reaction related to reduction potential for that reaction?
15. Determine the reduction potentials for each of the following half-reactions at 25°C.
    (a) $Fe^{3+} + e^- \rightleftharpoons Fe^{2+}$ ([$Fe^{3+}$] = 0.020 M; [$Fe^{2+}$] = 0.010 M)
    (b) $Al^{3+} + 3e^- \rightleftharpoons Al$ ([$Al^{3+}$] = 0.10 M; "concentration" of Al = 1.00)
    (c) $AgCl + e^- \rightleftharpoons Ag + Cl^-$ ([AgCl] = 1.00; "concentration" of Ag = 1.00; [$Cl^-$] = 0.050 M)
16. What is the "Nernst equation?" What are the various terms that appear in this equation? Why is this equation important when dealing with reduction-oxidation reactions?
17. Describe the standard hydrogen electrode. What half-reaction is used in this type of electrode? How is this electrode used as a reference in electrochemical cells?
18. Describe the silver-silver chloride electrode. What half-reaction is used in this type of electrode? What advantages does this electrode have over the standard hydrogen electrode for use as a reference in electrochemical cells?
19. Describe the saturated calomel electrode. What half-reaction is used in this type of electrode? What advantages does this electrode have over the standard hydrogen electrode for use as a reference in electrochemical cells.
20. Determine the cell potential for each of the following electrochemical cells:
    (a) $Fe^{3+} + e^- \rightleftharpoons Fe^{2+}$ (At Cathode; [$Fe^{3+}$] = 0.020 M; [$Fe^{2+}$] = 0.010 M)
        $Ce^{3+} \rightleftharpoons Ce^{4+} + e^-$ (At Anode; [$Ce^{3+}$] = 0.020 M; [$Ce^{4+}$] = 0.010 M)
    (b) $Hg^{2+} + 2e^- \rightleftharpoons Hg$ (At Cathode; [$Hg^{2+}$] = 0.015 M; "concentration" of Hg = 1.00)
        $Zn \rightleftharpoons Zn^{2+} + 2e^-$ (At Anode; "concentration" of Zn = 1.00; [$Zn^{2+}$] = 0.015 M)

(c) $AgCl + e^- \rightleftharpoons Ag + Cl^-$ (At Cathode; "concentration" of $AgCl = 1.00$; "concentration" of $Ag = 1.00$; $[Cl^-] = 0.20$ M)

$Ag \rightleftharpoons Ag^+ + e^-$ (At Anode; "concentration" of $Ag = 1.00$; $[Ag^+] = 0.20$ M)

21. Calculate the total change in free energy that is possible for each of the cells listed in Problem 20. Which of these are galvanic cells and which are electrolytic cells?

22. Describe what is meant by "chemical activity." How does this differ from the concentration of a chemical? When are the activity and concentration of a chemical about the same?

23. Fill in the following blanks with the terms "galvanic cell" or "electrolytic cell."
    (a) A car battery is a(n) _____ when it is used to start a car.
    (b) A car battery is a(n) _____ when it is being recharged.
    (c) A flashlight battery is a(n) _____ when is it being used to run a flashlight.
    (d) A positive change in free energy is a feature found in a _____.
    (e) A negative change in free energy is a feature found in a _____.

24. What is the difference between a "reduction potential" and a "standard reduction potential?" When are these the same? When are they different?

25. What are some industrial applications of galvanic cells? What are some industrial applications of electrolytic cells?

## For Class Discussion and Reports

26. An important industrial application of electrochemical reactions is their use in refining copper metal from the copper ore chalcopyrite ($CuFeS_2$). Find information on how this process is performed and on the reduction-oxidation reactions that are involved. Write a report on this process. Include information on the amount of copper that is refined each year by this technique and on the advantages and disadvantages of this refinement method.

27. Electrically powered automobiles have been a subject of interest for many years. These rely on some type of battery for supplying the power that is needed to run the automobile. Find information on this subject and report on the types of batteries that are currently used for this purpose. What are the current obstacles that must be overcome before these types of automobiles become competitive with those that run on fossil fuels? What are the potential advantages of using electrically powered automobiles instead of those that run on fossil fuels?

28. Reread CPW Box 12.3, where we defined a technique called a "titration." In that discussion, we also defined an indicator as a chemical that tells us when the end point of a titration has been reached. Some titrations utilize an electrochemical means of detecting an end point rather than an indicator. In an analytical chemistry book, look up "potentiometric titrations" and "amperometric titrations" and write a report on the end point detection methods in these techniques. Try to find at least one application for each that is performed by chemists and chemistry technicians in the workplace and give some details in your report.

# 17

# Biochemistry

## 17.1   Introduction

Chemical reactions are not only important in industry, but they are also responsible for the processes that go on in our bodies to sustain life. **Biochemistry** is the name given to the study of chemicals and chemical reactions that are present in living biological systems. In this chapter we will learn about some of the major groups of chemicals that are important in living organisms and we will examine the various roles that are played by each of these chemical groups in sustaining life.

## 17.2   Proteins and Peptides

The first general group of biological compounds that we will consider are the proteins and peptides. Proteins and peptides are the building blocks for many of the structural components of our bodies like skin, hair, and muscle. Proteins and peptides also act as the "engines" that perform most of the everyday reactions and tasks that take place within our cells. As we will see later, these tasks include such duties as transporting specific chemicals throughout the body, helping certain chemical reactions to occur within the body, fighting off invasion by foreign organisms, and sending chemical messages from one part of the body to another.

### 17.2.1   Amino Acids

In order to understand how proteins and peptides work, we first must look at their chemical structure. Proteins and peptides are made up of chains of **amino acids**. The basic structure of an amino acid is shown in Fig. 17.1. An amino acid always contains four types of functional groups that are connected to a central carbon atom. One of these groups is a hydrogen atom. Two of the other groups are a carboxylic acid and an amine group. It is the presence of these last two particular groups that gives rise to the name "amino acid." The general structure and properties of both carboxylic acids and amine groups were discussed earlier

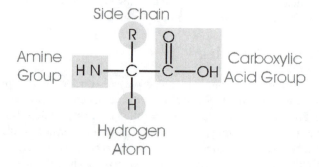

**FIGURE 17.1**  The basic structure of an amino acid. The symbol "R" represents the side chain of the amino acid.

in Sections 6.7.5 and 6.8.1 of Chapter 6. The hydrogen atom, amine group, and carboxylic acid group are present in all amino acids. It is these three groups that give all of the amino acids a set of common properties. For example, as we will see in Fig. 17.3, the presence of both an amine group and a carboxylic acid group in all amino acids is what makes it possible to connect amino acids together to form proteins and peptides.

The fourth type of group that is attached to the central carbon of an amino acid is different from one type of amino acid to the next. This fourth group, which is known as the amino acid's **side chain**, is represented by the symbol "R" in Fig. 17.1. Figure 17.2 shows the side chains for the 20 most common types of naturally occurring amino acids. The fact that each type of amino acid has a different chemical structure in its side chain means that each amino acid will have its own set of unique chemical and physical properties. For instance, the additional carboxylic acid groups that are present in the side chains

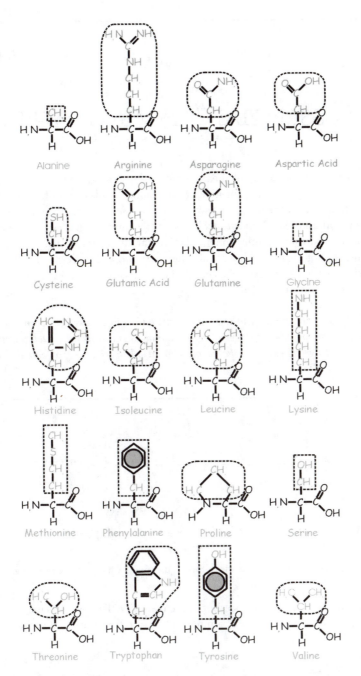

**FIGURE 17.2** Structures of the 20 most common types of amino acids. The portion of each amino acid in the dashed box is the amino acid's side chain.

of the amino acids glutamic acid and aspartic acid will make these act as better "acids" than other types of amino acids. On the other hand, the additional amine groups that can be found on the side chains of lysine, asparagine, histidine, and arginine make these amino acids act as better "bases" than other amino acids.

It is possible to sort all amino acids into one of two categories based on how they interact with water. Those amino acids with side chains that have relatively strong attraction to water are known as ***hydrophilic*** ("water-loving") amino acids. Most of the amino acids with oxygens or nitrogens in their side chains are in this category. The other group of amino acids are called ***hydrophobic*** ("water-fearing"). These amino

*Chemistry Professionals at Work*                                        *CPW Box 17.2*

# THE "ESSENTIAL" AMINO ACIDS

Our bodies can make many of the 20 amino acids that are shown in Fig. 17.2 for use in constructing proteins and peptides. But nine of these amino acids cannot be made by our bodies and must be obtained from the foods that we eat. The amino acids that we need to get from our diet are known as the **essential amino acids**. For adult humans, the essential amino acids are isoleucine, leucine, lysine, methionine, phenylalanine, threonine, tryptophan, valine, and possibly histidine. In infants, arginine is also essential. Chemistry professionals who work in the food and health industries need to know about the chemical content of foods in order to ensure that we get enough of these essential amino acids in the food we eat.

*For Homework: Look up the amino acid content of some typical foods, like beef, corn, beans, wheat, and cereal. Report on how knowledge of the essential amino acids is important in determining a good diet for young people in their growing years or for people who eat only specific types of food, such as vegetarians.*

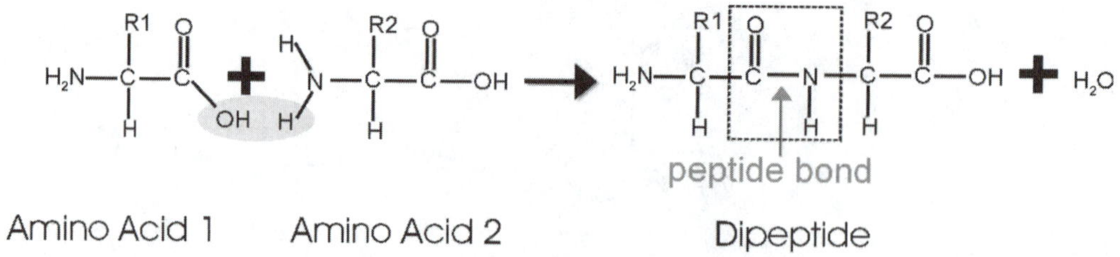

**FIGURE 17.3**   Combination of two amino acids to form a dipeptide, which is linked together by a peptide bond.

acids include all of those with simple hydrocarbon side chains (like alanine or leucine) and aromatic groups (like phenylalanine). Only one hydrophobic amino acid has a nitrogen in its side chain (tryptophan).

## Example 17.1

Determine whether each of the following amino acids have hydrophilic or hydrophobic side chains.
   (a)   Arginine
   (b)   Valine
   (c)   Glutamic acid

## Solution 17.1

   (a)   Hydrophilic, nitrogen present in side chain
   (b)   Hydrophobic, simple hydrocarbon side chain
   (c)   Hydrophilic, both nitrogen and oxygen present in side chain

*For Class Discussion: What properties do you think are responsible for making most amino acids hydrophilic when they have nitrogen or oxygen present in their side chains? Why are all amino acids with simple hydrocarbon side chains or aromatic side chains hydrophobic?*

## 17.2.2 Protein and Peptide Structures

The way in which amino acids are combined together to produce a protein or peptide is illustrated in Fig. 17.3. In this process, the amine group of one amino acid reacts with the carboxylic acid group of another amino acid. Water is one product of this reaction, which means this process is a type of **dehydration reaction**. The other product of this reaction is a chain of two amino acids that have joined together through the formation of an **amide bond**; this is also known in biochemistry as a **peptide bond**. The general properties of amide groups were discussed earlier in Chapter 6.

### Example 17.2
Write out the complete chemical structure for the following chain of amino acids where the amino acid at the far left of the chain (glycine) has an unreacted amine group after the chain has been formed, and the amino acid at the far right of the chain (arginine) has an unreacted carboxylic group after the chain has been formed.

Glycine-Serine-Arginine

### Solution 17.2
The structure of this molecule is shown below.

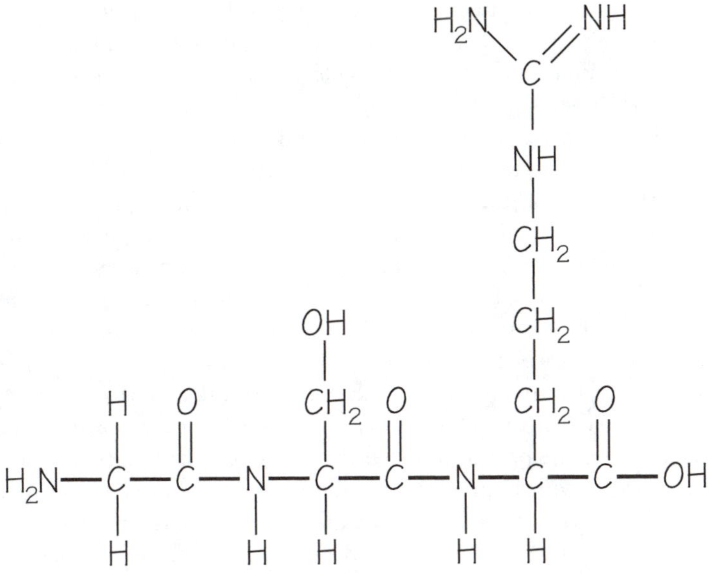

It is possible for the type of reaction in Fig. 17.3 to form a chain between any two amino acids. The same reaction can be used to link together even longer amino acid chains. Polymers of amino acids that are made in this way are called **polypeptides**. Common polypeptides can have sizes up to several hundred amino acids in length. Small polypeptides that have molecular weights less than about 5000 g/mol are often referred to as **peptides**. If the polypeptide has a molecular weight greater than 5000 g/mol, it is usually called a **protein**. Some proteins are so large that they have molecular weights of over 1,000,000 g/mol.

There are many different jobs that are performed by proteins and peptides in living organisms. Table 17.1 lists some of these functions. For example, proteins are important in creating various structural components

**TABLE 17.1** Some Functions of Proteins and Peptides in Our Bodies

| Function | Example |
| --- | --- |
| 1. Form structural features like skin, bones, tendons, and hair | Collagen |
| 2. Help in movement | Muscle proteins |
| 3. Catalyze reactions | Enzymes |
| 4. Transport chemicals within the body | Hemoglobin |
| 5. Protect body from disease and infection | Antibodies |
| 6. Transmit chemical messages throughout the body | Hormones |

in animals, such as skin, bones, cartilage, tendons, fingernails, and hair. The proteins present in muscle cells are essential in providing us with the ability to move. Other proteins, known as ***enzymes***, help catalyze many of the reactions that take place in living organisms. Some proteins, like ***hemoglobin***, are involved in the transport of chemicals throughout our bodies such as the movement of oxygen from the lungs to our cells and carbon dioxide from our cells to the lungs. Proteins and peptides also make up part of our body's defense system against disease and injury. For instance, ***antibodies*** are proteins that fight off bacteria and viruses that enter our bodies. ***Hormones*** are a group of proteins, peptides, and amino acid derivatives that act as chemical messengers that carry signals throughout our bodies for the control of various organs and tissues. Examples of protein and peptide-based hormones include insulin and human growth hormone.

As we can see from the list in Table 17.1, there is a large variety of jobs that are carried out by proteins and peptides in our bodies. But what is it that makes one protein or peptide different from another? The answer is the unique series of amino acids that are linked together to make up that protein or peptide. Based on only 20 different amino acids, it is possible to put together an enormous number of different combinations. For instance, there are 400 (or $20 \times 20$) combinations of amino acids that could be used to produce a simple two amino acid peptide. There are 8000 possible combinations ($20 \times 20 \times 20$) for peptides that contain three amino acids. And for a peptide that contains ten amino acids, there are $1 \times 10^{13}$ (or $20^{10}$) possible combinations! From this it is easy to see how our bodies can make so many types of peptides and proteins that have such different functions.

The sequence of amino acids that makes up a protein or peptide chain is known as the ***primary structure*** for that protein or peptide. An example is the structure of the three amino acid peptide that was given in Example 17.2. As was shown in this exercise, the primary structure for a protein or peptide is always written with the first amino acid being the one that has an unreacted amine group attached to its central carbon. This amino acid would be glycine in our example. This end of the protein or peptide is known as the ***N-terminal end*** or ***N-terminus***, because of the nitrogen atom that is present in the unreacted amine group. The last amino acid that is listed (arginine in Example 17.2) is the one that contains a central carbon with an unreacted carboxylic acid group. This end of the protein or peptide is called the ***C-terminal end*** or ***C-terminus***, because of the presence of the unreacted carboxylic acid.

When describing the primary structure of a protein or peptide, it is time-consuming to draw the full structures of all amino acids that are present in polypeptide chains. Instead, a type of shorthand is used to list the amino acids without having to provide all of the atoms or chemical groups that are present in them. This is done by using either a three letter code or one letter abbreviation for each amino acid. Both types of abbreviations are listed in Table 17.2. Based on this list, the primary structure for the peptide in Example 17.2 would be written in either of the following two ways.

| | |
| --- | --- |
| Using three letter amino acid codes: | Gly-Ser-Arg |
| Using one letter amino acid codes: | G S R |

As we saw before, the amino acid at the N-terminal end of the polypeptide chain is always written first (Gly or G, in this case) and the amino acid at the C-terminal end is written last (Arg or R). It is important for chemists to be familiar with these codes when they describe the primary structure of a peptide or protein to other scientists or when they look up the structure of a peptide or protein in the literature.

**TABLE 17.2**   Abbreviations for Common Amino Acids

| Full Name of Amino Acid | Three Letter Code | One Letter Code |
| --- | --- | --- |
| Alanine | Ala | A |
| Arginine | Arg | R |
| Asparagine | Asn | N |
| Aspartic acid | Asp | D |
| Cysteine | Cys | C |
| Glutamic acid | Glu | E |
| Glutamine | Gln | Q |
| Glycine | Gly | G |
| Histidine | His | H |
| Isoleucine | Ile | I |
| Leucine | Leu | L |
| Lysine | Lys | K |
| Methionine | Met | M |
| Phenylalanine | Phe | F |
| Proline | Pro | P |
| Serine | Ser | S |
| Threonine | Thr | T |
| Tryptophan | Trp | W |
| Tyrosine | Tyr | Y |
| Valine | Val | V |

## Example 17.3

Write the full amino acid names for each of the following primary structures.

(a)  H A N
(b)  Tyr-Arg-Lys
(c)  W Y R

## Solution 17.3

(a)  Histidine-Alanine-Asparagine
(b)  Tyrosine-Arginine-Lysine
(c)  Tryptophan-Tyrosine-Arginine

The sequence of amino acids in a protein or peptide not only determines that compound's primary structure, but also determines how the polypeptide chains in these compounds will fold and bend to form fairly well-defined three-dimensional shapes. This occurs because a given string of amino acids will tend to fold in its own particular way. We will see a few specific examples of this later when we talk more about particular types of proteins such as antibodies and enzymes.

Some special types of structures are produced when an oxygen on an amide group in one part of a polypeptide chain forms a hydrogen bond with a nitrogen in an amide group in a different part of the chain. This type of hydrogen bonding is shown below.

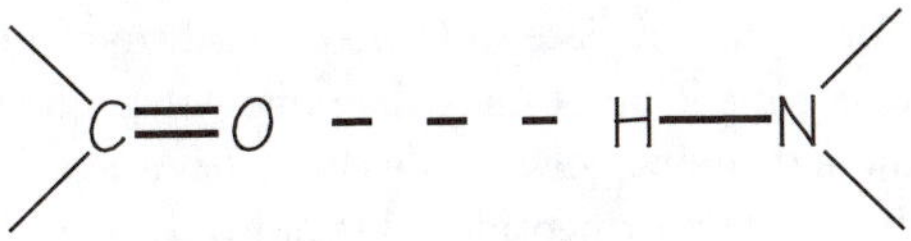

If many of these hydrogen bonds form within the same segment of a polypeptide chain, what is produced is a structure that looks like a spiral. This structure is known as an *α-helix* and it gives the protein or peptide both strength and a certain degree of springiness. α-Helices are quite common in proteins that make up structural features like hair and wool. Examples of proteins that contain α-helices are the α-keratins that make up an important part of horns and nails in animals. A ***pleated sheet*** (also known as a *β-sheet*) is another type of structure that can be formed by hydrogen bonding between amide oxygens and amide nitrogens in a polypeptide. This differs from an α-helix in that the hydrogen bonds now occur between different parts of a polypeptide that are usually quite distant from each other in the amino acid chain. An example of a protein that contains a lot of pleated sheets is β-keratin, which is found in silk. Both the α-helix and pleated sheet are formations that make up the ***secondary structure*** of a protein or polypeptide.

Another factor that causes a polypeptide or protein to take on a certain shape is determined by how well the different amino acids in the polypeptide interact with water. Because the hydrophilic amino acids are attracted to water but the hydrophobic amino acids are not, the polypeptide chain folds and arranges itself so that the water-loving amino acids are on the outside and the water-fearing amino acids are buried in the middle. In this way only those amino acids with side chains that are attracted to water are exposed to the water that surrounds the outside of a protein. It is this type of folding that gives rise to the final, overall shape of most proteins. The resulting shape is called the ***tertiary structure*** of the protein. In addition, it is possible for more than one type of polypeptide to bind together to give even larger formations known as the ***quaternary structure*** of a protein. Both the tertiary and quaternary structures are extremely important in determining the final purpose for which the protein or peptide will be used. Again, we will see some specific examples of this later when we consider how particular types of proteins work.

One last important feature that is present in many peptides and proteins is the presence of special covalent bonds that tie different parts of the protein or peptide together. These bonds occur when an -SH group on one cysteine side chain reacts with an -SH on another cysteine side chain to form a ***disulfide bond***. This takes place through the reduction-oxidation reaction that is shown below.

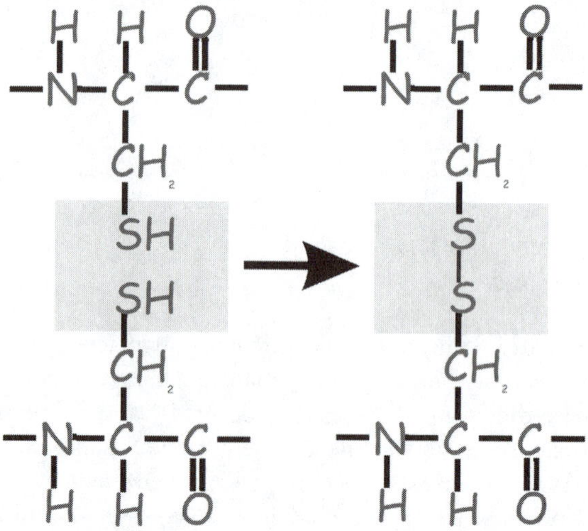

The −SH groups that are involved in linking polypeptides together might be present on cysteine amino acids that are on the same polypeptide chain or on entirely different polypeptide chains. These types of bonds help hold the structure of the protein or peptide in place and help the body to combine multiple types of polypeptides. An example of a protein that is made up of more than one polypeptide chain is the hormone insulin. Many other proteins also have more than one polypeptide chain that are tied together with disulfide bonds.

The particular way in which a protein or peptide folds not only gives that molecule its own unique shape, but it can also produce pockets in the protein or peptide that have just the right size and shape for binding to other molecules. An example is the group of proteins that are known as antibodies. Our bodies produce antibodies that bind to a particular virus or bacteria whenever we are exposed to these agents, like when we are vaccinated or when we become infected with them and get sick. The

*Chemistry Professionals at Work* | *CPW Box 17.3*

# INSULIN

One important type of protein is insulin. Insulin is a protein with a molecular weight of 5,734 g/mol that is produced in our bodies by the pancreas. The primary structure of human insulin is shown in Fig. 17.4. This structure consists of two polypeptide chains that are joined together by disulfide bonds. One of these polypeptide chains (the α-chain) has a string of 21 amino acids and the other (the β-chain) has a string of 30 amino acids. Two disulfide bonds tie these two chains together. There is also a third disulfide bond in insulin that ties together some amino acids that are in the α-chain. Insulin is a hormone that plays a vital role in determining how our bodies use sugars as a source of energy. If the correct levels of insulin are not maintained in the body, then a disease known as ***diabetes*** can result. Many people with diabetes must place extra insulin into their bodies in order for them to be able to properly handle sugars. Several drug companies produce insulin for this purpose. Most of this insulin is currently produced by using genetically altered bacteria that make insulin with the same primary structure and disulfide bonds as normal human insulin. A thorough understanding of protein and peptide structure is needed by the chemists who work at these companies in order for them to create insulin that will work properly in people with diabetes.

*For Homework: Look up the structure of the insulin that is produced by pigs and cows. These types of insulin, known as porcine insulin (from pigs) and bovine insulin (from cows), were once commonly used in treating people with diabetes. Compare the structures of these two types of insulin to the one that is shown for human insulin in Fig. 17.4. How are these structures the same? How are they different? Visit with a medical doctor and ask him or her what types of problems can be produced by these differences in structure when people with diabetes are treated with porcine or bovine insulin over long periods of time.*

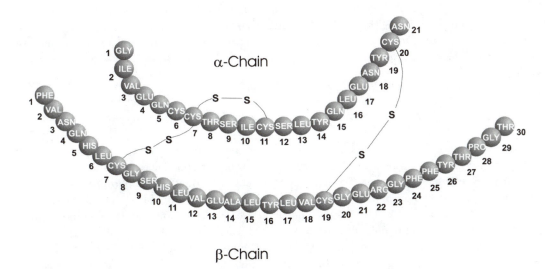

**FIGURE 17.4** The primary structure of human insulin.

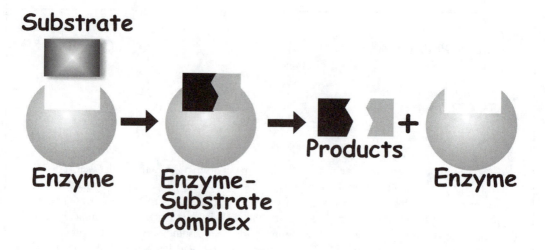

**FIGURE 17.5**   The basic reaction involved in the enzyme-catalyzed conversion of a substrate to a product.

structure of a typical antibody has two identical binding pockets that can form a good fit to the surface of the bacteria or virus to which the body is exposed. The structure of this binding pocket is determined by the amino acids that are present in the pocket and the way in which these amino acids fold to give a certain three-dimensional shape. Furthermore, the side chains on these amino acids are important in making sure that the foreign agent remains bound to the antibody. This is done through the formation of noncovalent bonds between the foreign agent and the amino acid side chains in the antibody's binding pocket. These types of bonds involve such things as ionic interactions, hydrogen bonding, and nonpolar or polar attractions. The result is a tight bond between the antibody and the invading species that helps our bodies to identify that species as being a foreign substance that must be destroyed or removed.

The same type of specific binding plays an important role in determining how many other proteins and peptides interact with chemicals. In special proteins that are called *enzymes*, this specific binding is used to help promote a particular type of chemical reaction. As we learned earlier, enzymes are protein-based catalysts (see Chapter 14 for a review of catalysts) that help make possible most of the reactions that happen within our bodies. They do this by first binding to the reactants (called *enzyme substrates*) in a specific pocket on the enzyme. This pocket also usually contains other substances (known as *cofactors* and *coenzymes*) that are required for the reaction, and may contain one or more metal ions to help increase the reactivity of these bound substances. An example of an enzyme-catalyzed reaction is shown below and in Fig. 17.5.

$$\text{Creatine} + \text{Adenosine Triphosphate} \xrightarrow{\text{Creatine Kinase}} \text{Creatine Phosphate} + \text{Adenosine Diphosphate}$$

This reaction is important in muscles and other tissues as a way of storing energy, as represented here by the product creatine phosphate. In this reaction, the enzyme creatine kinase binds to the substrates creatine and adenosine triphosphate. The enzyme then causes the transfer of a phosphate group from adenosine triphosphate to creatine. The creatine phosphate that is produced has high energy chemical bonds that can later be used by muscles as a source of reserve energy. After the desired products have been formed by this reaction, these products are released from the binding pocket of the enzyme and another set of reactant molecules are allowed to bind. The process is then repeated so that even more creatine phosphate can be made. It is in this way that only a small amount of enzyme can be used to produce a large amount of a particular biological product.

## Example 17.4

Identify the enzyme, enzyme substrates, and products in each of the following reactions.

(a) Aspartate + Oxoglutarate $\xrightarrow{\text{Aspartate Aminotransferase}}$ Oxaloacetate + Glutamate

(b) Lactate + NAD$^+$ $\xrightarrow{\text{Lactate Dehydrogenase}}$ Pyruvate + NADH + H$^+$

(c) Glucose + 1/2 O$_2$ $\xrightarrow{\text{Glucose Oxidase}}$ H$_2$O$_2$ + Gluconolactone

## Solution 17.4

(a) The enzyme is aspartate aminotransferase. The substrates are aspartate and oxoglutarate. The products are oxaloacetate and glutamate.

(b) The enzyme is lactate dehydrogenase. The substrates are lactate and NAD$^+$. The products are pyruvate, NADH, and H$^+$.

(c) The enzyme is glucose oxidase. The substrates are glucose and O$_2$. The products are H$_2$O$_2$ and gluconolactone.

Notice in Example 17.4 that the enzymes are shown above the reaction arrow since these act only as catalysts and are not used up by the reactions. This is one way of identifying the enzyme that is involved in the reaction. Another way is to look at the names of the chemicals that are listed. Most enzymes have names that end in "-ase," thus making them easy to identify. Some exceptions to this naming scheme include the enzymes trypsin, rennin, and pepsin.

# 17.3 Carbohydrates

A second general group of compounds that are important in our bodies are the **carbohydrates**. Carbohydrates are also known as **sugars**. The term carbohydrate comes from the fact that many of these compounds have the empirical formula $C_x(H_2O)_y$, which caused early chemists to think that these were made through simple hydration, or combination of water with carbon atoms. Carbohydrates are important in humans and animals as a source of energy. Plants also use carbohydrates as an energy source as well in the construction of such features as wood, bark, and stems.

## 17.3.1 Simple Carbohydrates

Carbohydrates can be divided into three general categories: **monosaccharides, disaccharides**, and **polysaccharides. Monosaccharides** are also known as simple sugars or **simple carbohydrates**. An example of a monosaccharide is glucose, whose structure is shown in Fig. 17.6. All monosaccharides contain a single chain of carbon atoms. For glucose, this chain is six carbon atoms long, but in other monosaccharides the length might be anywhere between three and nine atoms. One of the carbon atoms in a monosaccharide is bound to an oxygen atom to form an aldehyde or ketone group. In the structure of glucose that is shown in Fig. 17.6, this carbon atom occurs at the top of the carbon chain. All or most of the other carbon atoms are attached to an oxygen atom that is part of an alcohol group. These alcohol groups make the monosaccharide hydrophilic, or "water loving." The number of carbon atoms in the chain, the location of the aldehyde or ketone group, and the arrangement of the alcohol groups are all important features that determine how the properties of one monosaccharide will differ from the next.

## Example 17.5

Compare the structure of glucose in Fig. 17.6 to the structure that is shown below for fructose, another type of monosaccharide. How are these structures the same? How are they different?

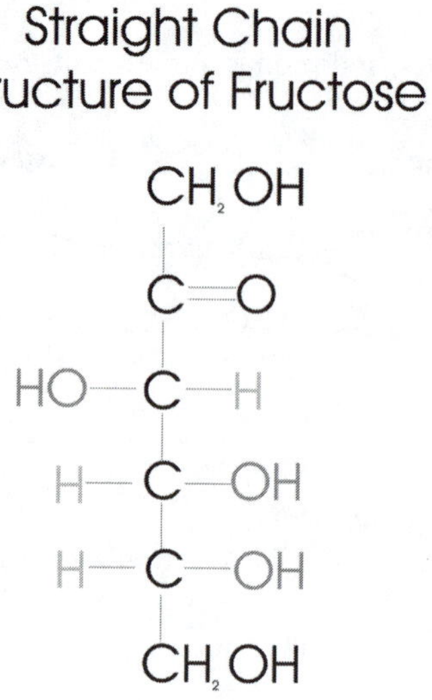

## Straight Chain Structure of Fructose

## Solution 17.5

Both glucose and fructose have chains containing six carbon atoms. Glucose has an aldehyde group on the first carbon in the chain, but fructose has a ketone group on the second carbon in the chain. Because of this, the location of the alcohol groups on the first and second carbons are just the opposite in glucose and fructose. All other carbons in these sugars have the same types of groups attached to them.

Monosaccharides are actually fairly flexible molecules that can easily rotate about their chemical bonds. Probably the most common way in which they orient themselves is into a shape that is a closed-ring, as illustrated in Fig. 17.7 for glucose. Monosaccharides produce these types of structures by having one of the alcohol groups at the end of the carbon chain react with the aldehyde or ketone group that is at or near the beginning of the same chain. This reaction occurs in water and forms a stable ring-type structure.

The ring that is produced by the **cyclization reaction** in Fig. 17.7 for a monosaccharide usually has a circle of five or six atoms (one oxygen, the rest carbon). For glucose, a six-membered ring is formed. This is a reversible process because this ring can go back to the straight chain form. But the greater stability of the ring form makes the closed-ring structure the preferred way that most monosaccharides actually exist in nature. Notice that there are two different types of rings that can form for glucose: one in which the alcohol group on the first carbon is drawn pointing up from the ring and another in which this alcohol group is drawn pointing down. We will see in Section 17.3.2 how this seemingly small difference can lead to big differences in the properties of more complex sugars that are made by linking together two or more molecules of glucose or other mono-saccharides.

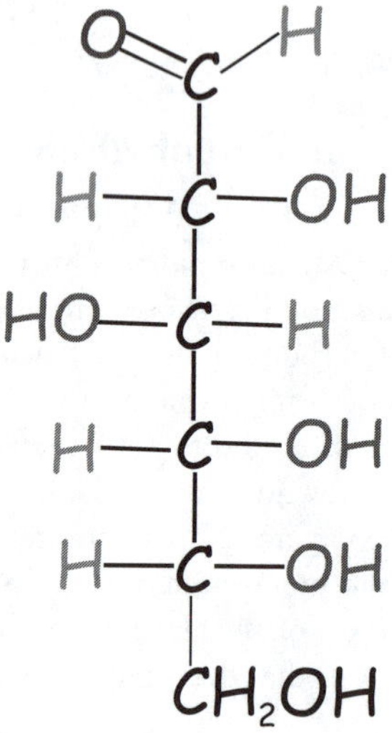

**FIGURE 17.6**   The structure of glucose, an example of a simple carbohydrate.

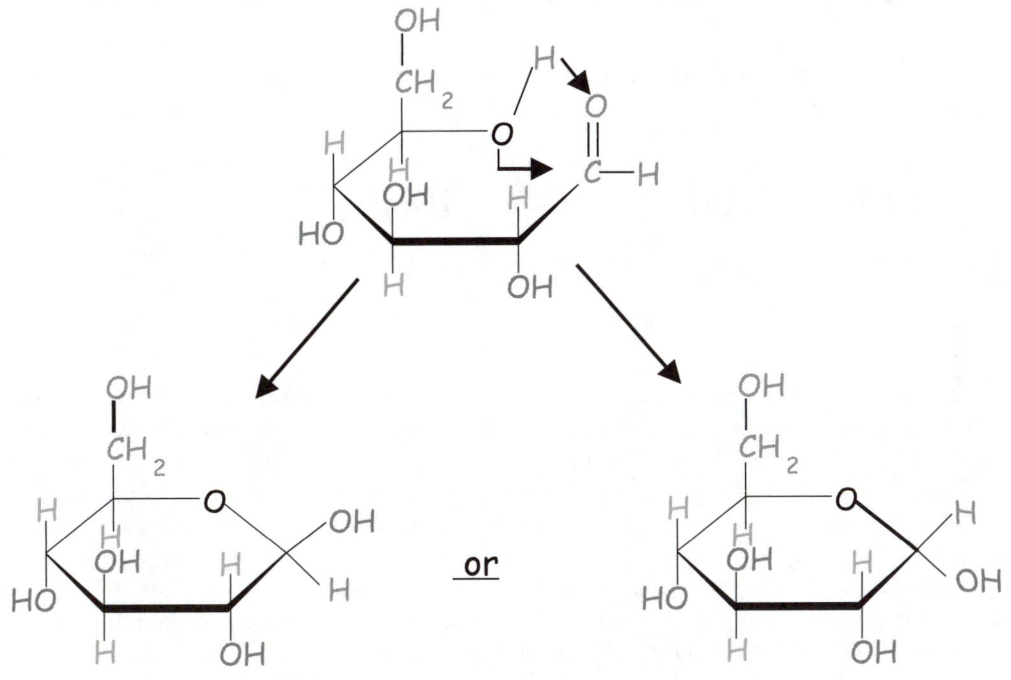

**FIGURE 17.7** Conversion of glucose from a straight chain to a closed-ring structure.

## Example 17.6

Draw a closed-ring structure that could form through the reaction of the alcohol group at the end of fructose and the ketone group on the second carbon atom of this monosaccharide.

## Solution 17.6

The result of the cyclization of fructose is a five-membered ring, as shown below.

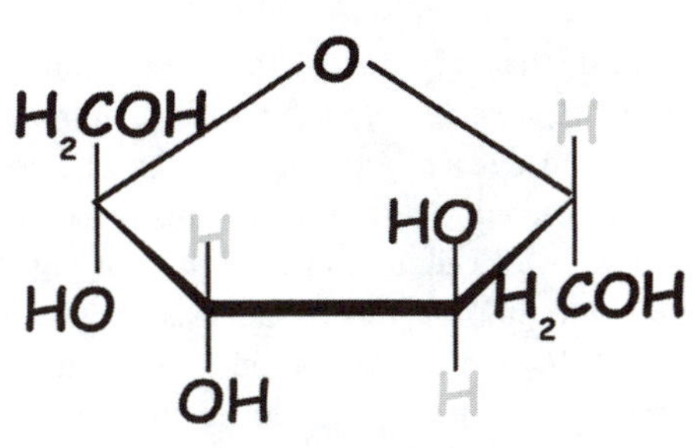

## 17.3.2 Disaccharides and Polysaccharides

Monosaccharides can be put together by living organisms to form larger structures called *complex carbohydrates*. If two monosaccharides are put together, then the product is referred to as a *disaccharide*. If more than two simple sugars are linked together, then the name *polysaccharide* is generally used for the complex carbohydrate that is produced.

# APPLICATIONS OF COMPLEX CARBOHYDRATES

The same general type of reaction that is shown in Fig. 17.8 for the production of sucrose can also be used to make long polymers of monosaccharides. Cellulose is a complex carbohydrate that makes up the cell walls in plants. It is the main component of wood and plant fibers. This makes it of great industrial importance in the manufacturing of building lumber, cotton, and paper. Cellulose is made up of a series of glucose molecules that are linked together in a straight chain, as indicated in Fig. 17.9.

Amylose is another type of complex carbohydrate that is found in plants. The structure of amylose is also shown in Fig. 17.9. Amylose and a related complex sugar, amylopectin, are the main components of starch. Like cellulose, amylose is a polymer that is made up of glucose molecules that are arranged in a straight chain. Amylopectin is similar to amylose but contains additional branches of glucose chains that are attached to the main string of glucose molecules. Amylose and amylopectin are important as a way of storing chemical energy in plants and as a food source for animals and humans. This makes these complex carbohydrates of great significance in the food and agricultural industries. An example of a good source of amylose and amylopectin is potatoes, which have a high starch content.

*For Homework: Look up information on some other types of complex carbohydrates. Report on how these are used in nature and/or in industrial applications.*

Sucrose is an example of a disaccharide that is formed by the combination of the simple sugars glucose and fructose. This process is shown in Fig. 17.8. Sucrose is formed when the oxygen on the alcohol group that is attached to the first carbon on glucose reacts with the second carbon on fructose to give a product that has two linked rings. Water is also created as a product by this reaction.

We can see by comparing the structures of cellulose and amylose in Fig. 17.9 that even though both of these complex carbohydrates are made up of chains of glucose, the glucose in these chains is linked together in slightly different ways. In the case of cellulose, the oxygen on the first carbon atom of each glucose ring to the left is drawn so that it is always pointed up from the chain, while in amylose it is drawn so that it is always pointed down. These two types of arrangements are known as a *β-linkage* (oxygen pointing up) and an *α-linkage* (oxygen pointing down). It is this seemingly small difference that is responsible for the very different properties and functions that cellulose and amylose have in plants. This difference is also the reason why humans can eat starch but not grass or wood, since the starch-digesting enzymes in our bodies can break down complex carbohydrates with α-linkages (like in amylose), but not those with β-linkages (like in cellulose).

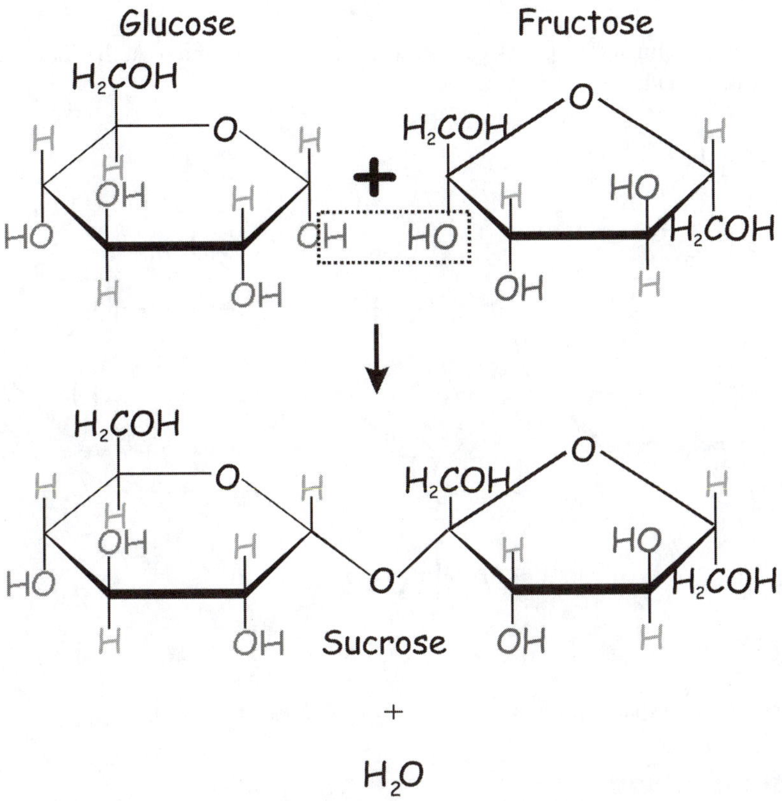

**FIGURE 17.8** Formation of the complex carbohydrate sucrose through the combination of the simple carbohydrates glucose and fructose.

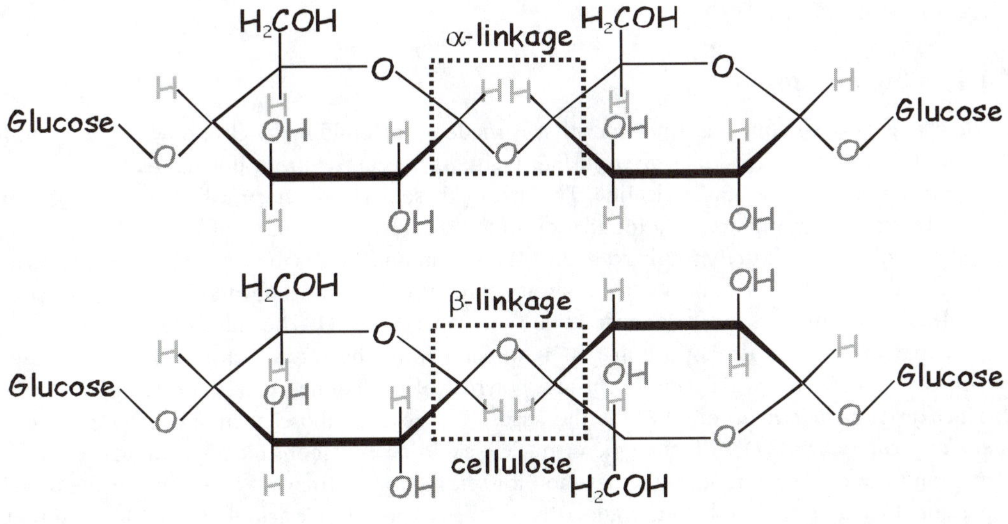

**FIGURE 17.9** The structures of cellulose and amylose, two examples of complex carbohydrates. The sections in the boxes show the type of linkage that is present between the glucose molecules that make up the chains in each of these complex carbohydrates.

*Example 17.7*

What type of linkage ($\alpha$-linkage or $\beta$-linkage) is present between the monosaccharides that make up the complex carbohydrate maltose.

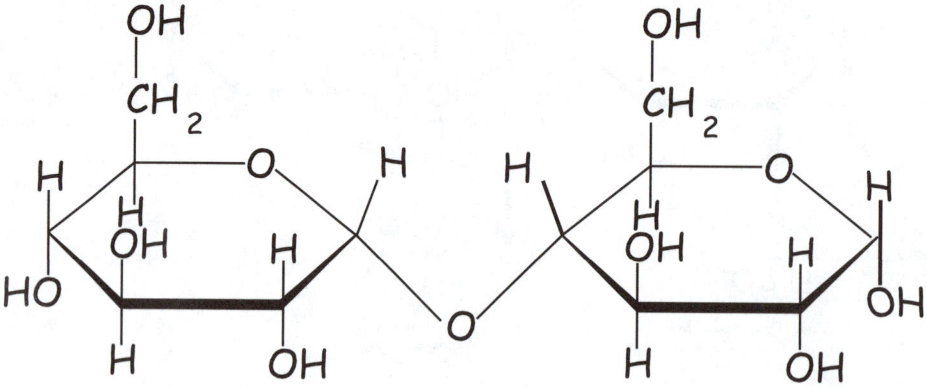

Maltose

*Solution 17.7*

Maltose has an $\alpha$-linkage that is present between two molecules of glucose.

# 17.4   Nucleic Acids

The nucleic acids are a third group of chemicals that are vital to all living organisms on our planet. Nucleic acids are responsible for acting as the "blueprint" when our bodies need to make a certain type of protein or peptide. Nucleic acids also serve as the means by which inherited traits are passed on from an organism to its offspring.

## 17.4.1   Nucleotides

A *nucleic acid* can be defined as a polymer that is made up of building blocks known as *nucleotides*. The basic structure of nucleotide is given in Fig. 17.10. As we can see from this figure, there are three main parts to the structure of a nucleotide. The first of these parts is a sugar, which consists of a five-membered ring. This group acts as a foundation onto which the other parts of the nucleotide can be attached. It also plays a role in the joining together of different nucleotides to form a nucleic acid polymer.

The sugar that is present in a nucleotide is always either the simple carbohydrate ribose or the simple carbohydrate deoxyribose. These carbohydrates are shown in Fig. 17.11. The only difference in these two sugars is that there is an alcohol group that is missing from the second carbon of deoxyribose. If deoxyribose is used as the sugar component in a polymer of nucleotides, then the resulting polymer is called *deoxyribonucleic acid*, or *DNA*. If the sugar is instead a ribose, then the polymer is called *ribonucleic acid*, or *RNA*. DNA is the type of nucleic acid that is responsible for passing on inherited information from one generation to the next and for serving as a permanent blueprint for the proteins and peptides that are produced by our bodies. RNA is the type of nucleic acid that is used by our bodies to make smaller copies of this blueprint and to act as a decoder for converting this information into a form that can be used for the construction of a protein or peptide.

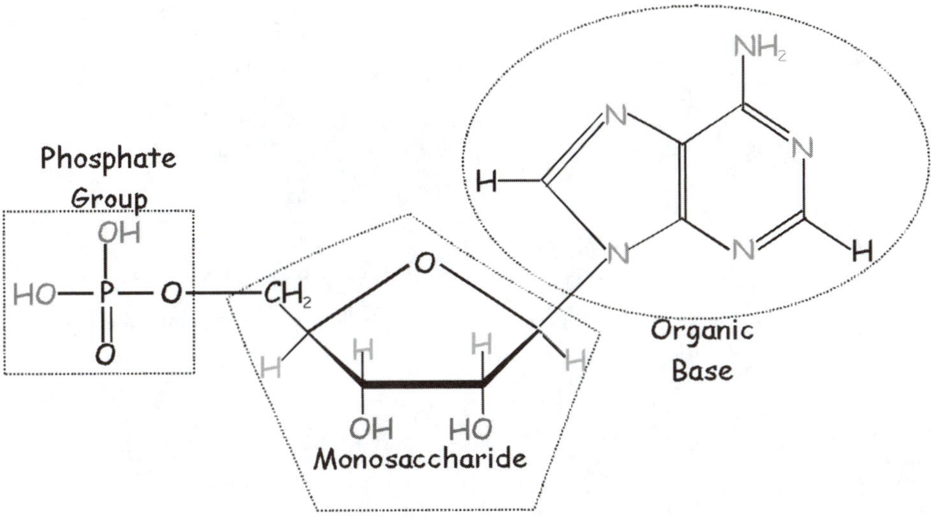

**FIGURE 17.10** Basic structure of a nucleotide.

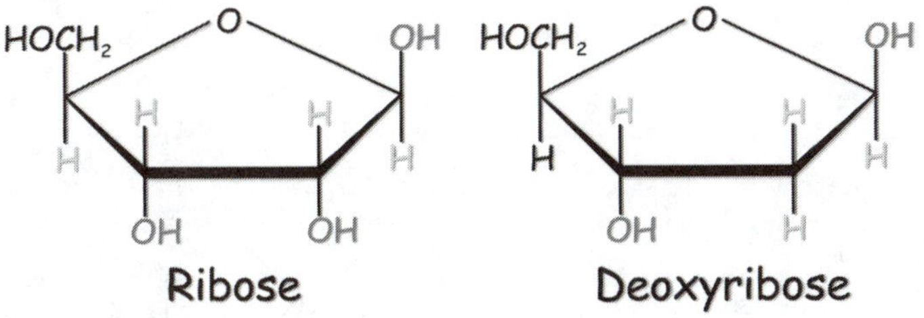

**FIGURE 17.11** Ribose and deoxyribose.

The second main part of a nucleotide is the nitrogen-containing aromatic ring that is attached to the deoxyribose or ribose carbohydrate. As we learned in Chapter 6, the presence of nitrogens in this type of ring will make this group act as an organic base. There are five main types of organic bases that are found in DNA and RNA. These are *cytosine (C)*, *thymine (T)*, *adenine (A)*, *guanine (G)*, and *uracil (U)*. The structures of these bases are shown in Fig. 17.12. All of these except uracil are found in DNA, while all but thymine are present in RNA. As we will see later, it is these organic bases that form the code by which DNA and RNA can be used to store the proper sequence of amino acids that should be used by the body in order to make a particular protein or peptide. The third part of a nucleotide is one or more phosphate groups that are attached to the deoxyribose or ribose sugar. These phosphate groups play a key role in linking different nucleotides together to form a larger polymer chain.

## Example 17.8

Identify each of the three main parts that are present in the following nucleotides. Identify the phosphate groups and state what type of sugar and organic base is present in each.

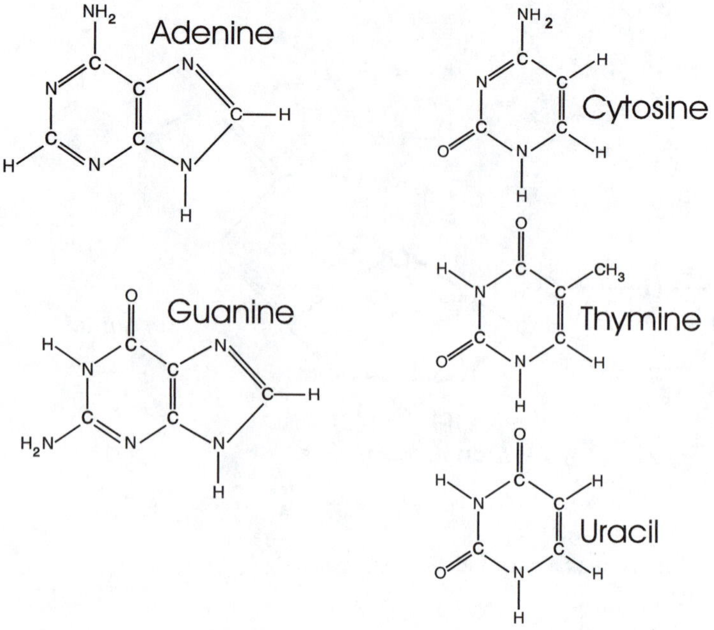

**FIGURE 17.12**  Structures of the organic bases cytosine, thymine, adenine, guanine, and uracil.

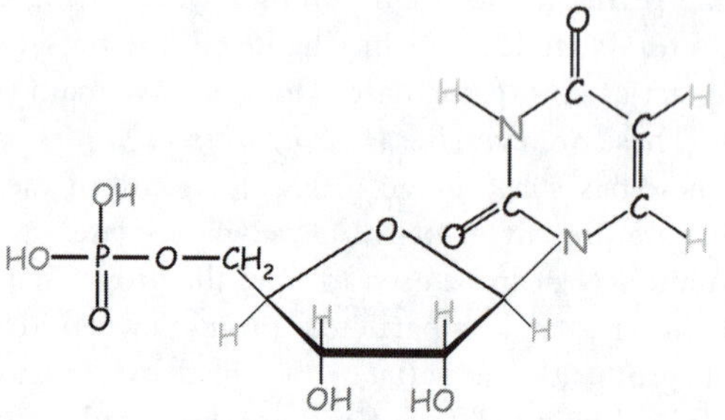

## Solution 17.8

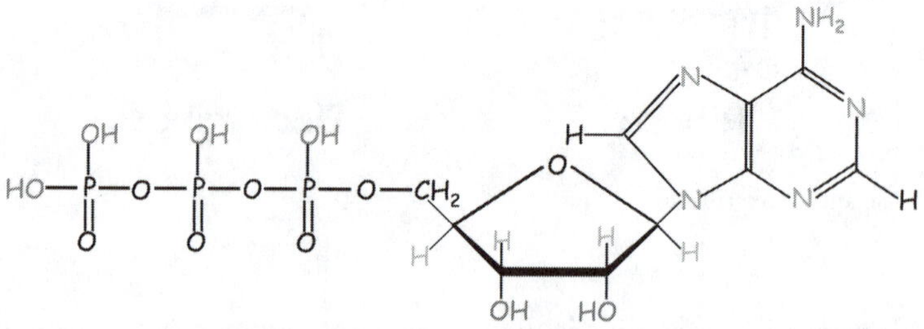

Each of the above molecules has a carbohydrate group, an organic base, and one or more phosphate groups.

(a) The carbohydrate is deoxyribose, the organic base is adenine, and there are three phosphate groups present. This nucleotide is known as deoxyadenosine triphosphate (dATP).

(b) The carbohydrate is ribose, the organic base is uracil, and one phosphate group is present. This nucleotide is known as uridine monophosphate (UMP).

---

# DNA TESTING

The fact that we each have our own unique set of nucleotide sequences in our DNA means that these sequences can be used to identify a particular person from a blood or cell sample. This approach, known as **DNA testing**, is becoming more and more common in identifying potential criminals or those who have somehow been associated with a suspected illegal activity. In addition, DNA testing is used in paternity testing to help identify the father of a child. DNA testing can also be used to help study how inherited diseases are passed from one generation to the next to determine which members of a family are at greatest risk for these diseases.

One way of determining the sequence of nucleotides that are present in a section of DNA is to use the technique of **gel electrophoresis**. This method separates a series of different nucleotide fragments of DNA based on how fast they travel through a porous gel and in the presence of an electric field. By combining the information that this provides on the size of each DNA fragment, it is possible to get a type of fingerprint pattern for that section of DNA. This can then be used to help identify the person who was the original source of the DNA.

*For Homework: Visit a police laboratory or a hospital laboratory in your area that performs DNA testing. Obtain information on how DNA testing techniques are used to help solve crimes or for the detection of disease. Write a report on your findings.*

---

## 17.4.2 DNA and the Genetic Code

We have already learned that DNA is the type of nucleic acid that is responsible for storing the set of instructions that our bodies use to produce proteins and peptides. This series of instructions is known as the **genetic code**. Each living organism has its own unique genetic code, and thus, its own particular set of proteins and peptides. To understand how this works, we first need to examine the structure of DNA in more detail.

In humans, DNA is located in the center portion of our cells in a region known as the **nucleus**. This DNA is present in 23 pairs of structures known as **chromosomes**. In each pair of chromosomes, we inherit one chromosome from our mother and the other from our father. This is the way in which DNA, and the proteins or peptides that it encodes, are passed on from parents to their children. Each chromosome contains a long string of DNA plus proteins and some RNA that help protect and package this DNA into a very compact form that can be stored within the nucleus of a cell.

The 23 types of chromosomes in humans each contain DNA that is responsible for producing a different set of proteins and peptides in our bodies. A **gene** is the part of DNA that acts as the code for making a single type of protein or peptide. There are a total of approximately 100,000 genes in human DNA. Because of the many types of proteins and peptides that must be produced by our bodies, we need to have a large number of genes that can act as blueprints for these proteins and peptides. The way in which a gene encodes the structure of a protein or peptide is through the sequence of nucleotides that are present in that gene. Sets of three nucleotides in a row are used to represent each amino acid. We will see some examples of these nucleotide codes later.

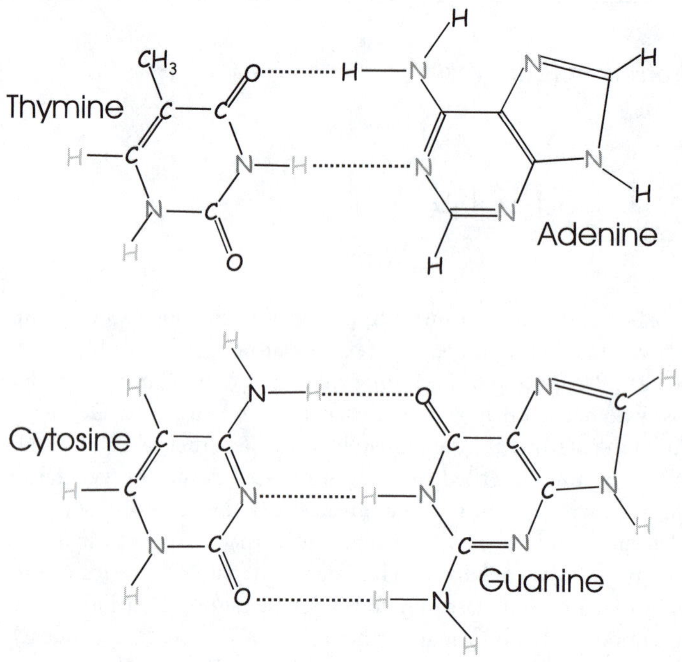

**FIGURE 17.13**    Hydrogen bonding between thymine and adenine (top), and cytosine and guanine (bottom).

An important property of each nucleotide is the ability of its organic base to form hydrogen bonds with an organic base in another nucleotide. In DNA, this hydrogen bonding is strongest when it occurs between the organic bases on a cytosine and a guanine, or between an adenine and a thymine. The way in which this hydrogen bonding occurs is shown in Fig. 17.13. The body uses this type of specific bonding to help keep the DNA in a stable form and to control when a particular gene is to be used for producing its protein or peptide. This occurs due to the DNA structure, which involves two strands that wind around each other and that have complementary nucleotides that can hydrogen bond between the strands. A double helix structure is what is formed by this process.

It is important that our cells be able to turn genes "on" and "off" so that the proteins and peptides that they make are produced only when they are needed. A gene is considered to be "on" when the cell has taken the DNA sequence for that gene and unwinds it into a nonhydrogen bonded single strand form that is now exposed and can be used for protein/peptide production. If the DNA is present in its normal, hydrogen-bonded and double helix form, then the gene is protected from exposure and cannot be used by the cell for protein or peptide production.

## 17.4.3   Messenger RNA and Transfer RNA

As we have already learned, much of the DNA in our cells is kept in the nucleus. (Note: some of our DNA is also present in structures known as ***mitochondria*** that are located outside of the nucleus). The nucleus can be thought of as a type of library for the cell, which contains all of the instructions for making the proteins that are needed by our bodies. But the place in a cell where most of the protein is actually made is outside of the nucleus and the DNA is much too large or precious for the cell to allow to leave this region. In other words, the DNA can be thought of as a "reference" that can only be used within the library and that cannot be checked out for external use. A cell overcomes this problem by making a copy of the part of the reference that it wants to use (that is, the gene for a particular protein or peptide). It then takes that copy to where it actually needs this material to perform its work. This overall process is summarized in Fig. 17.14.

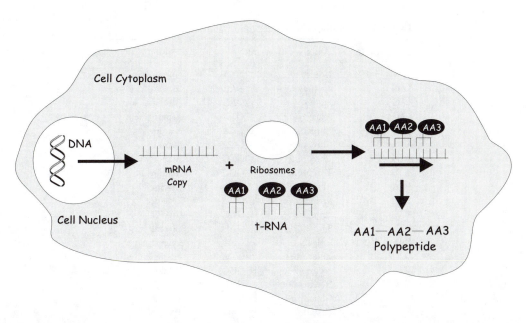

**FIGURE 17.14**  Basic processes involved in the production of a polypeptide by using DNA as the initial blueprint, mRNA as a copy of this blueprint, and tRNA plus ribosomes for construction of the polypeptide from this blueprint copy.

Copies of DNA are made by using a second type of nucleic acid known as *messenger RNA* (or *mRNA*, for short). mRNA is put together in the same general way as DNA, but with a few important differences. One of these differences is that RNA uses ribose sugars instead deoxyribose sugars in its nucleotide chain. Another difference between mRNA and DNA is that all types of RNA use the organic base uracil instead of thymine in their coding sequences. mRNA is also much smaller than DNA, since it contains the code for only one protein or peptide. And finally, mRNA is always present as a single nucleic acid chain, while DNA is normally present as a double-stranded chain that forms a *double helix*.

The way in which mRNA is used to make a copy of DNA begins when the cell causes the DNA in the region of the desired gene to unfold into two separate nonhydrogen bonding strands. When the gene is exposed, RNA-based nucleotides come in and form complementary hydrogen bonds to each of the exposed organic bases in the DNA. This hydrogen bonding occurs between adenine (DNA) and uracil (RNA), thymine (DNA) and adenine (RNA), cytosine (DNA) and guanine (RNA), and guanine (DNA) and cytosine (RNA). As these base pairs form, the RNA nucleotides are linked together by an enzyme. This RNA strand is then released for use. The result is that a copy of the original gene in the DNA is now represented by a sequence of mRNA.

## Example 17.9

Working from left to right, determine what amino acids are coded by the following sequence of mRNA: AUGCUCGUCUCG.

## Solution 17.9

According to Table 17.3, the above mRNA sequence represents the following amino acids: Methionine-Leucine-Valine-Serine.

Like in DNA, each amino acid is represented by a sequence of three specific nucleotides in the mRNA nucleotide chain. Table 17.3 shows the sequences of mRNA that form the codes for each amino acid.

**TABLE 17.3**   mRNA Codes for Amino Acids

| mRNA Code | Amino Acid | mRNA Code | Amino Acid |
|-----------|------------|-----------|------------|
| AAA | Lysine | GAA | Glutamic acid |
| AAC | Asparagine | GAC | Aspartic acid |
| AAG | Lysine | GAG | Glutamic acid |
| AAU | Asparagine | GAU | Aspartic acid |
| ACA | Threonine | GCA | Alanine |
| ACC | Threonine | GCC | Alanine |
| ACG | Threonine | GCG | Alanine |
| ACU | Threonine | GCU | Alanine |
| AGA | Arginine | GGA | Glycine |
| AGC | Serine | GGC | Glycine |
| AGG | Arginine | GGG | Glycine |
| AGU | Serine | GGU | Glycine |
| AUA | Isoleucine | GUA | Valine |
| AUC | Isoleucine | GUC | Valine |
| AUG | Methionine | GUG | Valine |
| AUU | Isoleucine | GUU | Valine |
| CAA | Glutamine | UAA | Stop code |
| CAC | Histidine | UAC | Tyrosine |
| CAG | Glutamine | UAG | Stop code |
| CAU | Histidine | UAU | Tyrosine |
| CCA | Proline | UCA | Serine |
| CCC | Proline | UCC | Serine |
| CCG | Proline | UCG | Serine |
| CCU | Proline | UCU | Serine |
| CGA | Arginine | UGA | Stop code |
| CGC | Arginine | UGC | Cysteine |
| CGG | Arginine | UGG | Tryptophan |
| CGU | Arginine | UGU | Cysteine |
| CUA | Leucine | UUA | Leucine |
| CUC | Leucine | UUC | Phenylalanine |
| CUG | Leucine | UUG | Leucine |
| CUU | Leucine | UUU | Phenylalanine |

Since there are only 20 common amino acids and there are 64 possible combinations of three nucleotides, some amino acids have more than one nucleotide sequence that codes for them.

After a gene sequence has been copied by the mRNA, the mRNA can leave the nucleus and move to the outer portions of the cell (known as the *cytoplasm*) where this sequence can be read and used to produce a protein or peptide. *Ribosomes* are the structures within a cell that are responsible for this production process. Another factor that is needed for this production is a small type of nucleotide called *transfer RNA* (or *tRNA*). tRNA is a special type of nucleic acid that has two main parts. The first part is a region where tRNA binds to an amino acid so that this can be taken to the ribosomes for protein or peptide construction. The second part of tRNA is a specific sequence of nucleic acids that is able to specifically hydrogen bond to the mRNA code for that same amino acid.

The binding of tRNA to the nucleotides in an mRNA strand allows the cell to translate a chemical message from a code written in nucleic acids to one that is represented by a string of amino acids. As the tRNA binds to the mRNA, the amino acids on neighboring tRNA molecules are joined together. As more of these amino acids are linked, they eventually form the entire protein or peptide of interest to the cell. After the protein/peptide has been formed, it is released from the ribosomes and the same sequence of mRNA can be used to make another copy of the same protein or peptide.

## 17.5  Lipids

The final class of biological compounds that we will consider is called the *lipids*. This category contains many types of biological compounds that all have the common property of being relatively insoluble in water, but easy to dissolve in an organic solvent like benzene. Four important groups of compounds that are found in this class are the fats or oils, phospholipids, waxes, and steroids.

### 17.5.1  Fats and Oils

This group of lipids is found in both plants and animals. These are used to provide a way of storing energy for long periods of time. In animals, fat also helps provide insulation to the body and helps to provide cushioning for various body parts. All fats and oils have the same basic structure that is given in Fig. 17.15. Fats and oils are formed by the reaction of glycerol with three fatty acids. Glycerol is a three-carbon-long hydrocarbon that contains one alcohol group on each carbon. A fatty acid consists of a carboxylic acid group that is attached to a long hydrocarbon chain. When the alcohol groups of glycerol react with the carboxylic acid groups of fatty acids, this reaction produces a series of ester bonds that link the glycerol and fatty acids together. If this new chemical is a solid at room temperature, it is called a *fat*. If this compound is a liquid at room temperature, it is called an *oil*.

The thing that makes one type of fat or oil different from another is the types of fatty acids that are added to the glycerol. The three fatty acids that are added to glycerol to make a fat or oil might all be the same or they might be different. There are many types of fatty acids that can be used. Some examples are shown in Table 17.4. Fatty acids can have hydrocarbon chains with different lengths. In addition, fatty acids can either have only single bonds present between all of the carbons in their chains, or they might have double bonds that are present. Fats or oils that have only single bonds in all of their fatty

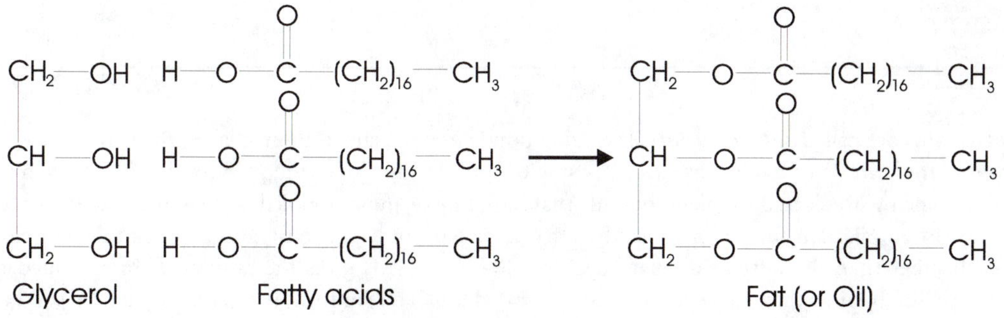

**FIGURE 17.15**  The reaction of glycerol with three fatty acids to form a fat or oil.

**TABLE 17.4**  Examples of Some Fatty Acids that Are Found in Fats and Oils

| Name of Fatty Acid | Melting Point ($^\circ$C) | Chemical Structure |
|---|---|---|
| Lauric acid | 44.2 | $CH_3(CH_2)_{10}CO_2H$ |
| Myristic acid | 53.9 | $CH_3(CH_2)_{12}CO_2H$ |
| Palmitic acid | 63.1 | $CH_3(CH_2)_{14}CO_2H$ |
| Stearic acid | 69.6 | $CH_3(CH_2)_{16}CO_2H$ |
| Palmitoleic acid | $-0.5$ | $CH_3(CH_2)_5CH{=}CH(CH_2)_7CO_2H$ |
| Oleic acid | 13.4 | $CH_3(CH_2)_7CH{=}CH(CH_2)_7CO_2H$ |
| Linoleic acid | $-5.0$ | $CH_3(CH_2)_4CH{=}CHCH_2CH{=}CH(CH_2)_7CO_2H$ |
| Linolenic acid | $-11.0$ | $CH_3CH_2CH{=}CHCH_2CH{=}CHCH_2CH{=}CH(CH_2)_7CO_2H$ |

---

*Chemistry Professionals at Work*                                    **CPW Box 17.6**

# SATURATED AND UNSATURATED FATS

W̲e hear almost every day about how we should limit the total amount of "fat" in our diet and try to replace saturated fats with unsaturated fats. The reason for reducing the overall amount of fats is related to weight control and the hardening of our arteries that occurs when such compounds are present for long periods of time in our blood. But why is it beneficial to use unsaturated fats instead of saturated fats? One answer is that saturated fats tend to raise our cholesterol levels more than unsaturated fats. Also, there are certain types of unsaturated fats, known as omega-3 polyunsaturated fats, that may even create a decreased risk of heart disease in certain people. Saturated fats are commonly found in animal-related foods, like meat and milk products. Unsaturated fats are found in such things as vegetable oils, nuts, and fish. The omega-3 polyunsaturated fats are also found in fish.

*For Homework: Go to the grocery store and look at the labels on items such as butter, margarine, cooking fat, or cooking oil. What types of fats and oils are present in each? Keep in mind that in many cases a mixture of various fatty acids might be present in the fat or oil. Look up the structures of some of these fatty acids, fats, and oils and report on how their structures are related to their effects on our health.*

---

acids chains are called **saturated fats**. If double bonds are present in these chains, then they are called **unsaturated fats**. Fatty acids that have only single bonds in their chains tend to produce a fat when they are combined with glycerol, while fatty acids that have one or more double bonds in their structure tend to give rise to oils. This occurs because the fatty acids with only single bonds can pack together closer to each other than the fatty acids that have double bonds. This gives the fatty acids in saturated fats higher melting points than those found in unsaturated fats. This, in turn, is what makes a fat a solid and an oil a liquid at room temperature.

## 17.5.2   Phospholipids and Waxes

Phospholipids and waxes are two other groups of lipids that are closely related to fats and oils. A phospholipid is produced by the reaction between glycerol and fatty acids, but now only two fatty acids combine with one glycerol molecule. The third alcohol group of the glycerol is combined in an ester bond with a phosphate group. Waxes are also formed through the reaction of a fatty acid with an alcohol to form an ester. But in waxes, a simple alcohol with only one –OH group is used in this reaction.

*Example 17.10*

State whether each of the following compounds is a fat/oil, phospholipid, or wax.

a)

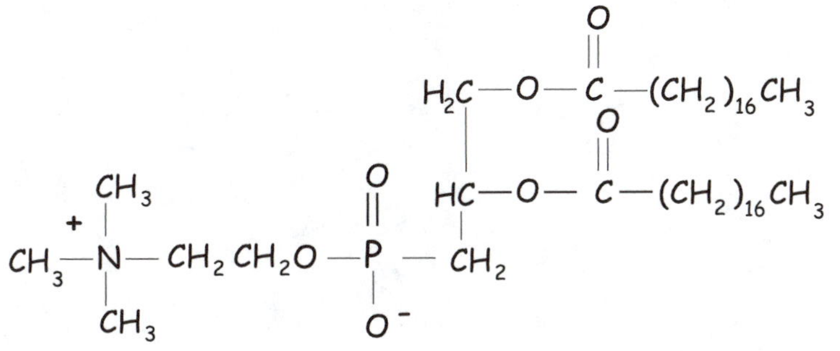

$$H_2C-O-\overset{\overset{\displaystyle O}{\|}}{C}-(CH_2)_{16}CH_3$$

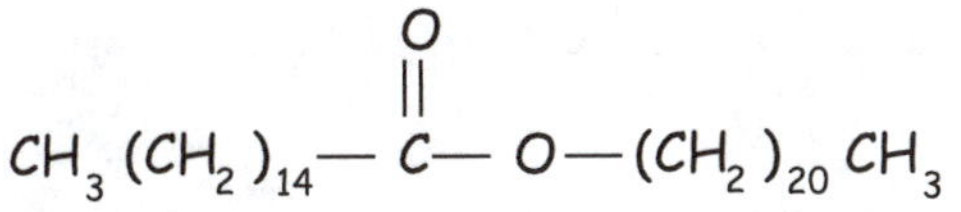

$$HC-O-\overset{\overset{\displaystyle O}{\|}}{C}-(CH_2)_{16}CH_3$$

$$CH_3-\overset{\overset{\displaystyle CH_3}{|}}{\underset{\underset{\displaystyle CH_3}{|}}{\overset{+}{N}}}-CH_2\ CH_2O-\overset{\overset{\displaystyle O}{\|}}{\underset{\underset{\displaystyle O^-}{|}}{P}}-CH_2$$

b)

$$CH_3\ (CH_2)_{14}-\overset{\overset{\displaystyle O}{\|}}{C}-O-(CH_2)_{20}\ CH_3$$

c)

$$H_2C-O-\overset{\overset{\displaystyle O}{\|}}{C}-(CH_2)_{18}\ CH_3$$

$$HC-O-\overset{\overset{\displaystyle O}{\|}}{C}-(CH_2)_{18}\ CH_3$$

$$H_2C-O-\overset{\overset{\displaystyle O}{\|}}{C}-(CH_2)_{18}\ CH_3$$

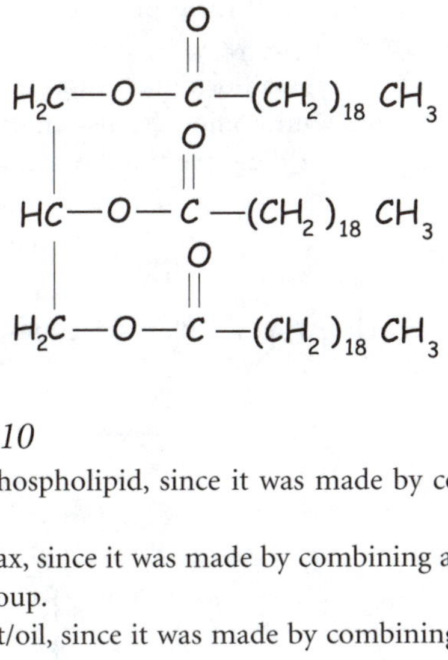

*Solution 17.10*

(a) This is a phospholipid, since it was made by combining two fatty acids and one phosphate group with glycerol.

(b) This is a wax, since it was made by combining a fatty acid with an organic compound that had only one alcohol group.

(c) This is a fat/oil, since it was made by combining three fatty acids with glycerol.

Waxes are valuable in plants and animals in helping to form a protective, water-proof coating. This is present as the outer layer on fruits, leaves, skin, and feathers. Phospholipids are important in cells in forming biological membranes that act as the boundary between the inside and outside of a cell. The presence of the long hydrocarbon chains from the fatty acids gives this part of the phospholipid a low solubility in water. In other words, this part of the molecule is hydrophobic, which we said earlier means "water-fearing." But the charged group that is present on the phosphate portion of this compound has very strong interactions with water and is hydrophilic, or "water-loving." Our cells make use of this property by having an outer boundary layer, known as the ***cell membrane***, that is two phospholipid molecules thick. The phosphate groups on each of these two molecules points away from the center of the membrane and towards the water that is present either within the cell or outside of the cell. The hydrophobic fatty acid chains on both phospholipids point inward toward each other and away from the water. Some proteins are also present in this membrane. It is through the formation and presence of this membrane that cells can control what enters and leaves them.

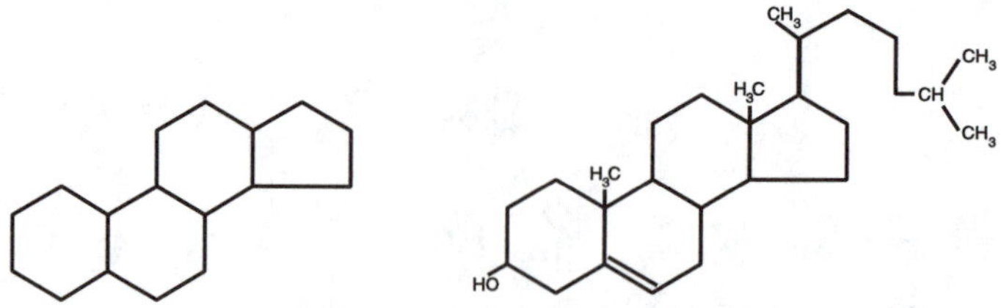

Basic Steroid
Ring Structure

Cholesterol

**FIGURE 17.16**   The basic backbone of a steroid (left) and the structure of cholesterol (right).

## 17.5.3   Steroids

The fourth general group of lipids are the steroids. These compounds all contain the hydrocarbon rings in the basic structure shown in Fig. 17.16. A common type of steroid is cholesterol. The structure of cholesterol is given in Fig. 17.16 next to the general form of a steroid. Cholesterol is needed by the body as a starting material for the synthesis of various hormones and bile salts. Hormones that are produced from cholesterol include the male hormone testosterone, the female hormone progesterone, and the female estrogen hormones. Our bodies also use cholesterol to make bile salts, which are produced by the liver and placed into our digestive tract to help us process fats and oils within our diet.

### Example 17.11

The structures of testosterone and progesterone are given below. How are these structures the same? How are they different?

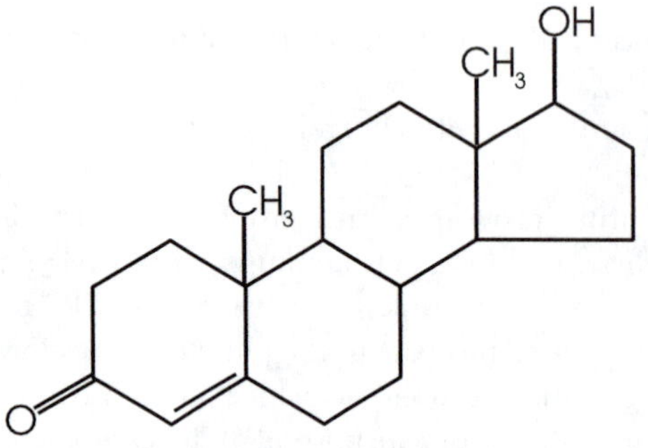

Testosterone

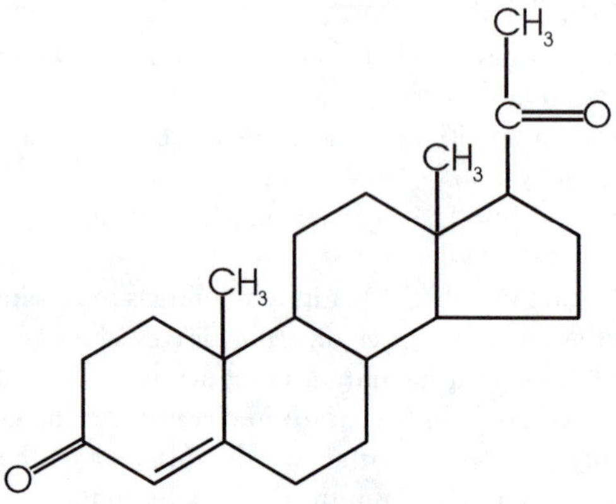

Progesterone

### Solution 17.11

Both testosterone and progesterone have the same basic ring structure that is shown in Fig 17.16. But they differ in the types of functional groups that are attached to this basic structure. In testosterone, there is a ketone group, two methyl groups ($-CH_3$), an alcohol group, and a double bond that have been added to the basic structure of a steroid. Progesterone contains most of these same groups, but has an extra ketone group ($H_3C-CO-$) that is present in place of the alcohol group in testosterone. It is this small change in chemical groups that gives rise to the very different effects that testosterone and progesterone have in humans.

## 17.6 Homework Exercises

1. What is an amino acid? Describe each main part of its structure.
2. What is a peptide? What is a protein? What is the difference between a protein and a peptide?
3. What is a peptide bond? How does this form?
4. Write the full chemical name for each amino acid in the following polypeptide chains.
   (a) I Q A N D W
   (b) His-Pro-Phe-Asp-Ala
   (c) Thr Val Gln Met Lys
   (d) S K I H G
5. Draw the full chemical structures for the following amino acid chains.
   (a) Lysine-Glutamate-Tyrosine
   (b) Cysteine-Serine-Cysteine
   (c) Tryptophan-Glycine-Isoleucine
6. Determine whether each of the following amino acids have hydrophilic or hydrophobic side chains.
   (a) Lysine
   (b) Threonine
   (c) Tyrosine
   (d) Isoleucine
   (e) Asparagine
7. Fill in the following blanks with the terms primary structure, secondary structure, tertiary structure, or quaternary structure.
   (a) The sequence of amino acids that make up a polypeptide is known as the _____ for that polypeptide.
   (b) Two or more polypeptides that are associated with each other gives rise to what is known as a _____.
   (c) An α-helix is an example of a _____.
   (d) Hydrophobic and hydrophilic amino acid side chains cause a protein to fold into a particular _____.
8. What is a disulfide bond? Why is this type of bond important to proteins and peptides?
9. What is the general function of an antibody? Describe the way in which an antibody performs this function. How is this process related to the antibody's structure?
10. What is the general function of an enzyme? Describe the way in which an enzyme performs this function. How is this process related to the enzyme's structure?
11. Identify the reactants, products, and enzyme that are involved in each of the following reactions.
    (a) 4-Nitrophenyl Phosphate $\xrightarrow{\text{Alkaline Phosphatase}}$ 4-Nitrophenoxide + Phosphate
    (b) Adenosine-5'-Monophosphate $\xrightarrow{\text{5'-Nucleotidase}}$ Adenosine + Phosphate
    (c) Triglyceride $\xrightarrow{\text{Lipase}}$ Lipase Glycerol + 3 Fatty Acids
12. Define each of the following terms:
    (a) Carbohydrate  (b) Simple carbohydrate  (c) Complex carbohydrate
    (d) Monosaccharide  (e) Disaccharide  (f) Polysaccharide

13. State whether each of the following compounds is a simple carbohydrate or a complex carbohydrate.

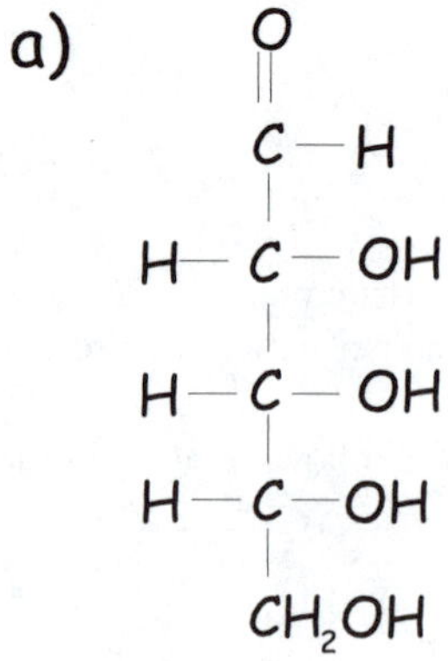

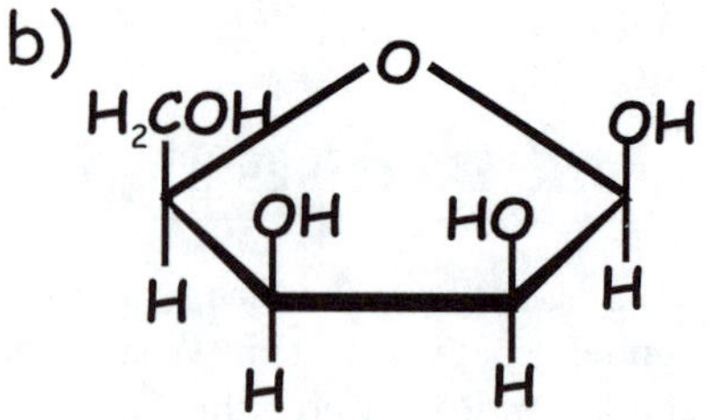

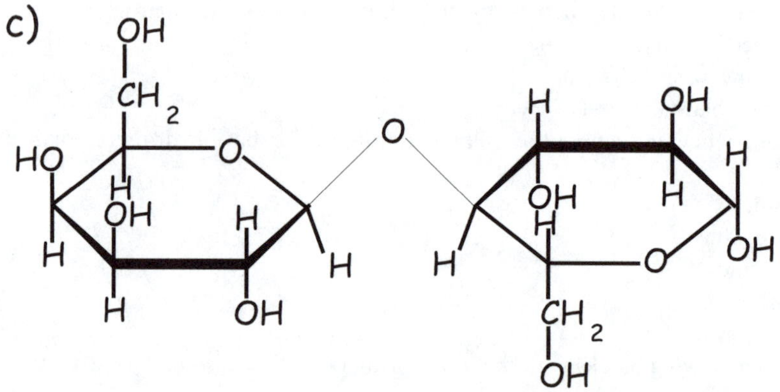

a)

$$
\begin{array}{c}
O \\
\parallel \\
C-H \\
\mid \\
H-C-OH \\
\mid \\
H-C-OH \\
\mid \\
H-C-OH \\
\mid \\
CH_2OH
\end{array}
$$

b)

c)

14. Compare the following structure to the one that was given for glucose in Fig. 17.9. How are these structures the same? How are they different?

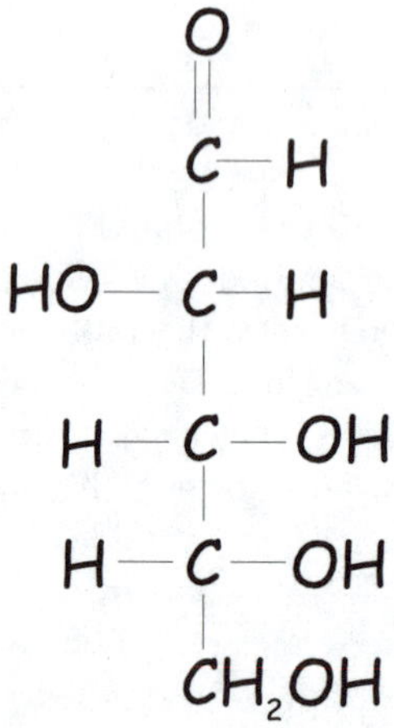

$$
\begin{array}{c}
O \\
\parallel \\
C-H \\
\mid \\
HO-C-H \\
\mid \\
H-C-OH \\
\mid \\
H-C-OH \\
\mid \\
CH_2OH
\end{array}
$$

15. Draw one possible cyclization product that might be obtained for the carbohydrate that is shown in Question 14.

16. State whether an α-linkage or β-linkage is present in the following carbohydrate.

17. Define or describe each of the following terms:
    (a) Nucleic acid      (b) Nucleotide          (c) DNA
    (d) RNA               (e) mRNA                (f) tRNA
    (g) Chromosome        (h) Gene                (i) Genetic code
    (j) Nucleus           (k) Ribosome

18. State at least two differences between DNA and RNA.

19. List the various types of organic bases that can be found in DNA and RNA. Which of these organic bases tend to form hydrogen bonds with each other?

20. Identify and name the organic base, sugar, and phosphate group that are present in the following nucleotide.

21. Working from left to right, determine what amino acids are being coded by the following mRNA sequences:
    (a) UUGUUCUCGGAUGAGCAA
    (b) AUCCUACCCCGUACGGCG
    (c) GUGGUAGCCACCAAAGGG

22. Define or describe each of the following terms:
    (a) Fat                (b) Oil                 (c) Saturated fat
    (d) Unsaturated fat    (e) Lipid               (f) Phospholipid
    (g) Wax                (h) Steroid             (i) Cell membrane

23. State whether each of the following compounds is a fat or oil, a phospholipid, or a wax.

a)

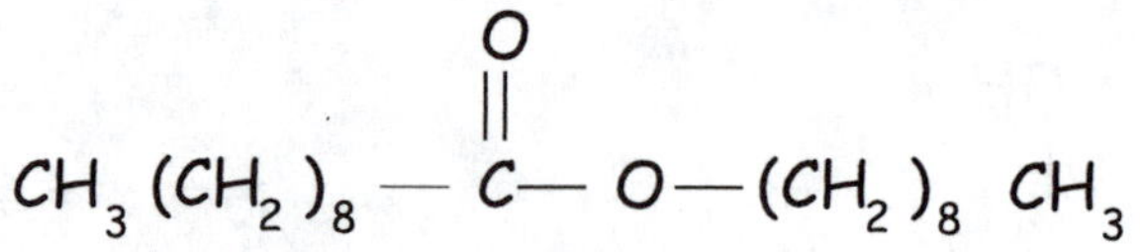

$$CH_3(CH_2)_8 - \overset{\overset{O}{\|}}{C} - O - (CH_2)_8 \ CH_3$$

b)

$$H_2C - O - \overset{\overset{O}{\|}}{C} - (CH_2)_{14} \ CH_3$$

$$HC - O - \overset{\overset{O}{\|}}{C} - (CH_2)_{12} \ CH_3$$

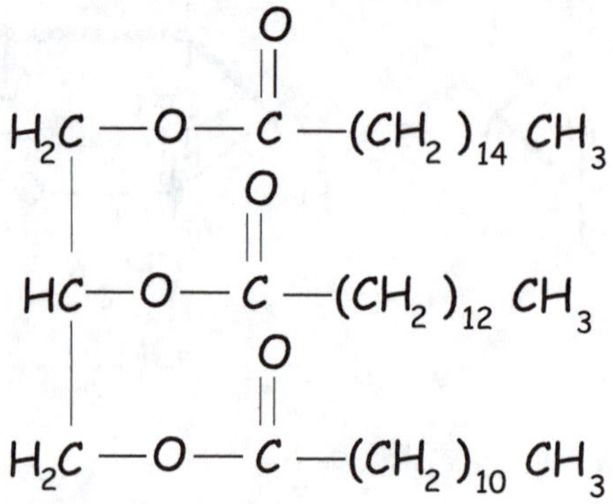

$$H_2C - O - \overset{\overset{O}{\|}}{C} - (CH_2)_{10} \ CH_3$$

c)

$$H_2C - O - \overset{\overset{O}{\|}}{C} - (CH_2)_{14} \ CH_3$$

$$HC - O - \overset{\overset{O}{\|}}{C} - (CH_2)_{16} \ CH_3$$

$$H_2C - O - \overset{O}{\underset{\underset{OH}{|}}{P}} - OH$$

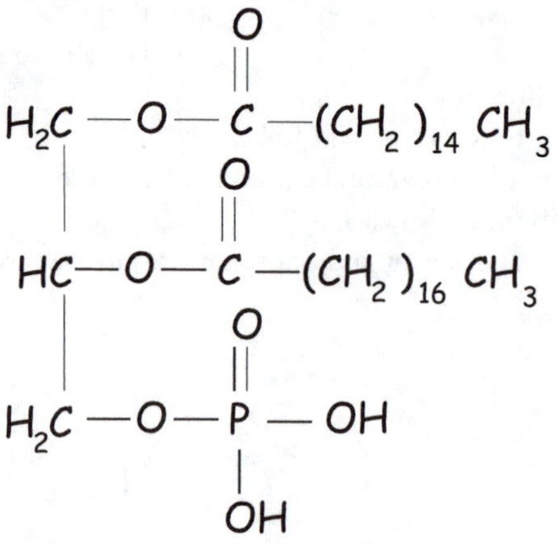

24. Compare the structures of the following steroid compounds to cholesterol (Fig. 17.16). How are they the same and how are they different?

a)

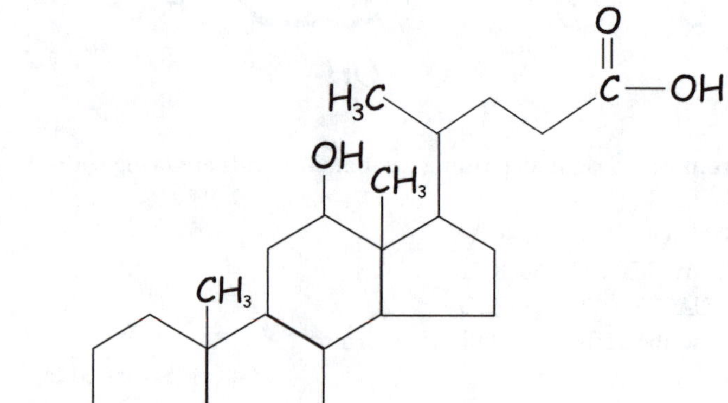

b)

25. Fill in the following blanks with the terms fat, oil, phospholipid, wax, or steroid.
    (a) A _____ is made by combining glycerol with two fatty acids and a phosphate-containing compound.
    (b) A _____ is made by combining glycerol with three fatty acids that contain mostly saturated hydrocarbon chains.
    (c) A _____ is made by combining glycerol with three fatty acids that tend to have some unsaturated bonds in their hydrocarbon chains.
    (d) A _____ makes up an important part of cell membranes.
    (e) A _____ is part of the outer coating of skin and leaves.
    (f) A _____ is sometimes a hormone, but it can also play a role as an agent to help our bodies digest fat.

## For Class Discussion and Reports

26. Anabolic steroids are a group of steroid compounds that are often used illegally to enhance the performance of athletes. Look up the structures of some of these compounds and report on the effects (both good and bad) that they can have on our bodies.
27. We often hear in the news about "good" and "bad" cholesterol. Look up information on these two types of cholesterol and report on how they are different from each other. Also, explain why one type is thought to be good for our bodies and the other is considered to be bad.
28. The Human Genome Project is an ongoing scientific project in the U.S. that is aimed at determining the entire sequence of nucleotides in human DNA. Obtain information on this project and on its goals. Discuss some of the technical challenges that this type of undertaking presents. What are some of the possible benefits that this type of knowledge might produce? What are some possible negative outcomes?
29. Bovine growth hormone has received a great deal of attention regarding its use in dairy cattle to increase their milk production. What is the primary structure of bovine growth hormone? How does this structure compare to that of human growth hormone? What are the actual effects of bovine growth hormone in cattle or of human growth hormone in humans? What is the basis for the controversy regarding the use of bovine growth hormone in cattle?

# Appendix

# Answers to Homework Questions

## A.1    Chapter 1

1. Composition, structure, properties, and changes.
2. Chemical properties involve chemical change; physical properties do not. Examples of physical properties: color, odor, texture, physical state, solubility, density, melting point, boiling point, freezing point. Examples of chemical change: iron rusts, wood burns, etc.
3. (a) chemical,   (b) physical,   (c) physical,   (d) physical,   (e) chemical,   (f) chemical.
4. Physical change is a change that occurs without any transformation of substance. The substance is the same substance both before and after the change. Examples include change of state (e.g., evaporation, condensation, freezing, melting, etc.), change of texture (e.g., the crushing of a granular solid to a powdery solid), etc.
5. Chemical change is a transformation of one substance into one or more others. Examples include the rusting of iron, the combination of hydrogen and oxygen to make water, etc.
6. (a) chemical change—there has been a transformation of substance.
   (b) physical change—sugar both before and after dissolving.
   (c) physical change—alcohol both before and after evaporation.
7. (a) physical change—water both before and after evaporation.
   (b) chemical change—there has been a transformation of substance.
   (c) physical change—salt both before and after dissolving.

   (d) chemical change—there has been a transformation of substance.

   (e) physical change—acetic acid before and after freezing.

8. A homogeneous mixture is a mixture in which the substances involved are mixed so well that one cannot tell by looking (even using a microscope) that it is in fact a mixture. It appears as just one substance. An example is a solution of sugar and water.

9. A heterogeneous mixture is a mixture in which the substances are not mixed very well, such that you can tell by looking that it consists of more than one substance. An example is sand and water mixed together in the same container.

10. (a) compound,   (b) heterogeneous mixture,   (c) homogeneous mixture,   (d) compound,
    (e) element,   (f) homogeneous mixture,   (g) element,   (h) heterogeneous mixture

11. Blanks from top to bottom: C, D, E, A, G, D, E, A, F, A

12. In a compound, elements are chemically combined. In a mixture, the components are not chemically combined, only physically mixed.

13. Because in a compound, the elements are chemically combined, meaning that it is a different substance with properties different from the elements before they were combined. In a mixture, the components are simply physically mixed, meaning the properties are retained.

14. (a) Entirely new properties   (b) New substance formed   (c) Atoms become chemically combined with other atoms forming molecules or formula units. Example: Carbon dioxide. Carbon atoms and oxygen atoms chemically combine forming molecules of carbon dioxide.

15. Two atoms of hydrogen and one atom of oxygen combined to form a molecule of water.

16. (a) $Na_3PO_4$: Sodium—3 atoms; phosphorus—1 atom; oxygen—4 atoms
    (b) $CCl_4$: Carbon—1 atom; chlorine—4 atoms
    (c) $Al_2(SO_4)_3$: Aluminum—2 atoms; sulfur—3 atoms; oxygen—12 atoms
    (d) $(NH_4)_2CO_3$: Nitrogen—2 atoms; hydrogen—8 atoms; carbon—1 atom; oxygen—3 atoms
    (e) $Ba(OH)_2$: Barium—1 atom; oxygen—2 atoms; hydrogen—2 atoms

17. (a) An atom is the smallest particle of an element that can exist and still be that element.
    (b) A molecule is the smallest particle of a molecular (or covalent) compound that can exist and still be that compound.
    (c) An ion is an atom or grouping of atoms that has developed an electrical charge.

18. (a) atom,   (b) molecule,   (c) ion,   (d) molecule,   (e) atom,   (f) molecule (Bromine exists as a diatomic molecule, $Br_2$, even though it is an element. This is true of several of the nonmetals—see Chapter 2, Section 2.4)

19. A formula unit is the simplest particle of an ionic compound that can exist and still be the compound. The term is used to differentiate between those compounds that are ionic from those that are molecular or covalent.

20. An atom is the fundamental unit of an element. A molecule is the fundamental unit of a molecular compound. A molecule is composed of atoms in chemical combination.

21. (a) Mass is the amount of material as measured by measuring its weight on the Earth's surface.
    (b) The atomic mass unit is one-twelfth the mass of an atom of the carbon-12 isotope.
    (c) The atomic weight of an element is the average weight of all atoms that exist for that element.
    (d) The atomic number of an element is the number of protons in the nucleus of its atom. It is the number that identifies or characterizes a given element.

22. See Table 1.1 in text.

23. See Table 1.1 in text.

24. An isotope of an element is an atom of that element that has a particular number of neutrons in its nucleus. One isotope of a given element differs from another in this number of neutrons.

25. Xe, 30, neutron, 19, 42, 24, protons, 20, C, electron.

26. Ag, 53, 28, 20, 95, 82, Al, atomic weight.

27. 127, 48, Br, Cr, 27, average atomic weight, 13, 28, 37.

28. (a) Atomic number is 6, mass number is 14, number of neutrons is 8.
    (b) Atomic number is 26, mass number is 58, number of neutrons is 32.
    (c) Atomic number is 83, mass number is 209, number of neutrons is 126.

29. (a) Atomic number is 93, mass number is 239, number of neutrons is 146.
    (b) Atomic number is 92, mass number is 235, number of neutrons is 143.
    (c) Atomic number is 82, mass number is 206, number of neutrons is 124.
    (d) Atomic number is 37, mass number is 87, number of neutrons is 50.
30. This is an isotope of hydrogen called "deuterium."
31. No. The different number of protons changes the identity of the element. An atom with nine protons is an atom of fluorine.
32. They are defined differently. The mass number is the sum of the protons and neutrons in a given atom. The atomic weight is the average weight of an atom expressed in atomic mass units, amu. They are approximately the same because one proton and one neutron each weigh approximately 1 amu while the weight of an electron is nearly negligible. The atomic weight is found in the periodic table.
33. Atomic weight is the average weight of all the isotopes of an element, not the weight of just one of the atoms.
34. Hydrogen-1 is the most abundant because the average atomic weight given in the periodic table is very close to 1.000 (1.008), indicating that hydrogen-2 and hydrogen-3 contribute very little to this average and are therefore rather scarce in nature.
35. Protons—53; neutrons—42; electrons—53
36. Radioactive elements have dangerous radiation emanating from them as a result of the breaking up of their atoms.
37. Families are the vertical columns of elements in the periodic table. See Fig. 1.7.
38. (a) Metal: metallic sheen, high electrical conductivity, high tensile strength, malleability, ductility, tend to form positively charged ions.  (b) Nonmetal: often gaseous, low or no electrical conductivity, tend to form negatively charged ions.  (c) Metalloids: can have properties of both metals and nonmetals.
39. Metals: Pu. Nonmetals: F, H. Metalloids: B, Si.
40. Cr—transition metal. Na—alkali metal or representative element. U—inner transition metal. Ca—alkaline earth metal or representative element. Cl—halogen or representative element. S—representative element. Ar—noble gas or representative element. Ge—representative element.
41. Many possible answers for each. Please refer to Figs. 1.6 and 1.7.
42. A monatomic ion is an ion that consists of just one atom. Examples: $Cl^-$, $Ca^{2+}$, etc.
43. Elements in this family achieve the same count as a noble gas by gaining an electron, not by losing any.
44. Nonmetals achieve the noble gas count of electrons more easily by gaining electrons.
45. Any three of the alkaline earth metals (beryllium, magnesium, calcium, strontium, barium, and radium) are possible answers here.
46. By losing two electrons, an atom of magnesium achieves the noble gas count. It does not achieve this count by losing one or three.
47. A polyatomic ion is an ion that consists of more than one atom. Examples include any of the "Big Six" in Table 1.2.
48. (a) $NO_3^-$,  (b) phosphate,  (c) $OH^-$,  (d) sulfate

# A.2   Chapter 2

1. A cation is a positively charged ion.
2. An anion is a negatively charged ion.
3. (a) CaO,  (b) $CaBr_2$,  (c) $K_3PO_4$,  (d) $Mg(NO_3)_2$,  (e) $Al_2S_3$,  (f) $(NH_4)_2CO_3$
4. (a) $K_2O$,  (b) $(NH_4)_2O$, (c) $Rb_2SO_4$,  (d) NaBr,  (e) $AlCl_3$,  (f) $Ca(NO_3)_2$,
   (g) $Mg_3(PO_4)_2$,  (h) BaS
5. (a) $CaF_2$,  (b) $Al_2O_3$,  (c) $K_2S$,  (d) MgO,  (e) $Rb_2CO_3$,  (f) $Li_3PO_4$,
   (g) $Sr(OH)_2$,  (h) $NH_4Cl$
6. $Ba_2SO_4$ cannot exist because the barium ion would need to have a charge of +1, and we know that being an alkaline earth metal, it must lose two electrons and acquire a change of +2 in order for it to have the same electron count as a noble gas.

7. The ionic compounds are those that contain a metal ion or the ammonium ion. These are $SrCO_3$, $MgBr_2$, $NH_4NO_3$, $Al_2(SO_4)_3$, $FeBr_2$, $K_2O$, and $MgCl_2$. The remaining ones are covalent.

8. A binary compound is a compound with two elements combined.

9. A ternary compound is a compound with three elements combined.

10. A hydride is a compound that has the hydride ion, $H^-$, in its formula.

11. The oxidation number of an element is a number assigned to that element that reflects how its electrons are involved in any chemical combination. It is useful to chemistry professionals in that it aids in the naming of some compounds and can also be used to track electrons in chemical reactions that involve electron transfer. In addition, it is useful in predicting chemical properties of substances.

12. The free elemental state of an element is the state the element is in if it is not chemically combined with any other element.

13. For most elements, this is merely the symbol of the elements. For some nonmetals, it is the diatomic molecule of that element. (a) Au,    (b) He,    (c) $O_2$,    (d) $Cl_2$

14. Assuming "normal" refers to what it is in most cases, we have: (a) $-2$,    (b) $+1$,    (c) $+1$,    (d) $+2$, (e) $+3$,    (f) $-1$.

15. $-1$

16. A peroxide is a compound that contains the peroxide ion, $O_2^{-2}$. The oxidation number of the oxygen is $-1$.

17. (a) $+1$,    (b) $+3$,    (c) $+5$,    (d) $0$,    (e) $-3$,    (f) $-3$,    (g) $+3$

18. (a) $-2$,    (b) $+6$,    (c) $0$,    (d) $+6$

19. (a) $+5$,    (b) $+3$,    (c) $+5$,    (d) $+3$

20. (a) $+2$,    (b) $0$,    (c) $-1$,    (d) $+4$,    (e) $-2$,    (f) $+2$,    (g) $+1$,    (h) $+4$,    (i) $+1$,    (j) tin: $+2$, nitrogen: $+5$

21. The three that are discussed in Chapter 2 are copper, iron, and tin. However, most transition metals and inner-transition metals can have more than one oxidation number.

22. (a) Acid: A compound that is a source of hydrogen ions and has hydrogen first in the formula.    (b) Base: A compound that consumes hydrogen ions, such as compounds that contain the hydroxide ion.    (c) Salt: A compound that is the product of an acid reacting with a base. These definitions explain how they are related.

23. (a) salt    (b) acid    (c) salt    (d) base    (e) salt    (f) base

24. (a) hydrobromic acid, hydrogen bromide
    (b) calcium oxide
    (c) diphosphorus pentoxide
    (d) potassium iodide
    (e) Ferrous chloride, Iron (II) chloride
    (f) sodium sulfide
    (g) sulfur trioxide
    (h) magnesium hydroxide
    (i) carbon tetrachloride
    (j) hydrochloric acid, hydrogen chloride
    (k) calcium bromide
    (l) ferrous oxide, iron (II) oxide
    (m) calcium sulfate
    (n) ferrous bromide, iron (II) bromide
    (o) sulfur hexafluoride
    (p) hydroiodic acid, hydrogen iodide
    (q) barium chloride
    (r) hydrosulfuric acid, hydrogen sulfide
    (s) magnesium oxide
    (t) cupric chloride, copper (II) chloride
    (u) phosphorus trichloride

25. (a) $CCl_4$  (b) $CuBr_2$  (c) $NH_4Cl$  (d) $H_2S$  (e) NaF  (f) $Fe_2O_3$  (g) $N_2O_3$  (h) $SnSO_4$
   (i) $CuI_2$  (j) $SF_6$  (k) HF  (l) $SnCl_4$  (m) $K_2O$  (n) NaOH  (o) CuO  (p) HCl
   (q) $P_2O_5$  (r) $MgCl_2$  (s) FeS  (t) $N_2O_5$  (u) $SnF_2$  (v) $AlBr_3$
26. (a) $MgBr_2$  (b) $H_3PO_4$  (c) $FeI_3$  (d) $K_2SO_4$  (e) $Cu_2CO_3$  (f) $SO_3$  (g) $H_2S$  (h) $N_2O$
27. (a) nitric acid  (b) carbonate tetrachloride  (c) aluminum bromide  (d) stannic oxide
   (e) ammonium carbonate
28. A chemical reaction is the transformation of one or more substances with known properties into one or more other substances with completely different properties.
29. A chemical equation is a statement of a chemical reaction that utilizes the convenient chemical shorthand of symbols and formulas. It is used by chemistry professionals to succinctly describe a chemical reaction.
30. (a) A reactant is a chemical present before a chemical reaction takes place and is placed to the left of the arrow in a chemical equation. (b) A product is a chemical that forms during a chemical reaction and is placed to the right of the arrow in a chemical equation.
31. The Law of Conservation of Mass is the law that states that matter can neither be created nor destroyed in a chemical reaction. The total number of atoms of each element present in the reaction container must be the same before the reaction begins and after the reaction takes place.
32. (a) $2Fe + O_2 \rightarrow 2FeO$
   (b) $4P + 5O_2 \rightarrow 2P_2O_5$
   (c) $2SO_2 + O_2 \rightarrow 2SO_3$
   (d) $H_3PO_4 + 3NaOH \rightarrow Na_3PO_4 + 3H_2O$
   (e) $BiCl_3 + H_2O \rightarrow BiOCl + 2HCl$
   (f) $2S + 3O_2 \rightarrow 2SO_3$
   (g) $4FeO + O_2 \rightarrow 2Fe_2O_3$
   (h) $4Ag + 2H_2S + O_2 \rightarrow 2Ag_2S + 2H_2O$
   (i) $3NO_2 + H_2O \rightarrow 2HNO_3 + NO$
   (j) $2HgO \rightarrow 2Hg + O_2$
   (k) $Mg + 2H_2O \rightarrow H_2 + Mg(OH)_2$
   (l) $2S + 3O_2 \rightarrow 2SO_3$
   (m) $PCl_5 + H_2O \rightarrow POCl_3 + 2HCl$
   (n) $3NaOH + H_3PO_4 \rightarrow Na_3PO_4 + 3H_2O$
33. (d) and (n), because one reactant is an acid (formula begins with "H") and one is a base (hydroxide).
34. (a) solid  (b) gas  (c) liquid  (d) a gas that is a product of a reaction and escapes the container
   (e) a precipitate  (f) heat added
35. (a) $CaCO_3$ and HCl are the reactants.
   (b) $H_2O$ is a liquid
   (c) $CO_2$ is a gas
36. $Mg(s) + HBr(aq) \rightarrow MgBr_2(aq) + H_2(l)$
37. "(s)" means that Zn is a solid, "(aq)" means that HCl and $ZnCl_2$ are in water solution, the up arrow means that $H_2$ is a gas that forms and escapes into the air.
38. $Pb(NO_3)_2(aq) + K_2CrO_4(aq) \rightarrow PbCrO_4(\downarrow) + KNO_3(aq)$

# A.3   Chapter 3

1. Dalton—first comprehensive atomic theory
   Crookes—electron
   Thomson—plum pudding model
   Millikan—value of electron charge
   Rutherford—the nucleus exists

Chadwick—neutron

Bohr—solar system model

Schrodinger—modern atomic theory

2. solar system model—c

nuclear atom—e

original atomic theory—b

quantum mechanical model—a

plum pudding model—d

neutron—g

electron—f

electron charge—h

3. (a) Dalton. See page 60 of the text.

(b) The electrons are particles in orbit around the nucleus much like the planets are in orbit around the sun.

(c) Most of the atom is empty space.

4. A small percentage of the alpha particles were deflected from their path and a very small percentage reflected back from the foil.

5. A very high percentage of the alpha particles came straight through the foil.

6. A positively charge particle consisting of 2 protons and 2 neutrons. It is the nucleus of a helium atom. It is one of the particles emitted from radioactive elements.

7. The boulder has the capacity to cause matter to move because it has potential energy. Potential energy can be converted to other forms of energy which then do have this capacity.

8. The different forms of energy mentioned in the text are: kinetic, potential, electrical, light, chemical, heat, and nuclear. Examples of transformation are: (1) kinetic is transformed into electrical in a dam on a river,   (2) potential is converted to kinetic when the boulder in #37 falls, (3) electrical is converted to light when a flashlight is switched on,   (4) light is converted to electrical in a solar battery,   (5) chemical is converted into electrical in a car battery,   (6) heat is converted to electrical in a steam turbine, and   (7) nuclear is converted to electrical in a nuclear power plant.

9. Light has properties of both particles and electromagnetic waves.

10. Light waves are electromagnetic, not mechanical, and as such, they do not require matter in order to exist.

11. Wavelength is the distance from the crest of one wave to the crest of the next wave in electromagnetic radiation. Long wavelength is low energy.

12. Gamma rays, ultraviolet waves, and x-rays. Visible and infrared light can cause eye damage if intense enough. Microwaves can indirectly cause damage through the generation of heat.

13. Red, orange, yellow, green, blue, violet.

14. From Fig. 3.3: 450—blue to violet; 600—yellow to orange; 550—green to yellow; 700—red; 350—ultraviolet.

15. If we solve Eq. (3.2) for $v$, we get:

$$v = \frac{c}{\lambda}$$

If we then substitute $c/\lambda$ into Eq. (3.3) for $v$, we get Eq. (3.4). A low frequency means a long wavelength. A long wavelength means a low energy. These equations indicate an inverse relationship between wavelength and both frequency and energy.

16. A line spectrum is a plot of emission intensity vs. wavelength that results from energized electrons. See Fig. 3.6 for examples.

17. Different colors of light represent different energies of light. According to the Bohr model, the different electron orbits represent different energies. Bohr theorized that electrons can gain or lose energy and move from one orbit to another. The energies that electrons can lose when jumping from a higher orbit to a lower orbit represent the different colors.

18. That the electron is a particle.

19. It is the Bohr model, not the quantum mechanical model, that describes the electrons that way.

20. (a) $2n^2$ = maximum number of electrons in principal level n. In this case, $2n^2 = 18$.
    (b) $n^2$ = the number of orbitals in principal level n. In this case, $n^2 = 9$
    (c) There can only be 6 electrons in any *p* sublevel.
    (d) There can only be 2 electrons in any orbital, whether it is *s, p, d*, or *f*.

21. $n^2$ = the number of orbitals in principal level n. If n = 4, 16 orbitals. If n = 5, 25 orbitals.

22. It is a *p*-orbital. It can have 3 orientations, each perpendicular to each other along the x, y, and z axes. The nucleus is at the center of the "figure 8."

23. The *s* orbital is spherical in shape. The lowest principal energy level in which this orbital is found is n = 1. It can have only one orientation.

24. The *f* sublevel has the most energy. The orbital in the *s* sublevel is spherical. The *d* sublevel can hold 10 electrons. The *p* sublevel has only 3 orbitals.

25. (a) Any one orbital can hold only 2 electrons regardless of type. (b) The 3*p* sublevel holds 6 electrons. (c) $2n^2 = 32$. (d) Any *s* sublevel has only 1 orbital, so just 2 electrons.

26. (a) orbital   (b) principal level   (c) sublevel

27. For *s*-orbital, see Fig. 3.8. For *p*-orbital, see Fig. 3.9(a). An *s*-orbital can have only one orientation, a *p*-orbital can have 3 orientations, as in Fig. 3.9(b).

28. (a) $2n^2 = 18$   (b) A single orbital of any type holds a max of 2 electrons.   (c) 6   (d) *s* sublevels hold 2 electrons.

29. For N-See Section 3.7

30. Some of these are listed in Section 3.7

31. Ni (28): $1s^2 2s^2 2p^6 3s^2 3p^6 4s^2 3d^8$
    As (33): $1s^2 2s^2 2p^6 3s^2 3p^6 4s^2 3d^{10} 4p^3$

32. (a) Seventeen electrons, so it is chlorine.   (b) Thirty electrons, so it is zinc.   (c) Ten electrons, so it is neon.

33. Because the n = 3 and the n = 4 levels overlap, the lowest sublevel in the n = 4 level, the 4*s*, actually lies below the highest sublevel in the n = 3 level, the 3*d*.

34. 1. Because principal levels overlap, some sublevels in higher principal levels get electrons before some sublevels in lower principal levels.
    2. Overlap places sublevels from different principal levels very close together in terms of energy. As a result we get the idiosyncracies exemplified by Cr and Cu in Fig. 3.12

35. A half-filled sublevel means that the sublevel has half the number of electrons that it can have and that each orbital within that sublevel has one electron. Sublevels that are half filled are lower in energy than those that are not. In order to achieve the favorable half-filled condition, some elements, notably chromium, "borrow" electrons from other sublevels.

36. (a) Same as argon, $1s^2 2s^2 2p^6 3s^2 3p^6$   (b) Same as krypton, $1s^2 2s^2 2p^6 3s^2 3p^6 4s^2 3d^{10} 4p^6$

37. Outermost electrons are those in the energy level that is the farthest from the nucleus. We pay attention in particular to them because outermost electrons are the ones affected by chemical combination.

38. This would be the energy level diagram shown in Fig. 3.14.

39. Electrons must first be elevated to a higher level, creating vacant levels below them, before they can drop back to a lower level.

40. The light emitted by an atom can only be the light represented by the energy difference between the energy levels involved.

41. FES:   energized by a flame, emission is measured, intensity of emission is plotted vs. concentration.

    FAAS and GFAAS:   energized by light, absorption is measured, absorbance is plotted vs. concentration.

    ICP:   energized by hot plasma source, emission is measured, emission intensity is plotted vs. concentration.

42. Atomic: atoms are energized, a flame, a hot graphite furnace, or a hot plasma source contains the species, and either absorbance or emission intensity is plotted, depending on the technique.

Molecular: molecules or ions are energized, a "cuvette" contains the species, absorbance is plotted.

## A.4　Chapter 4

1. The number of possibilities for each is large. Refer to Fig. 4.2.
2. (a) Manganese: d-block, (b) Radon: p-block, (c) P: p-block, (d) Ca: s-block, (e) Yb: f-block, (f) Yttrium: d-block, (g) Einsteinium: f-block, (h) carbon: p-block, (i) Se: p-block, (j) Ru: d-block, (k) Rubidium: s-block
3. The actinide series is the series of elements in the f-block period beginning with actinium.
4. The lanthanide series is the series of elements in the f-block period beginning with lanthanum.
5. Mendeleev's periodic table was based on properties. However, it is remarkably similar to the modern periodic table, which is based on electron configuration. Mendeleev's work thus shows that properties have much to do with electron configuration.
6. Properties have much to do with electron configuration.
7. See answer to #5.
8. Descriptive chemistry is the study of specific properties and other important facts about individual elements and compounds.
9. Alkali metals are highly reactive when exposed to air, water, and most other materials. Mineral oil is an exception. Because of this, these metals are often stored under mineral oil to preserve them in the free elemental state.
10. (a) See Section 4.4.2
    (b) Example: $Mg + 2H_2O \rightarrow Mg(OH)_2 + H_2$
11. See Section 4.4.3.
12. Hydrogen is explosively flammable and is dangerous to use in blimps. Helium, being a noble gas, is not reactive.
13. Allotropes are different forms of the same element. Examples are diamond and graphite, two allotropes of carbon. See Section 4.4.5 for more discussion.
14. See Sections 4.4.6 and 4.4.7.
15. It repeats or recurs periodically as atomic number is increased.
16. Properties repeat or recur periodically as atomic number is increased.
17. The Periodic Law states that the properties of the elements are periodic functions of their atomic numbers, meaning that properties recur periodically as atomic numbers increase. This is important because it recognizes the periodicity elemental properties and can be used to predict and explain why different elements behave the way that they do.
18. From left to right across a period in the periodic table, atomic size decreases. From top to bottom with a family, the atomic size increases.
19. More electrons in higher energy levels require more physical space. Therefore, atoms toward the bottom of a given family in the periodic table are larger than atoms of elements above them. Atoms of elements on the left end of a given period hold their electrons less tightly (because they tend to be given away such that cations form) and that means they are farther from the nucleus than are the electrons of those on the right end. Farther from the nucleus means a larger diameter, a larger atom.
20. Ionization energy increases as you go from left to right within a period in the periodic table. From top to bottom within a family in the periodic table, ionization energy decreases.

21. As you go from left to right within a period, the electrons are held more tightly by the atom because you are approaching a filled *p* sublevel and it is a more stable situation if atoms were to acquire more electrons rather than give them up. Hence more energy is required to remove an electron. Toward the bottom within a family, electrons in the outermost energy levels are held more loosely because they are farther from the nucleus. Thus it will take less energy to remove them.

22. As you go from left to right across a period, the affinity an atom has for electrons increases. As you go from top to bottom within a family, this affinity decreases.

23. See the answer to #21. A greater ionization energy also means a higher affinity for electrons. The explanations for this are the same as for ionization energy.

24. The screening effect is the effect the electrons in the lower energy levels (nearer the interior of the atom) have on those in the outermost levels. They, in effect, "screen" the outermost electrons from the pull of the nucleus because they are like charged. This results in lower ionization energy, lower electron affinity, and larger atoms near the bottom within a family.

25. Fluorine is at the top of the halogen family while bromine is nearer the bottom. Therefore, fluorine will have a higher electron than bromine because it has fewer electrons in the interior levels to screen other electrons from the pull of the nucleus.

26. (a) S, Cr, K, Rb;   (b) F, O, Ca, Cs;   (c) F, Ge, Fe, Ba;   (d) Fr, Hg, Te, S;   (e) Ba, Tc, Al, O;
    (f) Br, Mo, Sr, Cs;   (g) Ba, Ir, Au, S;   (h) Can't tell exactly where cobalt fits because it is both higher up and also to the left of the others. In terms of the other three, Pt, Pd, Ag;   (i) O, Zn, Ag, Rb;   (j) N, Mo, Ba;   (k) Sn, As, S, F

27. (a) Ca,   (b) S

28. An element with a low ionization energy would tend to form a cation because a low ionization energy means that it does not take much energy to remove an electron.

29. lower left

30. upper right

31. $K^{1+}$, because the one electron that K has in its outermost energy level is loosely held, so the atom is large. However, when that electron is lost, not only is the outermost energy level the next level down, but there are fewer electrons than protons, and all are more tightly held making for a smaller sphere.

32. Ionization energy is the energy required to remove an electron from an atom.

33. (a) nitrous acid,   (b) calcium sulfite,   (c) perchloric acid,   (d) potassium iodite
    (e) sodium dihydrogen phosphate,   (f) magnesium bicarbonate

34. (a) $H_2SO_3$,   (b) $K_3PO_4$,   (c) $NaHSO_4$,   (d) $KlO_3$,   (e) $Li_2HPO_4$,   (f) $Ca(BrO)_2$

35. (a) phosphorous acid,   (b) sodium monohydrogen phosphate,   (c) iodous acid,
    (d) calcium nitrite, (e) potassium bicarbonate

36. (a) $K_2SO_3$,   (b) $Ca(ClO_3)_2$,   (c) $HNO_2$,   (d) $HBrO$,   (e) $H_2CO_3$

37. (a) bromous acid,   (b) potassium bicarbonate,   (c) magnesium periodate,   (d) phosphoric acid,   (e) ferric nitrate, or iron (III) nitrate,   (f) sodium phosphite

38. (a) $H_2CO_3$,   (b) $K_2HPO_4$,   (c) $Fe(BrO)_2$,   (d) $HClO_3$,   (e) $HNO_2$,   (f) $CaSO_3$

39. (a) nitric acid,   (b) chloric acid,   (c) magnesium chlorite,   (d) sodium dihydrogen phosphate,
    (e) sodium bicarbonate

40. (a) HClO,   (b) $NaNO_2$,   (c) $KClO_3$

41. (a) potassium chromate,   (b) barium dichromate,   (c) potassium permanganate,
    (d) hydrogen cyanide,   (e) magnesium acetate,   (f) sodium oxalate,   (g) calcium thiosulfate

42. (a) KSCN,   (b) $NaC_2H_3O_2$,   (c) $K_2Cr_2O_7$,
    (d) $BaCrO_4$,   (e) $CaC_2O_4$,   (f) $Na_2S_2O_3$,
    (g) $(NH_4)_2Cr_2O_7$,   (h) NaCN,   (i) $NaMnO_4$

43. (a) $CaHPO_4$,   (b) $K_2HPO_4$,   (c) $Mg(H_2PO_4)_2$,   (d) $H_3PO_4$,   (e) $AlPO_4$,   (f) $Li_3PO_4$,
    (g) $(NH_4)_3PO_4$

# A.5    Chapter 5

1. With ionic bonding, ions form due to the outermost electrons being transferred from one chemical species to another. With covalent bonding, outermost electrons are not transferred; rather they are shared by the two atoms involved.

2. An ionic bond is the electrostatic attraction between oppositely charged ions. A covalent is a sharing of electrons between atoms.

3. KBr is an ionic compound because of the ionic bonding that is present. The important clue is the presence of a metal in the formula, because metals are only rarely found in covalent bonding situations. HBr is a covalent compound because there is no metal and a pair of electrons is being shared by the H and the Br.

4. (a) For an element, we show the outermost *s* and *p* electrons an atom has before bonding to another atom in any way. These electrons are either paired or unpaired according to that shown in the orbital diagram (outermost electrons only) for that element.

   (b) Simple ionic compound refers to formulas in which there are only monatomic positive and negative ions. These require no formal rules. Positive ions will have no electrons (they give them all to another chemical species), but will show its positive charge. Negative ions will have eight electrons, which is the number that the outermost *s* and *p* sublevels can hold, and will show its negative charge.

   (c) Simple covalent compounds can be done by slot filling. "Slot" refers to an open space in an orbital where there is already an unpaired electron. Thus, the two atoms share their electrons—an electron from one fills the slot of another.

   (d) These are drawn by following the steps in Table 5.1. They are too complicated to be done by slot filling.

   (e) This refers to ionic compounds that have one or more polyatomic ions present in their formula. These are also done by the steps in Fig. 5.1, remembering that ions are present and must have their charges shown.

5.

$$\cdot \ddot{C} \quad :\ddot{O}\cdot \quad \ddot{M}g \quad \cdot \ddot{A}l \quad \cdot \ddot{P}\cdot \quad :\ddot{Cl}\cdot \quad :\ddot{Ar}: \quad K\cdot$$

6. (a) $K_2O$                           (b) $H_2O$

$$\begin{array}{l} K^+ \quad {}_{\cdot\cdot}{}^{2-} \\ \quad :\ddot{O}: \\ K^+ \end{array}$$

$$\begin{array}{l} :\ddot{O}:H \\ \phantom{:}H \end{array}$$

   (c) Ca                              (d) $H_2CO_3$

   Ca:

$$\begin{array}{l} \ddot{O} \\ :: \phantom{..} \\ C:\ddot{O}:H \\ :\ddot{O}: \\ H \end{array}$$

   (e) $NO_3^-$                        (f) $AlCl_3$

$$\left( \begin{array}{l} \ddot{O} \\ :: \phantom{..} \\ N:\ddot{O}: \\ :\ddot{O}: \end{array} \right)^{-}$$

$$\begin{array}{l} :\ddot{Cl}: \\ Al^{3+} \; :\ddot{Cl}: \\ :\ddot{Cl}: \end{array}$$

(g) $Na_2SO_3$

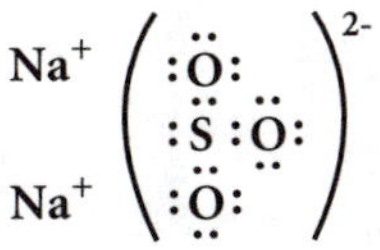

$$Na^+ \quad \begin{pmatrix} :\ddot{O}: \\ :\ddot{S}:\ddot{O}: \\ :\ddot{O}: \end{pmatrix}^{2-}$$
$$Na^+$$

(h) HCl

$$H:\ddot{\underset{..}{C}l}:$$

(i) S

$$:\overset{..}{\underset{.}{S}}\cdot$$

(j) $O_2$

$$\overset{.}{\underset{.}{O}}::\overset{.}{\underset{.}{O}}$$

(k) CO

$$:C:::O:$$

(l) $XeF_4$

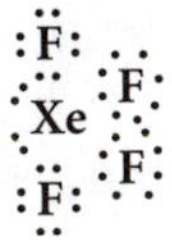

$$\begin{matrix} :\ddot{F}: & \\ \overset{..}{X}e & :\overset{.}{F}: \\ :\ddot{F}: & :\overset{..}{F}: \end{matrix}$$

7. (a) $IO_2^-$

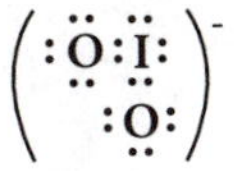

$$\begin{pmatrix} :\ddot{O}:\ddot{I}: \\ :\ddot{O}: \end{pmatrix}^-$$

(b) $Br_2$

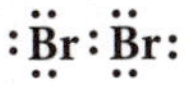

$$:\ddot{B}r:\ddot{B}r:$$

(c) Al

$$\cdot\ddot{A}l$$

(d) $K_2SO_3$

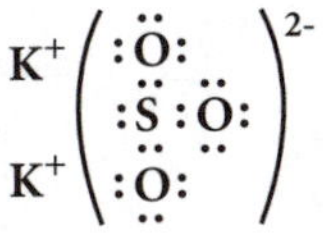

$$K^+ \quad \begin{pmatrix} :\ddot{O}: \\ :\ddot{S}:\ddot{O}: \\ :\ddot{O}: \end{pmatrix}^{2-}$$
$$K^+$$

(e) $BaCl_2$

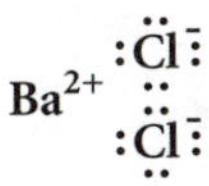

$$Ba^{2+} \quad \begin{matrix} :\ddot{C}l: \\ :\ddot{C}l: \end{matrix}$$

(f) $H_2S$

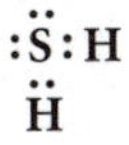

$$:\ddot{S}:H$$
$$\ddot{H}$$

(g) $HNO_3$

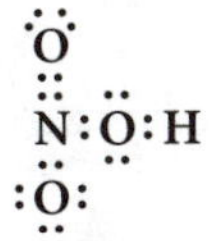

$$\overset{..}{\underset{..}{O}}$$
$$N:\ddot{O}:H$$
$$:\ddot{O}:$$

(h) $IF_5$

$$\begin{matrix} :\overset{.}{F}: & :\overset{.}{F}: \\ :\ddot{F}: & I & : \\ :\overset{.}{F}: & :\overset{.}{F}: \end{matrix}$$

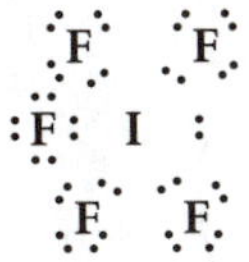

(i) $CN^-$

$$(:C:::N:)^-$$

(j) P

$$\cdot\ddot{P}\cdot$$

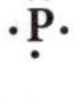

8. (a) $BrF_3$

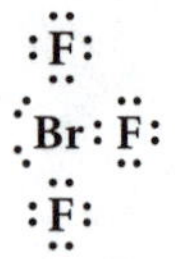

:F:
.Br:F:
:F:

(b) CO

:C:::O:

(c) $PH_3$

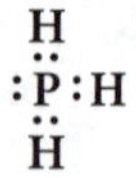

H
:P:H
H

(d) $SO_4^{2-}$

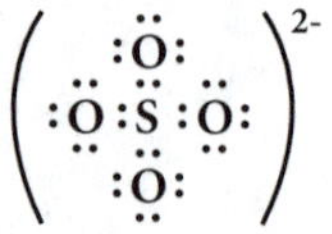

$$\left( \begin{array}{c} :O: \\ :O:S:O: \\ :O: \end{array} \right)^{2-}$$

(e) $CaCl_2$

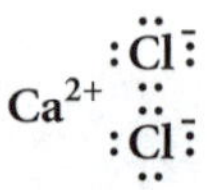

$Ca^{2+}$  :Cl:⁻  :Cl:⁻

(f) $HClO_3$

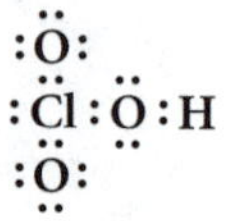

:O:
:Cl:O:H
:O:

(g) $KBrO_2$

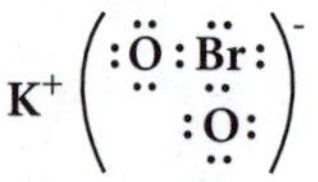

$K^+ \left( \begin{array}{c} :O:Br: \\ :O: \end{array} \right)^-$

(h) Al

·Al

(i) MgO

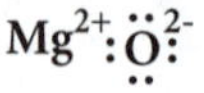

$Mg^{2+}$ :O:²⁻

(j) $SO_3$

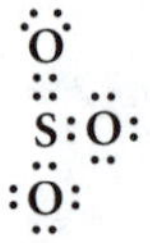

O
S:O:
:O:

(k) $H_2PO_4^-$

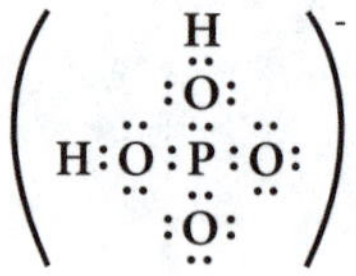

$$\left( \begin{array}{c} H \\ :O: \\ H:O:P:O: \\ :O: \end{array} \right)^-$$

9. (a) $Ca(NO_3)_2$

$$Ca^{2+} \left( \begin{array}{c} O \\ N:O: \\ :O: \end{array} \right)^- \left( \begin{array}{c} O \\ N:O: \\ :O: \end{array} \right)^-$$

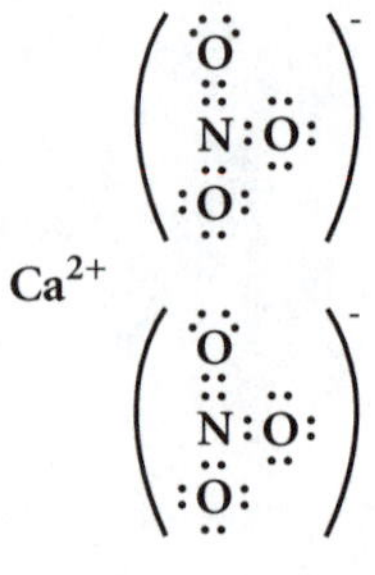

(b) $(NH_4)_2S$

$$\left( \begin{array}{c} H \\ H:N:H \\ H \end{array} \right)^+ \left( \begin{array}{c} H \\ H:N:H \\ H \end{array} \right)^+ :S:^{2-}$$

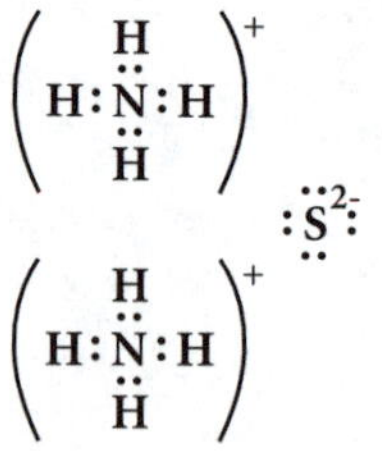

(c) $Al_2(SO_4)_3$

$$\left( \begin{array}{c} :\ddot{O}: \\ :\ddot{O}:S:\ddot{O}: \\ :\ddot{O}: \end{array} \right)^{2-}$$

$Al^{3+}$

$$\left( \begin{array}{c} :\ddot{O}: \\ :\ddot{O}:S:\ddot{O}: \\ :\ddot{O}: \end{array} \right)^{2-}$$

$Al^{3+}$

$$\left( \begin{array}{c} :\ddot{O}: \\ :\ddot{O}:S:\ddot{O}: \\ :\ddot{O}: \end{array} \right)^{2-}$$

(d) $(NH_4)_2CO_3$

$$\left( \begin{array}{c} H \\ H:N:H \\ H \end{array} \right)^{+}$$

$$\left( \begin{array}{c} H \\ H:N:H \\ H \end{array} \right)^{+}$$

$$\left( \begin{array}{c} :\ddot{O}: \\ C:\ddot{O}: \\ :\ddot{O}: \end{array} \right)^{2-}$$

10. The correct Lewis electron-dot structure is (b). $MgCl_2$ is an ionic compound (metal combined with nonmetal), one magnesium ion combined with two chloride ions as structure (b) indicates. Structure (a) would indicate that covalent bonds are present, which is inaccurate.

11. (a) The electron count is wrong. Three oxygen atoms contribute 18 electrons; one sulfur contributes six electrons; and two hydrogens contribute two electrons—a total of 26 electrons. The structure shows only 24. There should be no double bonds and there should be a nonbonded pair of electrons on the sulfur atom.

    (b) The nitrogen atom has only six electrons—it should have eight. There should be a double bond between the nitrogen atom and one of the oxygen atoms. One of the nonbonding pairs on the oxygen atom should be utilized for the second bond.

12. A hydrogen atom can only form one bond, since sharing one pair of electrons gives it the electrons needed to fill its n = 1 principal energy level. Thus it will always be found around the outside of a Lewis electron-dot structure.

13.

ethane:
$$\begin{array}{c} H \ \ H \\ H:C:C:H \\ H \ \ H \end{array}$$

ethylene:
$$\begin{array}{c} H \ \ \ \ \ H \\ C::C \\ H \ \ \ \ \ H \end{array}$$

acetylene: $H:C:::C:H$

When two hydrogen atoms are removed from ethane, an additional pair of electrons must be shared by the two carbon atoms, forming ethylene. When two hydrogens are removed from ethylene, an additional pair of electrons must be shared by the two carbon atoms forming acetylene.

14. A coordinate covalent bond is one in which both of the electrons of the pair of electrons being shared originated on only one of the two atoms involved. In the case of $NH_4^+$, both electrons of one of the pairs of electrons shared by the nitrogen and the hydrogens originated on the nitrogen. One of the hydrogens was a hydrogen ion (no electrons) before bonding. See Section 5.6.

15. See answer to #14.

16. Molecules are distinct units that exist independent of their surroundings. In ionic compounds, positive and negative ions exist in a 3-dimensional array, positive ions surrounded by negative ions and negative ions surrounded by positive ions, such that there is no distinct unit we can identify as a molecule.

17. Electron affinity is the attraction a free atom has for electrons. Electronegativity is the attraction a bonded atom has for its shared electrons. A number is assigned to the electronegativity of an element in order to compare one element to another and determine the polarity of a bond.

18. When the electronegativity values of two nonmetal atoms are different, the bond between the two is polar, and the larger the difference, the more polar the bond.

19. A polar bond is one in which the shared electrons are pulled more toward one atom in the bond than to the other, resulting in the imbalance of charge that is a polar bond. An example is the O–H bond in a water molecule.

20. A polar molecule is a molecule in which the polar bonds are not symmetrically arranged in space. An example is the water molecule because the polar O–H bonds are not symmetrically arranged in space.

21. The polar bonds in compound B are symmetrically arranged in space while the polar bonds in compound A are not.

22. (a) The O–H bond is a polar covalent bond. It is covalent and not ionic because it is between two nonmetals. It is polar because the electronegativities of oxygen and hydrogen are different (refer to Table 5.2).

    (b) The Br–Br bond is a nonpolar covalent bond. It is covalent and not ionic because it is between two nonmetals. It is nonpolar because the two atoms that are bonded are identical and thus have equal electronegativities.

    (c) The Ca–O bond is an ionic bond because a metal (calcium) is involved and metals tend to give up electrons and not share them.

    (d) The H–P bond is a nonpolar covalent bond. It is covalent and not ionic because it is between two nonmetals. It is nonpolar because the electronegativities of hydrogen and phosphorus are the same (refer to Table 5.2).

23. The carbon–chlorine bonds, while polar, are symmetrically arranged in space in $CCl_4$ (tetrahedral) and so the molecule is nonpolar.

24. The polar bonds must be symmetrically arranged in space.

25. An intermolecular force is the force of attraction one molecule has for another. The hydrogen bond is an example.

26. A hydrogen bond is an intermolecular force consisting of the attraction that a hydrogen atom bonded to an oxygen, nitrogen, or fluorine atom on one molecule has for a pair of electrons, usually on an oxygen, nitrogen, or fluorine atom on another.

27. Possible answers are high boiling point, significant expansion upon freezing, ice being less dense than liquid water, formation of ice causing water pipes and water hoses to burst.

28. The boiling point of carbon tetrachloride, $CCl_4$, is 78°C. The boiling point of chloroform, $CHCl_3$, is 61°C. Replacing a chlorine on the nonpolar $CCl_4$ molecule with hydrogen to form $CHCl_3$, creates a polar molecule that is lighter in weight. A polar molecule would mean a higher boiling point, but a lighter weight means a lower boiling point. It is apparent that the lighter weight has a greater effect in this case.

29. A hydrate is an ionic compound that has water molecules trapped within the crystal structure, a certain number of water molecules per formula unit. See Section 5.9.4 for some examples.

30. Since the water molecule retains its identity when combining with the ionic compound to form a hydrate, it is appropriate to show the water molecule intact in the formula of the hydrate. The use of the dot to separate the formula of the ionic compound from the formula of water is special symbolism to display this unique situation.

31. See Section 5.9.4 for the definitions.

32. The acronym VSEPR stands for the Valence Shell Electron Pair Repulsion theory. It is a theory by which electron pair repulsions are used to predict the geometric shape of a molecule.

33. The theory is that electron pairs, or, more correctly, regions of high electron density in the valence shell of the central atom repel each other and thereby get as far away from each other as they can get. This results in a particular geometric shape depending on how many total regions there are, how many bonding regions there are, and how many nonbonding pairs there are.

34. See Figs. 5.7(a) and 5.7(b). In terms of the total number of regions of high electron density, they are both the same—tetrahedral, i.e., the "general structure" of each is tetrahedral. In terms of the

specific molecular structure, for which only bonding regions are important, they are differ-ent—methane is tetrahedral and water is "angular."

35. See Fig. 5.8. The five structures in the left-most vertical column are the five general structures.

36. (a) angular,   (b) square planar,   (c) linear,   (d) octahedral,   (e) T-shaped,   (f) trigonal planar.

37. Answers are variable depending on which are selected.

38. The compound $XeF_4$ is nonpolar. The polar Xe–F bonds are symmetrically arranged in the square planar geometry. The compound $ICl_3$ is polar. The polar I–Cl bonds are not symmetrically arranged in the T-shaped geometry.

39. A molecular orbital is the region of space occupied by the bonding electrons in a covalent bond. It is formed by the overlap of the atomic orbitals that are participating in the bond.

40. A sigma bond is the end-to-end overlap between any two orbitals. A pi bond is the side-to-side overlap between two *p* orbitals. Any compound or ion that has a double bond or triple bond has one or two pi bonds, respectively. Examples of one pi bond include oxygen gas, the carbonate ion, the nitrate ion, and ethylene. Examples of two pi bonds include nitrogen gas, carbon monoxide, the cyanide ion, and acetylene.

41. (a) See Fig. 5.9,   (b) See Fig. 5.9,   (c) See Fig. 5.9,   (d) See Fig. 5.10

42. Situations (a), (b), and (c) produce a sigma bond. Situation (d) produces a pi bond.

43. All have one sigma bond; only "b" has just one pi bond; only "c" has two pi bonds; none have two sigma bonds; none have no sigma bond; only "a" has no pi bonds; none have three sigma bonds.

44. A pi bond is the side-to-side overlap between two p-orbitals. See Fig. 5.10

45. A triple bond has one sigma bond and two pi bonds.

46. (a) Any kind of orbital can be involved in a sigma bond. End-to-end.
    (b) Only p-orbitals are involved in a pi bond. Side-to-side.

47. (a) pi,   (b) sigma,   (c) pi, sigma,   (d) sigma

48. One sigma and one pi. The first bond is always a sigma bond. When two p-orbitals that have just one electron each line up side-to-side after that, then you will also have a pi bond.

# A.6   Chapter 6

1. Organic chemistry is the study of compounds derived from living systems, or systems that were once living. Organic chemistry is the study of the compounds of carbon. Most come from petroleum.

2. It refers to the fact that gasoline is derived from petroleum, or systems that were once living.

3. Carbon atoms are very prolific in their bonding capabilities and bond freely to one another, often forming long chains sometimes called backbones, or a network.

4. Hybridization, as defined in this module, is the rearrangement of electrons and orbitals such that carbon has four unpaired electrons rather than two, and can therefore bond four times.

5. Hybridization of a carbon atom means that the four electrons in the outermost energy level become unpaired, each in a separate atomic orbital. This results in carbon's ability to form four bonds with other atoms, including other carbon atoms, creating the possibility of "chains" and "networks" of carbon atoms, which in turn is responsible for the huge number of organic compounds that exist.

6. It is tetrahedral when there is nothing but single bonds attached. It is trigonal planar when there is a double bond attached. It is linear when there is a triple bond attached.

7. The four properties mentioned in Section 6.5 are liquids at room temperatures, flammable, odor, volatility.

8. Besides those listed in #7, specific gravity, density, viscosity, etc.

9. The boiling point will increase due to the increase in the weight of a molecule. More energy (higher temperature) is required to boil a liquid composed of heavier molecules.

10. A functional group is a structural feature that allows a compound to be placed into a particular classification.

11. Aromatic hydrocarbons have the benzene ring; aliphatic hydrocarbons do not.

12. A saturated hydrocarbon is one in which no more hydrogen atoms can be bonded to the carbons—it is saturated with hydrogen atoms. Saturated hydrocarbons are those with only single bonds between the carbon atoms in its structure. Hydrocarbons with double or triple bonds between carbons in the structure are not saturated with hydrogen atoms. More hydrogens can be added to the carbons involved in double or triple bonds because more hydrogens can attach to these atoms (eliminating the double or triple bond).

13. In each of the alkane molecules listed, there are "2n+2" hydrogen atoms bonded to "n" carbons. For example, in methane, $CH_4$, there is one carbon and "2(1) +2" or four hydrogen atoms.

14. In each of the alkene or cyclic molecules listed, there are two hydrogen atoms for every carbon atom. For example, in ethene, there are two carbon atoms and four hydrogen atoms.

15. In each of the alkyne molecules listed, there are "2n − 2" hydrogen atoms bonded to "n" carbon atoms. For example, in ethyne, there are two carbon atoms and "2n − 2" or two hydrogen atoms.

16. (a) alkene or cycloalkane (propylene or cyclopropane),   (b) alkyne (1- or 2-butyne),
    (c) aromatic (benzene),   (d) alkene (any isomer of octene, or cyclooctene), cyclooctene
    (e) probably aromatic (e.g., benzene ring with ethyl group replacing a hydrogen on the ring).

17. Simplest alkane is methane, Section 6.6.1. For the others, see Table 6.2.

18.
$$CH_3CH_2CH_2CH_2CH_3, \qquad CH_3CH_2\underset{\underset{CH_3}{|}}{C}HCH_3, \qquad CH_3\underset{\underset{CH_3}{|}}{\overset{\overset{CH_3}{|}}{C}}CH_3$$

19. All except (b) are isomers of the given hexane structure. Structure (b) is not an isomer because it is identical to the given structure—it has the six carbons in a chain. Twisting the chain to make it appear as shown in (b) does not make it an isomer.

20. A cyclohexane ring has all single bonds between the carbons in the ring. It represents an alkane. A benzene ring has pi bonding occurring with all the carbons in the ring (besides the sigma bonding), thus creating a unique situation in which pi electrons are delocalized around the entire ring.

21. The structures (b) and (d) are isomers. They both have the same formula ($C_4H_{10}O$), but a different structure.

22. The structures (b) and (e) are isomers. They both have the same formula ($C_5H_{10}O$), but a different structure.

23. See Table 6.2.

24. (a) ketone,   (b) carboxylic acid,   (c) aldehyde,   (d) alkene,   (e) aromatic hydrocarbon,
    (f) alcohol,   (g) alkyne,   (h) ketone

25. (a) acetic acid or ethanoic acid,   (b) diethyl ether,   (c) ethanol or ethyl alcohol,   (d) ethylene or ethene,   (e) methyl acetate,   (f) acetone or propanone,   (g) toluene   (h) isobutane

26. (a) methanol or methyl alcohol,   (b) butane,   (c) propylene or propene,   (d) formic acid or methanoic acid,   (e) dimethyl ether,   (f) methyl formate,   (g) cyclohexane,   (h) benzene

27. (a)       O       (b) $CH_3–OH$   (c) $CH_3CH=CH_2$
    $$\overset{\displaystyle ||}{H-C-OH}$$

(d)                  (e) See Table 6.2

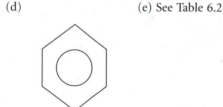

(f) $CH_3CH{\equiv}CHCH_3$

28. methane—D, acetone—G, formaldehyde—A, acetic acid—C, methanol—F, ethanol—B, isopropyl alcohol—E

29. acetylene—fuel for welder's torches
    ethyl alcohol—alcoholic beverages
    acetone—fingernail polish remover
    methane—natural gas
    polyethylene—plastic bottles
    toluene—airplane glue, whiteout
    methyl alcohol—windshield washer fluid
    diethyl ether—formerly used in hospitals as an anesthetic
    isopropyl alcohol—rubbing alcohol
    formaldehyde—biological specimen preserver
    acetic acid—vinegar

30. (a) See Table 6.2   (b) See Table 6.2   (c) See Table 6.2
    (d) See Table 6.2   (e)      O            (f) See Table 6.2
                                 ‖
                             H–C–H

    (g) See Table 6.2   (h)        O
                                   ‖
                            $CH_3$–C–O–$CH_3$

31. hexane:                        2-pentane:

    n-propyl alcohol:              ethyl methyl ether

    propanal                       acetic acid

32. (a) acetic acid,   (b) formaldehyde,   (c) methyl formate,   (d) ethyl methyl,   (e) ether,
    (f) ethanol or ethyl alcohol

33. See Sections 6.10 and 6.11.

34. The polymer is polyvinyl chloride. See Table 6.3 for the structure.

35. See Table 6.3.

36. Both take place by breaking one of the bonds of the double bond creating sites on these adjacent carbons where new substituents can attach. The new substituents in the case of a polymer are other monomer units that have undergone the same bond-breaking process.

37. A polymer is a material consisting of very large molecules that have formed by small "monomer" units attaching to one another to form a very long chain of such units. In an addition polymer, the monomer units consist of alkene molecules that link together through an "addition reaction." Examples are given in Table 6.3. In a condensation polymer, the monomer units consist of two

small molecules containing different functional groups (two each) that react with each other such that they link together. Examples are given in Table 6.4.

38. Since a condensation polymer forms because two functional groups react with each other, and since both ends of the molecule must link with other molecules in order to form the chain, both ends must have the functional group.

39. See Section 6.14.1 and Fig. 6.8.

40. The % transmittance is the percentage of the intensity of light that gets transmitted by a material on which it shines and through which it passes. It is determined by comparing the intensity of the light with nothing in its path to the intensity of the light with the material in its path. In infrared spectroscopy, the % transmittance measured at different wavenumbers of infrared light help to identify the structure of the material.

41. Wavenumber is another way to express the wavelength of light. It is the inverse of the wavelength when the wavelength is expressed in centimeters. The % transmittance measured at different wavenumbers of infrared light help to identify the structure of a material.

42. See the discussion in Section 6.14.4 and Example 6.1.

43. The spectrum displays the characteristic absorption pattern of an alcohol (broad absorption band centered at about 3300 $cm^{-1}$). It does not display the characteristic absorption pattern of a carbonyl group (sharp, strong absorption band near 1700 $cm^{1-}$), which methyl ethyl ketone possesses.

44. This spectrum has the strong, sharp absorption band at about 1700 $cm^{-1}$, which means that the compound tested possesses the carbonyl group. The carbonyl group is found in aldehydes, ketones, carboxyl acids, and esters.

# A.7    Chapter 7

1. Much of what occurs in modern chemistry laboratories is *chemical analysis* and this requires many different types of measurement, from basic weight and volume measurement to measurements with sophisticated instruments. For example, when analyzing a soil sample for moisture, the weight of the soil sample must be measured both before and after a drying step. When monitoring industrial products for quality, two common measurements are specific gravity and viscosity.

2. The fundamental rule of measurement is stated in Fig. 7.1. It is important because it helps us maximize the precision of the measurements we make and this can be quite important in real-world laboratories.

3. (a) 5.17 units    (b) 4.0 units

4. 24%, 22.0 mL

5. These rules are stated in Table 7.1.

6. (a) 3,   (b) impossible to say,   (c) 3,   (d) 5,   (e) 3,   (f) 1,   (g) impossible to say,   (h) 4,
   (i) 6,   (j) 4,   (k) 5,   (l) 4,   (m) impossible to say,   (n) 3,   (o) 4,   (p) 3,   (q) 4.

7. The number 4.00 is more precise than the number 4 because there are more significant figures.

8. "Rounding" is an operation in which digits are dropped from the end of a number in order to express the number to the correct precision.

9. The "even/odd" rule is a rule for rounding when the digit immediately to the right of the last digit to be retained is a 5 with no more digits to its right, or with only zeros to its right. In that case, if the last digit to be retained is odd, then we increase it by one. If it is even, then we keep it the same.

10. These rules are stated in Table 7.2.

11. (a) 0.000060,   (b) $1.58 \times 10^3$,   (c) 6.25,   (d) 103.6,   (e) 21.000,   (f) 0.037,
    (g) $1.24 \times 10^4$,   (h) 16.2,   (i) 11.31,   (j) $1.2 \times 10^2$,   (k) 0.0000767,   (l) 10.68,   (m) 0.50.

12. Dimensional analysis is a method by which unit conversions are carried out and/or checked for accuracy.

13. The basic rule of dimensional analysis is: When the same unit appears in both the numerator and denominator, these units cancel.

14. (a) An equality is an equation that gives the relationship between one unit and another. For example, "12 in = 1 ft" is an equality.
    (b) A conversion factor is one of two ratios that are derived from an equality.
    (c) A conversion is a calculation in which one or more conversion factors is (are) used to convert from one unit to another.

15. (a) 39.37 in. = 1 m,   (b) 1,000,000 $\mu$m = 1 m,   (c) 1000 mg = 1 g,   (d) 1.057 qt = 1 L,
    (e) 1000 mL = 1 L

16. (a) $\dfrac{39.37 \text{ in.}}{1 \text{ m}}$   $\dfrac{1 \text{ m}}{39.37 \text{ in.}}$   (b) $\dfrac{1,000,000 \text{ } \mu\text{m}}{1 \text{ m}}$   $\dfrac{1 \text{ m}}{1,000,000 \text{ } \mu\text{m}}$

    (c) $\dfrac{1000 \text{ mg}}{1 \text{ m}}$   $\dfrac{1 \text{ m}}{1000 \text{ mg}}$   (d) $\dfrac{1.057 \text{ qt}}{1 \text{ L}}$   $\dfrac{1 \text{ L}}{1.057 \text{ qt}}$

    (e) $\dfrac{1000 \text{ mL}}{1 \text{ L}}$   $\dfrac{1 \text{ L}}{1000 \text{ mL}}$

17. (a) $\dfrac{0.3048 \text{ m}}{\text{ft}}$   (b) $\dfrac{0.0001016 \text{ km}}{\text{yd}}$   (c) $\dfrac{453.6 \text{ g}}{\text{lb}}$

    (d) $\dfrac{946.1 \text{ mL}}{\text{qt}}$

18. (a) $\dfrac{3.281 \text{ ft}}{\text{m}}$   (b) $\dfrac{9843 \text{ yd}}{\text{km}}$   (c) $\dfrac{0.002205 \text{ lb}}{\text{g}}$

    (d) $\dfrac{0.001057 \text{ qt}}{\text{mL}}$

19. (a) deci   (b) micro   (c) milli   (d) centi
20. meter, gram, liter
21. 1 $cm^3$ = 1 mL
22. (a) $2.27 \times 10^3$ g,   (b) 6.224 m,   (c) 482 $cm^3$,   (d) 0.767 m
23. (a) 338 g,   (b) 0.00023 L,   (c) $5.4 \times 10^2$ cm,   (d) $3.380 \times 10^3$ mg
24. Density is the mass per unit volume. It can be measured by measuring the mass that a particular volume weighs and then dividing the mass by the volume.
25. Density is the mass per unit volume, not just the mass alone.
26. Density decreases with temperature.
27. 1.74 g/mL
28. Rectangular prisms, cylinders, and spheres
29. 0.835 g/mL
30. It can be measured by water displacement. The volume of water that an object displaces when completely immersed is equal to its volume.
31. 7.1 g/mL
32. 0.817 g/mL
33. 452 g
34. 1.71 mL
35. $2.0 \times 10^1$ mL
36. One gram of water is the same as one mL of water at 4°C.
37. Specific gravity is defined as the density of a material at a given temperature divided by the density of water at 4°C. One way to measure this is with a pycnometer as in Example 7.14. Numerically it is the same as density, but has no units.
38. 1.014

39. See Eqs. (7.7)–(7.10).
40. (a) 113.5°F   (b) 56.7°C   (c) 296.33 K   (d) 128.2
41. (a) 69°C   (b) 2.0 $\times 10^{10}$°C   (c) 365 K   (d) 57.7°F
42. Viscosity is a measure of how thick a liquid is. There are a number of ways to measure it, including the moving body method, the capillary method, and the rotational method as discussed in Section 7.8.
43. 1.527 cS/sec
44. Assuming the viscometer constant was determined with the same volume of a standard liquid, the viscosity is 13.82 cS.

# A.8   Chapter 8

1. Stoichiometry is the expression, measurement, and calculation of quantities of matter in compounds and in reactions.
2. The weight of the carbon-12 isotope includes the weight of the electrons and other subatomic particles, not just the weight of the protons and neutrons.
3. (a) Molecular weight is the average weight of a molecule of a covalent compound in amu. It is also the average weight of one mole of a covalent compound in grams.
   (b) Formula weight is the average weight of a molecule or formula unit of a compound in amu per molecule or formula unit, or grams per mole. For covalent compounds, it is the same as the molecular weight.
   (c) A mole is Avogadro's number of anything, just like the dozen is twelve of anything.
4. Formula weight is more general, encompassing both covalent and ionic compounds. Molecular weight is specific to covalent compounds. Formula weight acknowledges that not all compounds are composed of molecules.
5. (a) 74.551 grams per mole
   (b) 342.150 grams per mole
   (c) 116.315 grams per mole
6. Since the fundamental units of elements and compounds are so very small and weigh so little, it is important to use the concept of the mole so we can speak and work in terms of quantities that are measurable with the devices available.
7. (a) 2.6 $\times 10^3$ grams
   (b) 7.9 $\times 10^2$ grams
   (c) 19.37 grams
   (d) 8.33 $\times 10^3$ grams
8. (a) 0.0109 moles
   (b) 0.0003656 moles
   (c) 0.01253 moles
   (d) 0.0440 moles
9. The dozen is a group of twelve items. Use of the dozen makes it more convenient to count and work with a large number of such items. The mole is also a group of items and use of the mole is also more convenient, just like the dozen. They differ in terms of the number of items that make them up. The dozen is twelve and the mole is Avogadro's number, $6.022 \times 10^{23}$.
10. Avogadro's number is $6.022 \times 10^{23}$. This number is important because it is the number of atoms or molecules in one mole. The mole is used routinely in chemistry in many different ways.
11. (a) 4.9 $\times 10^{24}$ atoms
    (b) 5.97 $\times 10^{23}$ atoms
12. (a) 153 moles
    (b) 7.12 $\times 10^3$ moles
13. (a) 2.04 $\times 10^{23}$ molecules
    (b) 2.60 $\times 10^{25}$ formula units

14. (a) 15.32 moles
    (b) 4.35 $\times 10^3$ moles
15. (a) 3.7 $\times 10^{21}$ formula units
    (b) 827.9 grams
    (c) 1093 grams
    (d) 2.97 $\times 10^{22}$ molecules
16. (a) 0.108 moles of $CO_2$, 6.5 $\times 10^{22}$ molecules
    (b) 4.5 moles of $NH_3$, 2.7 $\times 10^{24}$ molecules
17. 0.416 moles of $CaCl_2$
18. 21.48% O
19. 20.93% water
20. 14.7% water
21. 18.401% S
22. (a) 13.854% N
    (b) 17.073% N
23. Percent composition is the weight percents of all elements in a compound expressed as one item.
24. (a) 7.8083% C, 92.192% Cl
    (b) 82.244% N, 17.756% H
25. 51.128% water
26. A "count" percent is not the same as a "weight" percent unless all items weigh the same. A hydrogen atom does not weigh the same as an oxygen atom.
27. The empirical formula of a compound has subscripts that represent the relative numbers of atoms of the given elements in a molecule or formula unit, not necessarily the *real* numbers. Molecular formulas give the real numbers.
28. This can be done by following the procedure given in Fig. 8.1
29. (a) $AlCl_3$
    (b) $Sr_2SiO_4$
    (c) $C_5H_{10}O_2$
    (d) $Fe_3O_4$
30. $Cu_2O$
31. $P_2O_5$
32. $C_6H_{12}O_3$
33. $C_2H_4O$, $C_4H_8O_2$
34. One atom of sulfur reacts with 3 molecules of fluorine to yield 1 molecule of sulfur hexafluoride. One mole of sulfur reacts with 3 moles of fluorine to yield one mole of sulfur hexafluoride.
35. (a) 0.00229 moles
    (b) 2.6 $\times 10^2$ grams
    (c) 0.00389 moles
    (d) 0.8383 grams
36. (a) 0.3727 moles
    (b) 10.55 moles
    (c) 12.0 grams
    (d) 6.925 grams
37. (a) 3.579 moles
    (b) 163.3 grams
38. (a) 0.110 moles
    (b) 8.59 grams
39. The limiting reactant is the reactant that is used up first in the process of a chemical reaction. It is described as "limiting" because when it is used up, the reaction stops, thus limiting the amount of a product that forms.

40. When carrying out a reaction, we are usually interested in how much product is going to form. Knowing which reactant is limiting allows us to predict this.
41. $I_2$
42. $Fe_3O_4$, 0.238 grams Fe
43. $Cu_2O$, 0.0809 grams CuO
44. See Sections 8.9 and 8.10.
45. 91.3%
46. 25.82%
47. 90.5%
48. 3.3 grams
49. HCl
50. 3.27 grams
51. 0.36 grams

# A.9   Chapter 9

1. See Section 9.2
2. See Section 9.3
3. See Section 9.3
4. When the tires get hot, heat energy is transformed into kinetic energy and the gas molecules move faster. When they move faster, they collide with the walls of the container more often. More collisions per unit time means more pressure.
5. Ether molecules move through air molecules because of the fact that gas molecules are far apart from one another and can accommodate other moving molecules with no opposing force.
6. (a) More air can be added because the air molecules of already inside the tire are far apart from one another and can accommodate more molecules in the same space.
   (b) More gas molecules in the confined volume of the tire means more collisions with the inner walls of the tire, which means more pressure.
7. (a) $\dfrac{1 \text{ atm}}{760 \text{ torr}}$   (b) $\dfrac{0.019337 \text{ psi}}{1 \text{ torr}}$   (c) $\dfrac{760 \text{ torr}}{1 \text{ atm}}$   (d) $\dfrac{14.696 \text{ psi}}{1 \text{ atm}}$

   (e) $\dfrac{1 \text{ torr}}{0.019337 \text{ psi}}$   (f) $\dfrac{1 \text{ atm}}{14.696 \text{ psi}}$

8. (a) 14.2 psi   (b) 0.963 atm
9. $9.80 \times 10^2$ torr, 19.0 psi
10. 861.8 torr, 16.67 psi
11. 0.978 atm, 14.4 psi
12. Equations (9.1) and (9.4) are expressions of Boyle's Law. Temperature is held constant while pressure is varied.
13. 82.6 mL
14. 33.2 mL
15. Equations (9.5) and (9.8) are expressions of Charles' Law. Pressure is held constant while temperature is varied.
16. Charles' Law, 249.8 mL
17. 7.26 L
18. Equation (9.9) is an expression of the Combined Gas Law. Both pressure and temperature are varied.
19. 3.01 L
20. 0.580 L

21. Standard temperature and pressure are specific temperature and pressure values that scientists use as the standard for the purpose of comparing gas volumes measured at conditions that can vary widely. Standard pressure is 760 torr (1 atm) and standard temperature is 0°C, or 273.15 K.
22. 61.9 mL
23. 0.184 L
24. See Eq. (9.15).
25. 0/082057 L atm/mole K
26. 2.01 moles
27. 11.3 L
28. 2.87 moles
29. 0.979 g/L
30. 2.11 g/L
31. 27.9 g/mole
32. Two statements of Avogadro's Law are given in Section 9.12.
33. 0.0327 moles
34. See Eq. (9.17) and the accompanying discussion.
35. Vapor pressure is the pressure exerted by the molecules or atoms of a gas in equilibrium with its liquid state in the same sealed container.
36. The normal boiling point is the temperature at which the vapor pressure equals the standard atmospheric pressure, 760 torr.
37. The vapor pressure of a liquid usually increases with temperature.
38. 743 mL
39. 2.02 L
40. 8.28 L
41. $1.25 \times 103$ mL
42. 365 mL
43. 0.171 L

# A.10   Chapter 10

1. (a) A solute is a solution component that is present in a lesser amount.
   (b) A solvent is the solution component that is present is the greatest amount.
   (c) A solution is a homogeneous mixture consisting of a solvent and one or more solutes.
2. (a) "Soluble" describes a solute, a small portion of which appears to dissolve freely in a quantity of solvent.
   (b) "Insoluble" describes a solute, a small portion of which does *not* dissolve in a quantity of solvent.
   (c) "Solubility" has both a qualitative definition and a quantitative definition. The "qualitative solubility" uses the words "soluble" and "insoluble" to describe a solute. The "quantitative solubility" uses a certain quantity (e.g., a certain number of grams per liter) to describe how much solute dissolves in a quantity of solvent or solution.
3. (a) "Saturated" describes a solution that has the maximum amount of solute possible dissolved at the given temperature. It is characterized by the presence of some undissolved solute.
   (b) "Unsaturated" describes a solution that has less than the maximum amount dissolved at the given temperature.
   (c) "Supersaturated" describes a solution that has more than the maximum amount possible dissolved at the given temperature. This condition is achieved by heating, dissolving more, then cooling back to the given temperature.
4. (a) "Miscible" describes a pair of liquids that appear to mix freely with each other. "Immiscible" describes a pair of liquids that appear not to mix freely with each other.

(b) "Dilute" describes a solution that does not have much solute dissolved. "Concentrated" refers to a solution that has a large quantity of solute dissolved. "Dilute" can describe a solution that has less than an amount that has been designated for a "concentrated" solution even though it may be a relatively large amount.

5. (a) T   (b) F   (c) F   (d) F   (e) T   (f) T   (g) F   (h) F   (i) T   (j) F   (k) T   (l) T
   (m) T   (n) T

6. (a) saturated,   (b) immiscible,   (c) solute,   (d) solubility,   (e) supersaturated,
   (f) soluble,   (g) concentrated

7. To say that a solute is insoluble is a qualitative description that is observed in exactly the manner by which this student observed it. So she is correct. However, all solutes dissolve to some extent, even though it may not be (cannot be) observed. The small amount that dissolves can be described quantitatively and referred to as the solubility. This is the manner by which the handbook lists the solubility, so the handbook is also correct.

8. Shaking, stirring, ultrasonic agitation.

9. Less surface area (large particles) in contact with the solvent means no contact of most of the solute with the solvent and therefore a slower rate of dissolving.

10. The negative end of polar water molecules are attracted to the positive sodium ions and the positive end of polar water molecules are attracted to the negative chloride ions. These forces are sufficient to pull the ions from the 3-D array of ions and dissolve the crystal.

11. I would refer him to Fig. 10.1 and tell him that there is considerable interaction between the ions in the crystal structure and the polar water molecule, causing the ions to be pulled from the crystal structure and carried off into the solution, or dissolved.

12. Soluble: NaBr, $K_2SO_4$, $NH_4Cl$, $AgNO_3$, $Ba(C_2H_3O_2)_2$ and $Na_3PO_4$. Insoluble: $Fe(OH)_3$, $Ag_2CO_3$.

13. Polar covalent compounds, such as acids. Examples: hydrochloric acid, sulfuric acid, sugar:

14. Molecules of nonpolar substances do not interact with polar water molecules and so there is no force that would pull apart water molecules held together by relatively strong intermolecular forces.

15. London forces are the very weak forces by which nonpolar solute and solvent molecules interact.

16. "Like dissolves like" refers to the fact that polar substances dissolve in other polar substances and nonpolar substances dissolve in other nonpolar substances, but polar substances and nonpolar substances don't dissolve in each other.

17. The presence of ions that are free to move is the condition that allows an electrical current to flow through a solution. A solution of NaCl has sodium ions and chloride ions that are free to move. Distilled water has no such ions.

18. Electrons do not actually flow through a solution. Rather, electrons are consumed at one electrode and generated at another, making it only *appear* that they flow through a solution.

19. Strong electrolytes are solutions that contain large numbers of ions or solutes that, when dissolved, provide large numbers of ions and thus conduct a relatively large electrical current. Examples are NaCl and HCl and their solutions. Weak electrolytes are solutions that contain only small numbers of ions or solutes that, when dissolved, provide only small numbers of ions and thus conduct only a relatively small electrical current. Examples are acetic acid and ammonium hydroxide and their solutions. Nonelectrolytes are solutions that contain no ions, like distilled water or other pure liquids or solutes such as ethyl alcohol or acetone that provide no, or very, very few ions when dissolved.

20. Covalent compounds would not provide ions, and thus would not conduct an electrical current when dissolved, as ionic compounds would.

21. The cathode is the electrode at which reduction takes place. For example, hydrogen ions can be reduced at a cathode and hydrogen gas can form. Also, metal ions can be reduced to the metal causing them to plate out on the cathode.

22. (a) Molarity is the number of moles dissolved per liter of solution.
    (b) Molality is the number of moles dissolved per kilogram of solvent.

(c) Normality is the number of equivalents dissolved per liter of solution.

(d) ppm is "parts per million" and refers to the number of milligrams dissolved per liter of solution.

23. 9.49%

24. Liquid volumes are not necessarily additive when mixed, meaning that the two solutions described may not be the same.

25. (a) 0.972%　(b) 0.981%　(c) 0.972%　(d) 0.981%

Weight percent is defined as the weight of solute per 100 g of solution. Thus, in (a) and (b) the two weights must be added together to give the weight of the solution. Weight percent is independent of the solute used, so (a) and (c) are the same and (b) and (d) are the same.

26. (a)1.169%

27. No. The one with 200.0 mL of water likely contains more water than the one with 200.0 mL of solution. In other words, the solute probably takes up some volume and so it takes less than 200.0 mL of water to give 200.0 mL of solution.

28. (a) 1.36 M　(b) 0.0127 M　(c) 0.562 M　(d) 0.02512 M

29. 0.141 M

30. (a) 2.52%　(b) 0.522 M

31. 0.163 m

32. 0.07457 m

33. (a) 27 ppm Fe　(b) 30.4 ppm Fe　(c) 36.4 ppm Fe　(d) 2.96 ppm Fe

34. (a) 33.7 ppm Cl　(b) 432 ppm Cl　(c) 84.2 ppm Cl　(d) 172 ppm Cl

35. 0.010 ppm; yes

36. (a) 1.79 N　(b) 1.35 N

37. 0.75 N

38. 2.0 N

39. 0.38 N

40. (a) 32.665 g/equiv　(b) 3.62 N

41. The dilution equation is Eq. (10.25). It is used to calculate the number of milliliters of a solution needed when preparing another solution by dilution.

42. (a) 7.29 mL　(b) 75.0 mL

43. (a) 9.72 mL　(b) 6.94 mL

44. (a) Dilute 6.3 mL of the 16 M $HNO_3$ to 500.0 mL of solution with water. Shake.
    (b) Dilute 35 mL of the 17 M acetic acid to 1.20 L of solution with water. Shake.

45. Dilute 0.25 L of the 12 M HCl to 2.00 L of solution with water. Shake.

46. (a) 15.0 mL　(b) 69.4 mL　(c) $1.0 \times 10^2$ mL

47. When a high degree of precision is important.

48. (a) 12.5 g of solute; 37.5 g of solvent
    (b) $1.80 \times 10^2$ g of solute; $1.020 \times 10^2$ g of solvent

49. (a) 45.0 mL　(b) 36.0 mL

50. 0.0075 L

51. 38 g

52. 50.0 g

53. 52 g

54. (a) 17 g　(b) 22 g

55. Weigh 8.4 g of the solid into the container, add water to dissolve, then add more water to the graduation line and shake.

56. First way: Weigh $4.0 \times 10^1$ grams of the solid into the flask, add water to dissolve, and then dilute to the mark with water and shake. Second way: Measure 68 mL of the 5.5 M solution into the flask, add water to the mark and shake.

57. (a) Weigh 24 g of the solid into a 500 mL flask, add water to dissolve, add more water to 500.0 mL and shake.
    (b) Place 44 mL of the 4.0 M solution into a 500 mL flask, add water to the 500.0 mL and shake.

58. Weigh 37 grams of the solid add water to dissolve, then add more water to the line and shake.
59. Weigh 8.6 grams of the solid and add the 400.0 grams of solvent. Shake.
60. 0.0013 g Cu
61. 0.032 g NaCl
62. Weigh 0.110 grams of $KH_2PO_4$ into a 500-mL volumetric flask, add water to dissolve, then add water to the 500-mL mark. Shake.
63. 1.6 g $H_3PO_4$
64. $9.0 \times 10^1$ g $NaH_2PO_4$

# A.11  Chapter 11

1.  (a) Vapor pressure is the pressure exerted by a gas when it is in equilibrium with its liquid in the same closed container. Thus equilibrium is an integral part of the definition.
    (b) A solute moves back and forth between the two liquids at the same time and at the same rate.
2.  It means that the reaction is a chemical equilibrium and the forward reaction is occurring at the same time at the same rate as the opposing reaction.
3.  The double arrow indicates that both the forward reaction and the reverse reaction are occurring at the same time and at the same rate.
4.  A chemical equilibrium is a reaction in which the forward reaction is occurring at the same time and at the same rate as the opposing reaction, with no net change.
5.  $2SO_2(g) + O_2(g) \rightleftharpoons 2SO_3(g)$
6.  The equilibrium constant is a constant that indicates the position of an equilibrium at a given temperature. It is the concentrations of the products of a chemical reaction raised to the power of their balancing coefficients, multiplied together and divided by the concentrations of the reactants raised to the power of their coefficients and multiplied together. See Eqs. (11.7) and (11.14).
7.  (a) $$K = \frac{[NOCl]^2}{[NO]^2[Cl_2]}$$

    (b) $$K = \frac{[NO]^2}{[N_2][O_2]}$$

    (c) $$K = \frac{[HCl]^4[O_2]}{[Cl_2]^2[H_2O]^2}$$

    (d) $$K = \frac{[SO_3^{2-}][H^+]^2}{[H_2SO_3]}$$

    (e) $$K = \frac{[H^+]^3[PO_4^{3-}]}{[H_3PO_4]}$$

    (f) $$K = \frac{[SO_3]^2}{[SO_2]^2[O_2]}$$

8.  A reaction in which a molecular or ionic species is partially converted to ions is described by an ionization constant, $K_i$. An example would be any weak acid or weak base ionization when dissolved in water.
9.  A reaction in which an ionic compound dissolves in water and partially is converted to ions is described by a solubility product constant, $K_{sp}$. An example is given in Eqs. (11.18) and (11.19).
10. (a) $Ag_2S \rightleftharpoons Ag^+ + S^{2-}$   $K_{SP} = [Ag^+]^2[S^{2-}]$
    (b) $PbI_2 \rightleftharpoons Pb^{2+} + 2I^-$   $K_{sp} = [Pb^{2+}][I^-]^2$
    (c) $Bi_2S_3 \rightleftharpoons 2Bi^{3+} + 3S^{2-}$   $K_{sp} = [Bi^{3+}]^2 + [s^{2-}]^3$

11. It is a very small number. Since the acid ionizes only slightly, the numerator is small compared to the denominator, which makes for a very small number after division.

12. If the equilibrium constant is small, the numerator is small compared to the denominator, which means that the $PO_4^{3-}$ concentration is small compared to the $H_3PO_4$ concentration.

13. (a)

$$K_{eq} = \frac{[H^+]^3[BO_3^{3-}]}{[H_3BO_3]}$$

(b) It can be $K_i$ because it is an ionization. It can be $K_a$ because it is a weak acid ionization.

(c) The number $1.6 \times 10^{-13}$ is very small. This would indicate only slight ionization because the numerator of the $K_{eq}$ expression, representing the products, is small compared to the denominator, which represents the reactants.

14. The quilibrium constant is small, so the numerator is small compared to the denominator. This means that the concentration of the $H^+$ is small at equilibrium.

15. (a) $HCHO \rightleftharpoons H^+ + CHO^-$

(b) $K = \dfrac{[H^+][CHO^-]}{[HCHO]}$

(c) It would be a small number because, with only slight ionization, the numerator will be small compared to the denominator.

16. Le Chatelier's Principle is a statement that says that if a stress is placed on a system at equilibrium, the equilibrium will respond to relieve the stress.

17. Le Chatelier's Principe is important in the description of chemical reactions because reaction conditions can represent a stress on a system and chemists must be able to predict the outcome.

18. An endothermic reaction is one in which heat is removed from the surroundings in order to make the reaction proceed from left to right.

19. The addition of heat would push an endothermic reaction to proceed or shift from left to right.

20. An exothermic reaction is one in which heat is added to the surroundings when the reaction proceeds from left to right.

21. The addition of heat would push an exothermic reaction to proceed or shift from right to left.

22. (a) increase,   (b) decrease,   (c) increase

23. (a) right, decrease;   (b) left, increase;   (c) left, increase

24. affect: a, c, f    not affect: b, d, e

25. (a) left, increase;   (b) left, increase;   (c) right, decrease

26. (a) left, decrease;   (b) neither, no change;   (c) left, decrease

27. The enthalpy change, $\Delta H$, for a reaction is the heat that is either absorbed or given off in the reaction.

28. An increase in temperature means that heat has been absorbed. A decrease in temperature means that heat has been given off. Thus the effects of temperature changes are really the effects of enthalpy changes.

29. (a) A negative $\Delta H$ means "heat" is on the right. Left, increase.
(b) neither, no change;   (c) right, decrease

30. (a) right, increase,   (b) left, decrease,   (c) left, decrease

31. (a) right, increase;   (b) left, decrease;   (c) right, increase

32. A high concentration of $H_2$ would be desirable because the equilibrium would shift toward the product in order to consume the added $H_2$. A high pressure would be desirable because the equilibrium would shift to the right where there is less gas. A low temperature would be desirable because the reaction is exothermic and the removal of heat would cause a shift to the right to replace this heat.

33. (a) $7.1 \times 10^{-7}$ M    (b) $5.0 \times 10^{-12}$ M
34. (a) $7.8 \times 10^{-5}$ M    (b) $4.8 \times 10^{-11}$ M
35. $3.0 \times 10^{-21}$
36. $3.7 \times 10^{-11}$
37. (a) $4 \times 10^{-17}$ M
    (b) 0.04 M
    (c) $4 \times 10^{-29}$ M
38. (a) $CaF_2$ is far less soluble in water.
39. (a) $1.5 \times 10^{-3}$
    (b) $8.4 \times 10^{-5}$ M
    (c) $8.8 \times 10^{-10}$ M
40. $s = 1.7 \times 10^{-11}$ M;   $5.5 \times 10^{-9}$ grams $Ca_3(PO_4)_2$
41. $K_{sp} = 4.6 \times 10^{-6}$
42. $1.9 \times 10^{-3}$ M
43. $8.8 \times 10^{-4}$ M
44. $3.07 \times 10^{-3}$ M

# A.12   Chapter 12

1. See Table 12.1.
2. A carboxylic acid is an organic acid with the following general structure: hydrochloric acid (HCl) is inorganic—it doesn't contain carbon and does not fit the above general structure.
3. The hydrogens in this formula are not "acidic" hydrogens. They are the hydrogens on the methyl group of the acetate ion and as such are not released or donated like the acidic hydrogens.
4. A strong acid is one that completely ionizes when dissolved in water. Example: hydrochloric acid, sulfuric acid, etc. A weak acid is one that is not completely ionized when dissolved in water. Example: acetic acid, carbonic acid.
5. A very small number. Weak means slight ionization, so the numerator of $K_a$ is small compared to the denominator.
6. $NaHCO_3$, $NaH_2PO_4$ and other partially neutralized acids.
7. Sodium hydroxide, NaOH, potassium hydroxide, KOH, and ammonium hydroxide, $NH_4OH$.
8. In concentrated solutions of $NH_4OH$, the following equilibrium is present:

$$NH_4OH \;\rightleftharpoons\; NH_3 + H_2O$$

Thus, ammonia, $NH_3$, is also present. In addition, the following equilibrium is present

$$NH_4OH \;\rightleftharpoons\; NH_4^+ + OH^-$$

and so the $NH_4OH$ is considered a weak base.
9. In both cases, "weak" means only slight ionization.
10. Because of the ingredient citric acid. Acids have a sour taste.
11. Acids have a sour taste, attack biological materials, are corrosive, ionize to give hydrogen ions in water, and turn litmus paper red.
12. Bases have a bitter taste, attack biological materials, are caustic and corrosive, feel slippery, ionize in water to give hydroxide ions, and turn litmus paper blue.
13. An indicator is a chemical that exhibits various colors in water solutions that vary in acidity level. For example, phenolphtalein is pink in basic conditions and colorless in acidic conditions.
14. pH paper is paper that is saturated with a mixture of various indicators and can thus detect specific acidity and basicity levels by simply displaying different colors.

15. It means that there is a higher concentration of hydrogen ions compared to hydroxide ions. It gives the solution acidic properties.

16. A neutralization reaction is the reaction of an acid with a base. A common example is the reaction of HCl with NaOH:

$$HCl + NaOH \rightarrow NaCl + H_2O$$

A salt and water are the products.

17. (a) damaging to skin, pungent odor,   (b) extremely damaging to skin, extreme heat when mixed with water,   (c) extemely damaging to skin,   (d) stifling odor, can cause skin burns,   (e) damaging to skin,   (f) heavy ammonia odor

18. (a) HF, hydrofluoric acid   (b) HCl, hydrochloric acid
    (c) $HNO_3$, nitric acid   (d) $H_2SO_4$, sulfuric acid
    (e) $NH_4OH$, ammonium hydroxide
    (f) HCl, hydrochloric acid   (g) $H_2SO_4$ sulfuric acid
    (h) $HC_2H_3O_2$, acetic acid

19. See Section 12.3.3. In addition, one should take special precautions to cool the container.

20. According to the Arrhenius Theory, an acid is a compound that donates hydrogen ions and a base is a compound that donates hydroxide ions. An example of an acid is HCl, (any compound whose formula begins with "H") and an example of a base is NaOH, or any hydroxide compound.

21. According to the Bronsted-Lowry Theory, an acid is a compound that donates hydrogen ions and a base is a compound that accepts hydrogen ions. An example of an acid is HCl (or any compound whose formula begins with "H"), and examples of bases are NaOH and $Na_2CO_3$ or any hydroxide or carbonate.

22. (1) a, c, d   (2) a, c, d   (3) f   (4) b, e, f, g

23. The hydronium ion is $H_3O^+$. It forms when water acquires a hydrogen. In acquiring a hydrogen, it behaves like a Bronsted-Lowry base, which is a substance that accepts hydrogens.

24. The Lewis Theory defines acids and bases in terms of the donation or acceptance of a pair of electrons rather than hydrogen ions and hydroxide ions. See Section 12.4.3.

25. HCN, $H_2O$, $HC_2H_3O_2$, $HPO_4^{2-}$.

26. A double replacement reaction is one in which two reactants "trade partners," meaning one element or polyatomic ion in one formula replaces an element or polyatomic ion in another and vice-versa.

Example: $HCl + NaOH \rightarrow NaCl + HOH$

A single replacement reaction is one in which an element reactants with a compound and replaces one of the elements in the formula.

Example: $HCl + Zn \rightarrow ZnCl_2 + H_2$

27. (a) Acid + Metal $\rightarrow$ Salt + $H_2$
    (b) Metal Oxide + Acid $\rightarrow$ Salt + Water
    (c) Nonmetal Oxide + Water $\rightarrow$ Acid
    (d) Metal Oxide + Water $\rightarrow$ Base

28. (a) $K_2O + H_2O \rightarrow 2KOH$
    (b) $2HCl + CaCO_3 \rightarrow CaCl_2 + CO_2 + H_2O$
    (c) $2NH_4OH + H_2SO_4 \rightarrow (NH_4)_2SO_4 + 2H_2O$
    (d) $NO_2 + H_2O \rightarrow HNO_3$ (requires special balancing procedure)
    (e) $2HCl + Mg \rightarrow MgCl_2 + H_2$
    (f) $CaO + 2HBr \rightarrow CaBr_2 + H_2O$

29. (a) $H_2SO_4 + K_2CO_3 \rightarrow K_2SO_4 + CO_2 + H_2O$
    (b) $2NO + H_2O \rightarrow 2HNO_3$
    (c) $2HNO_3 + Mg(OH)_2 \rightarrow Mg(NO_3)_2 + 2H_2O$
    (d) $2K + 2H_2O \rightarrow 2KOH + H_2$
    (e) $Ca + 2HCl \rightarrow CaCl_2 + H_2$

30. The violent reaction represented by the equation in #29(d) would occur.

31. $Mg(OH)_2 + 2HCl \rightarrow MgCl_2 + 2H_2O$

32. Acid rain is rain that has an acidic pH. It forms when rain falls through an atmosphere that contains oxides of nonmetals. An example is $HNO_3$, formed when rain falls through an atmosphere that contains $NO_2$.

33. See Eq. (12.32) for the definition. This is an important equation for water solutions because when hydrogen ions and hydroxide ions are out of balance because of acidic or basic conditions, the concentration of one may be calculated from the concentration of the other.

34. See Eq. (12.33) for the definition. This is an important equation because the pH of a solution may be calculated from the hydrogen ion concentration and vice versa.

35. (a) 2.299    (b) 4.032    (c) 7.419    (d) 10.312    (e) 12.218    (f) 2.910

36. (a) $7.60 \times 10^{-5}$ M    (b) $1.23 \times 10^{-3}$ M    (c) $3.56 \times 10^{-11}$ M    (d) $4.14 \times 10^{-13}$ M    (e) $5.90 \times 10^{-7}$ M

37. (a) 11.026    (b) 11.13    (c) 6.6    (d) 1.11

38. (a) $3.2 \times 10^{-10}$ M    (b) $5.32 \times 10^{-13}$ M    (c) $1.0 \times 10^{-7}$ M    (d) $5.1 \times 10^{-4}$ M

39. #35-acidic: a, b, f    basic: c, d, e
    #36-acidic: a, b, e    basic: c, d
    #37-acidic: c, d    basic: a, b
    #38-acidic: d    basic: a, b    neutral: c

40. (a) A pH meter is an instrument that, in combination with a pH electrode and a reference electrode, measures and displays the pH of solutions.    (b) A combination pH probe is a an electrode that develops a voltage proportional to the pH of a solution into which it is dipped, so, in combination with a pH meter, is useful for measuring pH of solutions. It consists of two concentric glass tubes, one the pH sensor and one a reference electrode.

# A.13    Chapter 13

1. See the discussion and figures in Section 13.3.2, in addition to your independent research.
2. $v = 9.22 \times 1014 \text{ s}^{-1}$
3. This includes the frequency range of $7.5 \times 10^{14} \text{ s}^{-1}$ to $6.3 \times 10^{14} \text{ s}^{-1}$.
4. The wavelength, $\lambda$, $= 5.51 \times 10^{-7}$ m $= 551$ nm.
5. $E = 2.45 \times 10^{-19}$ J. This is in the infrared region of the electromagnetic spectrum.
6. The energy of a mole of photons would be 147 kJ. The combustion of a mole of butane would release 2860 kJ/mol.
7. The change in energy is $1.5 \times 10^7$ kJ. This is many times greater than the energy released in the combustion of methane.
8. $\Delta H = -98$ kJ/mole
9. $\Delta H = -174$ kJ/mole
10. For the reaction in Exercise #8, $\Delta S^\circ = 0.126$ kJ/K and $\Delta G^\circ = -116.73$ kJ/mol. The reaction is spontaneous. For the reaction in Exercise #9, $\Delta S^\circ = -0.112$ kJ/K and $\Delta G^\circ = -141$ kJ/mol.
11. Ozone, $O_3$, is not in the standard state for oxygen. Energy is required to go from $O_2$ to $O_3$.
12. $\Delta H^\circ = -1440$ kJ/mole $C_2H_6$
13. $\Delta H^\circ = -1428$ kJ/mole $C_2H_6$
14. $\Delta H^\circ = -23.1$ kJ for 0.25 moles of $H_2$
15. (a) Increase
    (b) Based on entropy data, there is a slight increase.

16. (a) $\Delta S^\circ = +284.6$ J/K
    (b) $\Delta S^\circ = +21.1$ J/K
17. The entropy increases because the input of energy to melt the ice results in breaking some hydrogen bonds, thus allowing the water molecules to move more freely, and not be locked into specific positions in the ice crystal lattice.
18. The process is *not* spontaneous, because the entropy change of the *universe* is negative.
19. (c) The process tends to be spontaneous, but can be nonspontaneous under certain conditions.
20. (c) $\Delta S_{universe} > 0$
21. (a) $\Delta S > 0$, $\Delta H < 0$
22. (d) $\Delta G > 0$
23. The following sets of data represent spontaneous reactions:

| $\Delta H$(kJ) | $\Delta S$(J/K) | T(K) |
|---|---|---|
| a. $-250$ | $+50$ | 298 |
| c. $-250$ | 100 | 200 |
| e. $-53$ | $-30$ | 500 |

24. The reaction occurred at 192 K.
25. $\Delta G^\circ = -35.1$ kJ/mole, so the reaction is spontaneous.
26. The values agree.
27. $\Delta G^\circ = -2935$ kJ, which equals $-1467$ kJ/mol of $C_2H_6$. This reaction is spontaneous. It is a combustion reaction, which, like all combustion reactions, are highly thermodynamically favored.

# A.14   Chapter 14

1. The rate of a reaction tells us how quickly the reaction proceeds. Through this we can determine the concentration of reactants or products at any time. A variety of applications were discussed in the chapter, including medical and industrial uses.
2. (a) The rate of decomposition of $H_2O_2$ is $-7.04 \times 10^{-6}$ Ms$^{-1}$. The rate of production of $O_2$ is $\frac{1}{2}$ of that, $+3.52$ is $10^{-6}$ Ms$^{-1}$.
   (b) The rate of decomposition of $H_2O_2$ is $-3.52 \times 10^{-6}$ Ms$^{-1}$.
3. The rate of the reaction would increase substantially. As our benchmark, we say that the rate doubles for every 10°C increase in temperature. Therefore, as a first approximation, the new rate at 60°C would be $2^4 = 16$ times the rate at 20°C.
4. $k_1 = 9.76 \times 10^{-6}$ s$^{-1}$, $t_{1/2} = 71,000$ seconds.
5. After one week (8.518 half-lives), $[H_2O_2] = 0.00272$ M.
6. $6A \rightarrow 1B + 3C$
7. Rate $= 0.012$ M min$^{-1}$
8. $k_2 = 6.67 \times 10^{-4}$ M$^{-1}$ min$^{-1}$
9. The rate of production of $N_2 = 8.1 \times 10^{-5}$ M min$^{-1}$.
   The rate of production of $H_2$ is $2.44 \times 10^{-4}$ M min$^{-1}$.
10. (a) The rate of consumption of the barium ion would be 0.0500 Ms$^{-1}$.
    (b) The rate of production of $BaCrO_4$ would be 0.0250 Ms$^{-1}$.
11. (a) The reaction is first-order.
    (b) $k_1 = 0.046$ min$^{-1}$
12. (a) The reaction is second-order.
    (b) $k_2 = 1.34$ M$^{-1}$ min$^{-1}$.

13. (a) Rate = $k_1[PH_3]$

    (b) $k_1 = 0.020 \text{ s}^{-1}$, $t_{1/2} = 35$ seconds

14. (a) The reaction is second-order.

    (b) $k_2 = 0.025 \text{ M}^{-1} \text{ s}^{-1}$

    (c) [AB] = 0.28 M

15. Rate = $k_2[Ibr]^2$

16. The baking step would be rate-determining.

17. (a) Rate = $k[NOCl]^n$

    (b) The rate of disappearance (consumption) of $Cl_2$ is $1/2$ that of the rate of production of NOCl.

    (c) NO is consumed at the same rate, 4.5 M min$^{-1}$.

18. 0.43 M will be the concentration.

19. $k_1 = 0.547 \text{ dy}^{-1}$

20. It can sit on the shelf for 3.7 days.

21. 92% will have reacted.

22. $k_1 = 0.0575 \text{ yr}^{-1}$.

23. It would take 217 hours.

24. The concentration would be 0.482 M.

25. The half-life would be 0.00136 years, or 0.496 day.

26. (a) $t_{1/2} = 3759$ minutes

    (b) It will take 33,800 minutes for 90% of the sample to decompose.

27. After 2 weeks, $[NH_3] = 0.085$ M.

28. $t_{1/2} = 1160$ seconds

29. The intermediates are $NO_3$ and NO.

30. The reactant concentration will be 0.0125 M.

31. (a) $M^{-1} \text{ s}^{-1}$

    (b) 0.139 M of I will remain, so that 0.11 M will have been reacted. The mole ratio of I to $I_2$ is 2:1. Therefore, 0.050 M $I_2$ will have been produced.

    (c) $t_{1/2} = 5.7 = 10^{-10}$ seconds

    (d) It will take $1.1 \times 10^{-8}$ seconds

32. (a) $t_{1/2} = 4.2$ seconds

    (b) [NOBr] = 0.27 M

    (c) It will take 4200 seconds.

33. (a) It will take $8.0 \times 10^{-6}$ seconds.

    (b) $t_{1/2} = 2.0 \times 10^{-6}$ seconds.

    (c) It will take $1.1 = 10^{-5}$ seconds.

34. It was woven 2315 years before, or 315 B.C.E.

35. After 1.96 days (0.5 half-life) you will be left with 2.94 grams. After 3.82 days (one half-life) you will be left with 2.08 grams of $^{222}$Rn.

36. There are 11.3 grams remaining after 6.0 days.

37. It will take 404 seconds.

38. Among the reasons are the danger of the alpha-particle emissions of $^{239}$Pu and the expense.

39. These questions are discussed in Section 14.4.2.

40. Raising the temperature increases the motion and kinetic energy of the system particles, thus increasing the number and energy of the interactions.

41. There could be a number of explanations. Once experimental error is ruled out (by repeating the experiment several times), we could look at the mechanism being "mixed-order"—that is, a combination of first- and second-order, or, in fact, other orders. The answer can be established by doing a so-called "pseudo-first-order" experiment, in which one of the components is present in quite high amounts, allowing you to focus on the change in concentration of the other component.

# A.15   Chapter 15

1. The nuclide has 53 protons and 72 neutrons.

2.

| Symbol | Protons | Neutrons | Mass Number |
|--------|---------|----------|-------------|
| $^{130}_{55}$Cs | 55 | 75 | 130 |
| $^{89}_{35}$Br | 35 | 54 | 89 |
| $^{91}_{42}$Mo | 42 | 49 | 91 |

3. $^{224}_{88}$Ra + $^{0}_{-1}$e → $^{224}_{87}$Fr

4. $^{242}_{94}$Pu → $^{238}_{92}$U + $^{4}_{2}$He

5. (a) Gamma rays: These have no effect on the proton-to-neutron ratio.
   (b) Electron capture: This increases the neutron-to-proton ratio.
   (c) Alpha decay: This increases the neutron-to-proton ratio.
   (d) Positron emission: This increases the neutron-to-proton ratio.

6. (a) Gamma rays: These have no effect on the proton-to-neutron ratio.
   (b) Electron capture: This increases the neutron-to-proton ratio.
   (c) Alpha decay: This increases the neutron-to-proton ratio.
   (d) Positron emission: This increases the neutron-to-proton ratio.

7. (a) Gamma rays is the only one that does not affect this ratio.

8. Beta-particle emission

9. (a) ruthenium-95
   (b) arsenic-72
   (c) tungsten-182

10. (a) boron-13
    (b) neon-22
    (c) thorium-230

11. (a) The products are an alpha particle and protactinium-231.
    (b) The products are a beta-particle and protactinium-239.
    (c) The products are a positron and magnesium-21.

12. (a) $^{251}$No → $^{4}$He + $^{247}$Fm
    (b) $^{183}$Hf → $^{0}_{-1}$e + $^{183}$Ta
    (c) $^{212}$Po → $^{4}$He + $^{208}$Pb

13. (a) $^{19}_{7}$N → $^{19}$O + $^{0}_{-1}$e
    (b) $^{52}_{25}$Mn → $^{52}_{26}$Fe + $^{0}_{-1}$e
    (c) $^{214}_{84}$Po → $^{218}$Rn + $^{4}_{2}$He

14. (a) $^{71}$As + $^{0}_{-1}$e → $^{71}_{32}$Ge
    (b) $^{192}$Fr → $^{196}_{85}$At + $^{4}_{2}$He
    (c) $^{212}_{86}$Rn → $^{208}$Po + $^{4}_{2}$He
    (d) $^{191}_{80}$Hg → $^{191}$Au + $^{0}_{1}$e

15. Alpha-particles are quite hazardous because of their mass, and, therefore, their penetrating power.

16. Alpha-particle emission increases the neutron-to-proton ratio, which helps heavier nuclides toward the zone of stability. Beta-particle emission has just the opposite effect, decreasing the neutron-to-proton ratio, which is good for lighter nuclei that tend to be relatively neutron-rich.

17. $2.6 \times 10^{-10}$ J per atom ($1.6 \times 10^{14}$ J/mole)

18. $^{131}_{33}$0.0859 d$^{-1}$

19. $0.0282 \text{ s}^{-1}$

20. (a) $0.237 \text{ min}^{-1}$   (b) $0.389 \text{ s}^{-1}$   (c) $0.356 \text{ min}^{-1}$   (d) $4.6 \text{ s}^{-1}$

21. 63.6 minutes (2 half-lives)

22. $3.4 \times 10^7$ years (2 half-lives)

23. One day would be about 164 half-lives. The fraction left would be about $2^{-164} = 4 \times 10^{-50}$ of the original, which would be a fraction of an atom. Since that is not physically reasonable, the likelihood is that none would remain.

24. It would take about 8.6 half-lives, or about 45 minutes.

25. It would not have changed much—to 15.99 decays/g/minute, which, to 2 significant figures, is still 16 decays per minute.

26. A total of 0.625 grams remains.

27. The half-life is 14.64 days.

28. It would take 3.32 half-lives, or $1.48 \times 10^{10}$ years.

29. About 36 grams of $^{131}\text{I}$.

30. A total of about 285 half-lives have passed. None of the sample (actually, $10^{-86}$, which is "none") remains.

31. It will take a total of 18.8 years (6.65 half-lives).

32. About 2400 grams were initially present.

33. About 0.61 half-lives would have passed and about 16 grams of the cobalt-60 would remain.

34. The half-life of X is about 2.66 hours.

35. A total of 68.2 days.

36. About 2900 grams were in the original sample. Roughly 12.4 half-lives passed.

37. It will take 24 minutes.

38. The sample will need to be replaced in 0.415 half-lives, which is 12.45 years, or around January 2012.

39. The final product will be $^{206}\text{Pb}$.

40. The final product will be $^{207}\text{Au}$.

41. The nuclide is $^{224}\text{Pb}$.

42. This stable nuclide will be formed at the final step.

43. (a) $^{107}\text{Sn} + {}_{-1}^{0}\text{e} \rightarrow {}^{107}\text{In}$

    (b) 30.6% of the sample will remain after 5.00 minutes.

# A.16   Chapter 16

1. Oxidation is the process by which one chemical *loses* an electron to another; reduction is the process by which a chemical *gains* an electron from another. These processes occur simultaneously in a reduction-oxidation reaction and both are always present together in such reactions.

2. (a) An oxidizing agent is the reactant in a reduction-oxidation reaction that is being reduced.
   (b) A reducing agent is the reactant in a reduction-oxidation reaction that is being oxidized.
   (c) Oxidant is another name for an oxidizing agent (see definition in 2a).
   (d) Reductant is another name for a reducing agent (see definition in 2b).

3. Half-reactions are used by chemists for describing reduction-oxidation reactions. This is done by writing the redox process as separate reduction and oxidation reactions. Although both processes are actually occurring as part of the same net reaction, the use of half-reactions provides a more convenient means for seeing what is happening during this process. It is also useful as a tool in balancing reduction-oxidation reactions.

4. The standard reduction potential ($E^\circ$) for a given reduction half-reaction is a measure of that reaction's tendency to proceed under standard reaction conditions (that is, at a temperature of 25°C, a pressure of 1 atm for all gaseous reactants or products, a chemical activity of one for all solids, and 1 mol/L activity for all soluble reactants and products). In theory, the standard

reduction potential for any half-reaction is measured by comparing it against the standard hydrogen electrode; however, in practice, other types of reference electrodes can also be used for this purpose.

5. Chemicals that are reactants in reduction half-reactions with large, positive $E°$ values will be easily reduced and tend to act as good oxidizing agents, while chemicals that are products in reduction half-reactions with small or negative $E°$ values will be easily oxidized and will tend to act as good reducing agents.

6. The reduction half-reactions for both dichromate $(Cr_2O_7^{2-})$ and permanganate $(MnO_4^-)$ have high standard reduction potentials compared to the standard hydrogen electrode. This means that these compounds will be easily reduced and that they should act as good oxidizing agents.

7. Both elemental sodium (Na) and calcium (Ca) are products in reduction half-reactions that have low standard reduction potentials compared to the standard hydrogen electrode. This means that these compounds tend to act as good reducing agents.

8. This can be done by comparing the standard reduction potentials for these chemicals in Table 16.1. The ranking (from easiest to most difficult to reduce) is as follows: $Co^{3+}$, $Ce^{4+}$, $Ag^+$, $Na^+$, $Ca^{2+}$, and $K^+$.

9. This is done by comparing the standard reduction potentials for these chemicals in Table 16.1. The ranking (from easiest to most difficult to oxidize) is as follows: K, Ca, Na, Ce, Co, Ag.

10. (a) $2Fe^{3+} + Cu \rightleftharpoons 2Fe^{2+} + Cu^{2+}$
    (b) $2Ni(OH)_3 + Cd \rightleftharpoons 2Ni(OH)_2 + Cd(OH)_2$
    (c) $2Cl_2 + 2H_2O \rightleftharpoons 2HOCl + 2Cl^- + 2H^+$
    (d) $2MnO_4^- + 3S^{2-} + 4H_2O \rightleftharpoons 2MnO_2 + 3S + 8OH^-$

11. A galvanic cell is an electrochemical cell that is characterized by the production of electricity (or electron flow) as a result of the spontaneous reaction between an oxidizing agent and a reducing agent. An electrolytic cell is an electrochemical cell in which energy must be put into the system in order for a reaction to occur.

12. (a) The cathode is the electrode in an electrochemical cell where reduction takes place as electrons are passed from the electrode to a chemical.
    (b) An anode is the electrode in an electrochemical cell where oxidation takes place as electrons are released from a chemical and passed to the electrode.
    (c) A salt bridge is an intermediate solution or medium that is used in an electrochemical cell to help ions flow from one electrode to the other.
    (d) An electrolyte is a solution or medium that contains ions. This is used in an electrochemical cell to provide ions that can flow between the cathode and anode.

13. The anode and cathode are the places in the electrochemical cell at which oxidation and reduction of chemical species actually occurs. The salt bridge is used to connect the half-cells of these two electrodes together and the electrolyte is used to supply ions that can travel from one electrode to the other in order to complete the electrical circuit and maintain a balance of charge in the system.

14. This is done by using the Nernst equation, which shows how the measured reduction potential is related to the standard reduction potential as well as to the concentrations (or activities) of all chemical species that are involved in the given reduction-oxidation process.

15. The reduction potentials for the half-reactions at 25°C are as follows:
    (a) +0.79 volts
    (b) −1.68 volts
    (c) +0.30 volts

16. The basic form of the Nernst equation is shown below for a half-reaction where "a" moles of a chemical Ox receives "n" moles of electrons to be converted into "b" moles of another chemical, Red.

$$E = E° + 2.303 \, (R \cdot T)/(n \cdot F) \, \log([Ox]^a/[Red]^b)$$

In this equation, E is the measured reduction potential under nonstandard conditions, E° is the standard reduction potential (E°) for the reaction, R is the ideal gas law constant, T is the absolute temperature, and F is the Faraday constant; [Ox] and [Red] are the molar concentrations of Ox and Red in the system, (which are assumed here to be roughly equal to the chemical activities of Ox and Red). This equation is important in reduction-oxidation reactions in that it allows the potential of these reactions to be determined when working under nonstandard reaction conditions.

17. The standard hydrogen electrode (SHE) consists of a platinum electrode in contact with bubbling hydrogen gas (at 1 atm pressure) and in a solution that contains hydrogen ions at 1 mol/L activity (see Fig. 16.2). This makes use of the following half-reaction.

$$2H^+ + 2e^- \rightleftharpoons H_2$$

By convention, the reduction potential for this electrode is assigned an E° value of 0.00 volts. This is used as a reference in measuring the E° values for other half-reactions by using the half-reaction of interest as the cathode and an SHE as the anode.

18. The silver-silver chloride electrode (Ag/AgCl electrode) is a common type of reference electrode that is used in electrochemical cells. The general design of this electrode is given in Fig. 16.7. It is based on the following half-reaction.

$$AgCl + e^- \rightleftharpoons Ag + Cl^-    E° = +0.22 \text{ volts}$$

Although the standard reduction potential for this electrode is not equal to zero, the silver-silver chloride electrode is much easier to make and use than the standard hydrogen electrode.

19. The saturated calomel electrode (SCE) is another common type of reference electrode that is used in electrochemical cells. The general design of this electrode is given in Fig. 16.6. It is based on the following half-reaction.

$$Hg_2Cl_2 + 2e^- \rightleftharpoons 2 Hg + 2 Cl^-    E° = +0.27 \text{ volts}$$

Like the silver-silver chloride electrode, the value of E° for this electrode is not equal to zero, but it is still much easier to make and use than the standard hydrogen electrode.

20. (a) −0.80 volts
    (b) +1.61 volts
    (c) −0.50 volts

21. (a) +77 kilojoules, electrolytic
    (b) −311 kilojoules, galvanic
    (c) +48 kilojoules, electrolytic

22. Chemical activity is a type of concentration that takes into account the interactions that occur between neighboring molecules or ions in solution. Activity is expressed in the same units as ordinary concentrations and is closely related to these concentrations through a factor known as the activity coefficient. Chemical activities are a more accurate representation of the effective concentration of a compound and the ability of this compound to take part in chemical reactions. The chemical activity and the molar concentration of a compound are approximately the same when we are dealing with relatively dilute solutions, but these numbers can be very different for concentrated solutions.

23. (a) Galvanic cell
    (b) Electrolytic cell
    (c) Galvanic cell
    (d) Electrolytic cell
    (e) Galvanic cell

24. A standard reduction potential is one that has been determined under standard reaction conditions (that is, a temperature of 25°C, a pressure of 1 atm for all gaseous reactants or products, a chemical activity of one for all solids, and 1 mol/L activity for all soluble reactants and products). An ordinary reduction potential is that which is determined under any other set of conditions. These two terms are related to each other through the Nernst equation. If standard conditions are being used for an electrochemical cell, then the reduction potential for this cell will also be equal to its standard reduction potential.

25. Galvanic cells are commonly used in batteries to supply us with energy. Examples include the lead storage battery used in cars and smaller batteries like alkaline cells and nickel-cadmium batteries. Electrolytic cells are used in such industrial processes as chrome plating, the generation of magnesium metal from molten magnesium chloride, the refinement of copper from copper ore, and the production of aluminum metal from bauxite.

# A.17  Chapter 17

1. An amino acid is a chemical that contains four types of functional groups that are connected to a central carbon atom. These groups are (1) a hydrogen atom, (2) a carboxylic acid, (3) an amine group, and (4) a side chain. The hydrogen atom, amine group, and carboxylic acid group are present in all amino acids and give them a set of common properties such as the ability to act as an acid or a base or the ability to connect with other amino acids to form peptide bonds. The side chain varies in structure from one type of amino acid to the next and is what gives each amino acid its own set of unique chemical and physical properties.

2. Peptides and proteins are both polymers of amino acids, and are also known as polypeptides. Small polypeptides that have molecular weights less than about 5000 g/mol are often referred to as peptides. Polypeptides with molecular weights greater than 5000 g/mol are usually called proteins.

3. A peptide bond forms when the amine group of one amino acid reacts with the carboxylic acid group of another amino acid. This is a dehydration reaction that also forms water as a product.

4. (a) Isoleucine-Glutamine-Alanine-Asparagine-Aspartic Acid-Tryptophan
   (b) Histidine-Proline-Phenylalanine-Aspartic Acid-Alanine
   (c) Threonine-Valine-Glutamine-Methionine-Lysine
   (d) Serine-Lysine-Isoleucine-Histidine-Glycine

5.

c)

6. (a) Hydrophilic
   (b) Hydrophilic
   (c) Hydrophilic
   (d) Hydrophobic
   (e) Hydrophilic
7. (a) Primary structure
   (b) Quaternary structure
   (c) Secondary structure
   (d) Tertiary structure
8. A disulfide bond forms when an -SH group on one cysteine side chain of a protein or peptide reacts with an -SH on another cysteine side chain to form a covalent -S-S- bond. These types of bonds help hold the structure of the protein or peptide in place and help combine multiple types of polypeptides.
9. An antibody is a type of protein that is produced by our bodies to bind to a particular foreign agent, like a virus or bacteria. An antibody has specific binding pockets for this purpose. The structure of each binding pocket is determined by the amino acids that are present in the pocket and the way in which these amino acids fold to give a certain three-dimensional shape. The side chains on these amino acids are also important in making sure that the foreign agent remains bound to the antibody. This results in a tight bond between the antibody and the invading species that helps our bodies to identify that species as being a foreign substance that must be destroyed or removed.
10. An enzyme is a protein-based catalyst that helps promote a particular type of chemical reaction. Enzymes perform this function by first binding to the reactants in a specific pocket on the enzyme. This pocket also usually contains other substances that are required for the reaction. After the desired products have been formed by the enzyme, the products are released from the binding pocket and another set of reactant molecules are allowed to bind so that the process can be repeated.
11. (a) Reactant, 4-nitrophenyl phosphate; products, 4-nitrophenoxide and phosphate; enzyme, alkaline phosphatase.
    (b) Reactant, adenosine-5′-monophosphate; products, adenosine and phosphate; enzyme, 5′-nucleotidase.
    (c) Reactant, triglyceride; products, glycerol and fatty acids; enzyme, lipase.
12. (a) Carbohydrates, also known as sugars, are a group of compounds that have the empirical formula $C_x(H_2O)_y$. They are important in humans and animals as a sources of energy, and in plants as both an energy source and as a component of such features as wood, bark, and stems.
    (b) A simple carbohydrate, or a monosaccharide, is a carbohydrate that contains a single chain of carbon atoms. One of the carbon atoms in a monosaccharide is bound to an oxygen atom to form an aldehyde or ketone group. All or most of the other carbon atoms are attached to alcohol groups.
    (c) A complex carbohydrate is a group of two or more monosaccharides that are linked together.
    (d) A monosaccharide is another name for a simple carbohydrate (see definition in 12b).

(e) A disaccharide is a complex carbohydrate formed by linking two monosaccharides together.

(f) A polysaccharide is a complex carbohydrate formed by linking more than two monosaccharides together.

13. (a) Simple carbohydrate
    (b) Simple carbohydrate
    (c) Complex carbohydrate

14. This sugar has a chain of five carbons, while glucose has six. Both this sugar and glucose have an aldehyde group on their first carbon. The lower four carbons on each have alcohol groups arranged in the same fashion.

15.

16. This compound contains a β-linkage.

17. (a) A nucleic acid is a polymer made up of building blocks known as nucleotides.

    (b) A nucleotide is an organic compound that contains three main parts in its structure: a deoxyribose or ribose sugar, a nitrogen-containing organic base, and a chain of one or more phosphate groups.

    (c) DNA is a type of nucleic acid that contains deoxyribose as the sugar in each of its nucleotides.

    (d) RNA is a type of nucleic acid that contains ribose as the sugar in each of its nucleotides.

    (e) mRNA (or messenger RNA) is a special type of RNA that is used to make copies of DNA for use in the construction of a particular peptide or protein.

    (f) tRNA (or transfer RNA) is a special type of nucleic acid that has a region where it binds to an amino acid and a specific sequence of nucleic acids that is able to specifically hydrogen bond to the mRNA code for that same amino acid. tRNA is used along with mRNA and ribosomes to translate a chemical message from a code written in nucleic acids to one that is represented by a string of amino acids for the production of proteins and peptides.

    (g) A chromosome is a structure that contains a long string of DNA plus proteins and some RNA that help protect and package this DNA into a very compact form that can be stored within the nucleus of a cell.

    (h) A gene is the part of DNA that acts as the code for making a single type of protein or peptide.

    (i) The genetic code is the series of instructions, encoded by DNA, that our bodies use to produce all proteins and peptides.

    (j) The nucleus is the center portion of a cell that contains DNA.

    (k) Ribosomes are structures found within a cell that are responsible for making proteins and peptides.

18. DNA has deoxyribose as the sugar in each of its nucleotides, while RNA uses ribose. DNA contains the organic bases adenine, thymine, cytosine, or guanine, while RNA uses uracil instead of thymine. DNA is used to store the genetic code, while RNA is used to copy this code or to help translate it into protein and peptide structures. RNA structures are also usually much smaller than those that are usually observed for DNA.

19. The five types of organic bases found in DNA and/or RNA are cytosine, thymine, adenine, guanine, and uracil. Uracil is found only in RNA and thymine is found only in DNA. Cytosine and guanine form hydrogen bonds with each other. Adenine forms hydrogen bonds with both thymine and uracil.

20. The organic base is guanine, the sugar is deoxyribose, and the phosphate group is a disphosphate chain.

21. (a) Leucine-Phenylalanine-Serine-Aspartic Acid-Glutamic Acid-Glutamine
    (b) Isoleucine-Leucine-Proline-Arginine-Threonine-Alanine
    (c) Valine-Valine-Alanine-Threonine-Lysine-Glycine

22. (a) A fat is a type of lipid that is a solid at room temperature and that is formed by the reaction of glycerol with three fatty acids.
    (b) An oil is a type of lipid that is a liquid at room temperature and that is formed by the reaction of glycerol with three fatty acids.
    (c) A saturated fat is a fat or oil that has only single bonds in all of its fatty acid chains.
    (d) An unsaturated fat is a fat or oil that has one or more double bonds present in some or all of its fatty acid chains.
    (e) Lipids are a group of biological compounds that all have the common property of being relatively insoluble in water but easy to dissolve in an organic solvent like benzene.
    (f) A phospholipid is a type of lipid that is produced by the reaction of glycerol with two fatty acids. It also contains a phosphate group.
    (g) A wax is a type of lipid that is formed through the reaction of a fatty acid with an alcohol to form an ester. A simple alcohol with only one $-OH$ group is used in this reaction.
    (h) A steroid is a type of lipid that contains the basic hydrocarbon ring structure that is shown in Fig. 17.16.
    (i) The cell membrane is an outer boundary in our cells that contains a layer of two phospholipid molecules. Some proteins are also present in this membrane. It is through the formation and presence of this membrane that cells control what enters and leaves them.

23. (a) Wax
    (b) Fat or oil (probably a fat)
    (c) Phospholipid

24. Both of these compounds contain the basic ring structure that is found in any steroid. The way in which these differ from cholesterol is indicated below.
    (a) This compound contains two more alcohol groups than cholesterol and is missing the double bond that is present in cholesterol. The upper side chain is also different, since it ends in a carboxylic acid group instead of the branched alkane group that is found in cholesterol.
    (b) One difference is in the upper side chain, which has a short ketone group in this compound instead of the longer branched alkane group that is found in cholesterol. Also, the double bond in the lower ring is in a different location and the alcohol group found in cholesterol is now a ketone group.

25. (a) Phospholipid
    (b) Fat
    (c) Oil
    (d) Phospholipid
    (e) Wax
    (f) Steroid

# Index